Genomics, Proteomics, and Metabolomics in Nutraceuticals and Functional Foods

Genomics, Proteomics, and Metabolomics in Nutraceuticals and Functional Foods

Editors

Debasis Bagchi, Ph.D.

Pharmacological and Pharmaceutical Sciences Department
University of Houston, College of Pharmacy
Houston, Texas

Francis C. Lau, Ph.D.

Dept. of Research and Development
InterHealth Research Center
Benicia, California

Manashi Bagchi, Ph.D.

Dept. of Research and Development
InterHealth Research Center
Benicia, California

WILEY-BLACKWELL

A John Wiley & Sons, Ltd., Publication

Edition first published 2010
© 2010 Blackwell Publishing

Blackwell Publishing was acquired by John Wiley & Sons in February 2007. Blackwell's publishing program has been merged with Wiley's global Scientific, Technical, and Medical business to form Wiley-Blackwell.

Editorial Office
2121 State Avenue, Ames, Iowa 50014-8300, USA

For details of our global editorial offices, for customer services, and for information about how to apply for permission to reuse the copyright material in this book, please see our website at www.wiley.com/wiley-blackwell.

Library of Congress Cataloging-in-Publication Data

Genomics, proteomics, and metabolomics in nutraceuticals and functional
foods / [edited by] Debasis Bagchi, Francis Lau, Manashi Bagchi.
     p. ; cm.
  Includes bibliographical references and index.
  ISBN 978-0-8138-1402-5 (hardback : alk. paper)
  1. Functional foods.   2. Dietary supplements.   3. Genomics.   4. Proteomics.
5. Physiological genomics. I. Bagchi, Debasis. II. Lau, Francis (Francis C.)
III. Bagchi, Manashi.
  [DNLM: 1. Functional Food–analysis.   2. Dietary Supplements–analysis.
3. Genetic Techniques.   4. Genomics–methods.   5. Metabolism.
6. Proteomics–methods. QU 145.5 G335 2010]
  QP144.F85G46 2010
  613.2–dc22   2009049304

A catalog record for this book is available from the U.S. Library of Congress.

Set in 9.5/11 pt Times New Roman by Aptara® Inc., New Delhi, India
Printed in Singapore by Markono Print Media Pte Ltd

1   2010

Dedication

We dedicate this book to our beloved Teacher, Professor
Amareshwar Chatterjee, Ph.D.

Debasis and Manashi Bagchi

I dedicate this book to my family, with love and gratitude.

Francis C. Lau

Contents

Editors and Contributors   ix
Preface   xiii

Section 1   Introduction
1. Recent advances in nutraceuticals and functional foods   3
    Francis C. Lau, Debasis Bagchi, Shirley Zafra-Stone, and Manashi Bagchi
2. Novel omics technologies in nutraceutical and functional food research   11
    Xuewu Zhang, Wei Wang, and Kaijun Xiao

Section 2   Genomics
3. Nutrigenomics and statistical power: The ethics of genetically informed nutritional advice   23
    Ruth Chadwick
4. NutrimiRomics: The promise of a new discipline in nutrigenomics   35
    Chandan K. Sen
5. Genomics in weight loss nutraceuticals   45
    Debasis Bagchi, Francis C. Lau, Hiroyoshi Moriyama, Manashi Bagchi, and Shirley Zafra-Stone
6. Application of genomics and bioinformatics analysis in exploratory study of functional food   61
    Kohsuke Hayamizu and Aiko Manji
7. Genomics as a tool to characterize anti-inflammatory nutraceuticals   73
    Sashwati Roy
8. Application of nutrigenomics in gastrointestinal health   83
    Lynnette R. Ferguson, Philip I. Baker, and Donald R. Love
9. Genomics analysis to demonstrate the safety and efficacy of dietary antioxidants   95
    Nilanjana Maulik
10. Genomics applied to nutrients and functional foods in Japan: State of the art   127
    Yuji Nakai, Akihito Yasuoka, Hisanori Kato, and Keiko Abe

Color plate appears between pages 154 and 155.

11. Genomic basis of anti-inflammatory properties of *Boswellia* extract   155
     Golakoti Trimurtulu, Chandan K. Sen, Alluri V. Krishnaraju, and Krishanu Sengupta
12. Nutrigenomic Perspectives on Cancer Chemoprevention with Anti-inflammatory
     and Antioxidant Phytochemicals: NF-κB and Nrf2 Signaling Pathways as Potential
     Targets   175
     Hye-Kyung Na and Young-Joon Surh

Section 3   Proteomics

13. Proteomics analysis of the functionality of *Toona sinensis* by 2D-gel electrophoresis
     201
     Sue-Joan Chang and Chun-Yung Huang
14. Application of proteomics in nutrition research   213
     Baukje de Roos
15. Proteomics approach to assess the potency of dietary grape seed proanthocyanidins   225
     Hai-qing Gao
16. Proteomics and its application for elucidating insulin deregulation in diabetes   241
     Hyun-Jung Kim and Chan-Wha Kim

Section 4   Metabolomics

17. NMR-based-metabolomics strategy for the classification and quality control of
     nutraceuticals and functional foods   265
     Yulan Wang and Huiru Tang
18. Metabolomics: An emerging post-genomic tool for nutrition   271
     Phillip Whitfield and Jennifer Kirwan
19. Evaluation of the beneficial effects of phytonutrients by metabolomics   287
     Katia Nones and Silas G. Villas-Bôas

Section 5   Nutrigenomics in Human Health

20. Omics for the development of novel phytomedicines   299
     Kandan Aravindaram, Harry Wilson, and Ning-Sun Yang
21. Contribution of omics revolution to cancer prevention research   315
     Nancy J. Emenaker and John A. Milner

Index   329

Editors and Contributors

EDITORS

Debasis Bagchi
Department of Pharmacology and Pharmaceutical Sciences, University of Houston College of Pharmacy, Houston, TX, USA

Manashi Bagchi
Dept. of Research and Development, InterHealth Research Center, 5451 Industrial Way, Benicia, CA, USA

Francis C. Lau
Dept. of Research and Development, InterHealth Research Center, 5451 Industrial Way, Benicia, CA, USA

CONTRIBUTORS

Keiko Abe
Dept. of Applied Biological Chemistry, Graduate School of Agricultural and Life Sciences, The University of Tokyo, Tokyo, Japan

Kandan Aravindaram
Agricultural Biotechnology Research Center, Academia Sinica, No. 128, Sec. 2, Academia Road, Nankang, Taipei 115, Taiwan

Philip I. Baker
Dept. of Nutrition, University of Auckland, Private Bag 92019, Auckland Mail Centre, Auckland 1142, New Zealand

Ruth Chadwick
CESAGen Furness College, Lancaster University, LA1 4YG, UK

Sue-Joan Chang
Dept. of Life Sciences, National Cheng Kung University, No. 1, University Road, Tainan City 701, Taiwan

Nancy J. Emenaker
Nutritional Science Research Group, Division of Cancer Prevention, National Cancer Institute, 6130 Executive Blvd., Executive Plaza North 3158, Bethesda, MD 20892 USA

Lynnette R. Ferguson
Dept. of Nutrition, University of Auckland, Private Bag 92019, Auckland Mail Centre, Auckland 1142, New Zealand

Hai-qing Gao
Dept. of Geriatrics, Qi-Lu Hospital of Shandong University, 107 Wenhuaxi Road, Jinan 250012, Shandong Province, China

Kohsuke Hayamizu
Evaluation Technology Group, Core Technology & Research Division, FANCL Research Institute, FANCL Corp., Yokohama, Japan

Chun-Yung Huang
Dept. of Life Sciences, National Cheng Kung University, No. 1, University Road, Tainan City 701, Taiwan

Hisanori Kato
Dept. of Applied Biological Chemistry, Graduate School of Agricultural and Life Sciences, The University of Tokyo, Tokyo, Japan

Chan-Wha Kim
Graduate School of Life Sciences and Biotechnology, Korea University, 1, 5-ka, Anam-dong, Sungbuk-ku, Seoul 136-701, South Korea

Hyun-Jung Kim
Graduate School of Life Sciences and Biotechnology, Korea University, 1, 5-ka, Anam-dong, Sungbuk-ku, Seoul 136-701, South Korea

Jennifer Kirwan
Proteomics and Functional Genomics Research Group, Faculty of Veterinary Science, University of Liverpool, Crown Street, Liverpool L69 7ZJ, UK

Alluri V. Krishnaraju
Laila Impex R&D Centre, Unit-I, Phase-III, Jawahar Autonagar, Vijayawada, A.P., India

Donald R. Love
School of Biological Sciences, University of Auckland, Private Bag 92019, Auckland Mail Centre, Auckland 1142, New Zealand
LabPLUS, PO Box 110031, Auckland City Hospital, Auckland, New Zealand

Aiko Manji
Evaluation Technology Group, Core Technology & Research Division, FANCL Research Institute, FANCL Corp., Yokohama, Japan

Nilanjana Maulik
Dept. of Surgery, Molecular Cardiology and Angiogenesis Laboratory, University of Connecticut Health Center, 263 Farmington Avenue, Farmington, CT 06030, USA

John A. Milner
Nutritional Science Research Group, Division of Cancer Prevention, National Cancer Institute, 6130 Executive Blvd., Executive Plaza North 3158, Bethesda, MD 20892, USA

Hiroyoshi Moriyama
Laboratory of Pharmacotherapeutics, Showa Pharmaceutical University, 3-3165 Higashi-tamagawagakuen, Machida, Tokyo 194-8543, Japan

Hye-Kyung Na
Dept. of Food and Nutrition, College of Human Ecology, Sungshin Women's University, Seoul 136-742, South Korea

Yuji Nakai
Dept. of Applied Biological Chemistry, Graduate School of Agricultural and Life Sciences, The University of Tokyo, Tokyo, Japan

Katia Nones
School of Biological Sciences, Microbiology & Virology Research Group, The University of Auckland, Private Bag 92019, Auckland 1142, New Zealand AcurePharma & BMC, Uppsala University, SE-751 82 Uppsala, Sweden

Baukje de Roos
University of Aberdeen, Rowett Institute of Nutrition and Health, Greenburn Road, Bucksburn, Aberdeen AB21 9SB, UK

Sashwati Roy
Davis Heart & Lung Research Institute, The Ohio State University Medical Center, 473 West 12th Avenue, 511 DHLRI, Columbus, OH 43210, USA

Chandan K. Sen
Deputy Director, Laboratory of Molecular Medicine, Department of Surgery, The Ohio State University Medical Center, Columbus, OH 43210, USA

Krishanu Sengupta
Laila Impex R&D Centre, Unit-I, Phase-III, Jawahar Autonagar, Vijayawada, A.P., India

Young-Joon Surh
National Research Laboratory of Molecular Carcinogenesis and Chemoprevention, College of Pharmacy, Seoul National University, Seoul 151-742, South Korea

Huiru Tang
State Key Lab of Magnetic Resonance and Atomic and Molecular Physics Wuhan Centre for Magnetic Resonance Wuhan Institute of Physics and Mathematics, The Chinese Academy of Sciences, Wuhan, 430071, China

Golakoti Trimurtulu
Laila Impex R&D Centre, Unit-I, Phase-III, Jawahar Autonagar, Vijayawada, A.P., India

Silas Villas-Bôas
School of Biological Sciences, Microbiology & Virology Research Group, The University of Auckland, Private Bag 92019, Auckland 1142, New Zealand AcurePharma & BMC, Uppsala University, SE-751 82 Uppsala, Sweden

Wei Wang
College of Light Industry and Food Sciences, South China University of Technology, 381 Wushan Road, Guangzhou, 510640, China

Yulan Wang
State Key Lab. of Magnetic Resonance and Atomic and Molecular Physics Wuhan Centre for Magnetic Resonance Wuhan Institute of Physics and Mathematics, The Chinese Academy of Sciences, Wuhan, 430071, China

Phil Whitfield
Proteomics and Functional Genomics Research Group, Faculty of Veterinary Science, University of Liverpool, Crown Street, Liverpool L69 7ZJ, UK

Harry Wilson
Agricultural Biotechnology Research Center, Academia Sinica, No. 128, Sec. 2, Academia Road, Nankang, Taipei 115, Taiwan

Kaijun Xiao
College of Light Industry and Food Sciences, South China University of Technology, 381 Wushan Road, Guangzhou, 510640, China

Ning-Sun Yang
Agricultural Biotechnology Research Center, Academia Sinica, No. 128, Sec. 2, Academia Road, Nankang, Taipei 115, Taiwan

Akihito Yasuoka
Maebashi Institute of Technology, the University of Tokyo, Tokyo, Japan

Shirley Zafra-Stone
Department of Research and Development, InterHealth Research Center, 5451 Industrial Way, Benicia, CA 94590, USA

Xuewu Zhang
College of Light Industry and Food Sciences, South China University of Technology, 381 Wushan Road, Guangzhou, 510640, China

Preface

The relationship between food and health was well established nearly two and a half millennia ago as indicated by Hippocrates, who famously proclaimed "Let food be thy medicine and medicine be thy food." Hippocrates, regarded as the Father of Medicine, also prophesied the importance of individualized nutrition that "if we could give every individual the right amount of nourishment and exercise, not too little and not too much, we would have found the safest way to health." These immortal words became the tenet for nutrition science. Indeed, more than 2000 years later, Thomas Edison concurred with Hippocrates by stating that "the doctor of the future will no longer treat the human frame with drugs, but rather will cure and prevent disease with nutrition."

In this regard, functional foods and nutraceuticals have received considerable interest in the past decade largely due to increasing consumer awareness of the health benefits associated with food. A functional food is a foodstuff that provides health or medical benefits beyond the basic nutrients it contains. When a functional food facilitates the prevention or treatment of certain diseases or disorders, it is a nutraceutical. The founder of the Foundation for Innovation in Medicine, Dr. DeFelice, coined the term "nutraceutical," which combines the words nutrition and pharmaceutical emphasizing its therapeutic properties. Functional foods and nutraceuticals are often used interchangeably because a functional food to one consumer may serve as a nutraceutical to another.

Consumers in the past were mainly reactive to existing health problems, but today they are more proactive and increasingly interested in the health benefits of functional foods in the prevention of illness and chronic conditions. This combined with an aging population that focuses not only on longevity but also quality of life has created a market for functional foods and nutraceuticals. The global nutraceuticals market is defined as the cumulative sales of nutraceutical foods, beverages, and supplements fortified with bioactive ingredients. This market was worth $117.3 billion in 2007 and it is projected to increase to $176.7 billion in 2013.

It is estimated that the functional food and nutraceutical market is growing at a rate surpassing the traditional processed food market. As this market expands, so will the demand for the identification of new bioactive food ingredients and the discovery of beneficial components in existing foods. The completion of the Human Genome Project and the advances in genomics technologies have revolutionized the field of nutrition research. Nutritional genomics or nutrigenomics provides the means for a high-throughput platform for simultaneously evaluating the expression of thousands of genes at the mRNA (transcriptomics), protein (proteomics), and metabolites (metabolomics) levels.

This book is divided into five main sections starting with a brief introductory section (chapters 1 and 2) followed by the second section (chapters 3–12), which covers recent advances in nutrigenomics. The third section (chapters 13–16) focuses on nutriproteomics and the fourth section (chapters 17-19) examines nutrimetabolomics. The final section summarizes the application of nutrigenomics to the development of phytomedicine (chapter 20) and to the prevention of cancer (chapter 21). The intent of this book is to bring together current advances and comprehensive reviews of nutritional genomics by a panel of experts from around the globe, with emphasis on the nutrigenomics approach to functional foods and nutraceuticals.

Debasis Bagchi
Francis C. Lau
Manashi Bagchi

Section 1

Introduction

1

Recent Advances in Nutraceuticals and Functional Foods

Francis C. Lau, Debasis Bagchi, Shirley Zafra-Stone, and Manashi Bagchi

INTRODUCTION

Postgenomic advances have revolutionized nutrition research. Traditional nutrition science focused on the investigation of nutrient deficiencies and impairment of health. In the past few years, an emerging discipline of nutrition research, functional genomics, has provided new approaches and techniques to elucidate how nutrients modulate gene expression, protein synthesis, and metabolism [1]. It has become apparent that nutrients not only fuel our life but also participate in gene regulation [2]. Functional genomics as applied to nutrition research includes nutrigenomics and nutrigenetics. Whereas nutrigenomics investigates the impact of nutrients on gene regulation, nutrigenetics studies the effect of genetic variations on individual differences in response to specific food components [3].

Nutrigenomics and nutrigenetics offer the promise to unravel complex interactions among genes, gene products, genetic polymorphisms, and functional food components [4]. The ultimate goal is to devise strategies for personalized nutrition and dietary recommendations aiming at improving human health [5]. Indeed, nutrigenomic approaches based on ethnopharmacology and phytotherapy concepts have demonstrated the interaction of nutrients and botanicals with the genome to cause significant changes in gene expression [6]. This observation has resulted in the commercial launch of nutraceuticals and functional foods that can regulate health effects of individual genetic profiles [7].

Nutraceuticals or functional foods are bioactive food components that provide medical or health benefits [8]. In addition to essential nutrients such as carbohydrates, proteins, fatty acids, minerals, and vitamins, there are various nonessential bioactive food components capable of modulating cellular processes. These nutraceuticals or functional foods contribute to the prevention of diseases such as cancer, cardiovascular disease (CVD), obesity, and type II diabetes [9]. For instance, obesity and CVD are multifactorial diseases influenced by a number of environmental and genetic factors [10]. Nutrigenomics approach has begun to reveal that obesity and CVD may be susceptible to dietary interventions and these interventions may modulate the onset and progression of the disorders [9]. In the case of CVD, there is evidence for interactions between dietary fat and three common polymorphisms in the apolipoprotein (apo) E, apoAI, and peroxisome proliferator-activated receptor-gamma (PPARγ) genes [11]. Therefore, a clear understanding of how these genes affect the response of individuals to certain nutrients should facilitate the progression of personalized nutrition for people with high propensity for CVD.

Although the traditional Food Guide Pyramid by the U.S. Department of Agriculture (USDA) was based on the estimated average nutrient requirements for the U.S. population as a whole, USDA has recently updated the Food Guide Pyramid to MyPyramid, focusing on individual nutrition needs [12]. The 2005 Dietary Guidelines for Americans published by USDA emphasizes on meeting adequate nutrients within calorie needs [13]. The guidelines further provide the Dietary Approaches to Stop Hypertension (DASH) Eating Plan designed to integrate dietary recommendations into a healthy way of eating [13]. It is foreseeable that nutritional recommendations will be subpopulation-based with variations according to ethnicity, age, gender, disease susceptibility, and genetic polymorphisms [14].

NUTRITIONAL GENOMICS

In the past, nutrition research was conducted mainly to retrace the importance of a nutrient through its deficiency, which manifested as health-related problems. The notion of the interrelationship between diet and health was firmly established. However, it was only until recently that nutrition research evolved to focus on the direct nutrient–genome interactions. This is largely due to the wealth of genomic information generated by the Human Genome Project (HGP), the largest international scientific research endeavor [15]. The HGP spent 13 years to sequence the entire 3 billion bases of genetic information in every human cell and was officially completed in 2003 [16]. The completion of the HGP significantly facilitated the identification of single nucleotide polymorphisms (SNPs) within populations leading to differential responses to specific nutrients [7]. Thus, nutritional genomics has emerged as a result of the genomic revolution. Nutritional genomics includes nutrigenetics and nutrigenomics. Nutrigenetics evaluates gene-nutrition regulation by showing how genetic variations (i.e. SNPs) among individuals affect their responses to certain nutritional components. On the other hand, nutrigenomics investigates nutrition-gene regulation by demonstrating how nutrients interact with the genome and modify the expression of certain genes.

NUTRIGENETICS

There exists a small fraction (about 0.1%) of variation in the human genome sequence, which manifests in the form of differences in phenotypes and in an individual's response to certain food components [17]. These genetic polymorphisms create diversity within the human population. Ninety percent of all human genetic polymorphisms are variations of a single base within a DNA sequence known as SNPs. SNPs may contribute to the inconsistencies observed in epidemiological studies concerning the effects of diet on chronic diseases. Mounting evidence indicates that bioactive food components may disrupt cellular pathways through alteration of gene expression, thereby increasing risks for developing various chronic diseases such as CVD and type II diabetes [18]. Thus, genetic information on the differences in response to dietary factors may aid in identification of candidate genes with functional variations that alter nutrient metabolism.

The classical example of folate and methyltetrahydrofolate reductase (MTHFR) gene interaction demonstrated that SNP at position 677 of MTHFR gene results in two variants of MTHFR protein. MTHFR protein plays an important role in providing the essential amino acid methionine. The wild-type MTHFR protein metabolizes folate normally, whereas the thermal-labile variant of MTHFR protein exhibits a significant reduction in its activity giving rise to higher homocysteine levels, leading to increased risk of CVD and accelerated cognitive decline [19]. Therefore, individuals with the thermal-labile form of MTHFR would greatly benefit by personalized diet supplemented with folate to reduce excess homocysteine levels [20, 21].

Although a plethora of SNP information exists, it has been slow in converting SNP information into individual-based nutritional practices. This is because it takes tremendous time and effort to collect and catalogue population SNP information, to integrate and assimilate such information in nutrient-disease scenarios, and to develop specific diagnostic tools [7].

NUTRIGENOMICS

Nutrigenomics appears to exploit the omics revolution at a rapid pace providing an ever-growing body of information on nutrition-gene regulation due to advances in omics technologies such as genomics, transcriptomics, proteomics, and metabolomics [22].

Transcriptomics

Transcriptomics seems to be the most successful omics technology in nutrigenomics studies because of its efficiency and high throughput characteristics [22]. Transcriptome consists of the entire complement of mRNA or transcripts generated from genes being actively transcribed or expressed. Therefore, transcriptomics is a powerful tool for profiling gene expression patterns. A wide variety of bioactive food components can influence the expression of genes leading to altered biological processes including cell proliferation and differentiation, cell metabolism, and cell death. The imbalance of these cellular processes may lead to diseases such as diabetes and cancer. In this regard, genome-wide interrogation of gene expression by nutrients is particularly relevant in nutrigenomics research. Microarray technology markedly facilitates the simultaneous quantification of thousands of mRNA, thereby providing detailed profiles of gene expression in scenarios such as before and after exposure to certain food components [22, 23]. Interfaced with bioinformatics platforms, it is possible to construct the pathways for the observed gene expression profiles.

Transcriptomics has been used to investigate the effect of bioactive dietary components on gene expression in a variety of experimental paradigms including cell cultures, animals, and humans. Dietary intervention human clinical trials in obese subjects have been conducted to examine the effects of energy-restricted diets on gene expression in adipose tissue using transcriptomics technology [24, 25]. Other transcriptomics human studies include the evaluation of impact of high-protein and high-carbohydrate breakfasts on transcriptome of human blood cells and the investigation of the influence of *Lactobacillus GG* on gene expression profiles of duodenal mucosa [26, 27]. Transcriptomics has also been applied to food safety evaluations [22]. The rapid accumulation of nutritranscriptomic microarray data has prompted the establishment of a Web-based database infrastructure. This integrated database, built on an open-source database platform, ensures the efficient organization, storage, and analysis of the immense amount of microarray data generated from each nutritranscriptomic experiment [28].

Proteomics

Proteomics is the large-scale analysis of a proteome expressed by a genome. A proteome is the entire complement of proteins synthesized in a biological system at a given time and under defined conditions, reflecting the expression of a set of specific genes in the situation pertaining to that time point [29]. The proteome is dynamic and more complex than the genome. A proteome continuously changes in the temporal continuum according to cell type and functional state of the cell [30]. Whereas the human genome encodes about 25,000 functional genes, the human proteome comprises an order of magnitude more proteins (about 250,000) due to alternative splicing and posttranslational modifications [31]. In order to assess the complex proteome, new proteomics tools have been developed. Indeed, protein analysis has rapidly progressed from gel-based techniques to technologies such as mass spectrometry, multiple reaction monitoring, and multiplexed immunoassays in recent years [32]. Currently, proteomics allows for the high-throughput investigation of numerous proteins simultaneously in cells, tissues, or biological fluids [30]. Proteomics also enables the discovery of novel proteins. As an integral part of nutrigenomics, nutritional proteomics examines the effects of food components on protein expression and provides the potential to identify biomarkers sensitive to dietary interventions [33].

Several cell culture studies using nutritional proteomics demonstrated the effects of food components such as butyrate, flavonoid, and genistein on protein profiles [34–39]. Animal studies also showed the potential of proteomics in nutritional research. Proteomics analysis of brain homogenates from rats fed a grape seed extract (GSE) supplemented diet identified 13 candidate proteins [40]. Many candidate proteins were regulated by GSE-supplementation in opposite direction from previous findings for the same proteins in Alzheimer's disease and mouse models of neurodegeneration, indicating that these candidate proteins may be modulated by GSE to confer neuroprotective benefits [40]. Proteomic analysis of liver tissues of atherosclerosis-susceptible and atherosclerosis-resistant mice identified 30 proteins significantly altered by atherogenic diet [41]. The findings revealed a clear distinction in differential expression of proteins involved in oxidative stress and lipid metabolism between the two strains of mice in response to atherogenic diet, suggesting that the candidate proteins may contribute to differences in susceptibility to atherogenesis [41]. Combining nutritional transcriptomics and proteomics, the enzymes and transporters responsible for fatty acid metabolism, sequestration, and their transcriptional control in zinc-deficient rats were identified and pathways for the observed increase in hepatic lipid accumulation were constructed [42].

In a randomized cross-over human study, matrix-assisted laser desorption/ionization-time of flight mass spectrometry (MALDI-TOF MS) was used to isolate serum biomarkers in subjects taking control or cruciferous-supplemented diet [43]. Serum protein B-chain of α_2-HS glyco-protein was identified as a diet-related biomarker involved in insulin resistance and immune function [43]. With advances in proteomic methodology, nutritional proteomics has the potential to rapidly generate new knowledge pertaining to the complex interplay of nutrition–protein regulation, to identify novel biomarkers for nutritional status, and to devise new strategies for dietary prevention and intervention of diseases [44].

Metabolomics

Metabolomics is one of the newest omics technologies in nutritional research. The metabolome consists of the entire set of metabolites synthesized in a biological system. Metabolites are the end products of metabolic reactions, reflecting the interaction of the genome with its environment [45]. Metabolomics is the study of global metabolite profiles in a biological system under specific environmental conditions.

Nutritional metabolomics has the potential to provide insight into biochemical changes after dietary intervention and to impact food safety issues pertaining to genetically modified food [46]. Metabolomic techniques such as nuclear magnetic resonance (NMR) and MS combined with powerful bioinformatics platforms greatly enhance metabolomic approach to nutrition research [4, 47, 48]. The first nutritional metabolomic approach to determining biochemical modifications following dietary intervention showed that soy isoflavones induced changes in plasma components in healthy premenopausal women under controlled environmental conditions [49]. The plasma biochemical profiles showed strong variability in each subject, indicating the complex interaction of factors such as genetics, age, health status, diet, and lifestyle. Despite the individual variability, there were clear diet intervention-related differences in the plasma lipoprotein, amino acid, and carbohydrate profile, suggesting a soy-induced alteration in energy metabolism [49].

Metabolic responses to chamomile tea ingestion in human subjects were evaluated by high-resolution ^{1}H NMR spectroscopy coupled with chemometric methods [50]. Although metabolite profiles exhibited a high degree of variation among subjects, there was a clear differentiation in urinary excretion demonstrating an increased hippurate and glycine with decreased creatinine level after chamomile ingestion [50]. A similar study evaluated the effects of black and green tea intake on human urinary metabolites. The study showed that green tea consumption resulted in a stronger increase in several citric acid cycle intermediates suggesting an effect of green tea flavanols on human oxidative energy metabolism [51].

Although metabolomics has contributed significantly to the omics revolution, a global description of human metabolism is impossible at this point due to limitations in current technologies and diversity among individuals in terms of age, gender, diet, lifestyle, health status, and other internal and external factors [44]. Currently, the extent to which food components in the human diet induce changes in nutritional metabolic profiles is poorly understood. However, with technological advances, the challenges of applying metabolomics in nutrition research can be overcome.

OMICS IN FUTURE NUTRITION RESEARCH

Nutrition research has accelerated greatly by the omics revolution. Nutrigenomics has already contributed a vast amount of information to nutrition science. The major tasks of postgenomic nutrition research are: to understand how diet or food components affect the genome and how genetic variations affect individual response to food components. The ultimate goal of nutritional genomics is to personalize diets based on individual needs for the maintenance of health and prevention of diseases. Emerging disciplines branching from genomics such as RNomics, miRNomics, liponomics, fluxomics, toxigenomics, and the like will further facilitate nutritional genomics.

REFERENCES

1. Elliott, R., Pico, C., Dommels, Y., Wybranska, I., *et al.*, Nutrigenomic approaches for benefit-risk analysis of foods and food components: defining markers of health. *Br J Nutr* 2007, *98*, 1095–1100.
2. Corthesy-Theulaz, I., den Dunnen, J. T., Ferre, P., Geurts, J. M., *et al.*, Nutrigenomics: the impact of biomics technology on nutrition research. *Ann Nutr Metab* 2005, *49*, 355–365.
3. Gillies, P. J., Nutrigenomics: the Rubicon of molecular nutrition. *J Am Diet Assoc* 2003, *103*, S50–S55.
4. Trujillo, E., Davis, C., Milner, J., Nutrigenomics, proteomics, metabolomics, and the practice of dietetics. *J Am Diet Assoc* 2006, *106*, 403–413.
5. Stover, P. J., Nutritional genomics. *Physiol Genomics* 2004, *16*, 161–165.
6. Subbiah, M. T., Understanding the nutrigenomic definitions and concepts at the food-genome junction. *OMICS* 2008, *12*, 229–235.
7. Subbiah, M. T., Nutrigenetics and nutraceuticals: the next wave riding on personalized medicine. *Transl Res* 2007, *149*, 55–61.
8. Ferguson, L. R., Nutrigenomics approaches to functional foods. *J Am Diet Assoc* 2009, *109*, 452–458.
9. Gorduza, E. V., Indrei, L. L., Gorduza, V. M., Nutrigenomics in postgenomic era. *Rev Med Chir Soc Med Nat Iasi* 2008, *112*, 152–164.
10. Lovegrove, J. A., Gitau, R., Personalized nutrition for the prevention of cardiovascular disease: a future perspective. *J Hum Nutr Diet* 2008, *21*, 306–316.
11. Lovegrove, J. A., Gitau, R., Nutrigenetics and CVD: what does the future hold? *Proc Nutr Soc* 2008, *67*, 206–213.
12. USDA, *http://MyPyramid.gov*; 2009.
13. USDA, Dietary Guidelines for Americans 2005. *U.S. Department of Health and Human Services* 2005.
14. Hernell, O., West, C., Do we need personalized recommendations for infants at risk of developing disease? *Nestle Nutr Workshop Ser Pediatr Program* 2008, *62*, 239–249; discussion 249–252.
15. Venter, J. C., Adams, M. D., Myers, E. W., Li, P. W., *et al.*, The sequence of the human genome. *Science* 2001, *291*, 1304–1351.

16. Austin, C. P., The impact of the completed human genome sequence on the development of novel therapeutics for human disease. *Annu Review Med* 2004, *55*, 1–13.
17. El-Sohemy, A., Nutrigenetics. *Forum Nutr* 2007, *60*, 25–30.
18. Mutch, D. M., Wahli, W., Williamson, G., Nutrigenomics and nutrigenetics: the emerging faces of nutrition. *FASEB J* 2005, *19*, 1602–1616.
19. Frosst, P., Blom, H. J., Milos, R., Goyette, P., *et al.*, A candidate genetic risk factor for vascular disease: a common mutation in methylenetetrahydrofolate reductase. *Nat Genet* 1995, *10*, 111–113.
20. Astley, S. B., An introduction to nutrigenomics developments and trends. *Genes Nutr* 2007, *2*, 11–13.
21. Miyaki, K., Murata, M., Kikuchi, H., Takei, I., *et al.*, Assessment of tailor-made prevention of atherosclerosis with folic acid supplementation: randomized, double-blind, placebo-controlled trials in each MTHFR C677T genotype. *J Hum Genet* 2005, *50*, 241–248.
22. Kato, H., Nutrigenomics: the cutting edge and Asian perspectives. *Asia Pac J Clin Nutr* 2008, *17* Suppl 1, 12–15.
23. Garosi, P., De Filippo, C., van Erk, M., Rocca-Serra, P., *et al.*, Defining best practice for microarray analyses in nutrigenomic studies. *Br J Nutr* 2005, *93*, 425–432.
24. Clement, K., Viguerie, N., Poitou, C., Carette, C., *et al.*, Weight loss regulates inflammation-related genes in white adipose tissue of obese subjects. *FASEB J* 2004, *18*, 1657–1669.
25. Dahlman, I., Linder, K., Arvidsson Nordstrom, E., Andersson, I., *et al.*, Changes in adipose tissue gene expression with energy-restricted diets in obese women. *Am J Clin Nutr* 2005, *81*, 1275–1285.
26. Di Caro, S., Tao, H., Grillo, A., Elia, C., *et al.*, Effects of Lactobacillus GG on genes expression pattern in small bowel mucosa. *Dig Liver Dis* 2005, *37*, 320–329.
27. van Erk, M. J., Blom, W. A., van Ommen, B., Hendriks, H. F., High-protein and high-carbohydrate breakfasts differentially change the transcriptome of human blood cells. *Am J Clin Nutr* 2006, *84*, 1233–1241.
28. Saito, K., Arai, S., Kato, H., A nutrigenomics database—integrated repository for publications and associated microarray data in nutrigenomics research. *Br J Nutr* 2005, *94*, 493–495.
29. Trayhurn, P., Proteomics and nutrition—a science for the first decade of the new millennium. *Br J Nutr* 2000, *83*, 1–2.
30. Thongboonkerd, V., Proteomics. *Forum Nutr* 2007, *60*, 80–90.
31. Kussmann, M., Affolter, M., Proteomic methods in nutrition. *Curr Opin Clin Nutr Metab Care* 2006, *9*, 575–583.
32. de Roos, B., McArdle, H. J., Proteomics as a tool for the modelling of biological processes and biomarker development in nutrition research. *Br J Nutr* 2008, *99* Suppl 3, S66–71.
33. Fuchs, D., Winkelmann, I., Johnson, I. T., Mariman, E., *et al.*, Proteomics in nutrition research: principles, technologies and applications. *Br J Nutr* 2005, *94*, 302–314.
34. Tan, S., Seow, T. K., Liang, R. C., Koh, S., *et al.*, Proteome analysis of butyrate-treated human colon cancer cells (HT-29). *Int J Cancer* 2002, *98*, 523–531.
35. Herzog, A., Kindermann, B., Doring, F., Daniel, H., *et al.*, Pleiotropic molecular effects of the pro-apoptotic dietary constituent flavone in human colon cancer cells identified by protein and mRNA expression profiling. *Proteomics* 2004, *4*, 2455–2464.
36. Wenzel, U., Herzog, A., Kuntz, S., Daniel, H., Protein expression profiling identifies molecular targets of quercetin as a major dietary flavonoid in human colon cancer cells. *Proteomics* 2004, *4*, 2160–2174.
37. Fuchs, D., Erhard, P., Rimbach, G., Daniel, H., *et al.*, Genistein blocks homocysteine-induced alterations in the proteome of human endothelial cells. *Proteomics* 2005, *5*, 2808–2818.
38. Fuchs, D., Erhard, P., Turner, R., Rimbach, G., *et al.*, Genistein reverses changes of the proteome induced by oxidized-LDL in EA.hy 926 human endothelial cells. *J Proteome Res* 2005, *4*, 369–376.

39. Fuchs, D., de Pascual-Teresa, S., Rimbach, G., Virgili, F., *et al.*, Proteome analysis for identification of target proteins of genistein in primary human endothelial cells stressed with oxidized LDL or homocysteine. *Eur J Nutr* 2005, *44*, 95–104.

40. Deshane, J., Chaves, L., Sarikonda, K. V., Isbell, S., *et al.*, Proteomics analysis of rat brain protein modulations by grape seed extract. *J Agric Food Chem* 2004, *52*, 7872–7883.

41. Park, J. Y., Seong, J. K., Paik, Y. K., Proteomic analysis of diet-induced hypercholesterolemic mice. *Proteomics* 2004, *4*, 514–523.

42. tom Dieck, H., Doring, F., Fuchs, D., Roth, H. P., *et al.*, Transcriptome and proteome analysis identifies the pathways that increase hepatic lipid accumulation in zinc-deficient rats. *J Nutr* 2005, *135*, 199–205.

43. Mitchell, B. L., Yasui, Y., Lampe, J. W., Gafken, P. R., *et al.*, Evaluation of matrix-assisted laser desorption/ionization-time of flight mass spectrometry proteomic profiling: identification of alpha 2-HS glycoprotein B-chain as a biomarker of diet. *Proteomics* 2005, *5*, 2238–2246.

44. Ovesna, J., Slaby, O., Toussaint, O., Kodicek, M., *et al.*, High throughput 'omics' approaches to assess the effects of phytochemicals in human health studies. *Br J Nutr* 2008, *99* E Suppl 1, ES127–134.

45. Rochfort, S., Metabolomics reviewed: a new "omics" platform technology for systems biology and implications for natural products research. *J Nat Prod* 2005, *68*, 1813–1820.

46. Dixon, R. A., Gang, D. R., Charlton, A. J., Fiehn, O., *et al.*, Applications of metabolomics in agriculture. *J Agric Food Chem* 2006, *54*, 8984–8994.

47. German, J. B., Roberts, M. A., Watkins, S. M., Genomics and metabolomics as markers for the interaction of diet and health: lessons from lipids. *J Nutr* 2003, *133*, 2078S–2083S.

48. Hall, R. D., Brouwer, I. D., Fitzgerald, M. A., Plant metabolomics and its potential application for human nutrition. *Physiol Plant* 2008, *132*, 162–175.

49. Solanky, K. S., Bailey, N. J., Beckwith-Hall, B. M., Davis, A., *et al.*, Application of biofluid 1H nuclear magnetic resonance-based metabonomic techniques for the analysis of the biochemical effects of dietary isoflavones on human plasma profile. *Anal Biochem* 2003, *323*, 197–204.

50. Wang, Y., Tang, H., Nicholson, J. K., Hylands, P. J., *et al.*, A metabonomic strategy for the detection of the metabolic effects of chamomile (Matricaria recutita L.) ingestion. *J Agric Food Chem* 2005, *53*, 191–196.

51. van Dorsten, F. A., Daykin, C. A., Mulder, T. P., van Duynhoven, J. P., Metabonomics approach to determine metabolic differences between green tea and black tea consumption. *J Agric Food Chem* 2006, *54*, 6929–6938.

2

Novel Omics Technologies in Nutraceutical and Functional Food Research

Xuewu Zhang, Wei Wang, and Kaijun Xiao

ABSTRACT

The detection of physiological effects induced in the human body by the uptake of nutrients requires robust technologies to measure many parameters. New omics technologies including transcriptomics, proteomics, and metabolomics offer exciting opportunities to address complex issues related to human health, disease, and nutrition. Systems biology opens new doors to understanding the complex interaction network between nutrients and molecules in biological systems. It is expected that omics-based human nutrition research can provide recommendations for personalized medicine and nutrition.

INTRODUCTION

Many nutrients and nonnutrient components of foods have multiple functions. For example, fatty acids not only function as constituents of cell membrane phospholipids but also participate in numerous biochemical processes in a cell-specific and tissue-specific fashion, involving hundreds of genes, many signal transduction pathways, and a large number of biomolecules, such as transcription factors, receptors, hormones, apolipoproteins, and enzymes. Hence, the measurements of a single gene, single protein, or single metabolite do not provide us with sufficient and thorough information to elucidate mechanisms that underlie the beneficial or adverse effects induced in the human body by the uptake of dietary nutrients or components. In recent years, novel omics technologies including transcriptomics, proteomics, metabolomics, and systems biology are getting more attention due to their power in addressing complex issues related to human health, disease, and nutrition.

To study the molecular basis of health effects of specific components of the diet, nutritionists increasingly make use of these state-of-the-art omics technologies (Zhang *et al.*, 2008). The term "genomics" refers to the study of nucleotide sequences in the genome of an organism. Nutrigenomics refers to the study of the impact of specific nutrients or diets on gene expression. It should not be confused with another closely related discipline, "nutrigenetics," which investigates how genetic variability influences the body's response to a nutrient or diet. Thus, nutrigenomics and nutrigenetics approach the interplay of diet and genes from opposing start points. Transcriptomics measures the relative amounts of messenger RNAs (mRNAs) in a given organism for determining patterns and levels of gene expression. Proteomics is the study of proteins expressed in a cell, tissue, or organism, including all protein isoforms and posttranslational modifications. Metabolomics is defined as the comprehensive analysis of all metabolites generated in a given biological system, focusing on measurements of metabolite concentrations and secretions in cells and tissues. It is not to be confused with "metabonomics," which investigates the fingerprint of biochemical perturbations caused by disease, drugs, and toxins (Goodacre, 2007). Systems biology aims at simultaneous measurement of genomic, transcriptomic, proteomic, and metabolomic parameters in a given system under defined conditions. The vast amount of data generated with such omics technologies requires the application of advanced bioinformatics tools to obtain a holistic view of the effects of nutrients or nonnutrient components of foods and to identify a system of biomarkers that can predict beneficial or adverse effects of dietary nutrients or components for promoting health and preventing disease.

TRANSCRIPTOMICS IN NUTRITIONAL RESEARCH

The classical gene analysis approach, such as Northern blotting and real-time RT-PCR, can only analyze gene expression for a limited number of candidate genes at a time. DNA microarray technology allows for measuring the expression level of thousands of genes, or even entire genomes, simultaneously. A typical DNA microarray experiment includes a number of characteristic steps: (1) RNA extraction from a sample; (2) reverse transcription of the RNA to obtain complementary DNA (cDNA) and labeling of the cDNA with specific dyes (usually fluorophores like Cyanine 3 and 5), or reverse transcription of the cDNA to obtain cRNA and labeling of the cRNA; (3) hybridization of the labeled cDNA or cRNA onto the microarray under given conditions; (4) washing of the slides to remove nonhybridized labeled oligonucleotides; (5) use of an appropriate scanning device to detect signal; and (6) data analysis by bioinformatics tools.

There are more examples of DNA microarray technology being performed in cell culture systems or laboratory animals to identify the cellular responses to dietary constituents and their molecular targets. For example, green tea catechins (McLoughlin et al., 2004; Vittal et al., 2004), soy isoflavones (Herzog et al., 2004), polyunsaturated fatty acids (Kitajka et al., 2004; Lapillonne et al., 2004; Narayanan et al., 2003), vitamins D and E (Johnson and Manor, 2004; Lin et al., 2002), quercetin (Murtaza et al., 2006), arginine (Leong et al., 2006), anthocyanins (Tsuda et al., 2006), hypoallergenic wheat flour (Narasaka et al., 2006). In particular, very recently, Lavigne et al. (2008) used a DNA oligo microarray approach to examine effects of genistein on global gene expression in MCF-7 breast cancer cells. They found that genistein altered the expression of genes belonging to a wide range of pathways, including estrogen- and p53-mediated pathways. At physiologic concentrations (1 or 5 mu M), genistein elicited an expression pattern of increased mitogenic activity, whereas at pharmacologic concentrations (25 mu M), genistein generated an expression pattern of increased apoptosis, decreased proliferation, and decreased total cell number. Park et al. (2008) performed a comprehensive analysis of hepatic gene expression in a rat model of an alcohol-induced fatty liver using the cDNA microarray. They found that chronic ethanol consumption regulated mainly the genes related to the processes of signal transduction, transcription, immune response, and protein/amino acid metabolism. For the first time, this study revealed that five genes (including β-glucuronidase, UDP-glycosyltransferase 1, UDP-glucose dehydrogenase, apoC-III, and gonadotropin-releasing hormone receptor) were regulated by chronic ethanol exposure in the rat liver.

Furthermore, the number of microarray-based transcriptomics analyses for assessing the biological effects of dietary interventions on human nutrition and health is steadily increasing. van Erk et al. (2006) investigated the effect of a high-carbohydrate (HC) or high-protein (HP) breakfast on the transcriptome of human blood cells with RNA samples taken from eight healthy men before and 2 hours after consumption of the diets. About 317 genes for the HC breakfast and 919 genes for the HP breakfast were found to be differentially expressed. Specifically, consumption of the HC breakfast resulted in differential expression of glycogen metabolism genes, and consumption of the HP breakfast resulted in differential expression of genes involved in protein biosynthesis. Using GeneChip microarrays, Schauber et al. (2006) examined the effect of regular consumption of low-digestible and prebiotic isomalt and the digestible sucrose on gene expression in rectal mucosa in a randomized double-blind crossover trial with 19 healthy volunteers over 4 weeks of feeding. They revealed that dietary intervention with the low digestible isomalt compared with digestible sucrose did not affect gene expression in the lining rectal mucosa, although gene expression of the human rectal mucosa can reliably be measured in biopsy material. Mangravite *et al.* (2007) used expression array analysis to identify molecular pathways responsive to both caloric restriction and dietary composition within adipose tissue from 131 moderately overweight men. They found that more than 1000 transcripts were significantly downregulated in expression in response to acute weight loss. The results demonstrated that stearoyl-coenzyme A desaturase (SCD) expression in adipose tissue is independently regulated by weight loss and by carbohydrate and saturated fat intakes, and SCD and diacylglycerol transferase 2 (DGAT2) expression may be involved in dietary regulation of systemic triacylglycerol metabolism. Kallio *et al.* (2007) assessed the effect of two different carbohydrate modifications (a rye-pasta diet characterized by a low postprandial insulin response and an oat-wheat-potato diet characterized by a high postprandial insulin response) on subcutaneous adipose tissue (SAT) gene expression in 47 persons with the metabolic syndrome. They detected that the rye-pasta diet downregulated 71 genes (linked to insulin signaling and apoptosis) and oat-wheat-potato diet upregulated 62 genes (related to stress, cytokine-chemokine-mediated immunity, and the interleukin pathway). Using microarray analysis, Niculescu *et al.* (2007) investigated the effects of dietary soy isoflavones on gene expression changes in lymphocytes from 30 postmenopausal women. They indicated that isoflavones had a stronger effect on some putative estrogen-responsive genes in equol producers than in nonproducers. In general, the gene expression changes caused by isoflavone intervention are related to increased cell differentiation, increased cAMP signaling and G-protein-coupted protein metabolism, and increased steroid hormone receptor activity.

However, there are some problems or limitations for transcriptomics approaches in nutritional research. One major problem is nonreproducibility of gene expression profiles. Different conclusions could be drawn from the same experiment performed at different times or in different labs or platforms. Fortunately, for reducing errors or variations, standards for reporting microarray data have been established under MIAME (minimum information about a microarray experiment) (Brazma *et al.*, 2001). Barnes *et al.* (2005) evaluated the reproducibility of microarray results using two platforms, Affymetrix GeneChips and Illumina BeadArrays. The results demonstrated that agreement was strongly correlated with the level of expression of a gene, and concordance was improved when probes on the two platforms could be identified as likely to target the same set of transcripts of a given gene. Another major issue is the analysis of data sets and their interpretation. Analyses only providing gene lists with significant p-values are insufficient to fully understand the underlying biological mechanisms; a single gene significantly upregulated or downregulated does not necessarily have any physiological meaning (Kussmann *et al.*, 2008). The combination of statistical and functional analysis is appropriate to facilitate the identification of biologically relevant and robust gene signatures, even across different microarray platforms (Bosotti *et al.*, 2007). An additional and more specific limitation in human nutritional applications is that microarray studies require significant quantities of tissues material for isolation of the needed RNA, whereas access to human tissues is obviously limited, although it is not impossible to obtain biopsies from control subjects involved in a nutrition research. If using human blood cell instead of tissue material, large interindividual variation exists in gene expression profiles of healthy individuals (Cobb *et al.*, 2005), this makes it challenging to identify robust gene expression signatures in response to a nutrition intervention. On the other hand, sample handling and prolonged transportation significantly influences gene expression profiles (Debey *et al.*, 2004). Highly standardized protocol across different labs is needed. Whole-blood samples require depletion of globin mRNA to enable detection of low-abundance transcripts, which have not been employed in human nutritional studies until now.

PROTEOMICS IN NUTRITIONAL RESEARCH

In the last 2 decades, proteomics has developed into a technology for biomarker discovery, disease diagnosis, and clinical applications (Beretta, 2007; Lescuyer *et al.*, 2007; Zhang *et al.*, 2007a, 2007b). The workflow for proteomics analysis consists of sample preparation, protein separation, and protein identification.

For gel-based proteomics experiments, proteins are extracted from cell or tissue samples, separated by two-dimensional polyacrylamide gel electrophoresis (2D gel), and stained. To identify differences in protein content between protein samples, images of the spots on the gels can be compared. Subsequently, the protein spots of interest are excised and the proteins are digested. Lastly, resulting peptides can be identified by mass spectrometry (MS). However, 2D-gel technology has many inherent drawbacks (Corthesy-Theulaz *et al.*, 2005; Kussmann *et al.*, 2005): (1) bias toward the most abundant changes, giving poor resolution for low abundant proteins, which might generate erroneous conclusions due to the fact that subtle variation may lead to important changes in metabolic pathways; (2) inability to detect proteins with extreme properties (very small, very large, very hydrophobic, and very acidic or basic proteins); (3) difficulty in identifying proteins.

Instead of the gel approaches, chromatography-based techniques have been developed for protein/peptide separation, such as gas chromatography (GC) and liquid chromatography (LC). When these separation technologies are combined with MS or tandem MS (MS/MS), the superior power of MS in the proteomic analysis is greatly enhanced. The MS instruments most used for proteomics experiments are ESI-MS (electrospray ionization MS), MALDI-TOF-MS (matrix-assisted laser desorption ionization with a time-of-flight MS), and its variant SELDI-TOF-MS (surface-enhanced laser desorption ionization with a time-of-flight MS). In addition, FTICR-MS (Fourier transform ion cyclotron resonance MS) is an increasingly useful technique in proteomic research, providing the highest mass resolution, mass accuracy, and sensitivity of present MS technologies, although with relatively high costs (Bogdanov and Smith, 2005).

In previous years, there were exponentially increasing publications on the application of proteomic techniques to nutrition research (Griffiths and Grant, 2006), but many investigations were performed in animal models (Breikers *et al.*, 2006; de Roos *et al.*, 2005; Kim *et al.*, 2006). Limited proteomics analysis in humans was involved in identifying the molecular target of dietary components in human subjects. For example, proteomic analysis of butyrate-treated human colon cancer cells (Tan *et al.*, 2002), identification of molecular targets of quercetin in human colon cancer cells (Wenzel *et al.*, 2004), and identification of cellular target proteins of genistein action in human endothelial cells (Fuchs *et al.*, 2005). Recently, Smolenski *et al.* (2007) applied 2D gel and MALDI-TOF-MS identified 15 proteins involved in host defense. Batista *et al.* (2007) employed 2D gel and MS method to identify new potential soybean allergens from transgenic and nontransgenic soy samples. Similarly, a proteomic analysis method based on 2D-gel and MALDI-TOF-MS was used to characterize wheat flour allergens and revealed that nine subunits of glutenins are the most predominant IgE-binding antigens (Akagawa *et al.*, 2007). Fuchs *et al.* (2007) conducted the proteomic analysis of human peripheral blood mononuclear cells (PBMC) from seven healthy men after a dietary flaxseed intervention. The results showed that flaxseed consumption affected significantly steady-state levels of 16 proteins, including enhanced levels of peroxiredoxin, reduced levels of the long-chain fatty acid beta-oxidation multienzyme complex, and reduced levels of glycoprotein IIIa/II. PBMCs are an important sample for monitoring dietary interventions and are accessible with little invasive means. Vergara *et al.* (2008) have established a public 2-DE database for human peripheral blood mononuclear cell (PBMCs) proteins, which have the potential of PBMCs to investigate the proteomics changes possibly associated with food or drug interventions.

In any proteomic study aiming at biomarker discovery, a critical question is: "How much of a given protein is present at a given time in a given condition?" Recently, a number of quantitative proteomic techniques have been developed, such as 2D DIGE (difference gel electrophoresis),

ICAT (isotope-coded affinity tag), iTRAQ (isobaric tags for relative and absolute quantification), and proteolytic O-18-labeling strategies (Chen *et al.*, 2007a; Miyagi *et al.*, 2007). Wu *et al.* (2006) conducted the comparative study of three methods (DIGE, ICAT, and iTRAQ) and demonstrated that all three techniques yielded quantitative results with reasonable accuracy, although iTRAQ is more sensitive than DIGE and ICAT. Due to the fact that these methods displayed limited overlapping among the proteins identified, the complementary information obtained from different methods should provide a better understanding of biological effects of dietary intervention. However, there are some potential problems: the protein comigration problem for DIGE, cysteine-content bias for ICAT, susceptibility to errors in precursor ion isolation for iTRAQ. All quantification approaches discussed so far deliver relative quantitative information. More recently, absolute or stoichiometric quantification of proteome is becoming feasible, in particular, with the development of strategies with isotope-labeled standards composed of concatenated peptides. On the other hand, remarkable progress has also been made in label-free quantification methods based on the number of identified peptides (Gerber *et al.*, 2003; Kito and Ito, 2008; Old *et al.*, 2005). To date, few samples of quantitative proteomics analyses in nutritional research are available. For example, using DIGE and MALDI-MS/MS, Alm *et al.* (2007) performed proteomic variation analysis within and between different strawberry varieties. They found that the biological variation was more affected by different growth conditions than by different varieties, the amount of strawberry allergen varied between different strawberry varieties, and the allergen content in colorless (white) strawberry varieties was always lower than that in the red ones. However, only three proteins are the same among the proteins correlated with the allergen and the color, which means it is possible to breed a strawberry with a low amount of allergen. Thus, the proteomic-based method has the potential to be used for the variety improvement of fruits or vegetables.

Furthermore, protein microarray technology is a promising approach for proteomics, which can be used to detect changes in the expression and posttranslational modifications of hundreds or even thousands of proteins in a parallel way. Its advantages include high sensitivity, good reproducibility, quantitative accuracy, and parallelization. The details of protein microarray method are described in a recent review (Kricka *et al.*, 2006). Protein microarray platforms should open new possibilities for novel insight into the molecular mechanisms underlying nutrient–gene or nutrient–drug interactions (such as grapefruit–cyclosporine interaction). Puskas *et al.* (2006) applied the Panorama protein microarray to analyze the cholesterol diet-induced protein expression and found that a different phosphorylation pattern could be detected as well. This is of great interest in nutrition research.

METABOLOMICS IN NUTRITIONAL RESEARCH

Changes in mRNA concentration do not necessarily result in changes in cellular protein levels, and changes in protein levels may not always cause changes in protein activity. Metabolites represent the real endpoints of gene expression. Thus, alterations in the concentrations of metabolites may be better suited to describe the physiological regulatory processes in a biological system and may be a better measure of gene function than the transcriptome and proteome. Biological effects in nutrition cannot be reduced to the action of a single molecule but actually result from the modulation of many metabolic pathways at the same time, which is the product of a complex interplay between multiple genomes represented by the mammalian host and its gut microflora and environmental factors (e.g., food habits, diet composition, and other lifestyle components) (Nicholson *et al.*, 2004; Rezzi *et al.*, 2007a). Metabolomics in nutrition, although a recent introduction in the field of nutrition research, has already delivered interesting insights to understand metabolic responses of humans or animals to dietary interventions.

The workflow for metabolomics involves a tandem use of analytical chemistry techniques to generate metabolic profiles and various bioinformatics tools to extract relevant metabolic information. Currently, the widely used tool for metabolomics experiments in nutrition research is proton nuclear magnetic resonance (NMR) technology. For example, the determination of metabolic effect of vitamin E supplementation in a mouse model of motor neuron degeneration (Griffin *et al.*, 2002); the evaluation of biochemical effects following dietary intervention with soy isoflavones in five healthy premenopausal women (Solanky *et al.*, 2003); the detection of human biological responses to different diets (e.g., chamomile tea (Wang *et al.*, 2005) or vegetarian, low-meat, and high-meat diets (Stella *et al.*, 2006)); the characterization of the metabolic variability due to different populations (e.g., American, Chinese, and Japanese (Dumas *et al.*, 2006a) or Swedish and British populations (Lenz *et al.*, 2004)). Recently, Bertram *et al.* (2007) employed a NMR-based metabolomic method to investigate biochemical effects of a short-term high intake of milk protein or meat protein on 8-year-old boys. This was the first report to demonstrate the capability of proton NMR-based metabolomics in identifying the overall biochemical effects of consumption of different animal proteins. They found that the milk diet increased the urinary excretion of hippurate, whereas the meat diet increased the urinary excretion of creatine, histidine, and urea. Moreover, based on NMR analysis of serum, the results demonstrated that the milk diet slightly changed the lipid profile of serum, but the meat diet had no effect on the metabolic profile of serum. Fardet *et al.* (2007) investigated the metabolic responses of rats fed whole-grain flour (WGF) and refined wheat flour (RF) by an NMR-based metabolomic approach. The results showed that some tricarboxylic acid cycle intermediates, aromatic amino acids, and hippurate were significantly increased in the urine of rats fed the WGF diet.

Another exciting and powerful tool for metabolomics is MS-based proteomics technology. The main advantage of MS technique is its high sensitivity and rapid determination of mass or structure information. MS instruments in combination with some separation technologies (such as gas or liquid chromatography (GC or LC), or capillary electrophoresis (CE)) can quantitatively profile molecular entities like lipids, amino acids, bile acids, and other organic solutes at high sensitivity (Fiehn *et al.*, 2000; Watkins and German, 2002). Moreover, a recent introduction into the field of metabolomics is the HPLC systems using sub-2 μm packing columns combined with high operating pressures (UPLC technology). Compared with conventional HPLC-TOF-MS systems using 3–5 μm packing columns, UPLC-TOF-MS systems allow a remarkable decrease in analysis time, higher peak capacity and increased sensitivity. Currently, only limited applications of MS-based metabolomics to nutritional research are reported. For example, HPLC-TOF-MS-based study of changes of urinary endogenous metabolites associated with aging in rats (Williams *et al.*, 2005); a noninvasive extractive ESI-Q-TOF-MS for differentiation of maturity and quality of bananas, grapes, and strawberries (Chen *et al.*, 2007b); a combined GC-MS and LC-MS metabolic profiling for a comprehensive understanding of systems responsive to aristolochic acid intervention in rats (Ni *et al.*, 2007).

However, a major problem for metabolomics is that the experimental metabolic profile is influenced not only by the genotype but also by age, gender, lifestyle, nutritional status, drugs, stress, physical activity, etc. To minimize variations in studies with humans, some attempts were made, such as using standardized diet, avoiding any vigorous activity, excluding smokers, and so on. Unfortunately, even under the consumption of standard diet, metabolic variability remains. Using 1H NMR spectroscopy, Walsh *et al.* (2006) investigated the acute effects of standard diet on the metabonomic profiles of urine, plasma, and saliva samples from 30 healthy volunteers. There are important biochemical variabilities to be observed for all biofluids at both intra- and interindividual levels, and significant variations in creatinine and acetate for urine and saliva respectively exist. After the consumption of standard diet, a reduction in interindividual variation was observed in urine, but not in plasma or saliva. Indeed, different diets in different populations lead to different metabolic profiles (Rezzi *et al.*, 2007a): higher urinary levels in creatine, creatinine, carnitine, acetylcarnitine, taurine, trimethylamine-N-oxide (TMAO), and glutamine are the metabolic signature of high-meat diet; higher urinary excretion of p-hydroxyphenylacetate, a microbial mammalian cometabolite, and a decreased level in N,N,N-trimethyllysine are associated with the vegetarian diet; elevated β-aminoisobutyric acid and ethanol in Chinese urinary samples; increased urinary excretion in TMAO in the Japanese and Swedish populations due to the high dietary intake of fish diet; and usually high level of urinary taurine in the British population as a consequence of the Atkins diet. A recent report reveals a "natural," stable over time, and invariant metabolic profile for each person, although the existence of human metabolic variations resulting from various dietary patterns (Assfalg et al., 2008). This provides the possibility of eliminating the day-to-day "noise" of the individual metabolic fingerprint and opens new perspectives to metabolomic studies for personalized therapy and nutrition.

Another important issue in nutritional metabolomics is gut microbiota–host metabolic inter-actions. Such an interaction between the microbiome and humans makes the human become a "superorganism" (Goodacre, 2007). More than 400 microbial species exist in the large-bowel microflora of healthy humans, which produce significant metabolic signals so that the true metabolomic signals of nutrients in diet could be "swamped," and the metabolome of biofluids in human nutrition is altered. Dumas et al. (2006b) investigated the metabolic rela-tionship between gut microflora and host cometabolic phenotypes using the plasma and urine metabolic NMR profile of mouse. They found that the urinary excretion of methylamines from the precursor choline was directly related to microflora metabolism, demonstrating significant interaction between the mammalian host and microbiota metabolism. Recently, Rezzi et al. (2007b) performed the NMR analysis of plasma and urine metabolic profiles in 22 healthy male volunteers with behavioral/psychological dietary preference (chocolate desiring or choco-late indifferent). The results revealed that chocolate preference was associated with a specific metabolic signature, which is imprinted in the metabolism even in the absence of chocolate as a stimulus. This suggests that specific dietary preferences can influence basal metabolic state and gut microbiome activity that in turn may have long-term health consequences to the host.

SYSTEMS BIOLOGY IN NUTRITIONAL RESEARCH

To better understand the complex interplay between genes, diet, lifestyle, and endogenous gut microflora, and to understand how diet can be modified to maintain optimal health throughout life, the integrative use of various omics technologies–systems biology technology offer exciting opportunities to develop the emerging area of personalized nutrition and health care (Naylor et al., 2008; Zhang et al., 2008). There has been limited work in this arena.

Using an integrated reverse functional genomic and metabolic approach, Griffin et al. (2004) identified perturbed metabolic pathways by orotic acid treatment. In searching for correla-tions between the 60 most differentially expressed genes and the largest changed metabolite trimethylamine-N-oxide, they found that the most significant negative correlation is stearyl-CoA desaturase 1, and they highlight the relationship between transcripts and metabolites in lipid pathways. Herzog et al. (2004) performed proteome and transcriptome analysis of human colon cancer cells treated with flavone. About 488 mRNA targets were found to be regulated by flavone at least twofold. On the other hand, many proteins involved in gene regulation, detoxification, and intermediary metabolism, such as annexin II and apolipoprotein, were found to be altered by flavone exposure. Dieck et al. (2005) conducted the transcriptome and pro-teome analysis to identify the underlying molecular changes in hepatic lipid metabolism in zinc-deficient rats. The experimental findings provide evidence that an unbalanced gene tran-scription control via PPAR-α,thyroid hormone and SREBP-dependent pathways could explain most of the effects of zinc deficiency on hepatic fat metabolism. Mutch et al. (2005) used an integrative transcriptome and lipid-metabolome approach to underdstand the molecular mech-anisms regulated by the consumption of PUFA. They identified stearoyl-CoA desaturase as a target of an arachidonate-enriched diet and revealed a previously unrecognized and distinct role for arachidonate in the regulation of hepatic lipid metabolism. By combining DNA microarray, proteomics, and metabolomics platforms, Schnackenberg et al. (2006) investigated the acute

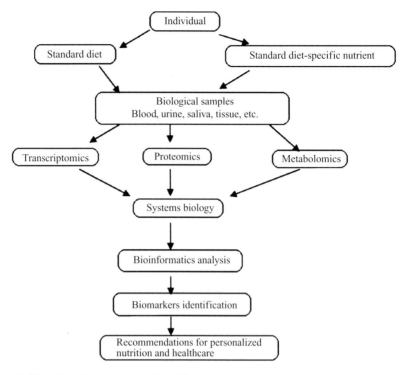

Figure 2.1 Flowchart for omics-based nutritional research.

effects of valproic acid in the liver and demonstrated a perturbation in the glycogenolysis pathway after administration of valproic acid.

CONCLUSIONS

The main goal of omics-based nutrition research is to understand the relationships between diet and disease and the relationships between diet and health, and finally to make recommendation for personalized nutrition or individualized diet (Figure 2.1, modified from Zhang *et al.* (2008)). To better understand the complex interplay that occurs between the individual in terms of genetics, physiology, health, diet and environment, comparative genetic, transcriptomic, proteomic, and metabolomic analyses for individuals and populations are highly required. In particular, systems biology, more than the simple merge of various omics technologies (transcriptomics, proteomics, and metabolomics), aims to understand the biological behavior of a cellular system in response to external stimuli, and opens the new road to understanding the complex interaction network between nutrients and molecules in biological systems. An era of personalized medicine and nutrition is coming.

ACKNOWLEDGMENTS

This work was partly supported by Guangdong Scientific Development Grant 2009B090300271 and 2007A020100001-4.

REFERENCES

Akagawa, M., Handoyo, T., Ishii, T., Kumazawa, S., *et al.*, Proteomic analysis of wheat flour allergens. *J Agric Food Chem* 2007, *55*(17), 6863–6870.

Alm, R., Ekefjard, A., Krogh, M., Hakkinen, J., *et al.*, Proteomic variation is as large within as between strawberry varieties. *J Proteome Res* 2007, *6*(8), 3011–3020.

Assfalg, M., Bertini, I., Colangiuli, D., Luchinat, C., *et al.*, Evidence of different metabolic phenotypes in humans. *Proc Natl Acad Sci USA* 2008, *105*(5), 1420–1424.

Barnes, M., Freudenberg, J., Thompson, S., Aronow, B., *et al.*, Experimental comparison and cross-validation of the Affymetrix and Illumina gene expression analysis platforms. *Nucleic Acids Res* 2005, *33*, 5914–5923.

Batista, R., Martins, I., Jenoe, P., Ricardo, C. P., *et al.*, A proteomic study to identify soya allergens - The human response to transgenic versus non-transgenic soya samples. *Int Arch Allergy Immunol* 2007, *144*(1), 29–38.

Beretta, L., Proteomics from the clinical perspective: many hopes and much debate. *Nat Methods* 2007, *4*, 785–786.

Bertram, H. C., Hoppe, C., Petersen, B. O., Duus, J. O., *et al.*, An NMR-based metabonomic investigation on effects of milk and meat protein diets given to 8-year-old boys. *Br J Nutr* 2007, *97*(4), 758–763.

Bogdanov, B., Smith, R. D. Proteomics by FTICR mass spectrometry: Top down and bottom up. *Mass Spectrom Rev* 2005, *24*(2), 168–200.

Bosotti, R., Locatelli, G., Healy, S., Scacheri, E., *et al.*, Cross platform microarray analysis for robust identification of differentially expressed genes. *BMC Bioinform* 2007, *8*(Suppl. 1), 5.

Brazma, A., Hingamp, P., Quackenbush, J., Sherlock, G., *et al.*, Minimum information about a microarray experiment (MIAME) – toward standards for microarray data. *Nat Genet* 2001, *29*, 365–371.

Breikers, G., van Breda, S. G. J., Bouwman, F. G., van Herwijnen, M. H. M., *et al.*, Potential protein markers for nutritional health effects on colorectal cancer in the mouse as revealed by proteomics analysis. *Proteomics* 2006, *6*(9), 2844–2852.

Chen, H. W., Sun, Y. P., Wortmann, A., Gu, H. W., *et al.*, Differentiation of maturity and quality of fruit using noninvasive extractive electrospray ionization quadrupole time-of-flight mass spectrometry. *Anal Chem* 2007b, *79*(4), 1447–1455.

Chen, X., Sun, L. W., Yu, Y. B., Xue, Y., *et al.*, Amino acid-coded tagging approaches in quantitative proteomics. *Expert Rev Proteomics* 2007a, *4*(1), 25–37.

Corthesy-Theulaz, I., den Dunnen, J. T., Ferre, P., Geurts, J. M. W., *et al.*, Nutrigenomics: The impact of biomics technology on nutrition research. *Ann Nutr Metab* 2005, *49*(6), 355–365.

Debey, S., Schoenbeck, U., Hellmich, M., Gathof, B. S., *et al.*, Comparison of different isolation techniques prior to gene expression profiling of blood derived cells: impact on physiological responses, on overall expression and the role of different cell types. *Pharmacogenomics* 2004, *4*, 193–207.

de Roos, B., Duivenvoorden, I., Rucklidge, G., Reid, M., *et al.*, Response of apolipoprotein E*3-Leiden transgenic mice to dietary fatty acids: combining liver proteomics with physiological data. *FASEB J* 2005, *19*(3), 813–838.

Dieck, H. T., Doring, F., Fuchs, D., Roth, H. P., *et al.*, Transcriptome and proteome analysis identifies the pathways that increase hepatic lipid accumulation in zinc-deficient rats. *J Nutr* 2005, *135*(2), 199–205.

Dumas, M. E., Barton, R. H., Toye, A., Cloarec, O., *et al.*, Metabolic profiling reveals a contribution of gut microbiota to fatty liver phenotype in insulin resistant mice. *Proc Natl Acad Sci USA* 2006b, *103*(33), 12511–12516.

Dumas, M. E., Maibaum, E. C., Teague, C., Ueshima, H., *et al.*, Assessment of analytical reproducibility of 1H NMR spectroscopy based metabolomics for large-scale epidemiological research: the INTERMAP Study. *Anal Chem* 2006a, *78*(7), 2199–2208.

Fardet, A., Canlet, C., Gottardi, G., Lyan, B., *et al.,* Whole-grain and refined wheat flours show distinct metabolic profiles in rats as assessed by a H-1 NMR-based metabonomic approach. *J Nutr* 2007, *137*(4), 923–929.

Fiehn, O., Kopka, J., Dormann, P., Altmann, T., *et al.,* Metabolite profiling for plant functional genomics. *Nat Biotechnol* 2000, *18*, 1157–1161.

Fuchs, D., Erhard, P., Rimbach, G., Daniel, H., *et al.,* Genistein blocks homocysteine-induced alterations in the proteome of human endothelial cells. *Proteomics* 2005, *5*(11), 2808–2818.

Fuchs, D., Piller, R., Linseisen, J., Daniel, H., *et al.,* The human peripheral blood mononuclear cell proteome responds to a dietary flaxseed intervention and proteins identified suggest a protective effect in atherosclerosis. *Proteomics* 2007, *7*, 3278–3288.

Gerber, S. A., Rush, J., Stemman, O., Kirschner, M. W., *et al.,* Absolute quantification of proteins and phosphoproteins from cell lysates by tandem MS. *Proc Natl Acad Sci USA* 2003, *100*, 6940–6945.

Goodacre, R., Metabolomics of a superorganism. *J Nutr* 2007, *137*(1), 259S–266S.

Griffin, J. L., Bonney, S. A., Mann, C., Hebbachi, A. M., *et al.,* An integrated reverse functional genomic and metabolic approach to understanding orotic acid-induced fatty liver. *Physiol Genomics* 2004, *17*(2), 140–149.

Griffin, J. L., Muller, D., Woograsingh, R., Jowatt, V., *et al.,* Vitamin E deficiency and metabolic deficits in neuronal ceroid lipofuscinosis described by bioinformatics. *Physiol Genomics* 2002, *11*(3), 195–203.

Griffiths, H. R., Grant, M. M., The use of proteomic techniques to explore the holistic effects of nutrients in vivo. *Nutr Res Rev* 2006, *19*(2), 284–293.

Herzog, A., Kindermann, B., Döring, F., Daniel, H., *et al.,* Pleiotropic molecular effects of the pro-apoptotic dietary constituent flavone in human colon cancer cells identified by protein and mRNA expression profiling. *Proteomics* 2004, *4*, 2455–2464.

Johnson, A., Manor, D., The transcriptional signature of vitamin E. *Ann NY Acad Sci* 2004, *1031*, 337–338.

Kallio, P., Kolehmainen, M., Laaksonen, D. E., Kekäläinen, J., *et al.,* Dietary carbohydrate modification induces alterations in gene expression in abdominal subcutaneous adipose tissue in persons with the metabolic syndrome: the FUNGENUT Study. *Am J Clin Nutr* 2007, *85*, 1417–1427.

Kim, H., Deshane, J., Barnes, S., Meleth, S., Proteomics analysis of the actions of grape seed extract in rat brain: Technological and biological implications for the study of the actions of psychoactive compounds. *Life Sciences* 2006, *78*(18), 2060–2065.

Kitajka, K., Sinclair, A. J., Weisinger, R. S., Weisinger, H. S., *et al.,* Effects of dietary omega-3 polyunsaturated fatty acids on brain gene expression. *Proc Natl Acad Sci USA* 2004, *101*(30), 10931–10936.

Kito, K., Ito, T., Mass spectrometry-based approaches toward absolute quantitative proteomics. *Curr Genomics* 2008, *9*(4), 263–274.

Kricka, L. J., Master, S. R., Joos, T. O., Fortina, P., Current perspectives in protein array technology. *Ann Clin Biochem* 2006, *43*, 457–467.

Kussmann, M., Affolter, M., Fay, L. B., Proteomics in nutrition and health. *Comb Chem High Throughput Screen* 2005, *8*, 679–696.

Kussmann, M., Rezzi, S., Daniel, H., Profiling techniques in nutrition and health research. *Curr Opin Biotechnol* 2008, *19*, 83–99.

Lapillonne, A., Clarke, S. D., Heird, W. C., Polyunsaturated fatty acids and gene expression. *Curr Opin Clin Nutr Metab Care* 2004, *7*(2), 151–156.

Lavigne, J. A., Takahashi, Y., Chandramouli, G. V. R., Liu, H. T., *et al.,* Concentration-dependent effects of genistein on global gene expression in MCF-7 breast cancer cells: an oligo microarray study. *Breast Cancer Res Treat* 2008, *110*(1), 85–98.

Lenz, E. M., Bright, J., Wilson, I. D., Hughes, A., Metabonomics, dietary influences and cultural differences: a H-1 NMR-based study of urine samples obtained from healthy British and Swedish subjects. *J Pharm Biomed Anal* 2004, *36*(4), 841–849.

Leong, H. X., Simkevich, C., Lesieur-Brooks, A., Lau, B. W., *et al.*, Short-term arginine deprivation results in large-scale modulation of hepatic gene expression in both normal and tumor cells: microarray bioinformatic analysis. *Nutrition & Metabolism* 2006, *3*, 37.

Lescuyer, P., Hochstrasser, D. F., Rabilloud, T., How shall we use the proteomics toolbox for biomarker discovery? *J Proteome Res* 2007, *6*, 3371–3376.

Lin, R., Nagai, Y., Sladek, R., Bastien, Y., *et al.,* Expression profiling in squamous carcinoma cells reveals pleiotropic effects of vitaminD3 analog EB1089 signaling on cell proliferation, differentiation, and immune system regulation. *Mol Endocrinol* 2002, *16*, 1243–1256.

Mangravite, L. M., Dawson, K., Davis, R. R., Gregg, J. P., *et al.,* Fatty acid desaturase regulation in adipose tissue by dietary composition is independent of weight loss and is correlated with the plasma triacylglycerol response. *Am J Clin Nutr* 2007, *86*(3), 759–767.

McLoughlin, P., Roengvoraphoj, M., Gissel, C., Hescheler, J., Transcriptional responses to epigallocatechin-3 gallate in HT 29 colon carcinoma spheroids. *Genes to Cells* 2004, *9*(7), 661–669.

Miyagi, M., Rao, K. C. S., Sekhar, K. C., Proteolytic O-18-labeling strategies for quantitative proteomics. *Mass Spectrom Rev* 2007, *26*(1), 121–136.

Murtaza, L., Marra, G., Schlapbach, R., Patrignani, A., *et al.,* A preliminary investigation demonstrating the effect of quercetin on the expression of genes related to cell-cycle arrest, apoptosis and xenobiotic metabolism in human COIIScolon-adenocarcinoma cells using DNA microarray. *Biotechnol Appl Biochem* 2006, *45*, 29–36.

Mutch, D. M., Grigorov, M., Berger, A., Fay, L. B., *et al.,* An integrative metabolism approach identifies stearoyl-CoA desaturase as a target for an arachidonate-enriched diet. *FASEB J* 2005, *19*(1), 599.

Narasaka, S., Endo, Y., Fu, Z. W., Moriyama, M., *et al.*, Safety evaluation of hypoallergenic wheat flour by using a DNA microarray. *Biosci Biotechnol Biochem* 2006, *70*(6), 1464–1470.

Narayanan, B. A., Narayanan, N. K., Simi, B., Reddy, B. S., Modulation of inducible nitric oxide synthase and related proinflammatory genes by the omega-3 fatty acid docosahexaenoic acid in human colon cancer cells. *Cancer Research* 2003, *63*(5), 972–979.

Naylor, S., Culbertson, A. W., Valentine, S. J., Towards a systems level analysis of health and nutrition. *Curr Opin Biotechnol* 2008, *19*(2), 100–109.

Ni, Y., Su, M. M., Qiu, Y. P., Chen, M. J., *et al.*, Metabolic profiling using combined GC-MS and LC-MS provides a systems understanding of aristolochic acid-induced nephrotoxicity in rat. *FEBS Lett* 2007, *581*(4), 707–711.

Nicholson, J. K., Holmes, E., Lindon, J. C., Wilson, I. D., The challenges of modeling mammalian biocomplexity. *Nat Biotechnol* 2004, *22*, 1268–1274.

Niculescu, M. D., Pop, E. A., Fischer, L. M., Zeisel, S. H., Dietary isoflavones differentially induce gene expression changes in lymphocytes from postmenopausal women who form equol as compared with those who do not. *J Nutr Biochem* 2007, *18*, 380–390.

Old, W. M., Meyer-Arendt, K., Aveline-Wolf, L., Pierce, K. G., *et al.*, Comparison of label-free methods for quantifying human proteins by shotgun proteomics. *Mol Cell Proteomics* 2005, *4*, 1487–1502.

Park, S. H., Choi, M. S., Park, T., Changes in the hepatic gene expression profile in a rat model of chronic ethanol treatment. *Food Chem Toxicol* 2008, *46*(4), 1378–1388.

Puskas, L. G., Menesi, D., Feher, L. Z., Kitajka, K., High-throughput functional genomic methods to analyze the effects of dietary lipids. *Curr Pharm Biotechnol* 2006, *7*(6), 525–529.

Rezzi, S., Ramadan, Z., Fay, L. B., Kochhar, S., Nutritional metabonomics: applications and perspectives. *J Proteome Res* 2007a, *6*, 513–525.

Rezzi, S., Ramadan, Z., Martin, F. P., Fay, L. B., *et al.,* Human metabolic phenotypes link directly to specific dietary preferences in healthy individuals. *J Proteome Res* 2007b, *6*, 4469–4477.

Schauber, J., Weiler, F., Gostner, A., Melcher, R., *et al.,* Human rectal mucosal gene expression after consumption of digestible and non-digestible carbohydrates. *Mol Nutr Food Res* 2006, *50*, 1006–1012.

Schnackenberg, L. K., Jones, R. C., Thyparambil, S., Taylor, J. T., *et al.,* An integrated study of acute effects of valproic acid in the liver using metabonomics, proteomics, and transcriptomics platforms. *OMICS* 2006, *10*(1), 1–14.

Smolenski, G., Haines, S., Kwan, F. Y. S., Bond, J., *et al.,* Characterisation of host defence proteins in milk using a proteomic approach. *J Proteome Res* 2007, *6*(1), 207–215.

Solanky, K. S., Bailey, N. J. C., Beckwith-Hall, B. M., Davis, A., *et al.,* Application of biofluid H-1 nuclear magnetic resonance-based metabonomic techniques for the analysis of the biochemical effects of dietary isoflavones on human plasma profile. *Anal Biochem* 2003, *323*(2), 197–204.

Stella, C., Beckwith-Hall, B., Cloarec, O., Holmes, E., *et al.,* Susceptibility of human metabolic phenotypes to dietary modulation. *J Proteome Res* 2006, *5*, 2780–2788.

Tan, S., Seow, T. K., Linag, R. C. M. Y., Koh, S., *et al.,* Proteome analysis of butyrate-treated human colon cancer cells (HT-29). *Int J Cancer* 2002, *98*, 523–531.

Tsuda, T., Ueno, Y., Yoshikawa, T., Kojo, H., *et al.,* Microarray profiling of gene expression in human adipocytes in response to anthocyanins. *Biochem Pharmacol* 2006, *71*(8), 1184–1197.

van Erk, M. J., Blom, W. A., van Ommen, B., Hendriks, H. F., High-protein and high-carbohydrate breakfasts differentially change the transcriptome of human blood cells. *Am J Clin Nutr* 2006, *84,* 1233–1241.

Vergara, D., Chiriaco, F., Acierno, R., Maffia, M., Proteomic map of peripheral blood mononuclear cells. *Proteomics* 2008, *8*(10), 2045–2051.

Vittal, R., Selvanayagam, Z. E., Sun, Y., Hong, J., *et al.,* Gene expression changes induced by green tea polyphenol (-)-epigallocatechin-3-gallate in human bronchial epithelial 21BES cells analyzed by DNA microarray. *Molecular Cancer Therapeutics* 2004, *3*(9), 1091–1099.

Walsh, M. C., Brennan, L., Malthouse, J. P. G., Roche, H. M., *et al.,* Effect of acute dietary standardization on the urinary, plasma, and salivary metabolomic profiles of healthy humans. *Am J Clin Nutr* 2006, *84*(3), 531–539.

Wang, Y. L., Tang, H. R., Nicholson, J. K., Hylands, P. J., *et al.,* A metabonomic strategy for the detection of the metabolic effects of chamomile (Matricaria recutita L.) ingestion. *J Agric Food Chem* 2005, *53*(2), 191–196.

Watkins, S. M., German, J. B., Toward the implementation of metabolomic assessments of human health and nutrition. *Curr Opin Biotechnol* 2002, *13*, 512–516.

Wenzel, U., Herzog, A., Kuntz, S., Daniel, H., Protein expression profiling identifies molecular targets of quercetin as a major dietary flavonoid in human colon cancer cells. *Proteomics* 2004, *4*, 2160–2174.

Williams, R. E., Lenz, E. M., Lowden, J. S., Rantalainen, M., *et al.,* The metabonomics of aging and development in the rat: an investigation into the effect of age on the profile of endogenous metabolites in the urine of male rats using 1H NMR and HPLC-TOF MS. *Mol Biosyst* 2005, *1*(2), 166–175.

Wu, W. W., Wang, G. H., Baek, S. J., Shen, R. F., Comparative study of three proteomic quantitative methods, DIGE, cICAT, and iTRAQ, using 2D gel- or LC-MALDI TOF/TOF. *J Proteome Res* 2006, *5*(3), 651–658.

Zhang, X. W., Li, L., Wei, D., Chen, F., Moving cancer diagnostics from bench to bedside. *Trends Biotechnol* 2007b, *25*(4), 166–173.

Zhang, X. W., Wei, D., Yap, Y. L., Li, L., *et al.,* Mass spectrometry-based "omics" technologies in cancer diagnostics. *Mass Spectrom Rev* 2007a, *26*, 403–431.

Zhang, X. W., Yap, Y., Wei, D., Chen, G., *et al.,* Novel omics technologies in nutritional research. *Biotechnol Adv* 2008, *26*, 169–176.

Section 2

Genomics

3

Nutrigenomics and Statistical Power: The Ethics of Genetically Informed Nutritional Advice

Ruth Chadwick

INTRODUCTION

Diet is increasingly a public policy issue, in light of increasing concerns about an obesity epidemic[1]. The ethics of nutritional advice requires attention in light of this. A variety of strategies is available for responding to the purported epidemic, on a spectrum from trying to change behavior patterns to developing pills that allow people to eat as much as they want without becoming obese. Specific possibilities include food labeling (e.g., the traffic light system), awareness campaigns such as the five-a-day program, through interventions such as stomach stapling and "obesity pills," to other forms of intervention such as controls on advertising of sugary snacks to children, including overfeeding of children in definitions of child abuse; and technological solutions such as functional foods and nutrigenomics (ngx). I will focus particularly on the significance of ngx in this context, with special reference to the extent of the ethical issues relating to the giving of advice and the relevance to ethics of statistics.

Whereas *nutrigenetics* is the study of the genetic differences between individuals that affect response to foods and food ingredients, *nutrigenomics* is genome-wide research into the genetic factors (including whole genes, single nucleotide polymorphisms (SNPs), and copy number variants (CNVs))[2] that are linked with these differential responses. Although there are differences between nutrigenetics and nutrigenomics, to some extent the terms are used interchangeably, and for the remainder of this article the latter term (ngx) will be used.

When any advice is given at population level, inevitably such differences between people are played down, if not ignored. Statistically, a given product may be effective in 80% of people, but for the remaining 20% it may be a waste of resources, whether public or private. Against this background, one of the promises of the postgenome era is the "personalizing" of advice.

In considering the coming together of public health concerns and the promises of genomics, the focus of social and ethical debates about biobanks is an additional factor that cannot be overlooked. It is in population-wide genomics research that the associations that underlie such personalization will be based. Although the tension between the interests of individuals and the common good has long been a focus of public health ethics, in discussions about public

---

[1] Some preliminary remarks are in order. First, the claim that there is an obesity "epidemic" or that it is a problem, is not universally accepted, but I shall not be investigating the evidence on this point. Second, "obesity" is not a category with clear boundaries, but I am not primarily concerned with its definition: my focus is on policy responses to the perceived problem. It is necessary, however, to take into account causal explanations of "obesity" when considering policy responses.

[2] The term "copy number variants" (CNBVs) refers to differences in the number of any given genetic factor that an individual has.

population research in relation to biobanks in particular there have been calls to rethink the relationship. Biobanks have been said, for example, by the World Health Organization, to be largely a force for the common good (World Health Organization, 2003). Ethical issues in relation to the promises of biobanks have been raised in relation to the process of collection of samples (including consent issues); access to information, including protection of privacy and confidentiality; the promises and hype surrounding them; whether population biobank research is a good thing in the first place given the opportunity costs; and whether there is a public mandate.

What has been underrepresented in biobank debates, I want to argue, is the relevance of statistical power, in relation to biobank ethics generally and subsequently in ngx in particular. Although it has long been recognized that the ethics of the use and abuse of statistics is a crucial issue, the implications of this for ngx and the population genetic research on which it rests have taken a back seat in relation to issues such as informed consent and choice. And yet these issues are central to the ethics of any *advice* that emanates from the research on biobanks, whether on diet or on any other type of health advice. Such personalized advice does not avoid the problems of statistical power that have traditionally attached to any kind of public health advice: it just shifts them to another place. On the other hand, arguably the ethics of advice is compromised by the *apparent* removal of some of the problems of statistical power. The relationship of advice-giving may be undermined by the way in which such advice is "marketed" through the rhetoric of personalization. It is crucial, therefore, to be clear about what "personalization" means.

PERSONALIZATION

What is meant by "personalization?" Arguably, the majority of discussion about pesonalization has taken place in relation to pharmaceuticals, although ngx has not been far behind. In the case of pharmacogenomics (pgx), in practice, it might mean no more than doing a genetic test to assign a person to one group or another – a poor or good responder to a drug, for example. If it is just allocation to a group, however, to use the term "personalization" is, to say the least, potentially misleading. The relevant difference is between this patient and the whole population, rather than between this patient and membership of *any* group. An ngx example can be found in the papers of the U.K. Advisory Committee on Novel Foods and Processes (ACNFP) where in ACNFP it was considered whether, in the light of the fact that 5–15% of the population have a genetically reduced activity of the enzyme responsible for converting folic acid to the active form, it would be more logical to administer folate in a different form to those individuals (ACNFP, 2007)

This example suggests that personalization can mean, secondly, particular action on the basis of genetic information – the choice between drug A and drug B; the selection of the appropriate dosage; the choice between drug A and no medication at all. In this sense, however, it could be argued that medicine has always been personalized in the sense of the application of professional judgment to the individual case. In this sense, again, to personalize and to individualize appear to be equivalent. In a third sense, to "personalize" might mean to treat the patient *as* a person, where a person is understood as a self-aware being with goals and life plans. It is in this sense that complementary therapies are sometimes said to be more concerned with the person than with the symptoms of the individual case. This is where a difference between individualization and personalization become apparent. To treat a patient as a person, rather than just as an (individual) case, is to have regard for their goals as well as symptoms.

The sense in which personalization is marketed as "tailoring," however, is not precisely equivalent to any of these meanings, although they are all significant from an ethical point of view. To have full tailoring to this individual/person before you, arguably, that person's whole genome needs to be looked at. Although the prospect of full genetic profiling is now firmly on the agenda for debate (Davies, 2007), the extent to which it is possible to make sense of all the information in a person's genome is still some way off. With those caveats in mind it may still be possible to come to some conclusions about the ethics of "personalized" advice.

NUTRIGENOMICS AND PHARMACOGENOMICS: THE DIFFERENCES

Although there are ethical issues that apply in the case of both pgx and ngx, there are also significant differences. These begin with the differences between drugs and foodstuffs. It is true that with the increasing prevalence of functional foods (Chadwick *et al.*, 2003), the boundary between drugs and foods is becoming fuzzier, but nevertheless it remains the case that although a drug is a well-defined substance that targets specific pathways in the body, this is rarely the case with foods, which tend to be complex and act upon multiple sites.

Beyond that, however, the moral argument for the importance of pgx is interestingly different from that for ngx. The moral argument for pgx has been put primarily in terms of avoiding harm (Chadwick, 2007). Statistics concerning the morbidity and mortality arising from adverse drug reactions are frequently quoted to support the moral urgency of developing genetically informed prescribing. It is not my purpose here to assess this argument in detail, which I have discussed elsewhere. The point is to draw a contrast with ngx.

The "do no harm" argument is, then, less prominent in discussions of ngx – on the contrary, despite the potential associations with prevention of diet-related disease (see Wallace, 2006), obesity, and allergy, there is a likelihood of a connection between ngx and enhancement[3]. Why is this? The context in which drugs, as opposed to food, are marketed and consumed may be an important consideration. Although both are associated with health claims, foods are marketed directly to consumers amid an increasingly varied choice of products, whereas drugs, apart from those that may be bought over the counter, are mediated by a professional's prescription. Food is consumed in a variety of social contexts and performs many functions apart from nutrition, including social bonding. Pharmaceuticals are typically sought in the context of an identified medical problem or at least a perceived symptom. Food is needed by all throughout life; medication is not. Any given food competes with a much larger array of alternatives than a particular drug. Given these differences, marketing foods in terms of their enhancing advantage may prove to be a beneficial marketing device. To say this is not to deny that pharmaceuticals may have enhancing properties that make them attractive – as in the case of the "better than well" effects of prozac, and amounts to the claim that the moral urgency claimed for the latter is not prominent in the case of the former.

Another reason for the lack of emphasis on moral urgency may be the number of alternatives to ngx – as indicated in the opening paragraph. Experience suggests, however, that strategies that attempt to change behavior are faced with particular difficulties in the case of food. Food is associated with powerful pleasures, for understandable biological reasons, and specific food preferences may also be embedded in particular peer group lifestyles or cultural practices.

SKEPTICISM

The fact that the moral argument for pgx has been put in terms of the moral urgency of avoiding harm is not the end of the story. The emphasis on individualism and personalization gives rise to the criticism that in so far as the promises of pgx can be delivered, they are likely to be available only in social contexts that can boast well-developed and well-funded health provision. In this case health inequalities worldwide may be exacerbated rather than reduced. The "boutique" model of personalization envisioned in this scenario has been challenged by Daar and Singer, who have argued that pgx could benefit developing countries, particularly by the resuscitation

---

[3] Personal communication, Human Genome Meeting Montreal, 2007.

of abandoned drugs, and that instead of individualization or personalization it is important to think in terms of variation between population groups (Daar and Singer, 2005).

In the ngx context, the promise of personal dietary advice, whether associated with enhancement or not, is a good marketing device. As the Food Ethics Council has pointed out, personalization serves both as a political and an economic project (Food Ethics Council, 2005). Personalization fits with prevailing political rhetoric in terms of the importance of "choice" and can also serve the ends of putting responsibility for health on the shoulders of individual consumers who decide what to buy and literally consume. There are equity concerns in this context too, of course, including differences in purchasing power, which set limits to food choices.

Additionally, whereas skepticism in relation to pgx has to a considerable extent been related to economic feasibility and the incentives of pharmaceutical companies, skepticism about ngx tends to focus on the potential usefulness or lack of it. Given how complicated diet is, why go down this route? (Müller and Kersten, 2003; Trivedi, 2007). Focusing on responses to specific foodstuffs may be useful for limited purposes, such as avoiding allergens, but do we need ngx for this? Is it likely to have an overall enhancing effect? And how do we assess the context of the whole diet, as opposed to specific foodstuffs? Although the claim that there is no such thing as a "healthy" or "unhealthy" food can be used as an excuse for not intervening in the market, it is true that overall diet gets short shrift in contemporary food safety legislation. In Europe, public policy concerning the *safety* of novel foods, for example, examines products on a case by case basis.

The suggestion that some foods might be "healthy" *per se* can be challenged in different ways. The first might be that it is indeed important to look at the whole diet, how much is consumed of a particular food and in relation to what; another is that it is relevant to take into account the genetic predispositions of the individual in question: it is the relation between the food and the individual rather than the relation between the food and other foods consumed that is important. This is where the advent of ngx becomes significant, as it explores the ways in which the optimum diet will be related to one's genetic make-up.

There are prior questions, however, such as how a food choice is characterized as "healthy" or not, and the ethics of the advice given. The fact that advice in this and related contexts is constantly changing. For example, how many glasses of wine can safely be consumed, according to whether one is a man or a woman, is a source of confusion.

The question less frequently addressed, however, is what basis there is for accepting dietary advice. The difficulty of establishing relevant (statistical) associations between genetic factors and dietary response in the first place is crucial. Beyond that, however, there is a problem about the meaning of that information for the individual. Let us take the five-a-day program. The health benefit of consuming five portions of fruit and/or vegetables a day has been established statistically at population level. For any given individual, it may not be true, and yet it is not clear that that is realized by those making food choices. Analogously in the pgx context, the degree of awareness that some drugs on the market for general use had established efficacy in as little as 30% of the population was limited until it became possible to promote the comparative benefits of personalized medicine. The issues surrounding awareness of the statistical evidence of efficacy are now replaced by those concerned with awareness of the degree to which the statistical associations have been or can be established between genetic factors and response in the phenotype.

Bearing these points in mind, is it possible to establish general principles related to the ethics of advice? It should be pointed out that it is not the purpose here to add to skepticism of ngx *per se*: the focus is the advice that emanates from it.

PREREQUISITES

As in the case of pgx, a prerequisite for ethically acceptable advice is quality control in relation to the following aspects. The first concerns establishing associations in the first place between genetic factors and dietary response. This requires association studies to be carried out in population groups. It does not need much thought to see that this is more complicated in ngx than in the case of pgx – even in pgx, the HUGO Ethics Committee has called for HUGO to consider establishing a body to oversee common standards for association studies (HUGO, 2007), as there has been concern over their reliability and replicability.

Although pgx may affect the conduct of drug trials, one thing is clear: in a drug trial there are fewer variables than there are in the case of dietary surveys. National dietary surveys predate genomics, of course, but they have always been subject to the problem that they rely on people keeping accurate diet diaries. These are notoriously unreliable. National dietary surveys pregenomics depended on making associations between dietary records and health status. Postgenomics, the aim is to identify underlying genetic factors affecting diet-related differences in health status. But only if a trial is carried out on a specific foodstuff or food ingredient, that is, approaching the specificity of a drug trial, would the same degree of specificity likely be obtained.

ETHICS AND ADVICE

The ethics of giving advice has a number of different aspects:

1. The basis on which advice is given (the evidence base) including the statistical power of the research. In addition to general points about the ethics of statistics, this requires thinking about statistical evidence specific to the context. It has to be acknowledged that advice may be given in a large number of settings, as in (2).
2. The authority of the advice giver to give advice and any associated agenda. What is at stake may be public health advice issued by a government department or nutritional advice given by a professional nutritionist or other health professional; advice issued by the media based on the latest (possibly unreplicated) research, including newspapers and magazines. Then, there is advice offered by commercial companies, including those that market genetic tests direct to the consumer. Although the latter has received a fair amount of attention, this is not my focus here because this introduces the issues of advertising rather than advice. My focus is on health advice in the public sector.
3. The extent to which both parties understand the parameters of what is going on – including the agenda of the advice giver and the extent to which the advice is individualized or statistical. Interestingly, as already mentioned above, it is only when the possibility of individualized or personalized advice appears on the horizon that the limitations of population advice are admitted to. This has been noticeable in the pgx context, where the statistics concerning adverse drug reactions are deployed to support the moral argument. It might be argued that prior to that there was no alternative to population-level advice, so what was the point in undermining what effectiveness it might have? But there is another important point not often alluded to, and that is that even with the advent of so-called personalization, the advice is still based on associations that are statistical.
4. The extent to which the advice is directive or nondirective. In the early 1990s, in discussion about the ethical dimensions of genetic counseling, much was made of the claim that genetic counseling should be nondirective, although the extent to which it *could* be nondirective was also hotly debated (Clarke, 1991). The reasons it was thought that nondirectiveness was problematic lay partly in the power relationships involved in the setting, but also in the fact that where termination is recognized as an option, there is already an in-built implication that in some cases that is *the* right option or at least acceptable option. The very offer of the service implies this.

While nondirectiveness, in so far as it is possible, may be suitable for the reproductive counseling setting, it should not be assumed without more, that it is necessarily appropriate for other contexts in achieving targets such as the reduction of adverse drug reactions or obesity.

Nevertheless, the rhetoric of choice, which is related to the ethos of nondirectiveness in counseling, *is* closely associated with the achievement of public health goals. This was noticeable in the U.K. Department of Health 2005 white paper, "Choosing Health." The focus of this white paper was not genomics. In this respect, it contrasts starkly with the earlier 2003 white paper, "Our Inheritance, Our Future," which addressed the challenges of postgenome health delivery and predicted a future when treatment and lifestyle advice would be tailored to the individual.

The point about the (im)possibility of nondirectiveness has a broader relevance outside the genetic counseling context. This point has been made by disability rights groups in relation to the social context in which reproductive decisions are made. In the context of a society that is not hospitable toward disability, the argument is that choices are inevitably subtly or not so subtly influenced by this. So, in the context of a society increasingly critical of obesity, food choices become increasingly subject to approval or disapproval.

The precedent of smoking is not reassuring in this regard. Although it might be argued that the increasing restrictions on the freedom to smoke are justified by the harm caused by passive smoking, and hence this is not (simply) an individual choice issue, there is not a sharp dividing line between smoking and choices to drink or to eat fatty food. "Binge drinking," for example, can lead to violence, which has the potential to harm people other than the drinker; certain eating choices lead to the use of health care resources, which leaves less for others.

Although the notion of "choice" is relied on in arguments on these issues, its meaning is not always clear. Different philosophical positions on "choice" are found in the arguments, with different implications for directiveness and the qualities of the information conveyed. It is important to be clear about what is involved. One position, identifiable with a "liberal" political philosophy, is that the job of government, in relation to food (analogous to its job in other areas) is to ensure that the food consumers eat is safe – but beyond that it is up to the individual what he or she eats. This appears to be the philosophy that underpins at least some current food policy in Europe. For example, novel foods are examined for safety but not for efficacy, but there is regard to issues of access and choice. The liberal philosophical position appears to be technology-friendly: it is of course important to give consumers information about the pros and cons of new developments, but in the name of "choice" it is up to them whether they avail themselves of these new possibilities.

This approach adopts a "thin" account of individual autonomy, where the concept of choice is limited to choosing for or against a new development on the basis of information provided. The provision of information is seen as potentially empowering, as is the technology itself. This position was reflected in the U.K. Department of Health white paper of 2003, "Our Inheritance, Our Future," where the vision of the future of health care includes individualized dietary advice, and the individualization will be made possible, it is predicted, on the basis of genetic information. On this position, then, advice will be ethical if the information is accurate and the advice giver has the authority appropriate to the context. Limitations on statistics would need to be transparent, however, for the information to be of the appropriate quality.

However, there are drawbacks in this position, both from the point of view of other interpretations of "choice" and concerning some practical issues. In the context of pgx, for example, there are worries about the implications for liability in the advent of genetically informed drug prescribing. What if someone wants to take a drug but not to have a genetic test? Can the physician prescribe it without fear of liability in the event of adverse reactions? Alternatively, what if someone has the genetic test, indications suggest he or she is a poor responder, but the person wants the drug anyway? Questions like these that show the pgx context are very different from the case of nondirective genetic counseling.

In the case of ngx, the situation may be different if, for example, the issue is whether enhancing effects, rather than avoidance of harm, will ensue. There are also considerations relating to whether the context is the commercial one, where the consumer pays for ngx advice.

Returning to underlying philosophical positions, when we look at "Choosing Health" it seems that what is envisaged is more like "advice" than providing information in a nondirective way. A second philosophical position, which seems to be encountered here, also operates with a notion of choice, but, a thicker one – instead of simply providing the information on which to make a choice, we find *encouragement* of healthy choice while still using the language of empowerment. The problems of what counts as healthy food and/or diet still remain.

Although this view still uses the rhetoric of choice, it implies individual responsibility for one's own health. Whereas the liberal view is in principle open to individuals, without disapproval, making any choice on the basis of information provided, on the second view, there is a barely hidden assumption that people given information will make certain sorts of choices. On this view, the autonomous agent is seen as a responsible agent, it being an essential part of responsibility to maintain the integrity and healthy functioning of the body, which is necessary for survival and future responsible action. When ngx is added to this mix, arguably, the responsible agent not only acts on ngx information when it is provided but takes action to know genetic information about him- or herself when the opportunity presents itself.

"Choosing Health" recognizes that children are a special case, but there is a wider problem than that. Clearly, in the case of food, the class of people who consume the food is not identical with the class of people who buy or prepare it. Many people in many different contexts do not buy and prepare their own food. This is probably true of the majority of people for some of the time and of a minority of people all of the time. In this sense, to encourage responsibility for one's own health can be beside the point.

Claims of responsibility for one's own health are also challenged by claims of a genetic determinist sort, including purported discoveries of "genes for" conditions such as obesity. There are different issues here: one concerns the status of factual claims about the causal effects of genetic factors; another concerns what follows from that. A third concerns people's reactions to those claims, which range from fatalism on the one hand to making extra efforts to counteract the genetic influence on the other.

The implementation of pgx in health care practice demands more responsibility on the part of *both* professional and patient. The professional has to be relied upon not to prescribe "off-label" to those whose genetic profile suggests they are likely to have an adverse reaction. The patient has to be relied upon to stick to the prescribed regime. For pgx may affect not only the choice of drug but also the dosage of the drug in question.

A third view relating to choice presupposes yet another notion of autonomy, according to which the agent's choice concerns a certain way of life. Here the agent himself is conceived of as making choices of what sort of life to live. To assume everyone is primarily concerned about health when choosing food is a mistake. Food choices are influenced by cultural factors, peer group pressure, and so on. From this perspective, it makes sense for stomach stapling to be preferred to changing diet so that obese children can continue to eat a diet of burgers with their peer group, only not in such great quantities.

Here we find the possibility of "personalization" being applied in the third sense described previously, of advice in relation to a person as a self-aware being with his or her own goals and life plan. What factors affect the kind of information relevant here? Nondirectiveness is implied, in so far as the advice has to be open to the advisee's goals rather than those of the adviser. One of the pertinent lifestyle choices may be a rejection of relevance of genetics, especially genetically modified foods, functional foods. On the other hand, there may be particular lifestyle choices such as sport that make the choice of ngx advice important.

It might be asked whether there is a real difference between the first position and the third, in relation to giving advice to support choice, on the grounds that even in the first position the advisee, given the information, is regarded as free to make any choice on the basis of that information without disapproval. The difference is that the third position has to have regard for different lifestyle choices and arguably make them *possible* by making options available. More is required than simply ensuring safety and then leaving choices up to the individual. To say there is respect for different lifestyles is empty if certain lifestyle choices are not possible.

In all cases, if the advisee thinks he or she is being treated as an individual and being given personalized (and more certain) advice rather than just statistical, it needs to be clear to what extent this is true. Otherwise, the benefit of supposed greater personal relevance is not the benefit it is claimed to be: the issues surrounding statistics have only been shifted to another place, to the reliability of the association study in question. The issues of statistical power need to be put firmly at the center of ethical debate.

CONCLUSION

Having looked at the issues surrounding the giving of genomically informed nutritional advice, is it possible to distill any conclusions? The following is a tentative list of points to consider:

A   The agenda of the adviser
D   The extent to which the conveying of information is done in a directive way, in relation to the relevant notion of choice
V   The value of the information, if any, to the advisee, in relation to personal goals
I   The adequacy and quality of the information conveyed
S   The statistical associations underlying the advice; the evidence base, which affects the extent to which it is personalized
E   Considerations of equity

Although all these are important, the underrepresentation of the ethics of statistics in this context needs to be addressed.

ACKNOWLEDGMENTS

The support of the Economic and Social Research Council (ESRC) is gratefully acknowledged. The work was part of the program of the ESRC Centre for Economic and Social Aspects of Genomics (CESAGen).

REFERENCES

Advisory Committee on Novel Foods and Processes (ACNFP), *www.acnfp.gov.uk*; 2007. Accessed on 21 September 2007.
Chadwick, R. *et al.*, *Functional Foods*. Springer-Verlag, 2003.
Chadwick, R., Nutrigenomics, individualism and public health. *Proceedings of the Nutrition Society*, 2004.
Chadwick, R., Pharmacogenomics. In R. Ashcroft *et al.* (eds.), *Principles of Health Care Ethics* (2nd ed.). Wiley, 2007.
Clarke, A., Is non-directive genetic counselling possible? *Lancet* 1991, *338*, 998–1001.
Daar, A., Singer, P. A., Pharmacogenetics and **geographical ancestry**: Implications for drug development and global health. *Nature Rev Genet.*
Davies, K., The drive for the $1000 dollar genome. *Bio.ITworld* 2007, *6*(4), 22–30.
Department of Health U.K., *Our Inheritance, Our Future*. London: Department of Health, 2003.
Department of Health U.K., *Choosing Health*. London: Department of Health, 2005.
Food Ethics Council, *Getting Personal: Shifting Responsibilities for Dietary Health*. London: Food Ethics Council, 2005.
Human Genome Organisation, *Statement on Pharmacogenomics, Solidarity and Equity*, 2007.
Müller, M., Kersten, S., Nutrigenomics: Goals and strategies. *Nature Reviews Genetics* 2003, *4*, 315–322.

Ries, N. M., Caulfield, T., First pharmacogenomics, next nitrigenomics: Genohype or genohealthy? *Jurimetrics* 2006, *46*, 281.

Trivedi, B., Hungry genes? *New Scientist* 2007, 20 January, 34–37.

Wallace, H., *Your Diet Tailored to Your Genes: Preventing Diseases or Misleading Marketing?* Buxton: Genewatch U.K., 2006.

World Health Organization, *Genetic Databases*. Geneva: World Health Organization, 2003.

4

NutrimiRomics: The Promise of a New Discipline in Nutrigenomics

Chandan K. Sen

INTRODUCTION

The field of nutrient–gene interaction has gradually unfolded during the last decade and established itself as a key cornerstone in nutritional research. Studies have focused on genes that encode proteins whose function are of interest in health and disease. Noncoding genes, which have been referred to as "junk DNA" [17,47], have been of no interest. Times have changed. What used to be "junk" is now under the spotlight center stage [1,27]. The central dogma in molecular biology ignored a significant part of the genetic code, which remained under veils for decades. Noncoding RNA (ncRNA) genes produce functional RNA molecules rather than encoding proteins. Several different systematic screens have identified a surprisingly large number of ncRNA genes. ncRNAs seem to be particularly abundant in roles that require highly specific nucleic acid recognition without complex catalysis, such as in directing posttranscriptional regulation of gene expression or in guiding RNA modifications. Although it has been generally assumed that most genetic information is transacted by proteins, recent evidence suggests that the majority of the genomes of mammals and other complex organisms are in fact transcribed into ncRNAs, many of which are alternatively spliced and/or processed into smaller products [43]. These RNAs (including those derived from introns) appear to comprise a heretofore hidden layer of internal signals that control various levels of gene expression in physiology and development, including chromatin architecture/epigenetic memory, transcription, RNA splicing, editing, translation, and turnover. This hidden layer of internal signals is now emerging to be of such critical significance that lack of consideration of that layer poses the serious risk of clouding our ability to understand the molecular basis of health and disease [18,43,50,67]. Posttranscriptional gene silencing (PTGS), which was initially viewed as an isolated regulatory mechanism in some plant species, now represents a major frontier in molecular medicine [15,50]. RNAi was first observed inadvertently in an experiment to increase the purple pigment in petunias. However, the experiment backfired when the gene introduced caused PTGS of the pigment production gene. Subsequent studies on *C. elegans* and the fruit fly Drosophila revealed that PTGS could be triggered by dsDNA. A similar phenomenon in fungus was termed "quelling" in 1992. Andrew Fire and Craig Mello (Nobel prize winners in physiology or medicine, 2006) are credited with the 1998 discovery of RNAi [16]. Earlier works had identified that both antisense [25] as well as sense [20] RNA could silence genes, although the results were inconsistent and the effects usually modest. In light of the observation that both sense and antisense RNA could cause silencing, Mello argued that the mechanism could not just be a pairing of antisense RNA to mRNA, and he coined the term RNAi for the unknown mechanism [53]. The discovery that short RNA is the effector of RNAi was rapidly followed by the identification of a class of endogenous RNA molecules of the same size in worms, flies, mice, and humans. This small RNA was called miRNA [30, 33, 51]. In all forms of life, ncRNA includes ribosomal RNA (rRNA), transfer RNA (tRNA), small nuclear RNA (snRNA), small nucleolar RNA (snoRNA), interference RNA (RNAi), short interfering RNA (siRNA), and micro RNA (miRNA or miR).

miRNAs are powerful regulators of gene expression [61]. It is estimated that about 3% of human genes encode for miRNAs [58]. An miRNA is approximately twenty-two ribonucleotides long, noncoding RNAs, with a potential to recognize multiple mRNA targets guided by sequence complementarity and RNA-binding proteins. Recent evidence suggests that the number of unique miRNA genes in humans exceeds 1000 [49]. The numbers of miRNAs and their targets turn out to be much greater than what we previously thought [52]. miRNAs are functionally versatile, with the capacity to specifically inhibit translation initiation or elongation, as well as induce mRNA destabilization, through predominantly targeting the 3′-untranslated regions of mRNA. Briefly, miRNA are transcribed in the nucleus by conventional mechanisms and are exported to the cytoplasm [73], where after biological processing they form the mature miRNA that can interact with matching mRNAs by RNA–RNA binding. This binding, with the assistance of the RNA-induced silencing complex (RISC), leads to modes of action, resulting in mRNA degradation or translational inhibition (see Fig. 4.1) [65]. This mechanism of action is termed posttranscriptional gene regulation (PTGS). In animals, in contrary to plants, there is not 100% nt match between miRNA and its target mRNA, leading to a mode of action causing mRNA translational inhibition and not mRNA degradation [7]. The interaction between the miRNA and its matching mRNA occurs between the 5′ UTR of the miRNA to the 3′ UTR region of the mRNA by a matching seed element in the miRNA. Utilizing this data, computational prediction

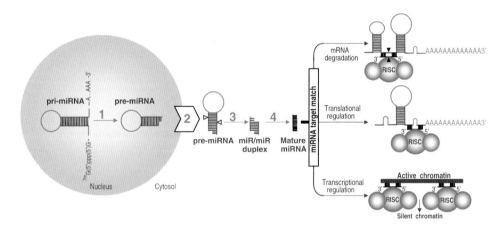

Figure 4.1 Overview of the major mechanisms involved in the generation and function of miRNA. Primary transcripts of miRNA (pri-miRNA) are generated by polymerase II and possess a 5′ 7-methyl guanosine cap and are polyadenylated. Processing of pri-miRNA in the nucleus is mediated by a microprocessor complex [1] including Pasha and Drosha. Drosha is an RNase III endonuclease that asymmetrically cleaves both strands of the hairpin stem at sites near the base of the primary stem loop thus releasing a 60- to 70-nucleotide pre-miRNA that has a 5′ phosphate and a 2-nucleotide 3′ overhang. Specific RNA cleavage by Drosha predetermines the mature miRNA sequence and provides the substrate for subsequent processing events. The pre-miRNAs are transported to the cytoplasm by Exportin-5 [2]. Once in the cytosol, a second RNase III endonuclease, Dicer [3] cleaves the pre-miRNA. Dicer releases a 22-nucleotide mature double-stranded miRNA with 5′ phosphates and a 2-nucleotide 3′ overhang. One strand of the miRNA duplex is subsequently incorporated into an effector complex termed RNA-induced silencing complex [4] or RISC that mediates target gene expression. *Reproduced with permission from [24].*

approaches estimate that miRNAs can target 30% of the human genome [29, 37, 58, 63]. Other estimates claim that more than 50% of human protein-coding genes might be regulated by miRNAs [70]. Furthermore, one miRNA can regulate hundreds of genes [70], and that one gene can be regulated by a number of miRNAs.

The first step in any gene expression is its transcription. Initially, it was believed that miRNA transcription is mediated by RNA polymerase III, because it transcribes most of the small RNAs. However, pri-miRNAs are sometimes several kilobases long and contain stretches of more than four uracils, which would have terminated transcription by pol III. Lee et al. [35] have concluded that miRNA transcription is accomplished by RNA polymerase II. The miRNA is first transcribed as a hundreds to thousands nt long miRNA precursor named a primary miRNA (pri-miRNA). Analysis of several, pri-miRNA precursors has shown that they all contain a 5′ 7-methyl guanosine cap and a 3′ poly-A tail. Therefore, this data indicate that pri-miRNAs are structurally analogous to mRNAs [12]. Following transcription, the miRNA goes through the first step of cleavage. It is initiated by the nuclear RNAs III Drosha, a double-stranded RNA (dsRNA)-specific endonuclease that introduces staggered cuts on each strand of the RNA helix [36]. It is responsible for nuclear processing of the pri-miRNAs into stem-loop (hairpin-shaped) precursors of ∼70 nucleotides named precursor miRNA (pre-miRNAs). It has been shown that RNA interference of Drosha results in the strong accumulation of pri-miRNA and the reduction of pre-miRNA and mature miRNA *in vivo* [32]. RNA stem-loops with a large, unstructured terminal loop (above 10 nt) are the preferred substrates for Drosha cleavage [75]. In the nucleus, Drosha functions as a large complex where it interacts with DGCR8, an essential cofactor for Drosha, which contains two dsRNA-binding domains [22, 72]. Recombinant human Drosha alone shows nonspecific RNase activity, but the addition of DGCR8 renders it specific for pri-miRNA processing [66]. The primary and secondary structure of miRNAs precursors is conserved as internal loops and bulges commonly appear in specific positions in the miRNA stem. This enables correct future processing by the following enzymes in the maturation of the miRNA [57].

Export of the pre-miRNA from the nucleus to the cytoplasm is mediated by Exportin 5 [73]. It is a nuclear export receptor for certain classes of dsRNA, including pre-miRNAs, viral hairpin RNAs, and some tRNAs [9]. It was demonstrated that the export of pre-miRNAs is sensitive to depletion of nuclear RanGTP, therefore mediated by it [4]. Once in the cytoplasm, there is a disassembly of the exported complex by GTP hydrolysis [42]. Besides the role of Exportin 5 in the nuclear export of pre-miRNA, there is evidence that it also has a role in preventing nuclear pre-microRNA degradation [74]. The second step of miRNA processing is confined to the cytoplasm [34]. The pre-miRNA goes through another cleavage step, conducted by Dicer. Dicer is a multidomain ribonuclease that processes the hairpin precursor to a ∼22-nt small dsRNA mature miRNA [28]. Dicer functions through intramolecular dimerization of its two RNase III domains, assisted by the flanking RNA-binding domains, PAZ and ds RNA-binding domains (dsRBD) that generate products with 2 nt 3′ overhangs [76]. PAZ domains are highly conserved domains of 130 amino acids that bind to RNA found only in Dicer and Argonaut proteins (discussed later) [6]. Following Dicer cleavage of the pre-miRNA, the mature miRNA is incorporated into a RISC complex whose diverse functions can include mRNA cleavage, suppression of translation, transcriptional silencing, and heterochromatin formation [3]. This complex functions as well in RNA interference. RISC is a multiple-turnover enzyme complex, meaning that miRNA can direct multiple rounds of target cleavage once incorporated. One strand of the ds miRNA is preferentially incorporated into RISC depending upon the thermodynamics of the duplex. Gregory RI [19] et al. have isolated a trimeric protein complex of ∼500 kDa that contains Dicer, human immunodeficiency virus transactivating response RNA-binding protein (TRBP), and Argonaute2 (Ago2) and have demonstrated that this complex is required for miRNA biogenesis. There is evidence that the complex forms prior to miRNA loading [41]. TRBP is a protein with three dsRBDs that has been shown to be essential for the miRNA prosecing [21]. Ago2 is a member of the Argonaute protein family and the only member in humans associated with both siRNA and miRNA silencing. It serves as the catalytic engine of RISC by a PIWI domain that contains a RNaseH-like structure for its endonucleolytic-slicer activity [45, 64]. It was found that Ago2 is essential for mouse development, and cells lacking it are unable to respond to siRNAs experiments. Moreover, mutations within its RNaseH domain inactivate RISC, proving its fundamental role in miRNA induced mRNA silencing.

Match between miRNAs and their target mRNAs inhibits translation. Therefore, RISC containing miRNA may directly interfere with translation initiation or elongation and perhaps target the mRNA to centers of degradation. These centers, which contain untranslated mRNAs, are sites of mRNA degradation previously observed in yeast and animal cells and are called processing (P) bodies [26]. Supporting this notion is the evidence of the presence of Argonaute family proteins in these p bodies, though it is not clear whether as a cause or as a consequence of inhibiting protein synthesis. Chu et al. [10] showed that RCK/p54 is the effector molecule in miRNA-RISC that represses translation. RCK/p54, the human homolog of yeast Dhh1p, is a P-body protein and a member of the ATP-dependent DEAD box helicase family. In human cells, RCK/p54 interacts in P-bodies with the translation initiation factor eIF4E. The overall result of the binding of mRNAs in the RISC complex by matching miRNAs will be inhibition of translation of the mRNAs, and therefore decreased protein levels of miRNAs target mRNAs. Thus, miRNAs play a key role in regulating functionality of coding genes. The mechanism of miRNA synthesis is summarized in Fig. 4.1.

miRomics: A New Cornerstone

Genome-wide analysis of miRNA expression provides powerful insight into the functional status of the coding genome. During the last 5 years, there has been a sharp improvement in the profiling technologies for miRome screening. What started in 2004 with laborious Northern blot analyses to screen 119 miRNA [59] has today matured into multiple technology platforms that can robustly screen more than a thousand miRNA [39, 48, 69]. One of the earliest versions of miR microchip contained oligonucleotides corresponding to 245 miRNAs from human and mouse genomes [40]. Contemporarily, 18–26 nucleotide RNAs were isolated from developing rat and monkey brains. From the sequences of these RNAs and the sequences of the rat and human genomes, the small RNAs likely to have derived from stem-loop precursors typical of miRNAs were selected. Using this approach, a microarray technology suitable for detecting 138 mammalian miRNAs was developed [44]. The short nature of the target sequences makes it difficult to achieve sufficient specificity with standard DNA oligonucleotide technologies. Direct random-primed cDNA synthesis on either chemically synthesized small RNAs (21–22 nucleotides) or gel-purified mature miRNAs from human cells can produce specific and sensitive full-length cDNA probes. Internally labeled cDNA probes are sensitive for detecting differential miRNA expression between control and test groups [62]. Although the goal of miRNA profiling remained the same, the technology has markedly improved over time, enabling more robust analyses. Some miRNAs differ from each other by as little as a single nucleotide, emphasizing the importance of good mismatch discrimination. To address this issue, in our laboratory, we utilize a locked nucleic acids (LNAs)-based approach [54]. LNAs are a class of conformationally restricted nucleotide analogs. The incorporation of LNA in an oligonucleotide increases the affinity of that oligonucleotide for its complementary RNA or DNA target by increasing the melting temperature of the duplex. Additionally, the Tm difference between a perfectly matched target and a mismatched target is substantially higher than that observed when a DNA-based oligonucleotide is used. These properties, high Tm and excellent mismatch discrimination, make LNA-modified probes ideal for analysis of short and similar targets like miRNAs. Furthermore, by adjusting the LNA content and probe length, it is possible to design Tm-normalized probes, allowing hybridization conditions that are optimal for all probes used on, for example, an array [5]. Recently, deep sequencing has been applied for miRNA profiling. Deep sequencing uses massively parallel sequencing, generating millions of small RNA sequence reads from a given sample. Profiling of miRNAs by deep sequencing measures absolute abundance and allows for the discovery of novel microRNAs that have eluded previous cloning and standard sequencing efforts. Public databases provide *in silico* predictions of miRNA gene targets by various algorithms. To better determine which of these predictions represent true positives, microRNA expression data can be integrated with gene expression data to identify putative microRNA:mRNA functional pairs [11]. miRNA expression profiling analysis can also be performed using a low-cost PCR-based assay platform. Primers associated with these miRNA assays were designed using a novel bioinformatics algorithm that has incorporated many primer selection features for assay specificity, sensitivity, and homogeneity [68].

NUTRIGENOMICS AND MIR

Successful conclusion of the Human Genome Project and emergence of the powerful "omics" tools have ushered in a new era of medicine and nutrition. Application of "omics" technologies to enhance the understanding of nutritional sciences has led to the development of nutrigenomics. "Nutrigenomics" was coined less than a decade ago to develop a subdiscipline that would address the application of high-throughput genomics tools in nutrition research. The overall goal was to understand how nutrition influences metabolic pathways and homeostatic control, how this regulation is disturbed in the early phase of a diet-related disease, and to what extent individual sensitizing genotypes contribute to such diseases [46]. Nutrition–health relationship is largely influenced by nutrient–gene interaction. The functional responsiveness of genes to nutritional elements consumed provides the fundamental basis of nutrigenomics. In humans, early-life metabolic imprinting has been evident in several epidemiological studies. Both in the uterus and during the first years of life, under and overfed mother–child units imprint gene changes that lead to chronic metabolic problems in later life [8]. During the last decade, nutrigenomics has emerged as a key driver of food commercialization. Range of products covered has been wide, spanning from functional food to individual nutrients. Development of miRNA biology has exposed a major crack in the armor of nutrigenomics. As of now, the field is limited to addressing coding genes only. In its current form, nutrigenomics fails to appreciate the key significance of noncoding genes in human health and disease. Current developments in the biology of noncoding genes, especially miRNA, provides an extraordinary opportunity to invigorate nutrigenomics by incorporating nutrimiromics as a key subcomponent. Nutrimiromics will help understand the relationship between food elements and the response of miRNA in specific body compartments. Such knowledge, taken together with information of nutrient-responsive coding genes, will help understand nutrigenomics as a whole. For example, let us assume that a nutrient induces a set of coding genes named CG1–100. Say, the same nutrient also induces a set of noncoding genes NG50–150 such that NG50–100 are miRs that specifically target CG50–100. In that case although expression profiling of coding genes would identify CG1–100 as candidate genes that are sensitive to the given nutrient, CG50–100 could be functionally inert because NG50–100 would silence them. Thus, when expression profiling of both coding as well as noncoding genes are taken into account, CG1–50 would emerge as the actual subset of coding genes sensitive to the given nutrients that are functionally active. This is an oversimplified example to communicate the broad point. A more specific example is provided below where obesity is discussed as a use case.

According to the Centers for Disease Control (CDC) and World Health Organization (WHO), in adults, a body mass index (BMI) of 25 or more is considered "overweight" and a BMI of 30 or more is considered "obese." In 2007, more than 1.1 billion adults worldwide were overweight and 312 million of them were obese [23]. The WHO estimates that by 2015, the number of overweight people worldwide will increase to 2.3 billion, and more than 700 million will be obese [2]. According to the World Health Organization, the United States ranks fifth among all countries for obesity-related deaths. Two-thirds of Americans are overweight, defined as having a BMI greater than 25. Americans spend close to $117 billion on obesity-related complications, with another $33 billion spent annually in attempts to control or lose weight. Nutritional supplements are considered a potentially valuable countermeasure to fight obesity [14, 31, 38]. Studies identifying nutritional supplement sensitive coding genes utilizing full-genome screening approaches are powerful in formulating hypotheses that would explain the mechanism of action of the supplement in question [55,56]. Obesity is the result of an imbalance between food intake and energy expenditure resulting in the storing of energy as fat. Microarray-based expression profiling studies have provided scientists with a number of new candidate genes whose expression in adipose tissue is regulated by obesity. Integrating expression profiles with genome-wide linkage and/or association analyses is a promising strategy to identify new genes underlying susceptibility to obesity [13]. This promise can only be realized when the significance of miRNA in adipose tissue is factored in. Because miRNA have a major say in the functional status of coding genes, upregulation of miRNA that target adipogenic coding genes is likely to provide productive solutions aimed at managing obesity. Support for this notion is provided by the recent observation that miRNAs induced during adipogenesis that accelerate fat cell development are downregulated in obesity. Ectopic expression of miR-103 or miR-143 in preadipocytes accelerated adipogenesis demonstrating that miRNA play a key role in obesity [71]. Angiogenesis is another factor that feeds adipose tissue growth. Recent studies demonstrate that angiogenesis is under the tight control of miRNAs [60]. Nutritional supplements aimed at managing obesity should be investigated for their ability to favorably regulate adipose tissue miRNA response minimizing storage and accelerating catabolism of cellular and tissue fat. Nutrimiromics represents a powerful tool in that regard and is likely to emerge as a major driver of the nutritional supplement industry in the near future.

REFERENCES

1. Transcriptomics: Rethinking junk DNA. *Nature* 2009, *458*, 240–241.
2. American Heart Association, Peripheral arterial disease statistics - 2008 update. *American Heart Association* 2008.
3. Andl, T., Murchison, E. P., Liu, F., Zhang, Y., *et al.*, The miRNA-processing enzyme dicer is essential for the morphogenesis and maintenance of hair follicles. *Curr Biol* 2006, *16*, 1041–1049.
4. Bohnsack, M. T., Czaplinski, K., Gorlich, D., Exportin 5 is a RanGTP-dependent dsRNA-binding protein that mediates nuclear export of pre-miRNAs. *Rna* 2004, *10*, 185–191.
5. Busch, A. K., Litman, T., Nielsen, P. S., MicroRNA expression profiling using LNA-modified probes in a liquid-phase bead-based array. *Nat Methods*, 2007, i–ii.
6. Carmell, M. A., Hannon, G. J., RNase III enzymes and the initiation of gene silencing. *Nat Struct Mol Biol* 2004, *11*, 214–218.
7. Carrington, J. C., Ambros, V., Role of microRNAs in plant and animal development. *Science* 2003, *301*, 336–338.
8. Chavez, A., Munoz de Chavez, M., Nutrigenomics in public health nutrition: short-term perspectives. *Eur J Clin Nutr* 2003, *57*(Suppl 1), S97–S100.
9. Chen, T., Brownawell, A. M., Macara, I. G., Nucleocytoplasmic shuttling of JAZ, a new cargo protein for exportin-5. *Mol Cell Biol* 2004, *24*, 6608–6619.
10. Chu, C. Y., Rana, T. M., Translation repression in human cells by microRNA-induced gene silencing requires RCK/p54. *PLoS Biol* 2006, *4*, e210.
11. Creighton, C. J., Reid, J. G., Gunaratne, P. H., Expression profiling of microRNAs by deep sequencing. *Brief Bioinform*, 2009.
12. Cullen, B. R., Transcription and processing of human microRNA precursors. *Mol Cell* 2004, *16*, 861–865.
13. Dahlman, I., Arner, P., Obesity and polymorphisms in genes regulating human adipose tissue. *Int J Obes (Lond)* 2007, *31*, 1629–1641.
14. Downs, B. W., Bagchi, M., Subbaraju, G. V., Shara, M. A., *et al.*, Bioefficacy of a novel calcium-potassium salt of (-)-hydroxycitric acid. *Mutat Res* 2005, *579*, 149–162.

15. Filipowicz, W., Jaskiewicz, L., Kolb, F. A., Pillai, R. S., Post-transcriptional gene silencing by siRNAs and miRNAs. *Curr Opin Struct Biol* 2005, *15*, 331–341.
16. Fire, A., Xu, S., Montgomery, M. K., Kostas, S. A., *et al.*, Potent and specific genetic interference by double-stranded RNA in Caenorhabditis elegans. *Nature* 1998, *391*, 806–811.
17. Flam, F., Hints of a language in junk DNA. *Science* 1994, *266*, 1320.
18. Goodrich, J. A., Kugel, J. F., Non-coding-RNA regulators of RNA polymerase II transcription. *Nat Rev Mol Cell Biol* 2006, *7*, 612–616.
19. Gregory, R. I., Chendrimada, T. P., Cooch, N., Shiekhattar, R., Human RISC couples microRNA biogenesis and posttranscriptional gene silencing. *Cell* 2005, *123*, 631–640. Epub Nov 2003.
20. Guo, S., Kemphues, K. J., par-1, a gene required for establishing polarity in C. elegans embryos, encodes a putative Ser/Thr kinase that is asymmetrically distributed. *Cell* 1995, *81*, 611–620.
21. Haase, A. D., Jaskiewicz, L., Zhang, H., Laine, S., TRBP, a regulator of cellular PKR and HIV-1 virus expression, interacts with Dicer and functions in RNA silencing. *EMBO Rep* 2005, *6*, 961–967.
22. Han, J., Lee, Y., Yeom, K. H., Kim, Y. K., *et al.*, The Drosha-DGCR8 complex in primary microRNA processing. *Genes Dev* 2004, *18*, 3016–3027.
23. Hossain, P., Kawar, B., El Nahas, M., Obesity and diabetes in the developing world – a growing challenge. *N Engl J Med* 2007, *356*, 213–215.
24. Ioshikhes, I., Roy, S., Sen, C. K., Algorithms for mapping of mRNA targets for microRNA. *DNA Cell Biol* 2007, *26*, 265–272.
25. Izant, J. G., Weintraub, H., Inhibition of thymidine kinase gene expression by anti-sense RNA: a molecular approach to genetic analysis. *Cell* 1984, *36*, 1007–1015.
26. Jabri, E., P-bodies take a RISC. *Nat Struct Mol Biol* 2005, *12*, 564.
27. Khajavinia, A., Makalowski, W., What is "junk" DNA, and what is it worth? *Sci Am* 2007, *296*, 104.
28. Kolb, F. A., Zhang, H., Jaronczyk, K., Tahbaz, N., *et al.*, Human dicer: purification, properties, and interaction with PAZ PIWI domain proteins. *Methods Enzymol* 2005, *392*, 316–336.
29. Kruger, J., Rehmsmeier, M., RNAhybrid: microRNA target prediction easy, fast and flexible. *Nucleic Acids Res* 2006, *34*, W451–W454.
30. Lagos-Quintana, M., Rauhut, R., Lendeckel, W., Tuschl, T., Identification of novel genes coding for small expressed RNAs. *Science* 2001, *294*, 853–858.
31. Lau, F. C., Bagchi, M., Sen, C. K., Bagchi, D., Nutrigenomic basis of beneficial effects of chromium(III) on obesity and diabetes. *Mol Cell Biochem* 2008, *317*, 1–10.
32. Lee, Y., Ahn, C., Han, J., Choi, H., *et al.*, The nuclear RNase III Drosha initiates microRNA processing. *Nature* 2003, *425*, 415–419.
33. Lee, R. C., Ambros, V., An extensive class of small RNAs in Caenorhabditis elegans. *Science* 2001, *294*, 862–864.
34. Lee, Y., Jeon, K., Lee, J. T., Kim, S., *et al.*, MicroRNA maturation: stepwise processing and subcellular localization. *Embo J* 2002, *21*, 4663–4670.
35. Lee, Y., Kim, M., Han, J., Yeom, K. H., *et al.*, MicroRNA genes are transcribed by RNA polymerase II. *Embo J* 2004, *23*, 4051–4060.
36. Lee, Y., Kim, V. N., Preparation and analysis of Drosha. *Methods Mol Biol* 2005, *309*, 17–28.
37. Lewis, B. P., Burge, C. B., Bartel, D. P., Conserved seed pairing, often flanked by adenosines, indicates that thousands of human genes are microRNA targets. *Cell* 2005, *120*, 15–20.
38. Li, J. J., Huang, C. J., Xie, D., Anti-obesity effects of conjugated linoleic acid, docosahexaenoic acid, and eicosapentaenoic acid. *Mol Nutr Food Res* 2008, *52*, 631–645.
39. Li, W., Ruan, K., MicroRNA detection by microarray. *Anal Bioanal Chem* 2009.
40. Liu, C. G., Calin, G. A., Meloon, B., Gamliel, N., *et al.*, An oligonucleotide microchip for genome-wide microRNA profiling in human and mouse tissues. *Proc Natl Acad Sci U S A* 2004, *101*, 9740–9744.
41. Maniataki, E., Mourelatos, Z., A human, ATP-independent, RISC assembly machine fueled by pre-miRNA. *Genes Dev* 2005, *19*, 2979–2990.

42. Matsuura, Y., Stewart, M., Structural basis for the assembly of a nuclear export complex. *Nature* 2004, *432*, 872–877.

43. Mattick, J. S., Makunin, I. V., Non-coding RNA. *Hum Mol Genet* 2006, *15* Spec No 1, R17–R29.

44. Miska, E. A., Alvarez-Saavedra, E., Townsend, M., Yoshii, A., *et al.*, Microarray analysis of microRNA expression in the developing mammalian brain. *Genome Biol* 2004, *5*, R68.

45. Miyoshi, K., Tsukumo, H., Nagami, T., Siomi, H., *et al.*, Slicer function of Drosophila Argonautes and its involvement in RISC formation. *Genes Dev* 2005, *19*, 2837–2848. Epub 2005 Nov 14.

46. Muller, M., Kersten, S., Nutrigenomics: goals and strategies. *Nat Rev Genet* 2003, *4*, 315–322.

47. Nowak, R., Mining treasures from 'junk DNA.' *Science* 1994, *263*, 608–610.

48. Pene, F., Courtine, E., Cariou, A., Mira, J. P., Toward theragnostics. *Crit Care Med* 2009, *37*, S50–S58.

49. Perera, R. J., Ray, A., MicroRNAs in the search for understanding human diseases. *BioDrugs* 2007, *21*, 97–104.

50. Racz, Z., Hamar, P., Can siRNA technology provide the tools for gene therapy of the future? *Curr Med Chem* 2006, *13*, 2299–2307.

51. Reinhart, B. J., Slack, F. J., Basson, M., Pasquinelli, A. E., *et al.*, The 21-nucleotide let-7 RNA regulates developmental timing in Caenorhabditis elegans. *Nature* 2000, *403*, 901–906.

52. Ro, S., Park, C., Young, D., Sanders, K. M., *et al.*, Tissue-dependent paired expression of miRNAs. *Nucleic Acids Res* 2007, *35*, 5944–5953.

53. Rocheleau, C. E., Downs, W. D., Lin, R., Wittmann, C., *et al.*, Wnt signaling and an APC-related gene specify endoderm in early C. elegans embryos. *Cell* 1997, *90*, 707–716.

54. Roy, S., Khanna, S., Hussain, S. R., Biswas, S., *et al.*, MicroRNA expression in response to murine myocardial infarction: miR-21 regulates fibroblast metalloprotease-2 via phosphatase and tensin homologue. *Cardiovasc Res* 2009, *82*, 21–29.

55. Roy, S., Rink, C., Khanna, S., Phillips, C., *et al.*, Body weight and abdominal fat gene expression profile in response to a novel hydroxycitric acid-based dietary supplement. *Gene Expr* 2004, *11*, 251–262.

56. Roy, S., Shah, H., Rink, C., Khanna, S., *et al.*, Transcriptome of primary adipocytes from obese women in response to a novel hydroxycitric acid-based dietary supplement. *DNA Cell Biol* 2007, *26*, 627–639.

57. Saetrom, P., Snove, O., Nedland, M., Grunfeld, T. B., *et al.*, Conserved microRNA characteristics in mammals. *Oligonucleotides* 2006, *16*, 115–144.

58. Sassen, S., Miska, E. A., Caldas, C., MicroRNA: implications for cancer. *Virchows Arch* 2008, *452*, 1–10.

59. Sempere, L. F., Freemantle, S., Pitha-Rowe, I., Moss, E., *et al.*, Expression profiling of mammalian microRNAs uncovers a subset of brain-expressed microRNAs with possible roles in murine and human neuronal differentiation. *Genome Biol* 2004, *5*, R13.

60. Sen, C. K., Gordillo, G. M., Khanna, S., Roy, S., Micromanaging vascular biology: tiny micrRNAs play big band. *J Vasc Res* 2009, in press.

61. Sen, C. K., Roy, S., miRNA: licensed to kill the messenger. *DNA Cell Biol* 2007, *26*, 193–194.

62. Sioud, M., Rosok, O., Profiling microRNA expression using sensitive cDNA probes and filter arrays. *Biotechniques* 2004, *37*, 574–576, 578–580.

63. Smalheiser, N. R., Torvik, V. I., Complications in mammalian microRNA target prediction. *Methods Mol Biol* 2006, *342*, 115–127.

64. Sontheimer, E. J., Carthew, R. W., Molecular biology. Argonaute journeys into the heart of RISC. *Science* 2004, *305*, 1409–1410.

65. Tang, G., siRNA and miRNA: an insight into RISCs. *Trends Biochem Sci* 2005, *30*, 106–114.

66. Tomari, Y., Zamore, P. D., MicroRNA biogenesis: drosha can't cut it without a partner. *Curr Biol* 2005, *15*, R61–R64.

67. Tomaru, Y., Hayashizaki, Y., Cancer research with non-coding RNA. *Cancer Sci* 2006, *97*, 1285–1290.

68. Wang, X., A PCR-based platform for microRNA expression profiling studies. *Rna* 2009, *15*, 716–723.
69. Wei, J. J., Soteropoulos, P., MicroRNA: a new tool for biomedical risk assessment and target identification in human uterine leiomyomas. *Semin Reprod Med* 2008, *26*, 515–521.
70. Wu, W., Sun, M., Zou, G. M., Chen, J., MicroRNA and cancer: Current status and prospective. *Int J Cancer* 2007, *120*, 953–960.
71. Xie, H., Lim, B., Lodish, H. F., MicroRNAs induced during adipogenesis that accelerate fat cell development are downregulated in obesity. *Diabetes* 2009, *58*, 1050–1057.
72. Yeom, K. H., Lee, Y., Han, J., Suh, M. R., *et al.*, Characterization of DGCR8/Pasha, the essential cofactor for Drosha in primary miRNA processing. *Nucleic Acids Res* 2006, *34*, 4622–4629.
73. Yi, R., Qin, Y., Macara, I. G., Cullen, B. R., Exportin-5 mediates the nuclear export of pre-microRNAs and short hairpin RNAs. *Genes Dev* 2003, *17*, 3011–3016.
74. Zeng, Y., Cullen, B. R., Structural requirements for pre-microRNA binding and nuclear export by Exportin 5. *Nucleic Acids Res* 2004, *32*, 4776–4785.
75. Zeng, Y., Yi, R., Cullen, B. R., Recognition and cleavage of primary microRNA precursors by the nuclear processing enzyme Drosha. *Embo J* 2005, *24*, 138–148.
76. Zhang, H., Kolb, F. A., Jaskiewicz, L., Westhof, E., *et al.*, Single processing center models for human Dicer and bacterial RNase III. *Cell* 2004, *118*, 57–68.

5

Genomics in Weight Loss Nutraceuticals

Debasis Bagchi, Francis C. Lau, Hiroyoshi Moriyama,
Manashi Bagchi, and Shirley Zafra-Stone

ABSTRACT

The field of nutritional science has been revolutionized by the emergence of genomics. Nutrigenomics, the study of interaction of diet and gene regulation, is now considered an integral part of nutritional research. According to the World Health Organization (WHO), more than 60% of the global disease burden will be due to obesity-related chronic disorders by the year 2020. In the United States, the prevalence of obesity has doubled in adults and tripled in children over the past 3 decades. In this regard, nutrigenomics may provide strategies for the design and development of safe and effective dietary interventions against obesity. Prompted by the ever-growing obesity epidemic and the increasing emphasis on physical beauty, numerous dietary supplements have been introduced to the billion-dollar weight loss industry. Many of these supplements make claims that lack scientific substantiation. However, there are a few natural dietary supplements that have undergone extensive safety and efficacy evaluation for their potential in weight management. (–)-Hydroxycitric acid (HCA) isolated from the dried fruit rind of *Garcinia cambogia*, and the micronutrient niacin-bound chromium(III) (NBC) are among products consistently shown to be safe and efficacious for use in weight management. Utilizing cDNA microarrays, we demonstrated for the first time that potassium/calcium-double salt of HCA (HCA-SX) supplementation modulated the expression of genes involved in fat metabolism from obese women. Similarly, we showed that NBC supplementation induced the expression of myogenic genes while it suppressed the expression of genes that are highly expressed in brown adipose tissue in diabetic obese mouse model. The molecular basis for the beneficial effects of these supplements as elucidated by state-of-the-art nutrigenomic technologies will be discussed in this chapter.

INTRODUCTION

Obesity has rapidly grown into a global epidemic. In the United States, for example, the number of overweight and obese U.S. adults surpassed that of normal weight U.S. adults for the first time in history [1,2]. It is projected that by the year 2015, about 75% of U.S. adults will be overweight or obese [3]. Obesity significantly decreases quality of life and life expectancy because it is a risk factor for other diseases such as type 2 diabetes, cardiovascular disease, and certain cancers including colon and breast cancer [4, 5]. It has been shown that a reduction of life expectancy by 7 years at age 40 was attributed to obesity [6]. The obesity epidemic has also imposed a tremendous burden on socioeconomic resources. In the United States, the direct obesity costs are estimated to be 5.7% of total national health expenditures. The expected lifetime costs for cardiovascular disease and its related risk factors increase by nearly 200% with severe obesity [7]. Nevertheless, evidence suggests that obesity-related risk factors are preventable and even convalescent through weight loss and long-term weight management programs [8–10]. Because increased food consumption combined with lack of exercise seem to be the major factors contributing to the obesity epidemic, long-term lifestyle changes toward a healthy diet and increased exercise may be a cost-effective alternative for obesity treatment [11,12]. Indeed, micronutrients and photochemicals have become increasingly popular as part of the dietary lifestyle intervention for the obesity epidemic [13]. Unfortunately, many of these weight management supplements are under scrutiny because they lack the scientific evidence to support their claims. However, the safety and efficacy of HCA-SX and NBC have been well studied in animal models and clinical trials as weight loss supplements [14,15]. This chapter focuses on the potential molecular mechanisms underlying the observed beneficial effects of the two supplements as elucidated by state-of-the-art nutrigenomic technologies.

OBESITY EPIDEMIC AND ITS CONSEQUENCES

Obesity is often assessed by body mass index (BMI, kg of body weight/m^2 of height). The WHO defines overweight as BMI greater than 25 and obese as BMI exceeding 30 [3]. Obesity is described by WHO as one of the most visible but at the same time most neglected public health crises [5]. It is estimated that globally 1.6 billion adults are overweight, whereas 400 million are obese [16]. In the United States, the estimated number of overweight and obese adults has recently surpassed the number of those of normal weight [3,17,18]. This unprecedented increase in the prevalence of obesity in the United States as well as in the rest of the developed Western world began as an unintended by-product of economic, social, and technological advancements achieved in the past several decades. These advances have created a fertile obesogenic environment wherein food supply is abundant and costs are low, leading to food overconsumption. And readily available machine-operated technologies as well as electronic devices greatly reduce physical activity, resulting in a sedentary lifestyle [19,20].

Obesity carries with it a national financial burden that, according to the Surgeon General, amounted to over $120 billion at the turn of the millennium from direct and indirect costs [21]. Furthermore, obesity significantly decreases quality of life and life expectancy while greatly increasing the risk for a number of diseases related to increased morbidity and mortality [6,22,23]. It is estimated that about 80% of type 2 diabetes, 70% of cardiovascular disease, and 42% of certain forms of cancers are linked to excess body fat [15,24–26]. In addition, 85% of children diagnosed with type 2 diabetes are classified as obese [15]. The correlation between type 2 diabetes and obesity is so strong that "diabesity" was coined for this phenomenon [27,28].

Lifestyle modification and weight management

Obesity and obesity-related risk factors are preventable and even ameliorable through weight loss and long-term weight management programs [8–10]. Because increased food consumption is a central component to the obesity epidemic, weight loss and weight management strategies should be designed to curb appetite, reduce food intake, and decrease fat absorption and increase fat oxidation [12,19]. There are three commonly used methods to treat obesity: (1) bariatric surgery, (2) pharmacotherapy, and (3) lifestyle interventions [7]. Bariatric surgery, also known as obesity surgery, is the most invasive but effective weight loss procedure for people suffering from severe clinical obesity. Typically, these are patients with a BMI of 40 and higher or patients with a BMI of 35 and more who also experience severe obesity-related comorbidities [29]. There are three categories of bariatric procedures: malabsorptive, restrictive, and the combination of both methods [30]. These approaches are designed to achieve a stable reduction of total caloric intake and/or assimilation. Pharmacotherapy involves the use of antiobesity drugs, such as rimonabant, phentermine, sibutramine, and orlistat [31, 32]. However, the use of these pharmaceuticals often results in significant adverse side effects. Combined with regular exercise, dietary invention may be the most cost-effective strategy for weight management. A plethora of diets have been proposed to treat obesity. Although some dietary approaches may result in short-term weight loss, most diets lack long-term compliance, leading to weight regain. Incorporating nonprescriptional antiobesity dietary supplements into lifestyle modification may facilitate weight loss. In 2005, American consumers spent more than $1.6 billion on weight loss supplements [33]. Because dietary supplements are defined and regulated as foods according to the Dietary Supplement Health and Education Act (DSHEA), they can be marketed without Food and Drug Administration (FDA) approval and without any scientific substantiation for their safety and efficacy [33].

There are two weight loss supplements whose safety and efficacy have been systematically and extensively evaluated by a number of biochemical, pharmacological, and toxicological studies. In this context, the discussion will be restricted to these two nutritional supplements: niacin-bound chromium(III) complex (ChromeMate®, NBC) and (−)-hydroxycitric acid conjugated to potassium and calcium (Super CitriMax®, HCA-SX).

SAFETY AND EFFICACY OF DIETARY SUPPLEMENTS

NBC SAFETY EVALUATIONS

The most commonly available chromium(III) dietary supplements are chromium(III) chloride ($CrCl_3$), chromium(III) picolinate (CrPic), and the oxygen-coordinated niacin-bound chromium(III) (NBC) [34, 35]. There exists a number of studies evaluating the safety and efficacy of these Cr(III) supplements.

Clinical records revealed no evidence of toxicity in patients receiving Cr(III) supplemented total parenteral nutrition (TPN) for more than 20 years [36]. A number of pharmacotoxicological safety studies have shown that there is no toxic effect associated with NBC [37, 38]. Acute oral test indicated that LD_{50} for NBC was greater 5000 mg/kg of body weight. Dermal toxicity as well as primary dermal and eye irritation studies did not produce any signs of toxicity or irritation induced by NBC. Ames bacterial reverse mutation and mouse lymphoma mutagenicity assays also showed that NBC was nonmutagenic. A 90-day subchronic toxicity study did not result in any toxicological effects of NBC. A chronic safety study demonstrated that rats orally administered a human equivalency dose of 1000 μg/day elemental Cr(III) in the form of NBC for a duration of 52 weeks exhibited no signs of NBC-induced toxicological effects [38]. Particularly, the parameters such as hepatic lipid peroxidation and DNA fragmentation, hematology and clinical chemistry, and histopathological parameters were not significantly changed [38].

However, evidence from an *in vitro* study indicated that CrPic was genotoxic and caused mutation at the hypoxanthine (guanine) phosphoribosyltransferase locus of the Chinese hamster ovary (CHO) cells [39, 40]. CrPic was also known to induce clastogenesis or chromosomal damage in CHO cells; however, treatment of the cells with the same physiologic doses of $CrCl_3$ or NBC did not result in any clastogenic effect [41]. Several clinical cases also linked CrPic to nephrotoxicity resulting in renal failure [42, 43].

NBC IN WEIGHT LOSS

Several lines of evidence indicate that NBC is a novel micronutrient capable of reducing body fat while increasing lean body mass [14, 35]. NBC is also able to effectively diminish the obesity-related risk factors of metabolic syndrome by promoting glucose–insulin sensitivity [44–46]. The importance of Cr(III) was revealed 50 years ago with the discovery that chromium(III) was the central component of the biologically active form of glucose tolerance factor (GTF) found in brewer's yeast [47, 48]. GTF prevented diabetes in experimental animals by potentiating the action of insulin and modulating protein, fat, and carbohydrate metabolism [49].

A human clinical trial was conducted to show the comparative clinical efficacy of NBC and CrPic in young obese women with or without exercise [50]. Combined with exercise training, NBC supplementation significantly promoted weight loss and lowered insulin response to an oral glucose load [50]. CrPic, on the other hand, resulted in a significant weight gain. This study suggests that combined with exercise, NBC supplementation facilitates weight loss and reduces risk factors associated with diabetes [50].

A subsequent randomized, double-blinded, placebo-controlled, crossover human clinical study further demonstrated the efficacy of NBC in weight loss [51]. Twenty overweight African-American women taking 200 μg NBC three times daily exhibited a significant fat loss without causing a reduction in lean body mass as compared to subjects receiving placebo [51].

Therefore, human clinical studies demonstrated evidence of efficacy for NBC supplementation in promoting weight loss and improving insulin sensitivity and/or blood lipid profiles [48].

HCA-SX SAFETY STUDIES

There are different forms of HCA available in the supplement market. However, their safety and efficacy are not the same. Super CitriMax® (HCA-SX) is a standardized, chemically analyzed extract of *Garcinia cambogia* containing 60% HCA [52]. The superiority of HCA-SX resides in the fact that HCA-SX is an HCA double salt complexed with calcium and potassium. The unique structural characteristics of HCA-SX make it water soluble and highly bioavailable, as well as tasteless, odorless, and colorless in solution [52–59]. Furthermore, HCA-SX is the only HCA-containing product on the market whose bioavailability profile in humans has been systematically demonstrated [60].

HCA-SX passed a battery of safety tests and was subsequently awarded the Generally Recognized As Safe (GRAS) status. Acute oral test showed that LD_{50} of HCA-SX was greater than 5000 mg/kg of body weight when gavaged gastrically to fasted male and female albino rats [58]. Acute dermal, primary dermal irritation and primary eye irritation toxicity studies have demonstrated the safety of HCA-SX. The dermal LD_{50} of HCA-SX was greater than 2000 mg/kg of body weight when applied once for 24 hours to the shaved, intact skin of male and female albino rabbits. There was no evidence of acute systemic toxicity among rabbits that were dermally administered HCA-SX at 2000 mg/kg of body weight. The eye irritation study indicated that HCA-SX was mildly irritating to the eye of rabbits. No gross toxicological findings were observed under the experimental conditions [58].

Subchronic safety of HCA-SX was evaluated in Sprague-Dawley rats on body weight, selected organ weights, hepatic lipid peroxidation and DNA fragmentation, hematology and clinical chemistry over a period of 90 days [55, 57]. The body weight and selected organ weights were calculated as a percentage of body weight and brain weight at 90 days of treatment. HCA-SX supplemented rats exhibited a significant reduction in body weight as compared to control counterparts. There was no significant difference in selected organ weights between the HCA-SX group and control group at the end of the 90-day treatment. In terms of hepatic DNA fragmentation, no statistically significant difference was observed between the HCA-SX-treated and control rats. Hematology and histopathological examinations also showed no significant difference for the two treatment groups [55, 57]. Ames bacterial reversal mutation test and mouse lymphoma tests revealed that HCA-SX did not induce genomic mutation [55].

Recently, studies elucidating the effect of HCA-SX supplementation on two-generation reproduction and teratogenicity were carried out in Sprague-Dawley rats [61, 62]. These studies assessed the reproductive systems of both males and females, the postnatal maturation and reproductive capacity of their offspring, and the possible cumulative effects through multiple generations. Rats were fed either none or one of three doses of HCA-SX at 100 mg, 300 mg, or 1000 mg/kg/day for 10 weeks prior to and during mating and, for females, through gestation and lactation, across two generations. The study findings indicated no reproductive or terato-genic toxicity in either the parents or offspring. No adverse effects on reproductive performance were observed as evaluated by sexual maturity, fertility and mating, gestation, parturition, litter properties, lactation, or development of offspring. At these high dose levels, HCA-SX did not induce any systemic toxicity in the parental rats or their offspring. Neither sperm motility nor sperm count was negatively affected in either the parental groups or the first-generation groups. No teratogenicity or birth defects were observed. These findings add to the extensive body of evidence showing the long-term safety of HCA-SX [61, 62].

Taken together, these results show that supplementation of HCA-SX results in a reduction in body weight and does not cause changes in major organs or in hematology, clinical chemistry, histopathology, reproduction, or teratogenicity.

HCA-SX EFFICACY IN WEIGHT LOSS

Numerous clinical trials were performed to assess the weight management potential of different (–)-hydroxycitric acid preparations. Following are the summaries of human clinical studies conducted on HCA-SX formulation.

A randomized, placebo-controlled, single-blind study was conducted to assess the effect of HCA-SX on energy intake and satiety in twenty-four overweight, healthy, dietary unrestrained subjects (12 males and 12 females; BMI: 27.5 ± 2.0 kg/m^2; Age: 37 ± 10 years) [63]. In this 6-week trial, subjects were instructed to take 100 ml tomato juice (placebo) three times per day for 2 weeks. After a 2-week wash-out period, the subjects were given 100 ml tomato juice containing 300 mg HCA-SX three times daily for another 2 weeks. At the conclusion of the supplementation, 24-hour energy intake, appetite profile, hedonics, mood and possible change in dietary restraint were evaluated. A 24-hour energy intake was significantly reduced by HCA-SX treatment as compared to the placebo treatment. Although no difference was observed in appetite profile, dietary restraint, mood, taste perception, and hedonics, a trend toward body weight decrease and satiety was sustained [63].

The efficacy of HCA-SX and a combination of HCA-SX, niacin-bound chromium, and *Gymnema sylvestre* extract (GSE) in weight management was investigated in randomized, placebo-controlled, double-blind pilot study [64]. Thirty overweight subjects (aged 21–50; BMI > 26 kg/m^2) were randomly divided into three groups (ten subjects/group): HCA-SX group was given 4667 mg HCA-SX, combination group was given a combination of HCA-SX 4667 mg, 400 μg elemental chromium as NBC and 400 mg GSE, and placebo group was given a placebo daily in three equally divided doses 30–60 minutes before meals. In addition, subjects were given a 2000 kcal/day diet and underwent a 30-minutes/day supervised walking program, 5 days/week for 8 weeks. Subjects receiving HCA-SX exhibited a reduction in body weight and BMI by 6.3%, a decrease in food intake by 4%. Total cholesterol, low-density lipoprotein (LDL) and triglyceride levels were reduced by 6.3%, 12.3%, and 8.3%, respectively, whereas high-density liproprotein (HDL) and serotonin levels increased by 10.7% and 40%, correspondingly. Serum leptin levels were decreased by 36.6%, and the enhanced excretion of urinary fat metabolites, including malondialdehyde (MDA), acetaldehyde (ACT), formaldehyde (FA), and acetone (ACON) increased by 125–258% in the HCA-SX group. Under identical conditions, the combination of HCA-SX, NBC, and GSE reduced body weight and BMI by 7.8% and 7.9%, food intake by 14.1%. Total cholesterol, LDL, and triglyceride levels were reduced by 9.1%, 17.9%, and 18.1%, respectively, whereas HDL and serotonin levels increased by 20.7% and 50%, respectively. Serum leptin levels decreased by 40.5%, and enhanced excretion of urinary fat metabolites increased by 146–281%. Subjects in the placebo group showed marginal reduction in body weight and BMI by 1.6% and 1.7%, respectively. There were no significant changes for other test parameters observed in the placebo group [64].

A subsequent follow-up randomized, placebo-controlled, double-blind study was conducted to corroborate the observed effects of HCA-SX and a combination of HCA-SX plus NBC and

GSE in the pilot study [56, 65]. Sixty obese subjects (ages 21–50, BMI > 26 kg/m^2) were randomly divided into three groups as in the pilot study. The study procedures were the same as those in the pilot study. At the end of an 8-week supplementation regimen, body weight and BMI were significantly decreased by 5–6% in both treatment groups as compared to those in the placebo group. In addition, food intake, total cholesterol, LDL, triglycerides, and serum leptin levels were significantly attenuated, whereas high-density lipoprotein levels and excretion of urinary fat metabolites were significantly increased in both treatment groups when compared to the placebo group. No significant effects were observed for all test parameters in subjects receiving placebo. Results demonstrate that HCA-SX, and to a greater degree, the combination of HCA-SX, NBC, and GSE, are effective in reducing body weight and promoting healthy cholesterol levels [56, 65].

Nutrigenomics in Weight Management Supplements

Nutritional genomics is a nascent field that rides on the waves of the successful genomic revolution. Nutritional genomics includes nutrigenomics and nutrigenetics. Whereas nutrigenomics is the study of the effect of nutrients or dietary components on the transcriptome, epigenome, proteome, and/or metabolome in cells or tissues, nutrigenetics is the retrospective investigation of how genetic variations such as single nucleotide polymorphisms respond differently to specific dietary components leading to variation in health and disease status among individuals [66, 67]. This chapter will focus on the application of nutrigenomics to evaluating weight management supplements. Specifically, the effect of these supplements on gene-expression patterns using the high-throughput transcriptomics technology will be discussed.

Pregenomic nutritional research has been focused mainly on epidemiological studies evaluating the correlation between food and disease incidences at the population level. With the advent of genomics, nutritional science has gradually shifted toward personalized nutrition as demonstrated by the transition of Food Guide Pyramid to MyPyramid by United States Department of Agriculture (USDA) emphasizing individualized nutritional needs. In fact, postgenomic studies have indicated that dietary components are involved in the modulation of gene expression [68]. Nutrigenomics is a subsidiary of the science of genomics that utilizes the powerful technology to study the effect of nutrients or dietary components on the structure, integrity, and function of the genome. Thus, nutrigenomic approaches may be utilized to gain new insights into the mechanisms of nutrient–gene interactions and eventually to develop evidence-based nutritional inventions to restore and/or prevent diet-related diseases such as obesity [69, 70].

NBC-INDUCED TRANSCRIPTOMIC CHANGES

Cr(III) was discovered as the central component of biologically active complex known as the glucose tolerance factor (GTF) found in brewer's yeast [48]. Diabetic laboratory animals supplemented with GTF showed improved insulin sensitivity [48]. Although there are several Cr(III) supplements on the market, not all of these supplements are equally safe and efficacious. Several lines of evidence indicate that NBC is safe, bioavailable, and efficacious in improving insulin sensitivity and increasing lean body mass [14, 35, 38]. However, the molecular mechanism modulating the genetic response to NBC supplementation was largely unknown. Here we summarize the use of a high-throughput screening technology to examine the effect of NBC-induced alteration in transcriptomic profiles in subcutaneous adipose tissue of type 2 diabetic obese mice [71].

Male obese mice were fed a control- (CON-) or an NBC-supplemented diet for a period of 10 weeks. Blood samples from the animals were drawn before (baseline) and after 6 weeks of supplementation. Blood glucose level as well as lipid profile parameters such as total cholesterol (TC), HDL, LDL, triglycerides, and the TC-to-HDL ratio were evaluated at sixth week. Oral glucose tolerance test (OGTT) was performed at eighth week by challenging the animals with

1.5 mg/g of body weight of glucose solution and measuring blood glucose levels at 30, 60, and 120 minutes after glucose loading [71]. At the conclusion of 10-week supplementation, mice were euthanized and subcutaneous fat was removed for transcriptomic analysis.

NBC-supplementation significantly attenuated the levels of triglycerides, TC, LDL, and TC-to-HDL ratio in the plasma of the obese mice. The plasma level of HDL was also significantly increased by NBC-supplementation at the sixth week. Furthermore, NBC supplementation significantly improved the rate of blood glucose clearance 60 to 120 minutes after glucose challenge [71]. These biophysiological results are in agreement with the previous findings that NBC supplementation improves glucose and lipid metabolism, which play an important role in the invention of obesity [71].

The effect of 10-week NBC supplementation on the transcriptome of the adipose tissue from the obese mice was assessed by cDNA microarrays comprised 41, 101-probe sets. The results from transcriptomic profiling indicated that NBC supplementation stimulated the expression of 161 genes while suppressing the expression of 91 genes. Among the NBC-stimulated genes are those that play an important role in glycolysis, muscle metabolism, and muscle development. For example, enolase-3 (ENO3) gene was significantly upregulated upon NBC supplementation as compared to CON supplementation. The gene product encoded by ENO3 is the β-enolase subunit, which is responsible for more than 90% of the enolase activity in adult human muscle [72]. Mutation in this gene results in deficiency of β-enolase leading to defective glycolysis and metabolic myopathies [72]. Glucose phosphate isomerase (GPI) gene was another NBC-induced gene involved in the glycolytic pathway. Both ENO3 and GPI were found to be suppressed in visceral fat tissue of morbidly obese patients [73]. However, NBC was able to stimulate these genes in the adipose tissue of obese mice, suggesting that NBC may be able to reduce the accumulation of visceral fat by upregulating the expression of these genes, which are markedly suppressed in obese patients. Genes involved in muscle metabolism such as calsequestrin and tropomyosin-1 were also upregulated by NBC supplementation. Calsequestrin controls calcium storage whereas tropomyosin-1 is involved in calcium-regulated muscle contraction. Enhanced expression of these genes over time was linked to diminished fat content in the fat tissue [74].

NBC supplementation also suppressed the expression of adipocyte-specific genes such as the tocopherol transfer protein (TTP), mitochondrial uncoupling protein 1 (UCP1), and cell death-induced DNA fragmentation factor (CIDEA). TTP gene product plays an important role in the transport of α-tocopherol from hepatocytes into peripheral tissues including adipose tissue [75]. Adipose tissues are the major storage for lipid-soluble α-tocopherol, which is a potent antioxidant. Downregulation of TTP gene may decrease the lipid-phase antioxidant defense in the adipose tissue, thus promoting the breakdown of adipose tissue [71]. UCP1 and CIDEA genes are highly expressed in brown adipose tissue (BAT). They are involved in the thermogenesis and energy expenditure of BAT [76, 77]. UCP1 protein uncouples the generation of adenosine triphosphate (ATP, a high-energy molecule) [78]. UCP1 is thought to be tightly linked to adipocyte growth [79]. Mice lacking the CIDEA gene are found to be resistant to diet-induced obesity and diabetes [77]. Therefore, suppression of UCP1 and CIDEA by NBC may restrict adipocyte growth and promote lipolysis.

Taken together, the results provide direct evidence that NBC may exert its weight loss effects through the regulation of specific genes in fat cells of obese diabetic mice. Specifically, NBC supplementation-induced genes are responsible for muscle metabolism whereas suppressed genes are involved in adipocyte growth and lipogenesis. Such a genomic approach to nutritional research should provide insight into future investigations to delineate the molecular basis of nutrient–gene interactions.

HCA-SX REGULATION OF TRANSCRIPTOMIC EXPRESSION

HCA-SX has been used in the weight loss market for decades. However, the underlying mechanisms for its weight loss property are not well understood. An unbiased genome-wide interrogation approach was used to investigate alterations in gene expression profile following HCA-SX treatment in Sprague-Dawley rats. Rat genome arrays containing 15,923 probe sets were used to profile changes in the transcriptome of abdominal fat in rats fed an HCA-SX supplemented or a control diet [80]. There were no behavioral changes observed in HCA-SX-supplemented rats compared to control rats. There was, however, a significant weight loss observed in the HCA-SX group starting from week 6 that continued until the end of the 8-week supplementation period [80].

Genomic analysis led to the identification of three candidate genes responsive to HCA-SX supplementation. These genes were prostaglandin D synthase (PDS), aldolase B (AldB), and lipocalin (LCN2). The expression pattern of these genes was confirmed by real-time RT-PCR, which was congruent to the results from DNA microarrays [80]. To further investigate the relevance of these genes in signaling pathways related to fat metabolism, several bioinformatic platforms were used to map these genes to known fat metabolism-related pathways. Pathway construction clearly linked the candidate genes to monoamine G-protein-coupled receptor cascade such that serotonin receptor expression was consistently stimulated by HCA-SX supplementation. This finding was consistent with the previous observation that HCA-SX possessed serotonergic potential to promote serotonin release from rat brain slices [81]. Hence, HCA-SX may regulate serotonergic pathway to suppress appetite, thereby promoting weight loss.

A subsequent *in vitro* genomic study was conducted to investigate the effect of HCA-SX on lipolysis in cultured human adipocytes because defects in the lipolytic processes in adipocytes have been implicated in human obesity [82, 83]. This study used subcutaneous preadipocytes isolated from obese, nondiabetic women. These preadipocytes were allowed to differentiate to adipocytes for 2 weeks in culture. The effects of HCA-SX on lipid metabolism and transcriptomic profile were examined by treating the cells with or without HCA-SX. Treatment with HCA-SX at the experimental conditions did not cause any cytotoxicity in the human adipocytes. Biochemical analysis revealed that HCA-SX treatment induced a significant increase in the fat droplet dispersion of adipocytes as compared to the control cells, suggesting that HCA-SX potentiated lipolysis through mobilization of triacylglycerol storage [83].

Unbiased genome-wide interrogation using human genome arrays containing a total of 54,676 probe sets per chip revealed that HCA-SX treatment of human adipocyte induced statistically significant changes in the transcriptome of a small cluster of genes. There were 348 genes upregulated, whereas 366 genes downregulated by HCA-SX treatment. Of these genes, a selected group of HCA-SX responsive candidate genes was confirmed by quantitative real-time PCR. Among these genes, tissue-type plasminogen activator (PLAT) gene was stimulated by HCA-SX. PLAT protein regulates fibrinolysis and facilitates cellular defense mechanism against thrombosis [84]. Elevated risk of arterial thrombosis is intimately linked to adiposity. It has been shown that the release of PLAT from the endothelium of obese and overweight adults was significantly decreased [84]. In this context, upregulation of PLAT gene by HCA-SX may restore endothelial fibrinolytic dysfunction associated with overweight and obesity.

The expression of leptin gene in the human adipocytes was significantly upregulated by HCA-SX treatment. Numerous studies have established that the adipocyte-derived hormone leptin together with its receptors are the major components in the maintenance of energy balance [85, 86]. Inactivation of the leptin receptor gene has been shown to cause obesity [87]. Because leptin plays an important role in fat oxidation, weight loss, and cardioprotection, upregulation of leptin by HCA-SX may be beneficial in promoting energy balance [88, 89].

The other HCA-SX-stimulated genes included a number of the matrix metalloproteinase (MMP) family of proteases. MMP proteases participate in matrix protein turnover and tissue remodeling. MMP1 and MMP10 proteases are involved in fibrillar collagen and proteoglycan metabolism [90]. Proteoglycan turnover is associated with adipocyte degradation and weight loss [91, 92]. MMP3 is capable of activating MMP1 resulting in impaired adipose tissue development [93, 94]. It has been suggested that low levels of stromelysins (such as MMP3 and MMP10) play an important role in obesity [93, 95]. In addition, it has been demonstrated that MMP10 suppresses angiogenesis and vascular remodeling via degradation of extracellular matrix [96]. Therefore, upregulation of these MMP proteases by HCA-SX may induce an antiangiogenic and antiadipogenic effect on the human adipocytes leading to the observed fat droplet dispersion as revealed by the biochemical analysis.

Furthermore, HCA-SX treatment suppressed several fat- and obesity-related genes. Of these genes, perilipin gene was especially interesting in the context of lipid droplet dispersion. Perilipin protein envelops the surfaces of lipid droplets and potentiates triacylglycerol storage in adipocytes through regulation of the rate of basal lipolysis [97]. HCA-SX may facilitate the dispersion of fat droplets as revealed by the biochemical analysis via suppression of perilipin gene expression.

The expression of peroxisome proliferator-activated receptor (PPAR) gamma coactivator 1-alpha (PGC1-alpha) gene was suppressed by HCA-SX. PGC1-alpha gene serves as a transcriptional coactivator protein that coordinately modulates a number of metabolic pathways, thereby contributing to pathogenic conditions such as obesity, diabetes, and cardiomyopathy [98]. It has been shown that PGC1-alpha gene is induced during mitochondrial biogenesis accompanied by adipogenesis to complement the requirement of ATP and acetyl-CoA for lipogenesis [99]. Peroxisomal trans-2-enoyl-CoA reductase (PECR) gene was also downregulated by HCA-SX treatment. PECR catalyzes the conversion of phytol to phytanoyl-CoA in peroxisomes. It has been suggested that phytanic acid, a derivative of phytol, induces adipocyte differentiation in human preadipocytes indicating that PECR may function as an activator of retinoid X receptors (RXR) or PPAR [100, 101]. Therefore, downregulation of PGC1-alpha and PECR by HCA-SX may act in concert to promote the antiadipogenic effect of HCA-SX resulting in healthy weight loss.

HCA-SX treatment also significantly decreased the expression of endothelial lipase (LIPG) gene in human adipocytes. Overexpression of LIPG is associated with reduced plasma level of HDL. Because HDL level is inversely related to the risk of atherosclerotic cardiovascular disease, decrease in plasma level HDL may result in adverse effects to increase the risk factors for cardiovascular diseases [102]. In addition, HDL hydrolysis by endothelial lipase (gene product of LIPG) activates PPARα, which in turn induces inflammation [103]. Thus, HCA-SX may be beneficial for lipoprotein metabolism and vascular health by reducing inflammation through downregulation of LIPG.

HCA-SX significantly reduced the expression of cytoplasmic epoxide hydrolase 2 (EPHX2) gene. EPHX2 protein is involved in the metabolism of arachidonic acid [104]. Polymorphisms in EPHX2 gene have been connected to the pathogenesis of coronary heart disease (CHD) and atherosclerosis; therefore, EPHX2 has been postulated as a potential cardiovascular disease-susceptibility gene [105, 106]. Because obesity is a major risk factor for CHD and atherosclerosis, suppression of EPHX2 by HCA-SX in human adipocytes may afford protective effect against these cardiovascular diseases.

Elongation of very-long-chain fatty acids-like3 (ELOVL3) protein is involved in fatty acid biosynthesis [107]. It has been shown that ELOVL3 gene expression is significantly stimulated during adipogenesis. ELOVL3 protein catalyzes the synthesis of very-long-chain fatty acids and triglycerides in the adipose tissues, thus facilitating the process of adipogenesis [108]. HCA-SX treatment significantly suppressed the expression of ELOVL3 gene in human adipocytes, which may inhibit adipogenesis in fat tissues [83].

Taken together, these nutrigenomic findings strongly support the antilipogenic and antiadipogenic effects of HCA-SX through modulation of genes involved in fat metabolism.

CONCLUSIONS

Although a large body of evidence suggests the safe and efficacious use of NBC and HCA-SX in weight loss, the molecular mechanisms for their observed weight management effects are poorly understood. In fact, nutritional research has suffered from the inability to correlate dietary benefits to nutrient–gene regulation. However, the advent of nutrigenomics has opened the door for nutrition science to evaluate diet–gene interaction at the molecular level. In this regard, the molecular basis for gene modulation by NBC and HCA-SX has been revealed for the first time by nutrigenomics approach.

Indeed, genomic technology provides an extremely efficient means to assess genome-wide alterations in the transcriptomes of biological systems in response to any given therapeutic regimen [109]. Consequently, physionutrigenomics holds the promise to facilitate the design and development of novel and safe nutraceuticals and functional foods to combat obesity and other metabolic disorders. Studies such as those mentioned in this chapter represent a revolutionized approach in nutritional research that not only elucidates the underlying molecular mechanisms but also evaluates the health and safety concerns related to the consumption of dietary supplements. These studies may pave the way for personalized nutrition as more associations between dietary components and gene regulations are identified by such functional genomics investigations. In addition, through further understanding of the biological impacts of nutrient–gene modulations, nutrigenomics will provide key insights into the pathogenesis and progression of diet-related disorders.

With respect to obesity, nutratherapeutics combined with a healthy lifestyle may be the most cost-effective and natural way to combat this crisis. The studies discussed here also offer unequivocal evidence that it is possible to use the powerful tool of nutrigenomics to delineate the molecular basis for the observed beneficial effects of dietary supplements.

REFERENCES

1. NIH, Clinical Guidelines on the Identification, Evaluation, and Treatment of Overweight and Obesity in Adults: The Evidence Report 2000.
2. http://www.cdc.gov/nchs/fastats/overwt.htm; July 2007.
3. Wang, Y., Beydoun, M. A., The obesity epidemic in the United States – gender, age, socioeconomic, racial/ethnic, and geographic characteristics: a systematic review and meta-regression analysis. *Epidemiol Rev* 2007, *29*, 6–28.
4. Shaw, D. I., Hall, W. L., Williams, C. M., Metabolic syndrome: what is it and what are the implications? *Proc Nutr Soc* 2005, *64*, 349–357.
5. Haslam, D. W., James, W. P., Obesity. *Lancet* 2005, *366*, 1197–1209.
6. Peeters, A., Barendregt, J. J., Willekens, F., Mackenbach, J. P., *et al.,* Obesity in adulthood and its consequences for life expectancy: a life-table analysis. *Ann Intern Med* 2003, *138*, 24–32.
7. Mctigue, K. M., Harris, R., Hemphill, B., Lux, L., *et al.,* Screening and interventions for obesity in adults: summary of the evidence for the U.S. Preventive Services Task Force. *Ann Intern Med* 2003, *139*, 933–949.
8. Anderson, J. W., Konz, E. C., Obesity and disease management: effects of weight loss on comorbid conditions. *Obes Res* 2001, *9*(Suppl 4), 326S–334S.
9. Bray, G. A., The missing link - lose weight, live longer. *N Engl J Med* 2007, *357*, 818–820.
10. Wadden, T. A., Butryn, M. L., Wilson, C., Lifestyle modification for the management of obesity. *Gastroenterology* 2007, *132*, 2226–2238.
11. Ross, R., Dagnone, D., Jones, P. J., Smith, H., *et al.,* Reduction in obesity and related comorbid conditions after diet-induced weight loss or exercise-induced weight loss in men. A randomized, controlled trial. *Ann Intern Med* 2000, *133*, 92–103.
12. Blackburn, G. L., Treatment approaches: food first for weight management and health. *Obes Res* 2001, *9*(Suppl 4), 223S–227S.
13. Pittler, M. H., Ernst, E., Dietary supplements for body-weight reduction: a systematic review. *Am J Clin Nutr* 2004, *79*, 529–536.
14. Bagchi, M., Preuss, H. G., Zafra-Stone, S., Bagchi, D., In D. Bagchi, H. G. Preuss (eds.), *Obesity: Epidemiology, Pathophysiology, and Prevention*. Boca Raton, FL: CRC Press, 2007, 339–347.

15. Zafra-Stone, S., Bagchi, M., Preuss, H. G., Grover, G. J., *et al.,* In D. Bagchi, H. G. Preuss (eds.), *Obesity: Epidemiology, Pathophysiology, and Prevention.* Boca Raton, FL: CRC Press, 2007, 349–370.

16. Ahima, R. S., Obesity: much silence makes a mighty noise. *Gastroenterology* 2007, *132*, 2085–2086.

17. Ordovas, J. M., The quest for cardiovascular health in the genomic era: nutrigenetics and plasma lipoproteins. *Proc Nutr Soc* 2004, *63*, 145–152.

18. Gibbs, W. W., Obesity: an overblown epidemic? *Sci Am* 2005, *292*, 70–77.

19. Pi-Sunyer, F. X., The obesity epidemic: pathophysiology and consequences of obesity. *Obes Res* 2002, *10*(Suppl 2), 97S–104S.

20. Caballero, B., The global epidemic of obesity: an overview. *Epidemiol Rev* 2007, *29*, 1–5.

21. http://www.surgeongeneral.gov/topics/obesity/calltoaction/CalltoAction.pdf; 2007.

22. Livingston, E. H., Fink, A. S., Quality of life: cost and future of bariatric surgery. *Arch Surg* 2003, *138*, 383–388.

23. Haslam, D., Obesity: a medical history. *Obes Rev* 2007, *8*(Suppl 1), 31–36.

24. Adamson, A. J., Mathers, J. C., Effecting dietary change. *Proc Nutr Soc* 2004, *63*, 537–547.

25. Joyal, S. V., A perspective on the current strategies for the treatment of obesity. *Curr Drug Targets CNS Neurol Disord* 2004, *3*, 341–356.

26. Walker, C. G., Zariwala, M. G., Holness, M. J., Sugden, M. C., Diet, obesity and diabetes: a current update. *Clin Sci (Lond)* 2007, *112*, 93–111.

27. From the NIH: Successful diet and exercise therapy is conducted in Vermont for "diabesity." *JAMA* 1980, *243*, 519–520.

28. Astrup, A., Finer, N., Redefining type 2 diabetes: 'diabesity' or 'obesity dependent diabetes mellitus'? *Obes Rev* 2000, *1*, 57–59.

29. Hubbard, V. S., Hall, W. H., Gastrointestinal surgery for severe obesity. *Obes Surg* 1991, *1*, 257–265.

30. Elder, K. A., Wolfe, B. M., Bariatric surgery: a review of procedures and outcomes. *Gastroenterology* 2007, *132*, 2253–2271.

31. Schnee, D. M., Zaiken, K., McCloskey, W. W., An update on the pharmacological treatment of obesity. *Curr Med Res Opin* 2006, *22*, 1463–1474.

32. Bray, G. A., Ryan, D. H., Drug treatment of the overweight patient. *Gastroenterology* 2007, *132*, 2239–2252.

33. Pillitteri, J. L., Shiffman, S., Rohay, J. M., Harkins, A. M., *et al.,* Use of dietary supplements for weight loss in the United States: results of a national survey. *Obesity (Silver Spring)* 2008, *16*, 790–796.

34. Porter, D. J., Raymond, L. W., Anastasio, G. D., Chromium: friend or foe? *Arch Fam Med* 1999, *8*, 386–390.

35. Zafra-Stone, S., Bagchi, M., Preuss, H. G., Bagchi, D., In J. B. Vincent (ed.), *The Nutritional Biochemistry of Chromium(III).* Amsterdam: Elsevier, 2007, 183–206.

36. Jeejeebhoy, K. N., The role of chromium in nutrition and therapeutics and as a potential toxin. *Nutr Rev* 1999, *57*, 329–335.

37. Shara, M., Yasmin, T., Kincaid, A. E., Limpach, A. L., *et al.,* Safety and toxicological evaluation of a novel niacin-bound chromium (III) complex. *J Inorg Biochem* 2005, *99*, 2161–2183.

38. Shara, M., Kincaid, A. E., Limpach, A. L., Sandstrom, R., *et al.,* Long-term safety evaluation of a novel oxygen-coordinated niacin-bound chromium (III) complex. *J Inorg Biochem* 2007, *101*, 1059–1069.

39. Coryell, V. H., Stearns, D. M., Molecular analysis of hprt mutations induced by chromium picolinate in CHO AA8 cells. *Mutat Res* 2006, *610*, 114–123.

40. Stearns, D. M., Silveira, S. M., Wolf, K. K., Luke, A. M., Chromium(III) tris(picolinate) is mutagenic at the hypoxanthine (guanine) phosphoribosyltransferase locus in Chinese hamster ovary cells. *Mutat Res* 2002, *513*, 135–142.

41. Hathcock, J. N., Vitamins and minerals: efficacy and safety. *Am J Clin Nutr* 1997, *66*, 427–437.

42. Wasser, W. G., Feldman, N. S., D'agati, V. D., Chronic renal failure after ingestion of over-the-counter chromium picolinate. *Ann Intern Med* 1997, *126*, 410.

43. Cerulli, J., Grabe, D. W., Gauthier, I., Malone, M., *et al.,* Chromium picolinate toxicity. *Ann Pharmacother* 1998, *32*, 428–431.

44. Lefavi, R. G., Anderson, R. A., Keith, R. E., Wilson, G. D., *et al.,* Efficacy of chromium supplementation in athletes: emphasis on anabolism. *Int J Sport Nutr* 1992, *2*, 111–122.

45. Mertz, W., Chromium research from a distance: from 1959 to 1980. *J Am Coll Nutr* 1998, *17*, 544–547.

46. Shapcott, D., Hubert, J., *Proceedings of the Symposium on Chromium in Nutrition and Metabolism, held in Sherbrooke, Canada on July 13–15, 1979.* Amsterdam: Elsevier/North-Holland Biomedical Press, 1979.

47. Mertz, W., Effects and metabolism of glucose tolerance factor. *Nutr Rev* 1975, *33*, 129–135.

48. Mertz, W., Chromium in human nutrition: a review. *J Nutr* 1993, *123*, 626–633.

49. Mertz, W., Toepfer, E. W., Roginski, E. E., Polansky, M. M., Present knowledge of the role of chromium. *Fed Proc* 1974, *33*, 2275–2280.

50. Grant, K. E., Chandler, R. M., Castle, A. L., Ivy, J. L., Chromium and exercise training: effect on obese women. *Med Sci Sports Exerc* 1997, *29*, 992–998.

51. Crawford, V., Scheckenbach, R., Preuss, H. G., Effects of niacin-bound chromium supplementation on body composition in overweight African-American women. *Diabetes Obes Metab* 1999, *1*, 331–337.

52. Soni, M. G., Burdock, G. A., Preuss, H. G., *et al.,* Safety assessment of (-)-hydroxycitric acid and Super CitriMax, a novel calcium/potassium salt. *Food Chem Toxicol* 2004, *42*, 1513–1529.

53. Preuss, H. G., Garis, R. I., Bramble, J. D., Bagchi, D., *et al.,* Efficacy of a novel calcium/potassium salt of (-)-hydroxycitric acid in weight control. *Int J Clin Pharmacol Res* 2005, *25*, 133–144.

54. Downs, B. W., Bagchi, M., Subbaraju, G. V., Shara, M. A., *et al.,* Bioefficacy of a novel calcium-potassium salt of (-)-hydroxycitric acid. *Mutat Res* 2005, *579*, 149–162.

55. Shara, M., Ohia, S. E., Schmidt, R. E., Yasmin, T., *et al.,* Physico-chemical properties of a novel (-)-hydroxycitric acid extract and its effect on body weight, selected organ weights, hepatic lipid peroxidation and DNA fragmentation, hematology and clinical chemistry, and histopathological changes over a period of 90 days. *Mol Cell Biochem* 2004, *260*, 171–186.

56. Preuss, H. G., Bagchi, D., Bagchi, M., Rao, C. V., *et al.,* Effects of a natural extract of (-)-hydroxycitric acid (HCA-SX) and a combination of HCA-SX plus niacin-bound chromium and Gymnema sylvestre extract on weight loss. *Diabetes Obes Metab* 2004, *6*, 171–180.

57. Shara, M., Ohia, S. E., Yasmin, T., Zardetto-Smith, A., *et al.,* Dose- and time-dependent effects of a novel (-)-hydroxycitric acid extract on body weight, hepatic and testicular lipid peroxidation, DNA fragmentation and histopathological data over a period of 90 days. *Mol Cell Biochem* 2003, *254*, 339–346.

58. Ohia, S. E., Opere, C. A., Leday, A. M., Bagchi, M., *et al.,* Safety and mechanism of appetite suppression by a novel hydroxycitric acid extract (HCA-SX). *Mol Cell Biochem* 2002, *238*, 89–103.

59. Asghar, M., Monjok, E., Kouamou, G., Ohia, S. E., *et al.,* Super CitriMax (HCA-SX) attenuates increases in oxidative stress, inflammation, insulin resistance, and body weight in developing obese Zucker rats. *Mol Cell Biochem* 2007, *304*, 93–99.

60. Loe, Y. C., Bergeron, N., Rodriguez, N., Schwarz, J. M., Gas chromatography/mass spectrometry method to quantify blood hydroxycitrate concentration. *Anal Biochem* 2001, *292*, 148–154.

61. Deshmukh, N. S., Bagchi, M., Yasmin, T., Bagchi, D., Safety of a novel calcium/potassium salt of hydroxycitric acid (HCA-SX): I. Two-generation reproduction toxicity study. *Toxicology Mechanisms and Methods* 2008, *18*, 433–442.

62. Deshmukh, N. S., Bagchi, M., Yasmin, T., Bagchi, D., Safety of a novel calcium/potassium salt of (-)-hydroxycitric acid (HCA-SX): II. Developmental toxicity study in rats. *Toxicology Mechanisms and Methods* 2008, *18*, 443–451.

63. Westerterp-Plantenga, M. S., Kovacs, E. M., The effect of (-)-hydroxycitrate on energy intake and satiety in overweight humans. *Int J Obes Relat Metab Disord* 2002 *26*, 870–872.

64. Preuss, H. G., Bagchi, D., Bagchi, M., Rao, C. V. S., *et al.,* Efficacy of a novel, natural extract of (-)-hydroxycitric acid (HCA-SX) and a combination of HCA-SX, niacin bound chromium and Gymnema sylvestre extract in weight management in human volunteers: a pilot study. *Nutr Res* 2004, *24*, 45.

65. Preuss, H. G., Rao, C. V., Garis, R., Bramble, J. D., *et al.,* An overview of the safety and efficacy of a novel, natural(-)-hydroxycitric acid extract (HCA-SX) for weight management. *J Med* 2004, *35*, 33–48.

66. Mutch, D. M., Wahli, W., Williamson, G., Nutrigenomics and nutrigenetics: the emerging faces of nutrition. *Faseb J* 2005, *19*, 1602–1616.

67. Kaput, J., Rodriguez, R. L., Nutritional genomics: the next frontier in the postgenomic era. *Physiol Genomics* 2004, *16*, 166–177.

68. Corthesy-Theulaz, I., Den Dunnen, J. T., Ferre, P., Geurts, J. M., *et al.,* Nutrigenomics: the impact of biomics technology on nutrition research. *Ann Nutr Metab* 2005, *49*, 355–365.

69. Elliott, R. M., Johnson, I. T., Nutrigenomic approaches for obesity research. *Obes Rev* 2007, *8*(Suppl 1), 77–81.

70. Afman, L., Muller, M., Nutrigenomics: from molecular nutrition to prevention of disease. *J Am Diet Assoc* 2006, *106*, 569–576.

71. Rink, C., Roy, S., Khanna, S., Rink, T., *et al.,* Transcriptome of the subcutaneous adipose tissue in response to oral supplementation of type 2 Leprdb obese diabetic mice with niacin-bound chromium. *Physiol Genomics* 2006, *27*, 370–379.

72. Comi, G. P., Fortunato, F., Lucchiari, S., Bordoni, A., *et al.,* Beta-enolase deficiency, a new metabolic myopathy of distal glycolysis. *Ann Neurol* 2001, *50*, 202–207.

73. Baranova, A., Collantes, R., Gowder, S. J., Elariny, H., *et al.,* Obesity-related differential gene expression in the visceral adipose tissue. *Obes Surg* 2005, *15*, 758–765.

74. Kocaefe, Y. C., Israeli, D., Ozguc, M., Danos, O., *et al.,* Myogenic program induction in mature fat tissue (with MyoD expression). *Exp Cell Res* 2005, *308*, 300–308.

75. Stocker, A., Molecular mechanisms of vitamin E transport. *Ann N Y Acad Sci* 2004, *1031*, 44–59.

76. Cinti, S., Adipocyte differentiation and transdifferentiation: plasticity of the adipose organ. *J Endocrinol Invest* 2002, *25*, 823–835.

77. Zhou, Z., Yon Toh, S., Chen, Z., Guo, K., *et al.,* Cidea-deficient mice have lean phenotype and are resistant to obesity. *Nat Genet* 2003, *35*, 49–56.

78. Cassard-Doulcier, A. M., Gelly, C., Bouillaud, F., Ricquier, D., A 211-bp enhancer of the rat uncoupling protein-1 (UCP-1) gene controls specific and regulated expression in brown adipose tissue. *Biochem J* 1998, *333*(Pt 2), 243–246.

79. Valladares, A., Porras, A., Alvarez, A. M., Roncero, C., *et al.,* Noradrenaline induces brown adipocytes cell growth via beta-receptors by a mechanism dependent on ERKs but independent of cAMP and PKA. *J Cell Physiol* 2000, *185*, 324–330.

80. Roy, S., Rink, C., Khanna, S., Phillips, C., *et al.,* Body weight and abdominal fat gene expression profile in response to a novel hydroxycitric acid-based dietary supplement. *Gene Expr* 2004, *11*, 251–262.

81. Ohia, S. E., Awe, S. O., Leday, A. M., Opere, C. A., *et al.,* Effect of hydroxycitric acid on serotonin release from isolated rat brain cortex. *Res Commun Mol Pathol Pharmacol* 2001, *109*, 210–216.

82. Langin, D., Dicker, A., Tavernier, G., Hoffstedt, J., *et al.,* Adipocyte lipases and defect of lipolysis in human obesity. *Diabetes* 2005, *54*, 3190–3197.

83. Roy, S., Shah, H., Rink, C., Khanna, S., *et al.,* Transcriptome of primary adipocytes from

obese women in response to a novel hydroxycitric acid-based dietary supplement. *DNA Cell Biol* 2007, *26*, 627–639.

84. van Guilder, G. P., Hoetzer, G. L., Smith, D. T., Irmiger, H. M., *et al.,* Endothelial t-PA release is impaired in overweight and obese adults but can be improved with regular aerobic exercise. *Am J Physiol Endocrinol Metab* 2005, *289*, E807–E813.

85. Friedman, J. M., Halaas, J. L., Leptin and the regulation of body weight in mammals. *Nature* 1998, *395*, 763–770.

86. Friedman, J. M., Leptin, leptin receptors, and the control of body weight. *Nutr Rev* 1998, *56*, s38–s46; discussion s54–s75.

87. Cohen, P., Zhao, C., Cai, X., Montez, J. M., *et al.,* Selective deletion of leptin receptor in neurons leads to obesity. *J Clin Invest* 2001, *108*, 1113–1121.

88. Wittert, G. A., Turnbull, H., Hope, P., Morley, J. E., *et al.,* Leptin prevents obesity induced by a high-fat diet after diet-induced weight loss in the marsupial S. crassicaudata. *Am J Physiol Regul Integr Comp Physiol* 2004, *286*, R734–R739.

89. Nijhuis, J., Van Dielen, F. M., Buurman, W. A., Greve, J. W., Leptin in morbidly obese patients: no role for treatment of morbid obesity but important in the postoperative immune response. *Obes Surg* 2004, *14*, 476–483.

90. Dannewitz, B., Edrich, C., Tomakidi, P., Kohl, A., *et al.,* Elevated gene expression of MMP-1, MMP-10, and TIMP-1 reveal changes of molecules involved in turn-over of extracellular matrix in cyclosporine-induced gingival overgrowth. *Cell Tissue Res* 2006, *325*, 513–522.

91. Figueroa, J., Vijayagopal, P., Debata, C., Prasad, A., *et al.,* Azaftig, a urinary proteoglycan from a cachectic cancer patient, causes profound weight loss in mice. *Life Sci* 1999, *64*, 1339–1347.

92. Obunike, J. C., Sivaram, P., Paka, L., Low, M. G., *et al.,* Lipoprotein lipase degradation by adipocytes: receptor-associated protein (RAP)-sensitive and proteoglycan-mediated pathways. *J Lipid Res* 1996, *37*, 2439–2449.

93. Maquoi, E., Demeulemeester, D., Voros, G., Collen, D., *et al.,* Enhanced nutritionally induced adipose tissue development in mice with stromelysin-1 gene inactivation. *Thromb Haemost* 2003, *89*, 696–704.

94. Sasaki, K., Takagi, M., Konttinen, Y. T., Sasaki, A., *et al.,* Upregulation of matrix metalloproteinase (MMP)-1 and its activator MMP-3 of human osteoblast by uniaxial cyclic stimulation. *J Biomed Mater Res B Appl Biomater* 2007, *80*, 491–498.

95. Lijnen, H. R., Van, H. B., Frederix, L., Rio, M. C., *et al.,* Adipocyte hypertrophy in stromelysin-3 deficient mice with nutritionally induced obesity. *Thromb Haemost* 2002, *87*, 530–535.

96. Chang, S., Young, B. D., Li, S., Qi, X., *et al.,* Histone deacetylase 7 maintains vascular integrity by repressing matrix metalloproteinase 10. *Cell* 2006, *126*, 321–334.

97. Garcia, A., Sekowski, A., Subramanian, V., Brasaemle, D. L., The central domain is required to target and anchor perilipin A to lipid droplets. *J Biol Chem* 2003, *278*, 625–635.

98. Lin, J., Handschin, C., Spiegelman, B. M., Metabolic control through the PGC-1 family of transcription coactivators. *Cell Metab* 2005, *1*, 361–370.

99. Kim, B. W., Choo, H. J., Lee, J. W., Kim, J. H., *et al.,* Extracellular ATP is generated by ATP synthase complex in adipocyte lipid rafts. *Exp Mol Med* 2004, *36*, 476–485.

100. Gloerich, J., Ruiter, J. P., Van Den Brink, D. M., Ofman, R., *et al.,* Peroxisomal trans-2-enoyl-CoA reductase is involved in phytol degradation. *FEBS Lett* 2006, *580*, 2092–2096.

101. Schluter, A., Yubero, P., Iglesias, R., Giralt, M., *et al.,* The chlorophyll-derived metabolite phytanic acid induces white adipocyte differentiation. *Int J Obes Relat Metab Disord* 2002, *26*, 1277–1280.

102. Jaye, M., Lynch, K. J., Krawiec, J., Marchadier, D., *et al.,* A novel endothelial-derived lipase that modulates HDL metabolism. *Nat Genet* 1999, *21*, 424–428.

103. Ahmed, W., Orasanu, G., Nehra, V., Asatryan, L., *et al.,* High-density lipoprotein hydrolysis by

endothelial lipase activates PPARalpha: a candidate mechanism for high-density lipoprotein-mediated repression of leukocyte adhesion. *Circ Res* 2006, *98*, 490–498.

104. Przybyla-Zawislak, B. D., Srivastava, P. K., Vazquez-Matias, J., Mohrenweiser, H. W., *et al.,* Polymorphisms in human soluble epoxide hydrolase. *Mol Pharmacol* 2003, *64*, 482–490.

105. Lee, C. R., North, K. E., Bray, M. S., Fornage, M., *et al.,* Genetic variation in soluble epoxide hydrolase (EPHX2) and risk of coronary heart disease: the Atherosclerosis Risk in Communities (ARIC) study. *Hum Mol Genet* 2006, *15*, 1640–1649.

106. Wei, Q., Doris, P. A., Pollizotto, M. V., Boerwinkle, E., *et al.,* Sequence variation in the soluble epoxide hydrolase gene and subclinical coronary atherosclerosis: interaction with cigarette smoking. *Atherosclerosis* 2007, *190*, 26–34.

107. Westerberg, R., Tvrdik, P., Unden, A. B., Mansson, J. E., *et al.,* Role for ELOVL3 and fatty acid chain length in development of hair and skin function. *J Biol Chem* 2004, *279*, 5621–5629.

108. Westerberg, R., Mansson, J. E., Golozoubova, V., Shabalina, I. G., *et al.,* ELOVL3 is an important component for early onset of lipid recruitment in brown adipose tissue. *J Biol Chem* 2006, *281*, 4958–4968.

109. Liu-Stratton, Y., Roy, S., Sen, C. K., DNA microarray technology in nutraceutical and food safety. *Toxicol Lett* 2004, *150*, 29–42.

6

Application of Genomics and Bioinformatics Analysis in Exploratory Study of Functional Foods

Aiko Kohsuke and Manji Hayamizu

INTRODUCTION

The wealth of genomic information, genomics-based technologies, and model systems available provide a spectrum of new tools for use in human nutrition and food science. These new technologies are being used to study the molecular basis of the interaction of individual food constituents with both the genome and the metabolism of the human consumer. For historical reasons, nutrition and food science are not well prepared to exploit genomics technologies, primarily because of the lack of appropriate teaching of human genetics, genomics, and molecular biology in most university programs.

However, these deficits have been recognized and, in response, numerous initiatives have been launched recently in Europe, Asia, and the United States under the heading of "Nutrigenomics." Although to some, nutrigenomics might represent just another "-omic," it will change the face of research in nutrition and food science by moving the genome into the center of all processes that essentially determine mammalian metabolism in health and disease [1, 2, 3].

According to Kussmann et al., "nutritional genomics" as the collective term that covers the three subdisciplines of transcriptomics, proteomics, and metabolomics describes the use of medium- to high-throughput profiling technologies to assess the response of a cell or organism to dietary treatment or particular foods or food constituents [4].

Microarray-based transcriptome analysis may be considered the first mature genome-wide profiling technology. Consequently, it is also used widely, and applications of transcriptomics in nutritional studies seem unlimited when it comes to basic and preclinical research in either cell culture systems or animal models. The mRNA profiling techniques have the potential to easily identify specific transcript changes that respond to a given nutrient, non-nutrient compound, treatment, or diet in a well-defined experimental setting. This might not mean that the changes in mRNA level can be taken as a causal marker; it might rather be a pattern of expressed mRNAs that changes in a characteristic and reproducible way. Because the technology has the character of a screening process covering thousands of potentially affected indicators of the metabolic status simultaneously, it also reveals often totally unexpected findings [4]. In the infrastructure of microarray experiments, commercial platforms such as those of Affymetrix and Agilent are high quality and are provided with a manual, annotation file and other services [5, 6].

For the food industry, particularly functional food or dietary supplements, these new technologies look attractive to their business. Most industries conduct research on their most important ingredient that is important to human health. They hope to find new and novel functions for their items. A comprehensive approach may give them clue as to new ideas. There is no reason we do not conduct microarray experiments. Actually, wet experiments such as labeling fluorescent dyes, hybridization, or scanning of spots are not difficult to do. However, most beginners may be stalled by the analysis of immense amounts of data. Therefore, bioinformatics/dry experiment technique as well as wet experiment technique are very important in study of functional foods.

In this chapter, we present useful data analysis tools of microarray-based transcriptome for beginners. In addition, we show a practical analysis with our original data.

ANALYSIS TOOLS

Microarrays have become a standard tool for gene expression measurement in biology, medicine, and nutrition. Although microarrays are widely used, a fundamental challenge is to cope with the immense amount of data generated. Therefore, special software packages have been developed that are capable of handling the analysis of microarray data.

The common challenge faced by researchers is translating lists of differentially regulated genes into a better understanding of the underlying biological phenomena. Generally, the analysis flow is explained by two steps.

The first step is making a "gene list." This step can be the translation of the list of differentially expressed genes into a functional profile able to offer insight into the cellular mechanisms relevant in the given condition.

Gene lists are made by two methods that are gene expression-based and bioinformatics-based. The former is extracted by experimental data, for example, threshold of fold ratio or change pattern of gene expression on time course or treatment conditions.

The latter is extracted by known bioinformatics technologies. In this case, the researcher must have target fields before analysis.

The second step is filtering by gene list A and gene list B (and lists C, D, etc.) that are made in the first step.

Each operation is simple but takes a lot of time. It is inefficient to use Microsoft Excel for making gene lists because computational load is large.

GENESPRING GX

The most typical software is GeneSpring GX (Agilent) [7]. Other statistic analysis software has developed add-in packages to support microarray data. However, these are not dedicated software for analysis of microarray data. GeneSpring GX is a gene expression software tool enabling analysis, comparisons, visualization, and management of gene expression data. The best feature of this software is its easy-to-use interface.

BIOCONDUCTOR

We recommend Bioconductor for the R-user [8]. The Bioconductor toolkit is among the most sophisticated free software for microarray data analysis, based on the R statistical programming language [9]. Most algorithms developed for microarray data analysis are available within this package. Unfortunately, Bioconductor is a text-driven command line tool and does not provide an easy-to-use graphical interface. Therefore, it offers advanced analysis methods and the possibility of easy extension only for professional users, and it is difficult to use for people unskilled in R.

OTHERS

Other tools, EXPANDER [10] and TM4 [11], are well known as installed local machine software. The Web-based tools Expression Profiler [12] and GEPAS [13] are widely used for microarray data analysis.

These programs share a focus on data analysis, but most of them lack tools for the interpretation of the result. That is the biggest issue for beginners.

INTERPRETATION TOOLS

After making a list of differentially expressed genes from analysis tools GeneSpring GX and Bioconductor, the researcher conducts interpretation analysis using biological information.

Biological information for interpretation is grouped into three categories, Gene Ontology (GO), Pathway, and Association Network. Outlines of the three categories are explained in the following.

GO ANALYSIS TOOLS

The Gene Ontology (GO) project is a collaborative effort to address the need for consistent descriptions of gene products in different databases [14]. The GO project has developed three structured controlled vocabularies (ontologies) that describe gene products in terms of their associated biological process, cellular component, and molecular function in a species-independent manner. These are called GO Term. There are three separate aspects to this effort: first, the development and maintenance of the ontologies themselves; second, the annotation of gene products, which entails making associations between the ontologies and the genes and gene products in the collaborating databases; and third, development of tools that facilitate the creation, maintenance, and use of ontologies.

In 2002, an automatic ontological analysis approach using GO was proposed to help with this task [15]. From 2003 to date, many tools, including Onto-Express [16], DAVID (Database for Annotation, Visualization, and Integrated Discovery) [17], BinGO (The Biological Networks Gene Ontology) [18], and AmiGO [19], have been proposed for this type of analysis, and more tools appear every day. Currently, this approach is the *de facto* standard for the secondary analysis of high-throughput experiments, and a large number of tools have been developed for this purpose. Although these tools use the same general approach, they differ greatly in many respects that influence the results of the analysis. GO mining tools and their detailed information are available at the GO project Web site. You can get GO analysis results using your gene list by typing UniGene, EntrezGene, or GeneBank as a query.

PATHWAY ANALYSIS TOOLS

Pathway maps, especially metabolic pathway maps, are familiar to researchers. Most researchers are confident interpreting their own data on a pathway map. Pathways focus on physical and functional interactions between genes rather than the gene-centered view of GO-based analyses. Therefore, it is intriguing to map the list of significantly regulated genes onto precompiled pathways to elucidate the whole chain of events observed in a microarray experiment. However, not all pathways are equally suitable for microarray analysis. For instance, metabolic pathways are controlled largely by protein-based events, which are not observable in microarrays because only steady-state levels of mRNAs are monitored. Kinase-based signaling cascades also do not necessarily involve changes in mRNA levels. The best case for microarray-based pathway analysis is that transcriptional signaling pathways are directly coupled to *de novo* transcription. Although they contain post-transcriptional steps, there is usually enough transcriptional feedback regulation of pathway-related genes to allow for the identification of the pathways via alterations in mRNA level. Most pathway-analysis tools relying on precompiled databases of pathways derived from large-scale literature analysis require constant updating because of the continuous growth of the literature.

Many software tools capable of analyzing microarray data within the context of biological pathways have been developed. Recently released commercial software packages including

PathwayAssist [20], *PathArt* [21], Ingenuity Pathways Analysis tool [22], and MetaCore [23] also compete in the field of pathway-based microarray analysis. These tools provide an assortment of interfaces for the visualization of gene networks, natural language processing (NLP) extracted or hand-curated biological pathway/association network databases, and they accept gene list-based data input. Each of these tools has one or more unique features that distinguishes it from the others. Some open-source or publicly accessible software, such as GenMAPP [24] and Pathway Processor [25] display microarray data within the context of pathways annotated in *the Kyoto Encyclopedia of Genes and Genomes (KEGG)* pathways [26] and provide statistical assessment of the reliability of each differentially expressed gene.

KEGG is a widely used database of biological systems, consisting of genetic building blocks of genes and proteins (*KEGG GENES*), chemical building blocks of both endogenous and exogenous substances (*KEGG LIGAND*), molecular wiring diagrams of interaction and reaction networks (*KEGG PATHWAY*), and hierarchies and relationships of various biological objects (*KEGG BRITE*). KEGG provides a reference base for linking genomes to biological systems and to environments by the processes of PATHWAY mapping and BRITE mapping.

ASSOCIATION NETWORK ANALYSIS TOOLS

Protein interaction information is essential for systems-level understanding of cellular behavior, and it is needed to place the molecular functions of individual proteins into their cellular context. The database Web tool STRING (Search Tool for Retrieval of Interacting Genes/Protein) [27] is developed to collect, predict, and unify most types of protein–protein associations, including direct and indirect associations. It is a metaresource that aggregates most of the available information on protein–protein associations, scores and weights, and results of automatic literature-mining searches.

The most advanced point of STRING is the result of network analysis; it understands researchers intuitively. STRING analysis is simple; one can get network analysis results by using a gene list and writing biological common IDs, UniGene, EntrezGene, GeneBank..., as queries.

We have explained three categories of biological information. Many convenient tools are also present that refer to you.

In addition, some analysis tools integrate interpretation tools, for example, GeneSpring GX bundles the GO analysis tool, GEPAT (Genome Expression Pathway Analysis Tool) combines GO and KEGG analysis functions. These analysis programs and databases are no more than tools. Other tools such as TM4 also bundle a protein–protein interaction analysis function.

The following is an example of how to apply these tools for the study of the effect of kale on alcohol metabolism.

APPLICATION EXAMPLE
KALE (BRASSICA OLERACEA L. VAR ACEPHALA DC)

Kale (*Brassica oleracea L. var acephala DC*) is a form of cabbage, green in color, in which the central leaves do not form a head [28]. It is more closely related to wild cabbage than most domesticated forms. Until the end of the Middle Ages, Kale was one of the most common green vegetables in Europe. Curly-leafed varieties of cabbage already existed along with flat-leafed varieties in Greece in the fourth century BC. These forms, which were referred to by the Romans as Sabellian Kale, are considered the ancestors of modern Kale. Today, one may differentiate among different varieties according to the low, intermediate, or high length of the stem, with varying leaf types.

The species *Brassica oleracea* (Crucieferae family) contains a wide array of vegetables including broccoli, cauliflower, collard greens, and brussels sprouts. Epidemiological studies suggest that cruciferous vegetable intake may lower overall cancer risk. One such family of

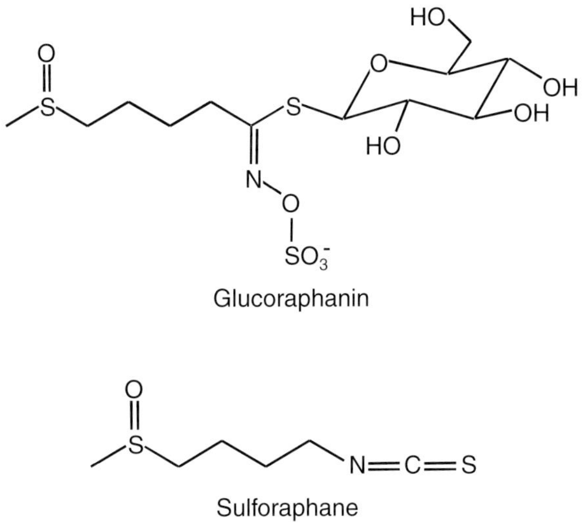

Figure 6.1 Structures of glucosinolate precursor glucoranin and its isothiocyanate hydrolysis product sulforaphane.

chemoprotective constituents are isothiocyanates (ITC), which are formed by glucosinolates [29]. Within the plant, glucosinolate content can vary greatly between and within members of the Crucieferae family depending on cultivation enviroment and genotype. There are more than 120 glucosinolates in the various varieties of cruciferous vegetables, each yielding different aglycone metabolic including ITC. The general structure of glucosinolate consists of a beta-D-thioglucose group, a sulfonated oxime group, and a variable side chain. Many of the anticancer effects observed from cruciferous vegetables have been attributed to the ITCs rather than to their parent glucosinolates. Two important and well-studied isothiocyanates derived from cruciferous vegetables are sulforaphane (SFN) and indole-3-carbinol.

The glucosinolate precursor to SFN, glucoraphanin, is abundant in kale, broccoli, cauliflower, and cabbage. Hydrolysis of glucoraphanin to its aglycone product SFN requires the activity of myrosinase enzymes released from the plant during consumption and other myrosinase enzymes present in human gut. The structures of glucoraphanin and SFN are shown in Fig. 6.1. The mechanisms of SFN chemoprevention have been well studied and reveal diverse responses depending upon the stage of carcinogenesis. SFN can function by blocking initiation via inhibiting Phase 1 enzymes that convert procarcinogens to proximate or ultimate carcinogens, and by inducing Phase 2 enzymes that detoxify carcinogens and facilitate their excretion from the body. Recently, Kale juice has become popular as a healthy vegetable juice in Japan. Kale juice is squeezed from fresh Kale leaves, and the slight bitter taste is caused by ITC. The studies of Kale as a functional food have shown beneficial properties such as antitumor activity, anti hypercholesterolemia, antihyperglycemia, improvement in constipation, antipigmentation and antiallergic effect. Incidentally, Kale is effective in preventing a hangover in popular folklore; however, scientific evidence of this was not reported. Therefore, we investigated the effect of Kale on alcohol and related metabolisms by microarray study.

ANIMAL STUDY AND DNA MICROARRAY ANALYSIS

Sixteen male ddY mice aged 3 weeks were used for animal experiments. Mice were divided into two groups of eight: control and kale group.

Kale was administered in the form of mixed food prepared by adding hot water extracted kale at 5% (w/w) in MF powder, a standard laboratory rodent diet (Oriental Yeast, Tokyo). The diet and drinking water were *ad libitum*. Fifteen weeks after kale supplementation, the mice were sacrificed and their livers were collected. There were no significant differences in body and liver weights between the kale treatment and control groups. Total RNA was extracted from the livers of eight mice fed on each diet, pooled and subjected to a DNA microarray analysis by whole mouse genome (Agilent). The number of probes consisted of 22,963 mouse genes. There were 17,472 genes that passed the selection criteria and subsequently used in the gene list for exploratory data analysis.

DATA ANALYSIS

EXPRESSION DATA

We conducted data analysis by filtering and selection. The analysis software used was Gene-Spring GX. At first, it required preparation of gene lists. We first identified 1682 genes whose expression was increased by twofold or decreased by <0.5 fold, and it was designated List A. We then extracted alcohol metabolism-related genes from Entrez Gene and STRING.

SEARCH BY ENTREZ GENE

An ethanol metabolism-related gene list was extracted by the search query "ethanol metabolism" limited to *Mus musculus*. A total of forty-four genes were identified by Entrez Gene search and annotated in a file named "Agilent whole mouse genome" (shown in Table 6.1).

SEARCH BY STRING

A gene list from STRING 8 was made with four queries, "alcohol dehydrogenase 1," "aldehyde dehydrogenase 2," "cyp2e1," and "catalase." The search condition was as follows:
   Required confidence (score): high confidence (0.07)
   Interactors shown: no more than 0 interactors
   Additional (white) nodes: 0
   or Network depth: 1
   The search result is illustrated as a network view. The search result of alcohol dehydrogenase 1 is shown in Fig. 6.2 as network view example.

FILTERING

Filtering was conducted by Venn diagram function on GeneSpring GX.
   Each gene list, extracted either from Entrez Gene or STRING 8, was written in common name, gene symbol, gene bank ID, and so on, for gene identity on GeneSpring GX. The result of filtering is shown in Fig. 6.3. We extracted fourteen genes showing significant changes related to alcohol metabolism by kale treatment. For interpretation of the fourteen genes, STRING search was used. The result of network analysis of remarkable genes is shown in Fig. 6.4.

RESULT

The identified genes were used to construct a network comprising alcohol metabolism, fatty acid metabolism, glycolysis/glucogenesis, tryptophan metabolism, and propanoate metabolism-related factors. The other gene network was MAPK signaling pathway-related factors.
   In general, the main pathway for ethanol oxidation involves hepatic alcohol dehydrogenase, a cytosolic enzyme that catalyzes the transformation of ethanol into acetaldehyde, which is in

Table 6.1 The list of ethanol metabolism genes extracted by Entrez Gene.

Gene Symbol	Genbank ID	Description
Adh1	NM_007409	Alcohol dehydrogenase 1 (class I)
Adh4	AK004863	Alcohol dehydrogenase 4 (class II), pi
Adh4	NM_011996	Alcohol dehydrogenase 4 (class II), pi
Aldh2	NM_009656	Aldehyde dehydrogenase 2, mitochondrial
Avpr1a	NM_016847	Arginine vasopressin receptor 1a
AI326910	NM_145429	Arrestin, beta 2
Bdnf	NM_007540	Brain-derived neurotrophic factor
Creb1	NM_025702	cAMP responsive element binding protein 1
Creb1	NM_009952	cAMP responsive element binding protein 1
Cat	NM_009804	Catalase
Cdc42	NM_009861	Cell division cycle 42 homolog (*S. cerevisiae*)
Cyp2e1	NM_021282	Cytochrome P450, 2, subfamily e, 1
Dbh	NM_138942	Dopamine beta hydroxylase
Gad1	NM_008077	Glutamic acid decarboxylase 1
Gad2	NM_008078	Glutamic acid decarboxylase 2
Gad2	AK018118	Glutamic acid decarboxylase 2
Gpt1	NM_182805	Glutamic pyruvic transaminase 1, soluble
Gpx1	NM_008160	Glutathione peroxidase 1
Il6	NM_031168	Interleukin 6
Lep	NM_008493	Leptin
Lpl	NM_008509	Lipoprotein lipase
Mapk14	NM_011951	Mitogen-activated protein kinase 14
Mapk8	NM_016700	Mitogen-activated protein kinase 8
Mapk8	AK047936	Mitogen-activated protein kinase 8
Maoa	NM_173740	Monoamine oxidase a
Nos1	NM_008712	Nitric oxide synthase 1, neuronal
Nos3	NM_008713	Nitric oxide synthase 3, endothelial cell
Ppar, PARalpha	NM_011144	Peroxisome proliferator-activated receptor alpha
Prkcd	NM_011103	Protein kinase c, delta
Rxra	NM_011305	Retinoid X receptor alpha
Spp1	NM_009263	Secreted phosphoprotein 1
Sirt1	NM_019812	Sirtuin 1 (silent mating type information regulation 2, homolog) 1
Slc18a2	AK035644	Solute carrier family 18 (vesicular monoamine), subfamily 2
Shh	NM_009170	Sonic hedgehog
Scd1	NM_009128	Stearoyl-Coenzyme A desaturase 1
Scd1	NM_009127	Stearoyl-Coenzyme A desaturase 1
Sod1	BC057592	Superoxide dismutase 1, soluble
Sod1	NM_011434	Superoxide dismutase 1, soluble
Sod1	AK080908	Superoxide dismutase 1, soluble
Trp53	NM_011640	Transformation related protein 53
Tnf	NM_013693	Tumor necrosis factor

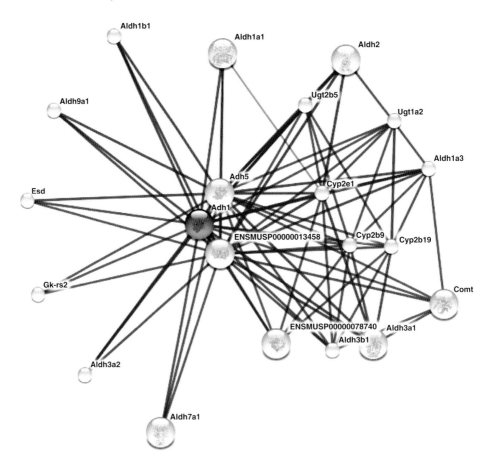

Figure 6.2 Gene networks analysis of alcohol dehydrogenase 1 by STRING 8.0. For color detail, please refer to the color plate section.

turn converted to acetic acid through aldehyde dehydrogenase. Ethanol can also be metabolized by peroxisomal catalase and an alcohol-inducible isoform of P450, CYP2e1, in the microsomal ethanol-oxidizing system [30].

Alcohol dehydrogenase 1(Adh1), Cyp2e1, and catalase (Cat) exhibited remarkable alteration in gene expression (fold ratios were 2.99, 5.91, 3.10, respectively). However, aldehyde dehydrogenase 2 (Aldh2), the main acetaldehyde dehydrogenase, was shown to be slightly changed (fold change: 1.6). Meanwhile, aldehyde dehydrogenase 1 (Aldh1), an isoform of Aldh2, was shown to change markedly (fold ratio: 3.05).

Approximately 60% of acetaldehyde was metabolized by Aldh2 and Aldh1 whereas 20% was metabolized by Cyp2e1. The pathway via Aldh1 and Cyp2e1 bypasses that expressed under excess alcohol intake. Taken together, we hypothesized that the mechanism of alcohol metabolism of kale is as follows. Kale upregulates Adh1 and alcohol is promoted by the conversion of ethanol to acetaldehyde. Acetaldehyde is converted to acetic acid by up-regulated Aldh1, Cyp2e1, and Cat supported to Aldh2 (Fig. 6.5). Fatty acid metabolism and glycolysis/glucogenesis are also known to relate to alcohol metabolism.

This study is preliminary and there are outstanding issues in proving the effect of kale on alcohol metabolism. For example, efficacy of kale on alcohol-loading animals, gene expression profile on time course (including early response gene expression), dose dependency, protein

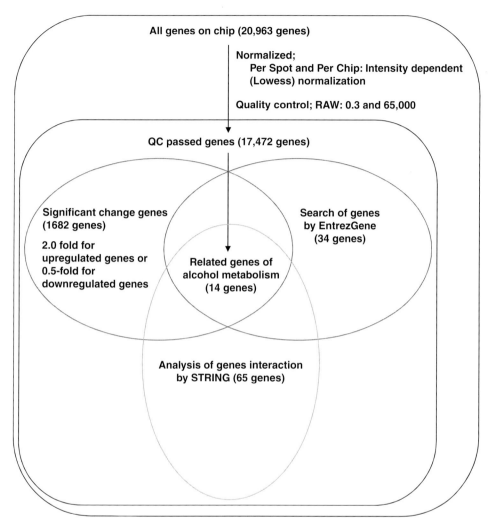

Figure 6.3 Scheme of identification of regulatory genes by *Brassica oleracea* var. *acephala* DC. For color detail, please refer to the color plate section.

expression analysis, and so on. Analysis of gene expression profile on time course may show an intriguing hint for understanding the mechanism of kale.

These results provided useful information for basic research of kale function. After this microarray study, we conducted a clinical study for evaluating alcohol metabolism improvement. In healthy subjects, the level of serum alcohol and acetaldehyde decreased when drinking alcohol with kale. Thus, the microarray study coupled with bioinformatics is efficient for exploratory study.

CONCLUSION

In this chapter, we introduced the use of bioinformatics databases for interpretation of microarray data for the novice. These methods may not be conventional or standard practice; however,

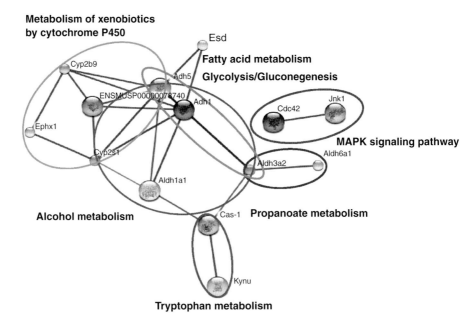

Figure 6.4 Genes network that shows remarkable change by *Brassica oleracea* var. *acephala* DC. For color detail, please refer to the color plate section.

we believe they are useful methods as an initial approach to microarray analysis. Moreover, construction of gene lists from databases is an efficient and effective method.

In our application example, we used Entrez Gene and STRING to obtain information not only related to alcohol metabolism but also related to other metabolisms. It is more efficient to combine the existing databases when analyzing microarray data.

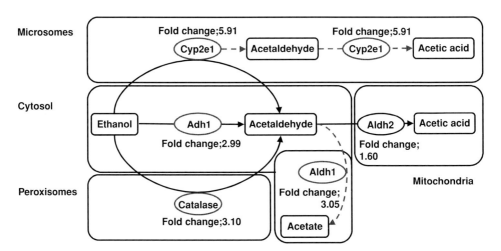

Figure 6.5 Effect of *Brassica oleracea* var. *acephala* DC on alcohol metabolism. For color detail, please refer to the color plate section.

REFERENCES

1. Muller, M., Kersten, S., Nutrigenomics: goals and strategies. *Nat Rev Genet* 2003, *4*, 315–322.
2. van der Werf, M. J., Schuren, F. H. J., Bijlsma, S., Tas, A. C., *et al.* Nutrigenomics: application of genomics technologies in nutritional sciences and food technology. *J. Food Sci* 2001, *66*, 772–780.
3. Rist, M. J., Wenzel, U., Daniel, H., Nutrition and food science go genomic. *Trends Biotechnol* 2006, *24*, 172–178.
4. Kussmann, M., Rezzi, S., Daniel, H., Profiling techniques in nutrition and health research. *Curr Opin Biotechnol* 2008, *19*, 83–99.
5. Affymetrix, *http://www.affymetrix.com/index.affx*.
6. Agilent Technologies, *http://www.home.agilent.com/agilent/home.jspx*.
7. Genespring GX Software, *http://www.chem.agilent.com/en-US/products/software/lifesciencesinformatics/genespringgx/*.
8. Bioconductor, *http://www.bioconductor.org/*.
9. The R project for Statistical Computing, *http://www.r-project.org/*.
10. Shamir, R., Maron-Katz, A., Tanay, A., Linhart, C., *et al.* EXPANDER – an integrative program suite for microarray data analysis. *BMC Bioinformatics* 2005, *6*, 232.
11. Saeed, A. I., Sharov, V., White, J., Li, J., *et al.*, TM4: a free, open-source system for microarray data management and analysis. *Biotechniques* 2003, *34*, 374–378.
12. Kapushesky, M., Kemmeren, P., Culhane, A. C., Durinck, S., *et al.*, Expression Profiler: next generation - an online platform for analysis of microarray data. *Nucleic Acids Res 2004, 32*, W465–470.
13. Herrero, J., Al-Shahrour, F., Díaz-Uriarte, R., Mateos, A., *et al.*, GEPAS: A web-based resource for microarray gene expression data analysis. *Nucleic Acids Res* 2003, *31*, 3461–3467.
14. The Gene Ontology, *http://www.geneontology.org/*.
15. Khatri, P., Draghici, S., Ostermeier, G. C., Krawetz, S. A., Profiling gene expression using onto-express. *Genomics* 2002, *79*, 266–270.
16. Khatri, P., Draghici, S., Ontological analysis of gene expression data: current tools, limitations, and open problems. *Bioinformatics* 2005, *21*, 3587–3595.
17. Dennis, G. Jr., Sherman, B. T., Hosack, D. A., Yang, J., *et al.*, DAVID: Database for Annotation, Visualization, and Integrated Discovery. *Genome Biol* 2003, *4*, P3.
18. Maere, S., Heymans, K., Kuiper, M., BiNGO: a Cytoscape plugin to assess overrepresentation of gene ontology categories in biological networks. *Bioinformatics* 2005, *21*, 3448–3449.
19. Carbon, S., Ireland, A., Mungall, C. J., Shu, S., *et al.*, AmiGO: online access to ontology and annotation data. *Bioinformatics* 2009, *25*, 288–289.
20. Nikitin, A., Egorov, S., Daraselia, N., Mazo, L., Pathway studio – the analysis and navigation of molecular networks. *Bioinformatics* 2003, *19*, 2155–2157.
21. PathArt, a product of Jubilant Biosys Ltd, *http://www.jubilantbiosys.com/pathart.html*.
22. Ingenuity Pathways Analysis tool, *http://www.ingenuity.com/products/pathways_analysis.html*.
23. MetaCore, a product of GeneGO Inc., *http://www.genego.com/*.
24. Dahlquist, K. D., Salomonis, N., Vranizan, K., Lawlor, S. C., *et al.*, GenMAPP, a new tool for viewing and analyzing microarray data on biological pathways. *Nat Genet* 2002, *31,* 19–20.
25. Grosu, P., Townsend, J. P., Hartl, D. L., Cavalieri, D., Pathway Processor: a tool for integrating whole-genome expression results into metabolic networks. *Genome Res* 2002, *12*, 1121–1126.
26. Kanehisa, M., Goto, S., Kawashima, S., Nakaya, A., The KEGG databases at GenomeNet. *Nucleic Acids Res* 2002, *30*, 42–46.
27. Jensen, L. J., Kuhn, M., Stark, M., Chaffron, S., *et al.*, STRING 8 - a global view on proteins and their functional interactions in 630 organisms. *Nucleic Acids Res* 2009, *37*(Database issue), D412–D416.

28. Olsen, H., Aaby, K., Borge, G. I., Characterization and quantification of flavonoids and hydroxycinnamic acids in curly kale *(Brassica oleracea* L. Convar. acephala Var. sabellica) by HPLC-DAD-ESI-MS(n). *J Agric Food Chem*, Mar. 2.(Epub ahead of print).

29. Clarke, J. D., Dashwood, R. H., Ho, E., Multi-targeted prevention of cancer by sulforaphane. *Cancer Lett* 2008, *269*, 291–304.

30. Lieber, C. S., Cytochrome P-4502E1: its physiological and pathological role. *Physiol Rev* 1997, *77*, 517–544.

7

Genomics as a Tool to Characterize Anti-inflammatory Nutraceuticals

Sashwati Roy

A rapid physiological response to tissue injury is to set the site on fire (Latin, *inflamatio*, to set on fire) by recruiting leukocytes from the blood into the injured tissues. A cascade of biochemical events propagates and matures the inflammatory response, involving the local vascular system, the immune system, and various cells within the injured tissue. Inflammation is viewed as an integral component of tissue repair or wound healing. The benefits of inflammation are only realized when inflammation is transient and resolves in a timely manner. Impairments in the resolution of inflammation lead to chronic inflammation, which is now known to be implicated in a wide range of disease processes. Prolonged inflammation, often associated with oxidative stress, may cause epigenetic changes and lead to a progressive shift in the type of cells present at the site of inflammation and is implicated in a variety of disease conditions. For example, chronic inflammation has been linked with an increased risk of type 2 diabetes and cardiovascular diseases. In this chapter, the broad-based implications of chronic inflammation in human health and disease are discussed. Next, the significance of nutraceuticals in managing chronic inflammation is addressed. Finally, a case of genomics as a tool in the characterization of the anti-inflammatory properties of nutraceuticals is discussed.

CHRONIC INFLAMMATION IN DISEASE

VASCULAR DISORDERS

Chronic inflammation is known to support angiogenesis, which in turn is expected to support tumor growth. Angiogenesis and unresolved inflammation have long been coupled together in many chronic inflammatory disorders with distinct etiopathogenic origin, including psoriasis, rheumatoid arthritis, Crohn's disease, diabetes, and cancer. Lately, this concept has been substantiated by the finding that several previously established noninflammatory disorders, such as osteoarthritis and obesity, display both inflammation and angiogenesis in an exacerbated manner. In addition, the interplay between inflammatory cells, endothelial cells, and fibroblasts in chronic inflammation sites, together with the fact that inflammation and angiogenesis can actually be triggered by the same molecular events, further strengthen this association [13]. Atherosclerosis is a chronic inflammatory disorder supported by inflammation that may be persistent for years and even decades [40].

RESPIRATORY DISORDERS

Inhaled environmental stressors damage the airways and lung parenchyma, producing irritation, recruitment of inflammatory cells, and oxidative modification of biomolecules. Oxidatively modified biomolecules, their degradation products, and adducts with other biomolecules can reach the systemic circulation, and when found in higher concentrations than normal they are considered to be biomarkers of systemic oxidative stress and inflammation. Metabolic stressors produced in the lung have a number of effects in tissues other than the lung, such as the brain, and they can also abrogate the mechanisms of immunotolerance [22]. Endogenous anti-inflammatory mediators and immune regulating mechanisms are important for the resolution of inflammatory processes. A disruption of these mechanisms can be causally related not only to the initiation of unnecessary inflammation but also to the persistence of several chronic inflammatory diseases. In asthma, chronic Th-2-driven eosinophilic inflammation of the airways represents one of the central abnormalities [55]. Oxidative and nitrative stress and severe inflammation account for the amplified inflammation in chronic obstructive pulmonary disease (COPD) [34]. A vicious cycle of airway obstruction, infection, and inflammation continues to cause most of the morbidity and mortality in cystic fibrosis (CF). The chronic inflammatory process damages and obstructs the airways and eventually claims the life of the patient [35].

GASTROINTESTINAL TRACT

Intestinal inflammation is implicated in inflammatory bowel disease (IBD) including Crohn's disease and ulcerative colitis. Enteric bacteria are a critical component in the development and prevention/treatment of chronic intestinal inflammation [57]. At the interface between the luminal content and host tissues, the intestinal epithelium must integrate pro- and anti-inflammatory signals to regulate innate and adaptive immune responses, that is, to control inflammation. However, under the influence of environmental factors, disturbance of the dialog between enteric bacteria and epithelial cells contributes to the development of chronic inflammation [11]. The genetic predisposition to deregulated mucosal immune responses and the concurrent prevalence of certain environmental triggers in developed countries are strong etiologic factors for the development of inflammatory bowel diseases in human subjects, including Crohn's disease and ulcerative colitis [10].

NEURODEGENERATIVE DISEASES

Recent studies demonstrate a strong link between neurodegeneration and chronic inflammation. The central nervous system (CNS) has limited regenerative capacity. Neural cell death occurs by apoptosis and necrosis. Necrosis in the CNS usually follows ischemic or traumatic brain injury [14]. Infection-triggered chronic inflammation may initiate a cascade of events leading to chronic inflammation and amyloid deposition in Alzheimer's disease (AD). Thus, anti-inflammatory therapy is considered to prevent dementia [33]. The immunohistochemical demonstration of reactive microglia and activated complement components suggests that chronic inflammation occurs in affected brain regions in Parkinson's disease (PD). Evidence from humans and monkeys exposed to 1-methyl-4-phenyl-1,2,3,6-tetrahydropyridine (MPTP) indicates this inflammation may persist many years after the initial stimulus has disappeared. Chronic inflammation can damage host cells and contributes to the pathogenesis of PD [30].

CANCER

Chronic inflammation, carcinogenesis, and cancer prognosis are tightly related. Inflammatory cells and cancer cells themselves produce free radicals and soluble mediators such as metabolites of arachidonic acid, cytokines, and chemokines, which act by further producing reactive species.

These, in turn, strongly recruit inflammatory cells in a vicious circle. Reactive intermediates of oxygen and nitrogen may directly oxidize DNA or may interfere with mechanisms of DNA repair. These reactive substances may also rapidly react with proteins, carbohydrates, and lipids, and the derivative products may induce a high perturbation in the intracellular and intercellular homeostasis until DNA mutation. The main substances that link inflammation to cancer via oxidative/nitrosative stress are prostaglandins and cytokines [19].

The proportion of total cancer deaths attributable to infectious agents is estimated to be about 20% to 25% in developing countries and 7% to 10% in more industrialized countries. Recurrent or persistent inflammation may induce, promote, or influence susceptibility to carcinogenesis by causing DNA damage, inciting tissue reparative proliferation, and/or creating a stromal "soil" enriched with cytokines and growth factors [48]. There is a proven association between carcinoma of the pancreas and both the sporadic and hereditary forms of chronic pancreatitis [31]. Kaposi's sarcoma (KS) is a complex cancer characterized by angioproliferative multifocal tumors of the skin, mucosa, and viscera. KS lesions comprised both distinctive spindle cells of endothelial origin and a variable inflammatory infiltrate. KS may result from reactive hyperproliferation induced by chronic inflammation [15]. Chronic inflammation is likely to have an important role in bladder carcinogenesis in developed countries [32].

RHEUMATIC DISEASES

Management of chronic inflammation represents the central mechanism targeted by most therapies addressing rheumatic diseases [46]. Rheumatoid arthritis (RA) is by definition a chronic disease with an autoimmune inflammatory attack on diarthrodial cartilaginous joints [25].

NUTRACEUTICALS IN THE MANAGEMENT OF CHRONIC INFLAMMATION

At present, monoclonal antibody-based therapeutics show clear promise in treating inflammatory disorders [21, 26]. The central role of TNFα in causing inflammation was initially provided by the demonstration that anti-TNFα antibodies added to *in vitro* cultures of a representative population of cells derived from diseased joints inhibited the spontaneous production of IL-1 and other proinflammatory cytokines. Systemic administration of anti-TNFα antibody or sTNFR fusion protein to mouse models of rheumatoid arthritis was shown to be anti-inflammatory and protective for joints. Clinical investigations in which the activity of TNFα in rheumatoid arthritis patients was blocked with intravenously administered infliximab, a chimeric anti-TNFα monoclonal antibody (mAB), has provided evidence that TNF regulates IL-6, IL-8, MCP-1, and VEGF production, recruitment of immune and inflammatory cells into joints, angiogenesis, and reduction of blood levels of matrix metalloproteinases-1 and –3 [20, 21]. The development of anti-TNF therapy is a key step forward in rheumatology as it is the first new therapy based on investigating the molecular mechanisms of this disease. Despite such major breakthroughs in investigative medicine, the fact remains that a vast population of individuals in developing countries suffering from inflammatory disorders do not benefit from monoclonal antibody-based therapy primarily because of excessive cost of acquisition and limited availability. For example, the cost associated with a single dose of infliximab is several thousand U.S. dollars [54].

The medical cost of rheumatoid arthritis averages $5919 per case per year in the United States and approximately £2600 per case per year in the United Kingdom. Current slow-acting antirheumatic drugs have limited efficacy and many side effects. Moreover, they do not improve the long-term prognosis of rheumatoid arthritis [9]. The use of medicinal plants to treat inflammatory disorders continues to be in practice worldwide [7, 8, 12, 18, 29, 38]. Medicinal plants often contain complex mixtures of phytochemicals that have additive or synergistic interactions. The anti-inflammatory properties of herbal preparations have been recognized in ancient Indian and Chinese medical literature. The gum resin of *Boswellia serrata*, known in the Indian Ayurvedic system of medicine as *Salai guggal*, contains Boswellic acids (BA), which have been shown to inhibit leukotriene biosynthesis. Compounds from the gum with proven anti-inflammatory effects are pentacyclic triterpenes of the BA type. Recently, the tetracyclic triterpene 3-oxo-tirucallic acid has been identified as a key active principle in *Boswellia* resin [6]. BA function as specific, nonredox inhibitors of leukotriene synthesis either interacting directly with 5-lipoxygenase or blocking its translocation [2–4]. Among the BA, acetyl-11-keto-beta-boswellic acid potently inhibits 5-lipoxygenase product formation with an IC_{50} of 1.5 μM. In contrast to the redox type 5-lipoxygenase inhibitor nordihydroguaiaretic acid, BA in concentrations up to 400 μM did not impair the cyclo-oxygenase and 12-lipoxygenase in isolated human platelets and the peroxidation of arachidonic acid by Fe-ascorbate. These data support that BA are specific, nonreducing-type inhibitors of 5-lipoxygenase [47]. In addition to their effects on the lipoxygenase system, certain BA inhibit elastase in leukocytes, inhibit proliferation, induce apoptosis, and inhibit topoisomerases of leukoma and glioma cell lines. A series of chronic inflammatory diseases are plausibly perpetuated by leukotrienes. In clinical trials promising results supporting the anti-inflammatory effects of *Boswellia* extract (BE) were observed in patients with rheumatoid arthritis, chronic colitis, ulcerative colitis, Crohn's disease, bronchial asthma, and peritumoral brain edema [1, 23, 24]. The enzymatic oxidation of arachidonic acid yields potent pathological agents by two major pathways, prostaglandin and lipoxygenase. The lipoxygenase pathway generates a new class of arachidonic acid oxygenation products, called the leukotrienes, which mediates inflammation. Unlike the prostaglandins, some of which play important roles as biological regulators, the action of the lipoxygenase products appear to be exclusively of a pathological nature [27]. Thus, the antilipoxygenase effects of BA are likely to have therapeutic implications.

Herbal medicines are widely used in the United States, with approximately one-quarter of adults reporting use of an herb to treat a medical illness within the past year. Of the ten most commonly used herbs in the United States, systematic reviews have concluded that only four are likely to be effective, and there is limited evidence to evaluate the efficacy of the approximately 20,000 other available products [5]. The emergent "omics" technology platform represents a powerful tool to examine the efficacy of herbals and nutraceuticals. Medicinal plants belonging to the *Burseraceae* family, including *Boswellia*, are especially known for their anti-inflammatory properties [16]. *Boswellia serrata* (frankincense) has been used in traditional medicine for treatment of inflammatory diseases since antiquity. BA have been studied for more than 30 years [17]. Acetyl-11-keto-beta-boswellic acid (AKBA) is a naturally occurring pentacyclic triterpene isolated from the gum resin exudate from the stem of the tree *Boswellia serrata*. AKBA has been recently identified as a novel, orally active, nonredox, and noncompetitive 5-lipoxygenase inhibitor that also inhibits topisomerase I and II *in vitro* [36, 37]. In humans, orally taken BE manifests as plasma KBA. The peak plasma levels of BE were reached at 4.5 hours. The plasma concentration attained a steady state after approximately 30 hours. BE has been proven to be safe and well tolerated on oral administration in humans. No adverse effects were seen with this drug when administered as a single dose in 333 mg [49].

GENECHIP™ AS A TOOL TO CHARACTERIZE THE ANTI-INFLAMMATORY PROPERTIES OF NUTRACEUTICALS

Almost 2 decades ago, TNF was identified as a protein produced by the immune system that suppressed tumor cell proliferation. Extensive research since then has revealed that TNFα is a major mediator of inflammation [51, 52, 56]. Endothelial cells are critical elements in the pathophysiology of inflammation. They participate through the synthesis and secretion of proinflammatory cytokines, including interleukin 1 (IL-1), IL-6, and IL-8, as well as M-CSF, G-CSF, GM-CSF, gro alpha, and MCP. They also express a series of cell-surface proteins and glycoproteins known as cell adhesion molecules that allow circulating leukocytes to bind to endothelial cells and allow endothelial cells to bind to matrix proteins [53]. TNFα potently induces inflammatory responses in endothelial cells [39]. Thus, the molecular basis of the anti-inflammatory effects of nutraceuticals may be tested in a system of TNFα-induced gene expression in human microvascular endothelial cells (HMEC) [43].

With the objective to identify sets of TNFα sensitive genes in HMEC, GeneChip[TM] analysis was performed in our own laboratories [41, 44, 45]. After 24 hours of seeding, HMEC cells were either treated with BE (50 μg/ml) or matching volume of dimethyl sulfoxide (DMSO) for 48 hours. As required, this was followed by treatment with recombinant human TNFα (50 ng/ml) for 6 hours. Cells were harvested after 6 hours of TNFα treatment and the total RNA was extracted using the RNeasy kit (Qiagen). Extracted RNA was treated with DNA-free (Ambion) to remove any possible DNA contamination present in the extracted RNA. The quality of RNA was assessed using Bioanalyzer 2100 (Agilent). To assess the quality of the labeled targets, the samples were hybridized for 16 hours at 45°C to GeneChip[TM] test arrays. Satisfactory samples were hybridized to the human genome arrays (U133 Plus 2.0) for the

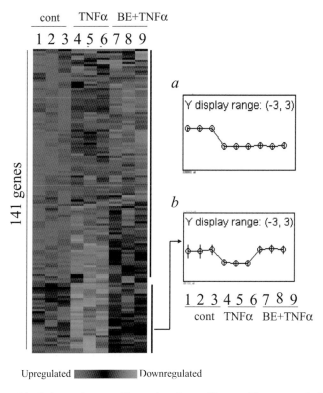

Figure 7.1 Hierarchical cluster images illustrating Boswellia-sensitive genes induced or down-regulated by TNFα in human microvascular endothelial cells. For clear graphic display of the specific clusters of genes showing an increase (A) or decrease (B) in expression following TNFα treatment to HMEC cells, a count percentage analysis was performed. Genes that were found up- or downregulated in 100% of replicates and all comparisons (nine out of nine) following TNFα treatment were selected. These select candidate genes were subjected to hierarchial clustering to identify clusters of genes induced/downregulated by TNFα and are sensitive to Boswellia. Red to green gradation in color represents higher to lower expression signal. Reproduced with permission from Roy *et al.*, *DNA Cell Biol* 2005, *24*, 244–255. For color detail, please refer to the color plate section.

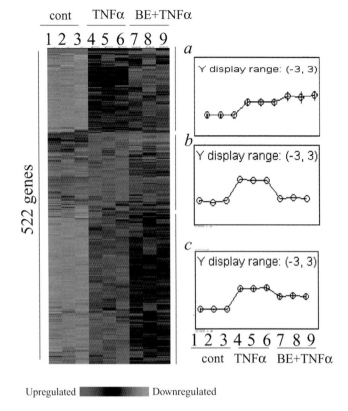

Figure 7.1 *(Continued)*

screening of more than 47,000 transcripts including the entire human genome. The arrays were washed, stained with streptavidin-phycoerythrin, and were then scanned with the high-resolution GeneChip scanner 3000 (Affymetrix) in our own facilities. To allow for statistical treatment, data were collected from three experiments. Raw data were analyzed using Affymetrix Microarray Suite 5.0 (MAS) and Data Mining Tool 2.0 (DMT) software. Additional processing of data was performed using dChip software [28]. We employed a stringent approach by taking a cutoff of 100%; that is, in a 3×3 comparison only those genes called increased in all nine of nine pair-wise comparisons were considered to increase in their expression. The magnitude of the change in gene expression (fold change) was reported for each comparison. These values were averaged to obtain an average fold change for each gene. Such conservative analytical approach limits the number of false-positive gene identifications. For data visualization and identification of BE-sensitive TNFα-inducible genes, genes filtered using the comparison analysis approach were subjected to hierarchical clustering using dChip software. This approach recognizes distinct clusters of transcriptome (Fig. 7.1). The genes that were significantly changed in BE and TNFα cotreated group compared to the group treated with TNFα alone were selected. Functional categorization and pathway construction were performed using the following software and Web resources: Gene Ontology Data Mining Tool (Affymetrix), *KEGG* (*Kyoto Encyclopedia of Genes and Genomes*), GenMAPP, DAVID (Database for Annotation, Visualization, and Integrated Discovery Verification), and LocusLink (Swiss-Prot). Acutely, TNFα induced 522 genes and downregulated 141 genes in 9 of 9 pair-wise comparisons. Of the 522 genes induced by TNFα in HMEC, 113 genes were clearly sensitive to BE treatment. Such genes directly related to

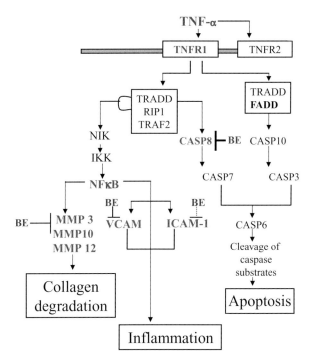

Figure 7.2 Boswellia-sensitive TNF-α induced signaling pathways in human microvascular endothelial cells. Pathway construction is based on GeneChipTM expression data and appropriate software resources (see Materials & Methods). To obtain insights on the effects of Boswellia on specific pathways induced by TNFα- in endothelial cells, the results of GeneChip analysis were mapped onto known pathways associated with inflammation, apoptosis, and collagen degradation. GenMAPP, *KEGG*, and Gene Ontology (GO) were used to develop the pathways. Genes shown in red are candidates identified using GeneChip assay that were upregulated following TNFα. Blunt arrow marked with BE marks the genes whose expression levels are fully (solid line) or partly (broken lines) normalized by BE pretreatment. TNFR, TNFα receptor; CASP, caspase; BE, Boswellia extract; TRADD, TNFR1-associated protein with death domain; NIK, NFκB-inducing kinase; IKK, Iκb kinase; MMP, matrix metalloproteinase; VCAM, vascular cell adhesion molecule; ICAM, intercellular adhesion molecule. Reproduced with permission from Roy *et al.*, *DNA Cell Biol* 2005, *24*, 244–255. For color detail, please refer to the color plate section.

inflammation, cell adhesion, and proteolysis. The robust BE-sensitive candidate genes were then subjected to further processing for the identification of BE-sensitive signaling pathways. The use of resources such as GenMAPP, *KEGG*, and Gene Ontology (GO) led to recognition of the primary BE-sensitive TNFα-inducible pathways (Fig. 7.2). BE prevented the TNFα-induced expression of matrix metalloproteinases. BE also prevented the inducible expression of mediators of apoptosis. Most strikingly, however, TNFα-inducible expression of VCAM-1 and ICAM-1 were observed to be sensitive to BE. Real-time PCR studies showed that although TNFα potently induced VCAM-1 gene expression, BE completely prevented it. This result confirmed our microarray findings and built a compelling case for the anti-inflammatory property of BE. In an *in vivo* model of carrageenan-induced rat paw inflammation, a significant anti-inflammatory property of BE was noted [43].

As follow-up to the previously mentioned GeneChip study, which recognized matrix metalloproteinases (MMP) as a key TNFα-inducible pathway that is BE-sensitive, the effects of BE on TNFα-inducible MMP expression in human microvascular endothelial cells was tested.

MMPs are a family of zinc-containing enzymes involved in the degradation and remodeling of extracellular matrix proteins. Under normal physiological conditions, the activities of these enzymes are well regulated by endogenous tissue inhibitors of metalloproteinases (TIMPs). Chronic stimulation of MMP activities, due to an imbalance in the levels of MMPs and TIMPs, has been implicated in the pathogenesis of a variety of diseases such as cancer, osteoarthritis, and rheumatoid arthritis. Thus, MMP inhibitors are expected to be useful for the treatment of these disorders. Because of their importance in a variety of pathological conditions, a number of small molecular weight MMP inhibitors have entered clinical trials in humans. However, the results of these trials have been disappointing [50]. In our study, to evaluate the significance of AKBA in the anti-inflammatory properties of BE, effects of BE containing either 3% (BE3%) or 30% (BE30%) were compared. Pretreatment of HMEC for 2 days with BE potently prevented TNFα-induced expression and activity of MMP3, MMP10, as well as MMP12. The Freund's adjuvant-induced rat paw edema experimental system was utilized to test the significance of the anti-inflammatory properties of BE *in vivo*. *In vivo*, BE protected against experimental arthritis. In all experiments, both *in vitro* and *in vivo*, BE30% was more effective than BE3%. These observations lend support to the GeneChip studies demonstrating that BE has potent anti-inflammatory properties both *in vitro* as well as *in vivo* [42]. Thus, the GeneChip screening and data mining approach adopted represents a powerful tool for hypothesis discovery.

REFERENCES

1. Ammon, H. P., Boswelliasauren (Inhaltsstoffe des Weihrauchs) als wirksame Prinzipien zur Behandlung chronisch entzundlicher Erkrankungen. *Wien Med Wochenschr* 2002, *152*, 373–378.
2. Ammon, H. P., Salai Guggal - Boswellia serrata: from a herbal medicine to a non-redox inhibitor of leukotriene biosynthesis. *Eur J Med Research* 1996, *1*, 369–370.
3. Ammon, H. P., Mack, T., Singh, G. B., Safayhi, H., Inhibition of leukotriene B4 formation in rat peritoneal neutrophils by an ethanolic extract of the gum resin exudate of Boswellia serrata. *Planta Medica* 1991, *57*, 203–207.
4. Ammon, H. P., Safayhi, H., Mack, T., Sabieraj, J., Mechanism of antiinflammatory actions of curcumine and boswellic acids. *J Ethnopharmacol* 1993, *38*, 113–119.
5. Bent, S., Ko, R., Commonly used herbal medicines in the United States: a review. *Am J Med* 2004, *116*, 478–485.
6. Boden, S. E., Schweizer, S., Bertsche, T., Dufer, M., *et al.*, Stimulation of leukotriene synthesis in intact polymorphonuclear cells by the 5-lipoxygenase inhibitor 3-oxo-tirucallic acid. *Mol Pharmacol* 2001, *60*, 267–273.
7. Borchers, A. T., Keen, C. L., Stern, J. S., Gershwin, M. E., Inflammation and Native American medicine: the role of botanicals. *Am J Clin Nutr* 2000, *72*, 339–347.
8. Chainani-Wu, N., Safety and anti-inflammatory activity of curcumin: a component of tumeric (Curcuma longa). *J Altern Complement Med* 2003, *9*, 161–168.
9. Choy, E. H., Panayi, G. S., Cytokine pathways and joint inflammation in rheumatoid arthritis. *N Engl J Med* 2001, *344*, 907–916.
10. Clavel, T., Haller, D., Bacteria- and host-derived mechanisms to control intestinal epithelial cell homeostasis: implications for chronic inflammation. *Inflamm Bowel Dis* 2007, *13*, 1153–1164.
11. Clavel, T., Haller, D., Molecular interactions between bacteria, the epithelium, and the mucosal immune system in the intestinal tract: implications for chronic inflammation. *Curr Issues Intest Microbiol* 2007, *8*, 25–43.
12. Cohen, S. M., Rousseau, M. E., Robinson, E. H., Therapeutic use of selected herbs. *Holist Nurs Pract* 2000, *14*, 59–68.
13. Costa, C., Incio, J., Soares, R., Angiogenesis and chronic inflammation: cause or consequence? *Angiogenesis* 2007, *10*, 149–166.

14. DeLegge, M. H., Smoke, A., Neurodegeneration and inflammation. *Nutr Clin Pract* 2008, *23*, 35–41.
15. Douglas, J. L., Gustin, J. K., Dezube, B., Pantanowitz, J. L., *et al.*, Kaposi's sarcoma: a model of both malignancy and chronic inflammation. *Panminerva Med* 2007, *49*, 119–138.
16. Duwiejua, M., Zeitlin, I. J., Waterman, P. G., Chapman, J., *et al.*, Anti-inflammatory activity of resins from some species of the plant family Burseraceae. *Planta Medica* 1993, *59*, 12–16.
17. el-Khadem, H., el-Shafei, Z. M., el-Sekeily, M. A., Rahman, M. M., Derivatives of boswellic acids. *Planta Medica* 1972, *22*, 157–159.
18. Ernst, E., Chrubasik, S.,Phyto-anti-inflammatories. A systematic review of randomized, placebo-controlled, double-blind trials. *Rheum Dis Clin North Am* 2000, *26*, 13–27.
19. Federico, A., Morgillo, F., Tuccillo, C., Ciardiello, F., *et al.,* Chronic inflammation and oxidative stress in human carcinogenesis. *Int J Cancer* 2007, *121*, 2381–2386.
20. Feldmann, M., Brennan, F. M., Paleolog, E., Cope, A., *et al.*, Anti-TNFalpha therapy of rheumatoid arthritis: what can we learn about chronic disease? *Novartis Found Symp* 2004, *256*, 53–69; discussion 69–73.
21. Feldmann, M., Maini, R. N., Anti-TNF alpha therapy of rheumatoid arthritis: what have we learned? *Ann Rev Immunol* 2001, *19*, 163–196.
22. Gomez-Mejiba, S. E., Zhai, Z., Akram, H., Pye, Q. N., *et al.*, Inhalation of environmental stressors & chronic inflammation: Autoimmunity and neurodegeneration. *Mutat Res* 2008, *31*, 62–72.
23. Gupta, I., Gupta, V., Parihar, A., Gupta, S., *et al.*, Effects of Boswellia serrata gum resin in patients with bronchial asthma: results of a double-blind, placebo-controlled, 6-week clinical study. *Eur J Med Res* 1998, *3*, 511–514.
24. Gupta, I., Parihar, A., Malhotra, P., Gupta, S., *et al.*, Effects of gum resin of Boswellia serrata in patients with chronic colitis. *Planta Medica* 2001, *67*, 391–395.
25. Holmdahl, R., Nature's choice of genes controlling chronic inflammation. *Ernst Schering Found Symp Proc* 2006, *4*, 1–15.
26. Kaplan, M., Eculizumab (Alexion). *Curr Opin Investig Drugs* 2002, *3*, 1017–1023.
27. Kuehl, F. A., Jr., Egan, R. W., Prostaglandins, arachidonic acid, and inflammation. *Science* 1980, *210*, 978–984.
28. Li, C., Wong, W. H., Model-based analysis of oligonucleotide arrays: Expression index computation and outlier detection. *Proc Natl Acad Sci USA* 2001, *98*, 31–36.
29. Long, L., Soeken, K., Ernst, E., Herbal medicines for the treatment of osteoarthritis: a systematic review.[see comment]. *Rheumatology* 2001, *40*, 779–793.
30. McGeer, P. L., McGeer, E. G., Inflammation and neurodegeneration in Parkinson's disease. *Parkinsonism Relat Disord* 2004, *10*(Suppl 1), S3–7.
31. McKay, C. J., Glen, P., McMillan, D. C., Chronic inflammation and pancreatic cancer. *Best Pract Res Clin Gastroenterol* 2008, *22*, 65–73.
32. Michaud, D. S., Chronic inflammation and bladder cancer. *Urol Oncol* 2007, *25*, 260–268.
33. Miklossy, J., Chronic inflammation and amyloidogenesis in Alzheimer's disease – role of Spirochetes. *J Alzheimers Dis* 2008, *13*, 381–391.
34. Mroz, R. M., Noparlik, J., Chyczewska, E., Braszko, J. J., *et al.*, Molecular basis of chronic inflammation in lung diseases: new therapeutic approach. *J Physiol Pharmacol* 2007, *58*(Suppl 5), 453–460.
35. Nichols, D., Chmiel, J., Berger, M., Chronic inflammation in the cystic fibrosis lung: alterations in inter- and intracellular signaling. *Clin Rev Allergy Immunol* 2008, *34*, 146–162.
36. Park, Y. S., Lee, J. H., Bondar, J., Harwalkar, J. A., *et al.*, Cytotoxic action of acetyl-11-keto-beta-boswellic acid (AKBA) on meningioma cells. *Planta Medica* 2002, *68*, 397–401.
37. Park, Y. S., Lee, J. H., Harwalkar, J. A., Bondar, J., *et al.*, Acetyl-11-keto-beta-boswellic acid (AKBA) is cytotoxic for meningioma cells and inhibits phosphorylation of the extracellular-signal regulated kinase 1 and 2. *Adv Exp Med Biol* 2002, *507*, 387–393.

38. Phillipson, J. D., 50 years of medicinal plant research – every progress in methodology is a progress in science. *Planta Medica* 2003, *69*, 491–495.

39. Pober, J. S., Endothelial activation: intracellular signaling pathways. *Arthritis Res* 2002, *4*(Suppl 3), S109–S116.

40. Rottenstrich, E., Rohana, H., [Atherosclerosis as a chronic inflammation]. *Harefuah* 1999, *136*, 785–788.

41. Roy, S., Khanna, S., Bentley, K., Beffrey, P., *et al.*, Functional genomics: high-density oligonucleotide arrays. *Methods in Enzymology* 2002, 487–497.

42. Roy, S., Khanna, S., Krishnaraju, A. V., Subbaraju, G. V., *et al.*, Regulation of vascular responses to inflammation: inducible matrix metalloproteinase-3 expression in human microvascular endothelial cells is sensitive to antiinflammatory Boswellia. *Antioxid Redox Signal* 2006, *8*, 653–660.

43. Roy, S., Khanna, S., Shah, H., Rink, C., *et al.*, Human genome screen to identify the genetic basis of the anti-inflammatory effects of Boswellia in microvascular endothelial cells. *DNA Cell Biol* 2005, *24*, 244–255.

44. Roy, S., Khanna, S., Wallace, W. A., Lappalainen, J., *et al.*, Characterization of perceived hyperoxia in isolated primary cardiac fibroblasts and in the reoxygenated heart. *J Biol Chem* 2003, *278*, 47129–47135.

45. Roy, S., Lado, B. H., Khanna, S., Sen, C. K., Vitamin E sensitive genes in the developing rat fetal brain: a high-density oligonucleotide microarray analysis. *FEBS Lett* 2002, *530*, 17–23.

46. Sacre, S. M., Drexler, S. K., Andreakos, E., Feldmann, M., *et al.*, Could toll-like receptors provide a missing link in chronic inflammation in rheumatoid arthritis? Lessons from a study on human rheumatoid tissue. *Ann Rheum Dis* 2007, *66*(Suppl 3), iii81–iii86.

47. Safayhi, H., Mack, T., Sabieraj, J., Anazodo, M. I., *et al.*, Boswellic acids: novel, specific, nonredox inhibitors of 5-lipoxygenase. *J Pharmacol Exp Ther* 1992, *261*, 1143–1146.

48. Schottenfeld, D., Beebe-Dimmer, J., Chronic inflammation: a common and important factor in the pathogenesis of neoplasia. *CA Cancer J Clin* 2006, *56*, 69–83.

49. Sharma, S., Thawani, V., Hingorani, L., Shrivastava, M., *et al.*, Pharmacokinetic study of 11-Keto beta-Boswellic acid. *Phytomedicine* 2004, *11*, 255–260.

50. Skiles, J. W., Gonnella, N. C., Jeng, A. Y., The design, structure, and clinical update of small molecular weight matrix metalloproteinase inhibitors. *Curr Med Chem* 2004, *11*, 2911–2977.

51. Strieter, R. M., Kunkel, S. L., Bone, R. C., Role of tumor necrosis factor-alpha in disease states and inflammation. *Crit Care Med* 1993, *21*, S447–S463.

52. Sullivan, K. E., Regulation of inflammation. *Immunol Res* 2003, *27*, 529–538.

53. Swerlick, R. A., Lawley, T. J., Role of microvascular endothelial cells in inflammation. *J Invest Dermatol* 1993, *100*, 111S–115S.

54. Valle, E., Gross, M., Bickston, S. J., Infliximab. *Expert Opin Pharmacother* 2001, *2*, 1015–1025.

55. Van Hove, C. L., Maes, T., Joos, G. F., Tournoy, K. G., Chronic inflammation in asthma: a contest of persistence vs resolution. *Allergy* 2008, *63*, 1095–1109.

56. Warren, J. S., Ward, P. A., Johnson, K. J., Tumor necrosis factor: a plurifunctional mediator of acute inflammation. *Mod Pathol* 1988, *1*, 242–247.

57. Werner, T., Haller, D., Intestinal epithelial cell signalling and chronic inflammation: From the proteome to specific molecular mechanisms. *Mutat Res* 2007, *622*, 42–57.

8

Application of Nutrigenomics in Gastrointestinal Health

Lynnette R. Ferguson, Philip I. Baker, and Donald R. Love

ABSTRACT

Nutrigenomics considers how common dietary chemicals affect health by altering the expression and/or structure of an individual's genetic makeup. Thus, it provides the scientific basis to underpin personalized nutrition and optimal health status (maintenance of homeostasis) through an extended lifespan. The gastrointestinal (GI) tract spans from the stomach to the anus, with the primary purpose of altering the composition of food to allow its digestion and absorption. The gastrointestinal epithelium also represents an initial defense against invading pathogens. Among a number of conditions classified as gastrointestinal diseases are various cancers, coeliac disease, irritable bowel syndrome, and inflammatory bowel diseases (Crohn's disease and ulcerative colitis). All of these diseases have a genetic susceptibility, which is strongly influenced by diet and environment. Genome-wide association studies provide an unprecedented understanding of the genetic basis of chronic disease, suggesting likely nutrigenomic approaches to intervention.

We develop the theme of Crohn's disease to highlight general principles applicable to gastrointestinal health. Although more than thirty genetic loci have now been identified as causal, many of these can be clustered with one another or alongside a parallel literature on diet, to suggest two main pathways important in maintaining gastrointestinal health. First, it is clear that the cells of the intestinal walls not only absorb nutrients but actively provide a protective barrier to the entry of pathogens, because several of the identified genes affect transport and barrier function. Secondly, the gut immune system provides a complex set of innate and adaptive defense mechanisms that discriminate between beneficial and pathogenic bacteria. Conspicuous among the risk loci are pattern recognition receptors and downstream genes affecting immune response. Related to these is a third mechanism whereby the resident microflora is able to compete for nutrients and for receptor sites on the gut wall, thereby reducing the likelihood that invading bacteria will be able to grow and reproduce. In this respect, the field of probiotics and prebiotics is an expanding area of active marketing of functional foods for gastrointestinal health. Finally, other -omics approaches can increase our understanding of mechanisms and validate dietary interventions to maintain optimal gastrointestinal health in individuals with Crohn's disease, as well as otherwise healthy individuals.

INTRODUCTION

Much of the focus of nutritional science throughout the twentieth century involved characterizing vitamins and minerals, defining their use, and seeking to prevent the deficiency diseases that they caused. However, to prevent disease development, it becomes important to understand not only how nutrients act at the molecular level, but also how nutrition can optimize and maintain homeostasis. Thus, a more modern interpretation of nutritional science extends this earlier knowledge into biochemistry and physiology, and also considers the way in which individual response to nutrients can differ according to genotype.

Nutritional genomics or "nutrigenomics" considers how common dietary chemicals affect health by altering the expression and/or structure of an individual's genetic makeup. Kaput and Rodriguez [1] developed five tenets of nutritional genomics, whose principles they stated as follows:

1. Common dietary chemicals act on the human genome, either directly or indirectly, to alter gene expression or structure;
2. under certain circumstances and in some individuals, diet can be a serious risk factor for a number of diseases;
3. some diet-regulated genes (and their normal, common variants) are likely to play a role in the onset, incidence, progression, and/or severity of chronic diseases;
4. the degree to which diet influences the balance between healthy and disease states may depend on an individual's genetic makeup; and
5. dietary intervention based on knowledge of nutritional requirements, nutritional status, and genotype (i.e., "individualized nutrition") can be used to prevent, mitigate, or cure chronic disease.

Interindividual differences in genotype profoundly affect metabolism and thereby phenotype. Much of the human variation described to date has focussed on single nucleotide polymorphisms (SNPs), but there is increasing recognition of the importance of more gross changes in chromosomal balance and/or of epigenetic effects [2]. Monogenic disorders causing strong pathological effects have been described, including a polymorphism in the gene for the hormone leptin, which results in morbid obesity [3]. Nevertheless, this disease provides a paradigm for gene–diet interactions, because even in individuals carrying the leptin variant, strictly controlled and supervised diets can overcome the effects of genetic susceptibility. Other polymorphisms have consequences for human nutrition, implying differences in required nutrient levels for such micronutrients as folate. In this context, the common polymorphism (C677T) in the gene encoding 5,10-methylenetetrahydrofolate reductase (MTHFR) significantly increases the required levels of dietary folate or related methyl donors required to protect against gastrointestinal diseases such as colon cancer [4]. Similar examples exist for selenium [5]. Many other SNPs may result in minor deviations in nutritional biochemistry, whose functional effects can be studied through changes in cellular gene expression (transcriptomics), proteins (proteomics), or metabolites (metabolomics).

Different diets may overcome the implications of variant genotypes through eliciting different patterns of gene and protein expression and metabolite production [6]. One of the approaches used in nutrigenomics involves identification of biomarkers of the early phase of diet-related diseases, at which intervention with nutrition can return the patient to health (maintenance of homeostasis) [7]. The promise of nutrigenomics is to use the burgeoning understanding of the relationship between genes and diets to develop personalized dietary recommendations to maintain optimal health throughout the lifespan. A schematic representation of approaches to achieving this goal is provided in Fig. 8.1. This review focuses on the application of this field to gastrointestinal health.

THE HUMAN GASTROINTESTINAL TRACT

In mammals, the gastrointestinal (GI) tract spans from the stomach to the anus, with the primary purpose of altering the physical and chemical composition of food so it can be absorbed and utilized by the body's cells [8]. The stomach and upper small bowel (ileum) are mostly involved

Case-control or family studies

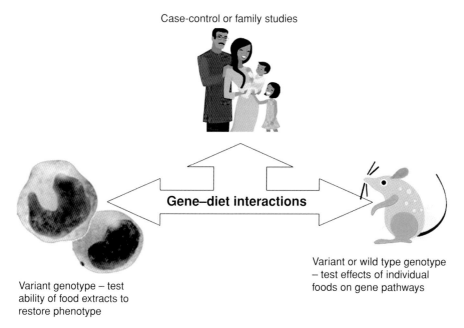

Variant genotype – test
ability of food extracts to
restore phenotype

Variant or wild type genotype
– test effects of individual
foods on gene pathways

Figure 8.1 Methods of studying nutrigenomics and gastrointestinal health, as utilized by Nutrigenomics New Zealand (www.nutrigenomics.org.nz). A fundamental part of studies recognizes human variation, usually studied as either single nucleotide polymorphism (SNPs) or copy number variants, and the effect that these have on the normal phenotype. In the case of gastrointestinal health, a challenge to homeostasis may be necessary to reveal the implication of variation on the phenotype. Once key genes are identified, then animal models may be utilized to study the effects of various foods and food components on cellular gene expression (transcriptomics), proteins (proteomics), or metabolites (metabolomics). It may also be appropriate to study cecal or fecal microbiota. *In vitro* or tissue culture models provide a high-throughput approach to test whether nutrients or food extracts can overcome the effect of the variant genotype. Evidence from all three lines of study can be integrated to select foods for human clinical trials, which involve stratifying subjects according to genotype and using a biomarker approach to relate genotype to the trial endpoint. Such approaches provide a pathway to develop personalized approaches in optimizing foods or diets to ameliorate the effects of a variant genotype. For color detail, please refer to the color plate section.

in food digestion, and the remaining small and large intestine (caecum and colon) are involved in the absorption of nutrients, water salvage, and the excretion of waste products. Goblet cells of the epithelia are responsible for secreting mucus, forming a lubricating surface for the passage of food and an attractive physical habitat for the microflora. The secretion of mucus is partly regulated by hormones, neurotransmitters, prostaglandins, and nitric oxide (NO) produced by the host [9].

The gastrointestinal epithelium represents an initial defense against invading pathogens through three main mechanisms. First, cells of the intestinal wall not only absorb nutrients, but have a number of active functions that provide a protective barrier to the entry of harmful substances or pathogens. Secondly, the gut immune system provides a complex set of innate and adaptive defense mechanisms that discriminate between beneficial and pathogenic bacteria [10–13]. Third, the resident microflora in a healthy individual is able to compete for nutrients and for receptor sites on the gut wall, thereby reducing the likelihood that invading bacteria will grow and reproduce. The resident flora may also generate an adverse environment for pathogens, for example, through maintaining a low pH level.

The lower small and large intestine are largely anaerobic environments [13]. pH is low in the stomach, increasing to between 6 and 7 in the terminal ileum, decreasing to ~5.7 in the caecum, and increasing again to ~6.7 in the rectum [14]. Lower pH levels have been observed clinically in some gastrointestinal diseases such as ulcerative colitis (UC) [15]. The implications of pH shift for the microbial composition are unclear, although it might be predicted that this shift will favor certain phylogenies over others less adapted to lower pH.

THE HUMAN GASTROINTESTINAL ECOSYSTEM

The complex structural and physiological characteristics of the human gut ensure an abundance of niche habitats for specialist microbes. Specialization of bacteria relative to these niche habitats confers evolutionary advantages to the host [16]. Experiments using narrow-spectrum antibiotics have shown that, within the enteric microbial ecosystem, differences in the competitive ability of bacteria should be considered relative to the matrix of competitive interactions of all bacterial species within the same niche, rather than competition between two species alone [17].

The gastrointestinal tract is sterile at birth, but rapidly develops an intestinal microflora whose composition varies according to mode of birth, infant nutrition, antibiotic use, diet, and age. Determining the microbial composition of the human gut has taken many years, because many of the relevant species are facultative anaerobes that cannot be cultured and characterized in a laboratory setting [18]. Recent advances in the development of molecular methodologies that rely on variability in the conserved 16S rRNA gene, rather than culture-dependent methodologies, have significantly expanded our knowledge of the microflora composition of the healthy and diseased human gut. Evidence from such studies suggests that each individual has a characteristic microflora composition that is stable over time [19,20]. It appears that approximately 500 species of bacteria normally reside in the human gut, but only 30-40 bacterial species comprise 99% of the bacterial population [19,21].

The intestinal microflora has been described as having the highest cell density of any ecosystem known to science [22–24]. Two divisions, the Firmicutes (low GC gram-positive obligate anaerobes) and Bacteroidetes (gram-negative obligate anaerobes) account for up to 90% of microbial mass [22,25]. Proteobacteria (gram-negative facultative anaerobes and sulfate reducers) are also important [22]. One member of the Archaea, *Methanobrevibacter smithi*, is also abundant, whereas other less abundant divisions include the Actinobacteria, Cyanobacteria, Fusobacteria, and Verrucomicrobia [19].

Ley and coworkers [26] demonstrated that altered human physiology, in the form of obesity, alters gut microbial ecology. They analyzed 5088 bacterial 16S rRNA gene sequences from the cecal microbiota of genetically obese mice, and various related mice, all fed the same diet. Compared with lean mice and regardless of parentage, obese animals exhibited a 50% reduction in the abundance of Bacteroidetes and a proportional increase in Firmicutes. The authors noted that, even though the Firmicutes and Bacteroidetes have shared the same gut environment for millennia, allowing plenty of time for the homogenization of the gene pool through lateral gene transfer, the genomic content of the Firmicutes has a significantly lower GC content than that of the Bacteroidetes [26]. This difference may suggest a functional retention of their complementary yet distinct roles in the microbial ecosystem, which allow them to codominate. It may also suggest that these phyla will differ in the chemical composition of their species, and therefore differentially stimulate the pathogen receptors of the innate immune system. These authors suggested that intentional manipulation of community structure may be useful for maintaining GI health.

Goodacre [27] suggested that humans should be considered human–microbe hybrids, whose health will be affected by intrinsic properties including genetics, diurnal cycles, and age. More importantly, health will also be affected by extrinsic factors that manifest as lifestyle choices (food, drink, and drug intake) and the acquisition of a stable healthy gut microflora (the microbiome).

GENE–DIET INTERACTIONS IN GASTROINTESTINAL HEALTH

By implication, gastrointestinal health means an absence of gastrointestinal disease. A wide range of conditions are classed as gastrointestinal diseases, including various cancers (colorectal, gastric, and esophageal), coeliac disease, irritable bowel syndrome (IBS), and inflammatory bowel diseases (IBD), classed as Crohn's disease (CD) and ulcerative colitis (UC). There is increasing evidence for all of these diseases having some degree of genetic susceptibility, which is influenced by diet and impacted by the microbial flora to varying extents.

IBD and IBS are chronic inflammatory conditions of the intestine that affect about 0.5% of the populations in the Western world. There is good evidence that these diseases arise in some people due to a lack of tolerance to gut bacteria. Through the rest of this review, we will focus on CD as a relevant example of immune intolerance within the context of nutrigenomics.

GENE-DIET INTERACTIONS IN CROHN'S DISEASE

Concordance of CD is approximately 36% in monozygotic twins and 4% in dizygotic twins, suggesting a strong underlying genetic basis of the disease [22, 30]. A diversity of gene variants that may predispose CD have been recognized, many of which have been identified or confirmed through genome-wide association studies (GWAs). A comprehensive meta-analysis of GWAs [28] confirmed eleven previously known susceptibility loci for CD, with strong evidence that a further twenty-one genes are likely to be important. Although evidence for genetic susceptibility is strong, discordance is found in 64% of identical twins, suggesting the importance of environmental factors, among which diet is likely to be highly significant [30]. We consider two main groups of genetic susceptibility alleles in the following sections.

CROHN'S DISEASE: INFLUENCE OF TRANSPORTER AND BARRIER FUNCTIONS

The sodium-dependent organic cation/carnitine transporters, SLC22A4 and SLC22A5 (alternatively called OCTN1 and OCTN2), function primarily in the transport of L-carnitine and the elimination of cationic drugs in the intestine. These genes are located within the IBD5 locus, which has been commonly associated with susceptibility to adult onset CD [29]. Peltekova and coworkers [30] suggested that SNPs in the SLC22A4 and SLC22A5 genes have a functional effect: 1672C/T in exon 9 of SLC22A4 (rs1050152) causes the amino acid substitution L503F, a conservative amino acid change, whereas a G to C transversion in the SLC22A5 promoter -207G/C disrupts a heat shock transcription factor binding element. The evidence for CD having susceptibility variants at the IBD5 locus in Caucasians is strong.

Although some authors [30, 31] believed that these OCTN1 and OCTN2 variants are causal variants, subsequent studies [32] have suggested that either the 1672C/T and -207G/C variants are not involved in CD pathogenesis, or that other IBD5 variants also confer risk for CD. We note, however, that a more recent meta-analysis [28] confirmed that CD-associated SNPs are associated with decreased OCTN2 mRNA expression. This data may confirm that the disease-associated variants in the IBD5 region, including a coding variant in the neighboring OCTN1 gene, are those most associated with changed OCTN2 expression.

CROHN'S DISEASE: INNATE AND ADAPTIVE IMMUNE RESPONSES TO MICROBIAL PATHOGENS

It seems that bacteria are recognized by several cellular mechanisms, and SNPs in several genes implicated in these mechanisms increase susceptibility to CD.

Autophagy has been defined as "a cellular homeostasis mechanism, involving the formation of autophagosomes within the cell to sequester intracellular waste products, and bacteria for lysosomal degradation" [33]. This mechanism is an essential component of the recognition response against bacteria and viruses. Included among the key SNPs strongly associated with CD are those in three autophagy genes [28]. The first autophagy gene identified was autophagy-related 16-like 1 (ATG16L1) [34–36], which acts in concert with a second autophagy protein, ATG5. These two proteins comprise an essential component of the biology of the Paneth cell within the ileal epithelium. Paneth cells are specialized epithelial cells whose function includes the secretion of antimicrobial peptides and other proteins that regulate the intestinal environment. CD patients who are homozygous for the ATG16L1 CD risk allele show abnormalities in Paneth cell function similar to those seen in autophagy-protein-deficient mice [36].

GWAs have also shown two nonfunctional SNPs, rs4958847 and rs13361189, in immunity-related GTPase – M (IRGM) gene, which is strongly associated with CD risk, especially of ileal disease [37,38]. IRGM is expressed in peripheral blood leukocytes and monocytes, as well as in tissues of the colon and ileum [38]. The IRGM gene product is required during the early stage initiation of IFN-y-induced autophagy and the facilitation of phagosome maturation in response to *Mycobacterium tuberculosis* infection [39]. These authors suggested that the mode of action of the gene product may be the inhibition or reduction of response to IFN-y-induced autophagy, resulting in high bacterial loading of enterocytes and peripheral immune cells.

The third autophagy gene confirmed by meta-analysis as strongly associated with CD [28] is the SNP rs11175593 located in the leucine-rich repeat kinase-2 (LRRK2) gene. This gene has recently been associated with the induction of autophagy in an *in vitro* neuronal cell transfection model of a common variant associated with Parkinson's disease. Initial *in vitro* studies suggest pathogenic variants of this gene result in a gain-of-function associated with increased kinase activity [40,41].

Intelectin-1 (ITLN-1): The SNP rs2274910 in this gene is also strongly associated with CD [28]. The gene product recognizes galactofuranosyl residues in microbial cell walls. Immunofluoresence studies show the gene to be expressed in Paneth and goblet cells of the mammalian small intestine and suggest that the protein is highly concentrated at the brush border membrane of enterocyte cells. It has been suggested to play an important role in innate immunity at the brush border membrane [42].

Pattern recognition receptors (PRRs) are germ-line encoded, and recognize pathogens by epithelial cells, dendritic cells, and macrophages. The PRRs include mannose receptors, complement receptors, lectins, scavenger receptors, nucleotide-binding oligomerization domain-containing (NOD) proteins, and toll-like receptors (TLRs) [11, 43–45]. Nucleotide-binding oligomerization domain-2 (NOD2) is a member of a family of more than twenty cytosolic proteins characterized by the presence of a conserved NOD domain. NOD2 is expressed in monocytes and in the intestinal epithelial cells, with the highest expression found in paneth cells of the ileum. Unlike toll-like receptors that recognize bacteria on the surface of mucosal cells, NODs are found in the cytoplasm [46, 47]. NOD2 was the first gene that was unequivocally associated with risk of CD [48].

Each group of PRRs has evolved to recognize a limited number of distinct and highly conserved products produced by the microflora, termed pathogen-associated molecular patterns (PAMPs). PRRs are a key determinant of host recognition between self and nonself, because they do not recognize host-derived products. The net immunomodulatory effect of the interaction of PAMPs with PRRs is an effector T-cell response, mediated through a number of proinflammatory pathways, including the nuclear factor-kappa B (NF-κB) pathway. For example, the NOD2 protein is thought to provide protection against invasive bacteria by eliminating intracellular pathogens in epithelial cells at the gastrointestinal mucosa barrier. It functions as an intracellular sensor of peptidoglycan components of bacterial cell walls (muramyl dipeptide or MDP). The gene is primarily expressed in macrophage cells and especially in intestinal Paneth cells and is involved in modulating the activity of the immune-related transcription factor, NFkB. When stimulated by MDP in wild-type individuals, NOD2 activates proinflammatory gene transcription via the NF-κB transcription factor and the MAPK-signaling pathways.

The NF-κB and MAPK pathways are thought to work synergistically to upregulate the expression of proinflammatory genes that stimulate innate and adaptive immunity. Examples of genes affected in downstream effects of this pathway, and in which variants are strongly associated with CD, include:

Chemokine receptor 6 (CCR6) encodes a member of the G protein-coupled chemokine receptor family, and is expressed by immature dendritic cells and memory T cells. It plays a key role in B-cell differentiation and tissue-specific migration of dendritic and T cells during inflammatory responses [49].

The IL-12–IL-23 pathway is essential for the effective resolution of chronic intestinal inflammation, and its role in CD is supported by GWAs, as well as strong functional evidence from mouse models of colitis [50–53]. The IL12B gene encodes the p40 subunit, a constituent of both interleukins IL-12 and IL-23 [54].

Signal transducer and activator of transcription 3 (STAT3) and Janus kinase 2 (JAK2) interact in signal transduction, downstream of cytokine, and growth factor signals. Signals are transmitted from cell surface receptors to the nucleus and thereby modify the transcription of various genes, especially in hematopoietic cells [55]. The role of these two genes in CD is strongly implicated by the meta-analysis of several GWAs [28].

CROHN'S DISEASE: DIETARY MODULATION IN RESPONSE TO SPECIFIC GENETIC VARIANTS

SNPs in OCTN1 and OCTN2 may affect the function of an intact intestine, and it is not surprising that a number of dietary chemicals are able to modulate this transport function. For example, a recent study by Le Borgne *et al.* [56] studied the effect of various hormones and nutrients, which are known to alter muscle metabolism, on L-carnitine import using a cell model system. They found that L-carnitine uptake was increased by thyroid hormones, but decreased by iron, whereas insulin had no effect.

Autophagy acts as an adaptation to cellular starvation, through processing intracellular organelles and proteins, leading to the generation of amino acids during nutrient deficiency [57–59]. This pathway is activated according to the nutritional status of the cell in response to a number of cellular stressors including nutrient deprivation, bacterial infection, and oxidative damage to cellular organelles. This suggests that accumulative intracellular bacterial concentrations, coupled with poor nutritional status, may act together to increase the risk of CD. It is recognized that elemental diets are often successful in attenuating inflammation in active Crohn's disease [2].

Dietary modulation of immune response may be particularly important for CD patients carrying SNPs in pathways for immune and adative response. We have previously summarized the various points at which key nutrients can intervene [60]. Immunomodulatory nutrients include amino acids such as glutamine or arginine, lipids such as the long-chain omega-3 polyunsaturated fatty acids, DHA or EPA, or various novel carbohydrates. A number of phytochemicals appear to affect signal transduction, including polyphenols such as epigallocatechin-3 galate or curcumin, or isothiocyanates such as beta-phenylethyl isothiocyanate.

Modulation of the resident microflora may provide one of the most constructive dietary approaches with some potential for controlling or reducing the symptomology of CD. A number of food products have been developed that can modify the intestinal microflora and have been claimed to benefit GI health. These include probiotics, prebiotics, and synbiotics (a combination of probiotics and prebiotics). Probiotics are "living micro-organisms that, upon ingestion in sufficient quantities, exert health benefits on the host beyond basic nutrition" [61]. Prebiotics are "non-digestible food ingredients that beneficially affect the host by selectively stimulating the growth and/or activity of one or a limited number of bacteria in the colon, and thus improve host health" [62]. They are found naturally in some foods (e.g., onions, artichokes, bananas, chicory, leeks) or can be added in other foods (e.g., breads, biscuits). They may also interact with probiotics in maintaining optimal gastrointestinal function [63,64].

As well as IBD, colon cancer is an area in which a significant number of animal studies link prebiotic and/or probiotic intake to risk reduction. For example, a combination of pre- and probiotics (BeneoOrafti's Synergy1 plus *Lactobacillus* GG and *Bifidobacteria*) appears to favorably shift the populations of fecal bacteria, with larger populations of protective bacteria and reduced numbers of cancer-promoting bacteria [65].

The importance of colonic microflora to GI health needs further investigation. However, research indicates an influence on constipation, diarrhea, the immune system, cancer, and the absorption of minerals. Thus, it is highly likely to reduce the risk of gastrointestinal disease. The extent of risk reduction will be influenced by other genetic and dietary factors.

-OMICs APPROACHES TO OPTIMIZING GASTROINTESTINAL HEALTH

Animal models provide a means of studying the effects of defined nutrients, foods, and food components on the expression of genes associated with gastrointestinal health. Although various different models can be used, an efficient study design measures food or food component-induced changes in gene and protein expression in gastrointestinal tissues, collected from 2x2 mouse experiments [66]. Transcriptomics (using microarrays), proteomics, and metabolomics techniques can be used to test the effects of the food in question on various cellular pathways. Two examples relevant to CD follow.

Dommels and coworkers [67] characterized intestinal inflammation and identified gene expression changes in mdr1a(-/-) mice to test whether these mice provided an appropriate model of CD. It is known that multidrug resistance-targeted mutation (mdr1a (-/-)) mice spontaneously develop intestinal inflammation, and this study considered the time for this to occur as well as the characteristic pattern of gene expression associated with inflammation. The inflammatory lesions resembled human CD in being transmural and discontinuous. Genes involved in inflammatory response pathways were upregulated whereas genes involved in biotransformation and transport were downregulated in colonic epithelial cell scrapings of inflamed mdra1 (-/-) mice at 25 weeks of age compared to noninflamed control mice.

Nones et al. [68] extended these studies to consider the effects of dietary incorporation of two flavonoids, curcumin and rutin, on colonic inflammation and gene expression in the same model. Dietary curcumin significantly reduced colonic inflammation. Microarray and pathway analyses suggested that the effect could occur through an upregulation of xenobiotic metabolism and a downregulation of proinflammatory pathways. The authors claimed that these results indicated the importance of global gene expression and pathway analyses in mechanistic studies of the effect of foods in modulating colonic inflammation.

As well as testing whether foods identified as potentially beneficial to gut health will actually act this way *in vivo*, such studies may also enable the identification of biomarkers [69], which can be used as surrogate endpoints to study effects of dietary interventions in small animal models of human metabolism or limited human clinical trials.

A third dimension will be required in subsequent studies. As discussed previously, it is increasingly recognized that colonic microflora plays a key role in disease susceptibility and progression. Fecal samples may be collected in both animal and human studies, and cecal samples in animal models, to study this important dimension [63].

CONCLUSIONS

Gastrointestinal health provides several important examples relevant to nutrigenomics, whereby personalized diets may overcome the detrimental effect of variant genotypes. The gastrointestinal tract provides an ecological habitat for a complex microbiota whose maintenance is critical to preventing gastrointestinal diseases including various cancers (colorectal, gastric, and esophageal), coeliac disease, IBS, as well as CD and UC. CD provides a good example where key genetic variants are becoming recognized through GWAs. These variants fall into a small number of key pathways, for which at least hypothetically, specific dietary interventions could be beneficial. Such diets can be studied by considering the patterns of gene and protein expression and metabolite production, as well as fecal or cecal microflora in animal models. The identification of biomarkers provides early evidence of diet-related diseases and/or evidence of protection against these by different diets. Personalized nutrition may allow the patient to maintain a state of health (maintenance of homeostasis). Therefore, nutrigenomics provides an approach to studying the relationships between genes and diets to enable personalized dietary recommendations to maintain optimal gastrointestinal health throughout the lifespan.

ACKNOWLEDGMENTS

Nutrigenomics New Zealand is a collaboration between AgResearch Ltd., Crop & Food Research, HortResearch, and The University of Auckland, with funding through the Foundation for Research Science and Technology.

REFERENCES

1. Kaput, J., Rodriguez, R. L., Nutritional genomics: the next frontier in the postgenomic era. *Physiol Gen* 2004, *16*, 166–177.
2. Ferguson, L. R., Shelling, A. N., Browning, B. L., Huebner, C., *et al.*, Genes, diet and inflammatory bowel disease. *Mutat Res* 2007, *622*, 70–83.
3. O'Rahilly, S., Farooqi, I. S., Human obesity: a heritable neurobehavioral disorder that is highly sensitive to environmental conditions. *Diabetes* 2008, *57*, 2905–2910.
4. Guerreiro, C. S., Carmona, B., Goncalves, S., Carolino, E., *et al.*, Risk of colorectal cancer associated with the C677T polymorphism in 5,10-methylenetatrahydrofolate reductase in Portugese patients depends on the intake of methyl-donor nutrients. *Am J Clin Nutr* 2008, *88*, 1413–1418.
5. Hesketh, J., Nutrigenomics and selenium: gene expression patterns, physiological targets, and genetics, *Ann Rev Nutr* 2008, *28*, 157–177.
6. Kaput, J., Ordovas, J.M., Ferguson, L., van Ommen, B., *et al.*, The case for strategic international alliances to harness nutritional genomics for public and personal health. *Br J Nutr* 2005, *94*, 623–632.
7. Williams, C.M., Ordovas, J.M., Lairon, D., Hesketh, J., *et al.,* The challenges for molecular nutrition research 1: Linking genotype to healthy nutrition. *Genes and Nutrition* 2008, *3*, 41–49.
8. Thibodeau, G.A., *Anatomy & Physiology*. St. Louis, MO: Elsevier Mosby, 2007.
9. Wapnir, R.A., Teichberg, S., Regulation mechanisms of intestinal secretion: implications in nutrient absorption, *J Nutr Biochem* 2002, *13*, 190–199.
10. Eckmann, L., Kagnoff, M.F., Fierer, J., Intestinal epithelial cells as watchdogs for the natural immune system. *Trends Microbiol* 1995, *3*, 118–120.
11. Braat, H., Peppelenbosch, M.P., Hommes, D.W., Immunology of Crohn's disease. *Ann NY Acad Sci* 2006, *1072*, 135–154.
12. Janeway, Jr., C.A., Medzhitov, R., Innate immune recognition. *Ann Rev Immunol* 2002, *20*, 197–216.
13. Atuma, C., Strugala, V., Allen, A., Holm, L., The adherent gastrointestinal mucus gel layer: thickness and physical state in vivo. *Am J Physiol- Gastrointest Liver Physiol* 2001, *280*, 922–929.
14. Fallingborg, J., Intraluminal pH of the human gastrointestinal tract. *Dan Med Bull* 1999, *46*, 183–196.

15. Nugent, S.G., Kumar, D., Rampton, D.S., Evans, D.F., Intestinal luminal pH in inflammatory bowel disease: possible determinants and implications for therapy with aminosalicylates and other drugs. *Gut* 2001, *48*, 571–577.

16. Ley, R.E., Peterson, D.A., Gordon, J.I., Ecological and evolutionary forces shaping microbial diversity in the human intestine. *Cell* 2006, *124*, 837–848.

17. Kirkup, B.C., Riley, M.A., Antibiotic-mediated antagonism leads to a bacterial game of rock-paper-scissors in vivo. *Nature* 2004, *428*, 412–414.

18. Hooper, L.V., Midtvedt, T., Gordon, J.I., How host-microbial interactions shape the nutrient environment of the mammalian intestine. *Ann Rev Nutr* 2002, *22*, 283–307.

19. Eckburg, P.B., Bik, E.M., Bernstein, C.N., Purdom, E., *et al.,* Diversity of the Human Intestinal Microbial Flora. *Science* 2005, *308*, 1635–1638.

20. Ott, S.J., Musfeldt, M., Wenderoth, D.F., Hampe, J., *et al.,* Reduction in diversity of the colonic mucosa associated bacterial microflora in patients with active inflammatory bowel disease. *Gut* 2004, *53*, 685–693.

21. Suau, A., Bonnet, R., Sutren, M., Godon, J.-J., *et al.,* Direct analysis of genes encoding 16S rRNA from complex communities reveals many novel molecular species within the human gut. *Appl Environ Microbiol* 1999, *65*, 4799–4807.

22. Backhed, F., Ley, R.E., Sonnenburg, J.L., Peterson, D.A., *et al.,* Host-bacterial mutualism in the human intestine. *Science* 2005, *307*, 1915–1920.

23. Whitman, W.B., Coleman, D.C., Wiebe, W.J., Prokaryotes: the unseen majority. *Proc Natl Acad Sci* 1998, *95*, 6578–6583.

24. Hugenholtz, P., Goebel, B.M., Pace, N.R., Impact of culture-independent studies on the emerging phylogenetic view of bacterial diversity. *J Bacteriol* 1998, *180*, 4765–4774.

25. Ley, R.E., Turnbaugh, P.J., Klein, S., Gordon, J.I., Microbial ecology: human gut microbes associated with obesity. *Nature* 2006, *444*, 1022–1023.

26. Ley, R.E., Backhed, F., Turnbaugh, P., Lozupone, C.A., *et al.,* Obesity alters gut microbial ecology. *Proc Natl Acad Sci* 2005, *102*, 11070–11075.

27. Goodacre, R., Metabolomics of a superorganism. *J Nutr* 2007, *137*, 259S–266S.

28. Barrett, J.C., Hansoul, S., Nicolae, D.L., Cho, J.H., *et al.,* Genome-wide association defines more than 30 distinct susceptibility loci for Crohn's disease. *Nat Genet*, advanced online publication, 2008.

29. Rioux, J.D., Daly, M.J., Silverberg, M.S., Lindblad, K., *et al.,* Genetic variation in the 5q31 cytokine gene cluster confers susceptibility to Crohn's disease. *Nat Genet* 2001, *29*, 223–228.

30. Peltekova, V.D., Wintle, R.F., Rubin, L.A., Amos, C.I., *et al.,* Functional variants of OCTN cation transporter genes are associated with Crohn's disease. *Nat Genet* 2004, *36*, 471–475.

31. Newman, B., Gu, X., Wintle, R., Cescon, D., *et al.,* A risk haplotype in the Solute Carrier Family 22A4/22A5 gene cluster influences phenotypic expression of Crohn's disease. *Gastroenterol* 2005, *128*, 260–269.

32. Fisher, S.A., Hampe, J., Onnie, C.M., Daly, M.J., *et al.,* Direct or indirect association in a complex disease: the role of SLC22A4 and SLC22A5 functional variants in Crohn's disease. *Hum Mutat* 2006, *27*, 778–785.

33. Levine, B., Kroemer, G., Autophagy in the pathogenesis of disease. *Cell* 2008, *132*, 27–42.

34. Hampe, J., Franke, A., Rosenstiel, P., Till, A., *et al.,* A genome-wide association scan of nonsynonymous SNPs identifies a susceptibility variant for Crohn's disease in ATG16L1. *Nat Genet* 2007, *39*, 207–211.

35. Rioux, J.D., Xavier, R.J., Taylor, K.D., Silverberg, M.S., *et al.,* Genome-wide association study identifies new susceptibility loci for Crohn disease and implicates autophagy in disease pathogenesis. *Nat Genet* 2007, *39*, 596–604.

36. Cadwell, K., Liu, J.Y., Brown, S.L., Miyoshi, H., *et al.,* A key role for autophagy and the autophagy gene Atg16l1 in mouse and human intestinal Paneth cells. *Nature* 2008, *456*, 259–263.

37. Roberts, R.L., Hollis-Moffatt, J.E., Gearry, R.B., Kennedy, M.A., *et al.,* Confirmation of

association of IRGM and NCF4 with ileal Crohn's disease in a population-based cohort. *Genes Immun* 2008, *6*, 561–565.

38. Parkes, M., Barrett, J.C., Prescott, N.J., Tremelling, M., *et al.,* Sequence variants in the autophagy gene IRGM and multiple other replicating loci contribute to Crohn's disease susceptibility. *Nat Genet* 2007, *39*, 830–832.

39. Singh, S.B., Davis, A.S., Taylor, G.A., Deretic, V., Human IRGM induces autophagy to eliminate intracellular mycobacteria. *Science (New York, N.Y.)* 2006, *313*, 1438–1441.

40. West, A.B., Moore, D.J., Biskup, S., Bugayenko, A., *et al.,* Parkinson's disease-associated mutations in leucine-rich repeat kinase 2 augment kinase activity. *Proc Natl Acad Sci U S A* 2005, *102*, 16842–16847.

41. Smith, W.W., Pei, Z., Jiang, H., Dawson, V.L., *et al.,* Kinase activity of mutant LRRK2 mediates neuronal toxicity. *Nat Neurosci* 2006, *9*, 1231–1233.

42. Wrackmeyer, U., Hansen, G.H., Seya, T., Danielsen, E.M., Intelectin: a novel lipid raft-associated protein in the enterocyte brush border. *Biochem* 2006, *45*, 9188–9197.

43. Cario, E., Bacterial interactions with cells of the intestinal mucosa: Toll-like receptors and Nod2. *Gut* 2005, *54*, 1182–1193.

44. Cash, H.L., Whitham, C.V., Behrendt, C.L., Hooper, L.V., Symbiotic bacteria direct expression of an intestinal bactericidal lectin. *Science* 2006, *313*, 1126–1130.

45. Hugot, J.-P., CARD15/NOD2 mutations in Crohn's disease. *Ann NY Acad Sci* 2006, *1072*, 9–18.

46. Bonen, D., Cho, J., The genetics of inflammatory bowel disease. *Gastroenterol* 2003, *124*, 521–536.

47. Van Heel, D., McGovern, D., Jewell, D., Crohn's disease: genetic susceptibility, bacteria, and innate immunity. *Lancet* 2001, *357*, 1902–1904.

48. Ogura, Y., Bonen, D.K., Inohara, N., Nicolae, D.L., *et al.,* A frameshift mutation in NOD2 associated with susceptibility to Crohn's disease. *Nature* 2001, *411*, 603–606.

49. Salazar-Gonzalez, R.M., Niess, J.H., Zammit, D.J., Ravindran, R., *et al.,* CCR6-mediated dendritic cell activation of pathogen-specific T cells in Peyer's patches. *Immunity* 2006, *24*, 623–632.

50. Hue, S., Ahern, P., Buonocore, S., Kullberg, M.C., *et al.,* Interleukin-23 drives innate and T cell-mediated intestinal inflammation. *J Exp Med* 2006, *203*, 2473–2483.

51. Kullberg, M.C., Jankovic, D., Feng, C.G., Hue, S., *et al.,* IL-23 plays a key role in Helicobacter hepaticus-induced T cell-dependent colitis. *J Exp Med* 2006, *203*, 2485–2494.

52. Uhlig, H.H., McKenzie, B.S., Hue, S., Thompson, C., *et al.,* Differential activity of IL-12 and IL-23 in mucosal and systemic innate immune pathology.[see comment]. *Immunity* 2006, *25*, 309–318.

53. Yen, D., Cheung, J., Scheerens, H., Poulet, F., *et al.,* IL-23 is essential for T cell-mediated colitis and promotes inflammation via IL-17 and IL-6. *J Clin Invest* 2006, *116*, 1310–1316.

54. Oppmann, B., Lesley, R., Blom, B., Timans, J.C., *et al.,* Novel p19 protein engages IL-12p40 to form a cytokine, IL-23, with biological activities similar as well as distinct from IL-12. *Immunity* 2000, *13*, 715–725.

55. Mudter, J., Neurath, M.F., The role of signal transducers and activators of transcription in T inflammatory bowel diseases. *Inflam Bowel Dis* 2003, *9*, 332–337.

56. Le Borgne, F., Rigault, C., Georges, B., Demarquoy, J., Hormonal and nutritional control of L-carnitine uptake in myoblastic C2C12 cells. *Muscle & Nerve* 2008, *38*, 912–915.

57. Scherz-Shouval, R., Shvets, E., Fass, E., Shorer, H., *et al.,* Reactive oxygen species are essential for autophagy and specifically regulate the activity of Atg4 [see comment]. *EMBO J* 2007, *26*, 1749–1760.

58. Kadowaki, M., Karim, M.R., Carpi, A., Miotto, G., Nutrient control of macroautophagy in mammalian cells. *Mol Asp Med* 2006, *27*, 426–443.

59. Reis e Sousa, C., Immunology. Eating in to avoid infection [comment]. *Science* 2007, *315*, 1376–1377.

60. Ferguson, L.R., Philpott, M., Cancer prevention by dietary bioactive components that target the immune response. *Curr Cancer Drug Targets* 2007, *7*, 459–464.
61. Fooks, L.J., Gibson, G.R., Probiotics as modulators of the gut flora. *Br J Nutr* 2002, *88*, S39–S49.
62. Gibson, G.R., Roberfroid, M.B., Dietary modulation of the human colonic microbiota: introducing the concept of prebiotics. *J Nutr* 1995, *125*, 1401–1412.
63. Lim, C.C., Ferguson, L.R., Tannock, G.W., Dietary fibres as "prebiotics": implications for colorectal cancer. *Mol Nutr Food Res* 2005, *49*, 609–619.
64. Teitelbaum, J.E., Walker, W.A., Nutritional impact of pre- and probiotics as protective gastrointestinal organisms. *Ann Rev Nutr* 2002, *22*, 107–138.
65. Rafter, J., Bennett, M., Caderni, G., Clune, Y., *et al.*, Dietary synbiotics reduce cancer risk factors in polypectomized and colon cancer patients. *Am J Clin Nutr* 2007, *85*, 488–496.
66. Park, E.I., Paisley, E.A., Mangian, H.J., Swartz, D.A., *et al.*, Lipid level and type alter stearoyl CoA desaturase mRNA abundance differently in mice with distinct susceptibilities to diet-influenced diseases. *J Nutr* 1997, *127*, 566–573.
67. Dommels, Y.E., Butts, C.A., Zhu, S., Davy, M., *et al.*, Characterization of intestinal inflammation and identification of related gene expression changes in mdr1a(-/-) mice. *Genes Nutr* 2007, *2*, 209–223.
68. Nones, K., Dommels, Y.E., Martell, S., Butts, C., *et al.*, The effects of dietary curcumin and rutin on colonic inflammation and gene expression in multidrug resistance gene-deficient (mdr1a- / -) mice, a model of inflammatory bowel diseases. *Br J Nutr* 2008, *2*, 1–13.
69. Ferguson, L.R., Han, D.Y., Huebner, C., Petermann, I., *et al.*, Single nucleotide polymorphisms in IL4, OCTN1 and OCTN2 genes in association with Inflammatory Bowel Disease phenotypes in a Caucasian population in Canterbury, New Zealand. *The Open Gastroenterology Journal* 2008, *2*, 50–56.

9
Genomics Analysis to Demonstrate the Safety and Efficacy of Dietary Antioxidants

Nilanjana Maulik

INTRODUCTION

The modern era of genomics and its understanding has opened up many opportunities to study the interactions between diet, gene expression, and genetic variability in health and disease. Genomics can be highly promising in reinvigorating several stages of drug discovery and development. In addition, the application of genomics toward drug safety assessment has been recently helpful to minimize late-stage drug attrition by pharmaceutical companies to predict toxicity and analyze the safety risks (Ulrich and Friend 2002; Suter, Babiss *et al*. 2004). The use of functional genomic approaches in relation with dietary nutrients or antioxidant research will present the possibility of an entirely new research field, which seeks to identify a global view of the biological effect of food (Trayhurn 2000; Elliott and Ong 2002). The ultimate aim of this kind of approach is to help people modify their diets, to avoid or include certain nutrients and antioxidants to stay healthy. If used carefully and sensibly along with the other nutritional and toxicological research tools available, this will elevate the comprehensive understanding of how different dietary nutrients or drugs promote health.

Nutrigenomics aims to estimate the influence of dietary nutrients on the genome, proteome, and metabolome, and hence can alter phenotype and balance between health and disease. It also identifies the differences in cellular and genetic response of the biological system to nutritional stimulus (Ordovas and Mooser 2004; Mutch, Wahli *et al*. 2005), which helps to better understand how nutrition can affect the regulation of gene expression and impact metabolic pathways (Gillies 2003; Muller and Kersten 2003; Gatzidou and Theocharis 2008). The major aim of the nutrigenomic approach is to discover the biomarkers and to investigate the nutrition based mechanisms (Afman and Muller 2006). Biomarker discovery will help to reveal the early indicators for disease disposition and early metabolic dysregulation along with alteration in the homeostasis modulated by diet and the efforts of the body to maintain it (Kussmann, Raymond *et al*. 2006). Identification of nutrition-based mechanism of dietary compounds would provide us with more detailed molecular information, which in turn increases the understanding of the interaction between nutrition and genome.

Dietary nutrients and antioxidants are the major sources of energy metabolism and play an important role in growth and development of structural components of the body. They also act as essential cofactors for enzymes that are critical and involved in various signaling pathways of metabolism. They can influence the expression of genes by three mechanisms either directly or indirectly: (a) bind to or directly activate or suppress specific transcription factors to alter the transcription rate of genes (Kaput and Rodriguez 2004), (b) convert to primary or secondary metabolic pathway products thereby altering the concentration of substrates or intermediates of critical pathways, (c) bind to receptors at cell surface, resulting in either induction or suppression of signal transduction cascade that ultimately facilitates DNA–transcription factor interactions (Kaput and Rodriguez 2004). Hence, nutrients can also be defined as signaling molecules, which can utilize the mechanisms mentioned previously to transmit and translate these dietary signals to alter the expression of genes, proteins, and metabolite expression (Gatzidou and Theocharis 2008). Nutritional genomics involves the studies of using "-omics" approaches to explore the impact of diet nutrients or drugs upon gene stability and gene expressions (Ovesna, Slaby *et al.* 2008). Identification of nutritionally responsive genome activity involves the study design, type of antioxidant or diet, genomic platform used, and data interpretation and evaluation, which involves numerous samples (Galitski 2004).

FUNCTIONAL GENOMICS

Functional genomics seeks to understand the multistep process of gene expression for which many modes of regulations have been recognized. There are forty-six chromosomes arranged in twenty-three pairs in every nucleus of a human cell. A chromosome is nothing but a long strand of supercoiled DNA, and genes are a part of DNA that codes for specific proteins. The gene expression process begins with transcription of a gene whereby information stored in the DNA sequence of a gene is transmitted into a complementary molecule of a particular mRNA, which may then be translated to a protein (Knoers and Monnens 2006). The synthesized protein may further undergo posttranslational modifications such as proteolytic cleavage, phosphorylation, glycation etc. The activity and amount of a gene product depends on the rate and efficiency of each of these processes, localization of the product, and its rate of degradation along with its regulations. DNA microarray, proteomics, and metabolomics can be used to detect the changes in each of these processes. Quantitative real-time RT-PCR is another technique that supports these functional genomics techniques by generating more accurate and sensitive data to validate the data provided by functional genomic approaches.

Recently, whole genome studies are becoming more popular and important because the number of DNA sequences available to researchers has been rising exponentially. To understand the entire genome with up to approximately 30,000 genes, it is very important to measure the mRNA levels for a complete set of transcripts of an organism. DNA microarray can be utilized to analyze the genome-wide changes associated with drug responses (Schena, Shalon *et al.* 1995). DNA microarray (DNA chip technology), a high-throughput method, is the most common standard tool for measuring the expression profile of a large number of genes in a single assay. Even though there is a great advantage in the traditional approach to assess gene expression changes (e.g., gene expression induced by pharmaceutical compounds), which includes Northern blotting of mRNAs, reverse transcription coupled polymerase chain reaction (RT-PCR) of mRNAs, Western blotting of proteins, etc., these methods allow only a few or at the maximum 100 genes in a study to be analyzed. Whereas DNA microarray technology allows us to study the expression of thousands of genes to the entire genome, which may be nutritionally (nutrigenomics), pharmacologically (pharmacological genomics), or toxicologically (toxicological genomics) related to drug administration (Nuwaysir, Bittner *et al.* 1999; Aardema and MacGregor 2002; Irwin, Boorman *et al.* 2004; Shioda 2004; Luhe, Suter *et al.* 2005; Lettieri 2006; Inadera, Uchida *et al.* 2007; Ju, Wells *et al.* 2007). Several different types of DNA chips are available, which are produced by photolithograph, spotted oligonucleotides, or cDNA arrays. But the most common and widely used arrays are the commercially available high-throughout arrays (Affymetrix, Illumina etc) as compared with the house-prepared arrays because they exhibit more advantages such as number of probes, accuracy, reproducibility, etc. (Ovesna, Slaby *et al.* 2008). High-throughput microarrays have been used in several pioneer studies on the changes in gene expression (Christen, Olano-Martin *et al.* 2002; Swanson 2008).

ARRAY PLATFORMS

Array platforms can be classified into three groups: macroarrays, microarrays, and high-density oligonucleotide arrays (gene chips) mainly based upon the material used to construct the matrix (platform), type of probes, array size, type of labeling, etc. (Vrana, Freeman *et al.* 2003). Macroarrays are the most basic forms of hybridization arrays and their main features are the deposition of probes on the membranes or plastic and radioactivity used as the labeling tool. In addition, they do not require specialized equipment. These arrays are generally capable of detecting 200-8000 genes simultaneously and are available commercially. The advantage of these types of arrays is that they do not require specialized equipment and can be performed in a normal laboratory, which can perform studies at the mRNA levels. The major disadvantage is the use of two different membranes for the sample and control, which can lead to experimental variability and hybridization errors. The second type of arrays used is microarrays, which differ from the macroarrays in three different aspects: glass as the matrix; fluorescence (usually Cy3 or Cy5) technology as the labeling method, and a greater number of probes than macroarrays. Usually, samples are labeled with fluorescence dyes to provide competitive hybridization where the main advantage is that it allows both the samples and the control to be hybridized on the same glass slide. The third group of arrays, high-density oligonucleotide arrays, also called gene chips, is the most widely used array recently and has several advantages with the other two types in relation to the highest number of genes. It allows analysis of the entire genome of a particular species. The most common arrays available now are from Affymetrix, Agilent, NimbleGen, Applied Biosystems, Febit, Illumina etc. (Hardiman 2004). The major limitation of these types of arrays is that they cannot be homemade and require several specialized equipment, which makes processing very expensive.

STEPS INVOLVED IN PERFORMING A MICROARRAY EXPERIMENT

Several steps are involved in performing a successful microarray experiment, which include (a) experimental design; (b) microarray platform selection and hybridization procedure; (c) image analysis and data extraction; (d) differential expression; (e) validation by real-time RT PCR.

EXPERIMENTAL DESIGN

Gene expression analysis by microarray experiments remains critical for production of high-quality and relevant data and is often compared between two samples of RNA, that is, a control sample compared with a treated sample or a normal sample compared with diseased samples. The first problem in the experimental design is the selection of the appropriate sample for the experiment along with its control. Complex experimental design involving more groups might be useful in determining the gene expression changes more directly but involves the difficulty of normalization, comparison, and data processing and hence should be kept in mind before designing a complex experiment. Variabilities in the microarray experiments are almost unavoidable, and hence proper experimental design is important in a microarray experiment (Weeraratna, Nagel *et al.* 2004). Relative expression levels of thousands of genes are compared based upon a single array, and this might lead to a large number of false positive and false negative. Hence it is important to repeat the experiments with replicates or triplicates of the array. Even though the use of replicates increases the cost to the experiment, it becomes unavoidable because it will decrease the variability in the experiment. In addition, the data extraction and processing should be carefully planned because there are numerous statistical tools and software available.

HYBRIDIZATION

Hybridization process involves reanealing single-stranded DNA with the target sequence printed or synthesized on the arrays. Successful hybridization depends on several factors such as the purity and quality of the RNA isolated from the sample, labeling procedure, and the type of microarray used. High-throughput microarrays or synthesized microarrays usually involve the use of biotinylated probes followed by staining with streptavadin conjugated to phycoerythrin, whereas the other type of microarrays involve either radioactive or fluorescence labeling methods (Lennon and Lehrach 1991; Schena, Shalon *et al*. 1995).

IMAGE ANALYSIS AND DATA EXTRACTION

Data analysis poses several challenges and is a critical step in microarray experiments due to large amount of data generated in an experiment. Microarray experiments are prone to false positive and false negative results, and hence the approach for the data analysis should be carefully determined from the available large number of methods (Brazma and Vilo 2000; Hess, Zhang *et al*. 2001). Different research groups analyze their data in a variety of ways using combinations of various microarray-based analysis tools available (Peterson 2002; Ringner, Peterson *et al*. 2002; Carr, Bittner *et al*. 2003; Ringner and Peterson 2003). There is no specific method available that could be universally accepted for microarray experiments.

Microarray experiment results are usually images of hybridized arrays with the samples scanned using an optical scanner. The images obtained should be subjected to background correction before performing any analysis or comparison to adjust for nonspecific binding, fluorescence from other chemicals on the slide, etc., and permitting comparison of specific signals. The data should be normalized to adjust the differences, which occur due to technology (e.g., dye effects) and are not biological in nature, so that the normalized data can be combined into a single value that represents the level of gene expression. Regardless of the type of arrays used, the normalization procedure is very important in an experiment to bring the signal range into acceptable confidence interval and adjust the signals on each filter to approximate a normal distribution pattern. It is also important to eliminate artifacts. In general, the normalization procedure should be used according to the microarray platform (Schadt, Li *et al*. 2000; Schadt, Li *et al*. 2001; Quackenbush 2002; Yang, Dudoit *et al*. 2002; Bolstad, Irizarry *et al*. 2003), and generally it is quite different between the cDNA arrays and affymetrix arrays. Linear normalization methods often miss obscuring variations, and hence nonlinear methods should be preferred (Yang, Dudoit *et al*. 2002; Bolstad, Irizarry *et al*. 2003).

DIFFERENTIAL EXPRESSION

The main goal of most microarray studies is to determine which genes were changed in their expression levels by the experimental conditions. Several methods are available for this, but there are no universally accepted standard methods (Tusher, Tibshirani *et al*. 2001; Zhang, Wang *et al*. 2002; Zhou and Abagyan 2002). Differential expression analysis involves two main issues such as determining a proper method to assess the extent of differential expression (statistical approaches and fold change) and a method for multiple comparisons of thousands of genes being studied. In general, normalized data or signals for a single gene are converted to a ratio compared between the treated and the untreated group called the expression ratio. Although this method is straightforward, it is not based on statistical principles and is dependent on the original signal intensities. Hence, more advanced methods that involve higher statistical techniques should be employed for the analysis, such as clustering analysis, which helps to identify groups of genes that display similar changes in the expression. Eisen *et al*. (Eisen, Spellman *et al*. 1998) were one of the first to apply the clustering analysis to microarray analysis, and it produces readily visualized patterns of coordinately regulated genes and is supported by software programs (Clusteru, Treeview). There are several clustering methods available such as k- means, self-organizing maps, hierarchical techniques, principal components analysis etc. (Alon, Barkai *et al.* 1999; Ben-Dor, Shamir *et al.* 1999; Tamayo, Slonim *et al.* 1999; Raychaudhuri, Stuart *et al.* 2000). Heat map represents a common approach to present gene expression data. This is an array where, typically, genes index the rows and chips index the columns. Regardless of statistical and clustering methods applied, it is obvious that only 5%-10% of the genes in general undergo significant changes in the expression patterns and therefore when running a microarray experiment it is necessary to replicate the experiment and further validate the experiment results using other techniques such as real-time PCR (Carlisle, Prabhu *et al.* 2000; Lee, Kuo *et al.* 2000; Coombes, Highsmith *et al.* 2002).

VALIDATION BY REAL-TIME PCR

Eventhough microarray analysis has several advantages including the ability to evaluate the expression of thousands of genes at the same time along with the impact of its results, it has several limitations, including potential incorrect gene annotation, quenching of signals resulting in loss of sensitivity to detect changes, intra- and interarray variations leading to misinterpretation of expression data. In addition, it is difficult to compare the results between two different array platforms eventhough the experimental condition is maintained the same (Kuo, Jenssen *et al.* 2002). Gene annotation problems might also occur due to error in probe development and spotting on the microchip. To overcome these, we could sequence verify every cDNA probe on the chip or alternately use another method of gene expression to validate the results. Several methods are available to measure the gene expression, including Northern analysis, RPA assay, real-time PCR. Real-time PCR (qPCR-quantitative PCR) is the most sensitive and flexible for RNA quantitation, hence it remains the gold standard to validate the microarray analysis results (Canales, Luo *et al.* 2006).

ANTIOXIDANTS AND GENE EXPRESSION

Oxidative stress is a condition caused when the balance between the production of reactive oxygen species (ROS) and a biological system's ability to readily detoxify the reactive intermediates fails. Cellular oxidant/antioxidant equilibrium plays a key role in redox-dependent signal transduction pathways both *in vitro* and *in vivo*. Several reports in the past decades have already shown that acute toxic effects of the ROS can be reduced by endogenous antioxidants (Gebhart 1974; Sarma and Kesavan 1993; Zimmermann and Kimmig 1998). Also, plant-derived foods have been reported to have a large number of compounds (flavonoids, polyphenolic compounds, allylsulfides, hydroxycinnamic acids, vegetable oils, specific ingredients of spices, etc.), which helps in protection against oxidative stress (Lopez-Velez, Martinez-Martinez *et al.* 2003; Masuda, Kikuzaki *et al.* 2004; Anagnostopoulou, Kefalas *et al.* 2005; Kim, Chang *et al.* 2006; Li, Tsao *et al.* 2006; Somparn, Phisalaphong *et al.* 2007). Vitamin C is known to be one of the most powerful antioxidants, which exists both endogenously and exogenously and acts as a scavenger of ROS and has been shown to be effective against superoxide radical anion, H_2O_2, hydroxyl radical, and singlet oxygen (Diplock, Charleux *et al.* 1998). Fruits, especially citrus fruits, cherries, melons, and vegetables such as tomatoes, leafy greens, broccoli, and cabbage are the most common dietary source of vitamin C (Diplock, Charleux *et al.* 1998). Vitamin C also functions as a cofactor for several important biological enzymes (Levine, Conry-Cantilena *et al.* 1996; Diplock, Charleux *et al.* 1998). Vitamin E is also an important antioxidant present both exogenously and endogenously, and its main function is to inhibit lipid peroxidation and scavenge lipid peroxyl radicals to yield lipid hydroperoxides and a tocopheroxy radical (Rock, Jacob *et al.* 1996). It is also reported to interact with peroxynitrite and in quenching singlet oxygen. The main dietary sources of vitamin E are vegetable oils from soybean, maize, cottonseed, and products from these oils such as margarine and mayonnaise, along with some leafy vegetables (Parker 1989). The next most important dietary antioxidants are the class of carotenoids that scavenge and peroxyl radicals (Palozza and Krinsky 1992). Carotenoids also belong to the group of lipophilic antioxidants like vitamin E and are present in lipoproteins such as low density lipoprotein (LDL) and high density lipoprotein (HDL). The major dietary sources of carotenoids are carrots, tomatoes, citrus fruits, spinach and maize (Mangels, Holden *et al.* 1993). Flavonoids, a large group of polyphenolic antioxidants commonly available in fruits, vegetables, tea, and wine, are also efficient antioxidants and are capable of scavenging radical species such as peroxyl radicals, hydroxyl radicals, and $O_2^{\cdot}$ – forming a phenoxy radical. It also has a number of subgroups, including flavanols, flavanones, flavones, isoflavones, or anthocyanidins (Rice-Evans and Miller 1996). Along with these, several other dietary constituents might also be involved in the antioxidant defense either directly or indirectly.

OXIDATIVE STRESS, ANTIOXIDANTS AND CELL SIGNALING

Oxidative stress can lead to several important changes in the gene expression patterns along with protein synthesis and posttranslational modifications. Some of the genes, which are transcriptionally regulated by oxidative stress, are shown in Table 9.1 (Knasmuller, Nersesyan *et al.* 2008). Endogenous and exogenous antioxidants can influence redox-dependent protein–protein interactions or interact with cell receptors to trigger or modify the key enzymes such as phosphatase and kinases in the cell signaling (Mathers, Fraser *et al.* 2004). This can further lead to changes in the transcription factor activity, which in turn affects the mRNA and protein levels. Extensive research in molecular and cell biology has helped us understand how the antioxidants can mediate and regulate cell signaling. Several such antioxidants, which help in maintaining the cellular oxidant/antioxidant equilibrium and/or serve as signaling molecules, are available both exogenously and endogenously.

Vitamin E is the most common example of endogenous antioxidant. Several studies have been done towards understanding the antioxidant capacity of vitamin E, the most important of which is the study by Azzi *et al.,* who first described that it can also function as a cell-signaling molecule in which they showed regulation of protein kinase C activity in smooth muscle cells (Boscoboinik, Szewczyk *et al.* 1991). Vitamin E has been found to exhibit many cellular functions independent of its antioxidant properties (Rimbach, Minihane *et al.* 2002). It is well established now that several antioxidants not only scavenge the free radicals but also act as cell-signaling molecules (Maulik and Das 2002), which may contribute to the beneficial effect of understanding and prevention of several disease conditions such as atherogenesis, carcinogenesis, neurodegeneration, etc.

Growth factor signaling triggers ROS production and mediates ROS as a signaling molecule and activates intracellular receptor-mediated signaling such as mitogen-activated protein kinases. Vascular endothelial cell growth factor (VEGF) is one of the most important growth factors found to trigger MAPK signaling (Maulik, Sato *et al.* 1998; Maulik and Das 2002; Maulik 2006). MAPKs are important signaling mediators that play an important role in the conversion of extracellular signals into intracellular responses by a series of phosphorylation cascades (Cobb and Goldsmith 1995). MAPK activation results in the appropriate gene expression based upon the signal received for the extracellular signaling molecules. MAPKs can activate either cell protection or cell death signaling depending upon the signal received and the cell environment. Three different MAP kinase pathways have been identified and well studied in most of the fields (Cano and Mahadevan 1995).

MAP kinase signaling cascades are capable of activating mechanisms, which result in the crosstalk between signaling pathways and the common result is the transduction of specific cellular gene expression to various extracellular stimuli form oxidative stress, pharmacological drugs, dietary antioxidants, etc. The most important response of MAPKs modulated by ROS is the activation of several transcription factors, which can regulate the gene expression related toward cell protection, cell death, etc. (Knasmuller, Nersesyan *et al.* 2008).

Cell signaling involves several redox transcription factors such as NFκB, AP-1, Nrf-1, SP-1, and Hif1α, which regulate the expression of several important genes. AP1 is an ubiquitous regulatory protein complex that interacts with its binding sites of target genes to regulate

Table 9.1 Examples for genes that are transcriptionally regulated by oxidative stress. Adapted from Knasmuller *et al.*

Gene name [Gene bank ID]	Protein function and occurrence
[GPX1 [NM_000581], GPX2 NM_002083], GPX3 [NM_002084], GPX4 [NM_002085], GPX5 [NM_001509], GPX6 [NM_182701], GPX7 [NM_015696]	Different forms of glutathione peroxidase that detoxify H_2O_2 OS*: GPX2 mainly in the GI tract and liver, GPX4 – testes, GPX5 – epidymis, GPX6 – embryonal form LO†: GPX1, GPX2 cytoplasm, GPX3, GPX5, GPX6 – secreted protein GPX4 – mitochondria, cytoplasm
GSTZ1 [NM_145870]	Maleylacetoacetate isomerase: bifunctional enzyme catalyzing glutathione- conjugation with specific substances – has also GPx activity OS: primary in liver and kidney LO: cytoplasm
PRDX1 [NM_002574], PRDX2 [NM_005809], PRDX3 [NM_006793], PRDX4 [NM_006406], PRDX5 [NM_012094], PRDX6 [NM_004905]	Peroxiredoxins reduce peroxides, PRDX1, PRDX2 may be involved in cell-signaling cascades of growth factors and TNFa by regulating the intracellular concentrations of H_2O_2. PRDX3 and PRDX4 are involved in MAP3K13, which triggers the regulation of NFkB, PRDX6 is involved in the detoxification of H2O2, fatty acids, and phospholipids LO: PRDX1, PRDX2, PRDX4 – cytoplasm, PRDX3 – mitochondria, PRDX5 – mitochondria, cytoplasm, peroxisomes, PRDX6 – cytoplasm, lysosome
CAT [NM_001752]	Catalase: detoxification of H_2O_2, one of the most important detoxifying enzymes found in many eukaryotic and prokaryotic cells LO: Peroxisomes
MGST3 [NM_004528]	Microsomal glutathione S-transferase 3: functions as a glutathione peroxidase OS: heart, skeletal muscle LO: microsome
SOD1 [NM_000454], SOD2 [NM_000636], SOD3 [NM_003102]	Superoxide dismutase, which detoxifies $O_2^{\cdot -}$ LO: SOD1 is found intracellulary (cytoplasm), SOD3 (Cu–Zn dependent) is found in extracellular fluids (plasma, plasma, lymph, and synovial fluid). SOD2 (Mn – form) is found in mitochondria
TXNRD1 [NM_003330], TXNRD2 [NM_006440]	Thioredoxin reductase contributes to oxidative stress resistance, maintains thioredoxin in a reduced state OS: TXNRD2 is highly expressed in the prostate, ovary, liver, testes, uterus, colon, and small intestine LO: TXNRD1 – cytoplasm, TXNRD2 – mitochondria
MT3 [NM_005954]	Metallothionein-3: binds heavy metals and is known to be also involved in oxidative defense LO: Abundant in a subset of astrocytes in the normal human brain

(Continued)

Table 9.1 (*Continued*)

Gene name [Gene bank ID]	Protein function and occurrence
NOS2A [NM_000625]	Nitric oxide synthase (inducible) produces NO, which is a messenger molecule with diverse functions OS: expressed in liver, retina, bone cells, and epithelial cells of the lung, not expressed in platelets
AOX1 [NM_001159]	Aldehyde oxidase, catalyses the formation of carboxylic acids from aldehydes (aldehyde $+ H_2O + O_2 =$ carboxylic acid $+ H_2O_2$) LO: abundant in liver and muscle
EPHX2 [NM_001979]	Epoxide hydrolase 2 – substrates are hydroxyl perepoxides such as alkene oxides, oxiranes, catalyses also the detoxification of toxic xenobiotics LO: cytoplasm, peroxisomes
FOXM1 [NM_202002]	Forkhead box protein M1 transcriptional activation factor: may play a role in the control of cell proliferation OS: Expressed in thymus, testis, small intestine, colon, followed by ovary; appears to be expressed only in adult organs containing proliferating/cycling cells or in response to growth factors LO: nucleus
GLRX2 [NM_016066]	Glutaredoxin-2 (mitochondrial form): glutathione-dependent oxidoreductase that maintains mitochondrial redox homeostasis upon induction of apoptosis by oxidative stress OS: Widely expressed in different organs (brain, heart, skeletal muscle, colon, etc.) LO: Isoform 1: mitochondria. Isoform 2: nuclei
FDX1 [NM_004109]	Ferredoxin-1: participates in the synthesis of thyroid hormones; electron transport intermediate of cytochrome P450 LO: mitochondria, mitochondrial matrix
CCNA2 [NM_001237]	Cyclin-A2 – cell cycle regulation LO: In contrast to cyclin A1, which is present only in germ cells, this cyclin is expressed in a broad variety of tissues
JUND [NM_005354]	Transcription factor jun-D: binds to AP-1 site and upon cotransfection stimulates the activity of a promoter that bears an AP-1 site, binds DNA as a dimmer LO: nucleus
HMOX1 [NM_002133]	Heme oxygenase 1 – cleaves the heme ring at the alpha methane bridge to form biliverdin, which is subsequently converted to bilirubin by biliverdin reductase. Under physiological conditions, the activity of heme oxygenase is highest in the spleen, where senescent erythrocytes are sequestrated and destroyed LO: microsomes

Table 9.1 (*Continued*)

Gene name [Gene bank ID]	Protein function and occurrence
HSPA1A [NM_005345], HSPA1B [NM_005346]	In cooperation with other chaperones, Hsp70s stabilizes preexistent proteins against aggregation and mediates the folding of newly translated polypeptides in the cytosol as well as within organelles. These chaperones participate in these processes through their ability to recognize nonnative conformations of other proteins
GCLM [NM_002061]	8-Glutamyl-cysteinyl ligase – ATP + L-glutamate + L-cysteine = ADP + phosphate + gamma-L-glutamyl-L-cysteine LO: in all tissues examined; highest levels in skeletal muscles
TOP2A [NM_001067], TOP2B [NM_001068]	DNA topoisomerase: control of the topological states of DNA by transient breakage and subsequent rejoining of DNA strands; topoisomerase II causes double-strand breaks LO: TOP2A: generally located in the nucleoplasm, TOP2B: in the cytoplasm
DDIT3 [NM_004083]	DNA damage-inducible transcript 3: inhibits the DNA-binding activity of C/EBP and LAP by forming heterodimers that cannot bind DNA LO: nucleus

*OS, Organ Specificity; †LO, Location.

transcription in response to stimuli (Angel and Karin 1991). AP1 plays an important role in cell growth, differentiation, and apoptosis, which has been reported to be induced by metals and H_2O_2 (Pinkus, Weiner *et al.* 1996). Contrary to AP-1, the activation of NFκB (DNA binding protein) induced by ROS stimuli occurs mainly by the dissociation of inhibitory protein IkB (Schreck, Rieber *et al.* 1991; Schreck, Meier *et al.* 1992), which in turn allows NFκB to translocate to nucleus and trigger expression of several genes involved in inflammatory responses, transformation, and angiogenesis (Maulik, Sato *et al.* 1998; Maulik and Das 2002; Amiri and Richmond 2005; Maulik 2006). A large number of studies from different laboratories including ours have shown the differential expression of several antioxidant genes during ischemic preconditioning (Addya, Shiroto *et al.* 2005; Thirunavukkarasu, Addya *et al.* 2008). Results from our previous studies have demonstrated nuclear translocation and activation of NFκB in response to preconditioning (Thirunavukkarasu, Juhasz *et al.* 2007). Increased binding of NFκB was dependent on both tyrosine kinase and p38 MAP kinase (Maulik, Sato *et al.* 1998). The NFκB appears as a critical regulator for gene expression induced by diverse stress signals which includes mutagenic, oxidative, and hypoxic stress (Maulik, Sato *et al.* 1998). Activation of NFκB is likely to be involved in the induction of gene expression associated with the ischemic adaptation, because this transcription factor has been found to play a crucial role in the regulation of ischemia/reperfusion-mediated gene expression (Yamazaki, Seko *et al.* 1993). Substantial evidence exists to support the notion that oxygen-derived free radicals are generated during the reperfusion of ischemic myocardium resulting in the development of oxidative stress (Baker, Felix *et al.* 1988). Due to its pivotal role in several redox-related signaling, several studies had been focused on identifying compounds or antioxidants that regulate NFκB activity. Several studies with phytochemicals have shown a protective mechanism with these processes (Surh, Han *et al.* 2000; Surh, Chun *et al.* 2001; Carluccio, Siculella *et al.* 2003; Kundu and Surh 2004;

Shen, Jeong *et al*. 2005; Surh, Kundu *et al*. 2005; Davis, Polagruto *et al*. 2006; de Sousa, Queiroz *et al*. 2007; Kundu and Surh 2007; Lee, Kang *et al*. 2007; Yoon and Liu 2007; Gopalakrishnan and Tony Kong 2008; Surh 2008). Resveratrol, capsaicin, and curcumin have been reported to inhibit NFκB, AP-1, and beta catenin TCF signaling via interaction with the upstream signaling pathways such as phosphorylation of AKT, MAPK and IKK (Kundu and Surh 2005).

CONVENTIONAL METHODS AVAILABLE TO MEASURE OXIDATIVE STRESS

Oxidative stress-mediated damage can be monitored by several different techniques, and it helps to identify the effect and mode of action of dietary antioxidant treatment (Knasmuller, Nersesyan *et al*. 2008). Some of the most common techniques employed are trapping reactive species by electron spin resonance, measurement of total antioxidant capacity, free radical quenching methods, measurement of LDL oxidation, TEAC assay, which utilizes spectrophotometic tests (oxidation of 2,2-azinobis(3-ethylbenzothiazoline-6 sulfonic acid (ABTS) by ROO), 2,2-diphenyl-1-picrylhydrazyl (DPPH) assay, measurement of malondialdehyde, isoprostanes, 8-Oxo-dg, breath hydrocarbons, etc. These methods are discussed in detail elsewhere (Smith and Anderson 1987; Pratico, Reilly *et al*. 1997; Prior and Cao 1999; Dizdaroglu, Jaruga *et al*. 2002; Rahman and Kelly 2003; Halliwell and Whiteman 2004; Rahman and Biswas 2004; Swartz 2004; Villamena and Zweier 2004; Cash, Pan *et al*. 2007; Somogyi, Rosta *et al*. 2007; Knasmuller, Nersesyan *et al*. 2008; Michel, Bonnefont-Rousselot *et al*. 2008).

Also, measurement of antioxidant enzyme activities have become more important in studies involving drug treatment toward ROS toxicity because they are a part of the endogeneous defense system against ROS. The most common antioxidant enzymes studied in extent are SOD, GPx, catalase, etc. Several studies have been reported to influence these enzymes by dietary treatment (Nelson, Bose *et al*. 2006). MnSOD has been reported to promote mitochondrial H_2O_2 production and in turn stimulates EC sprouting and neovascularization in the chorioallantoic membrane (CAM) assay (Abid, Tsai *et al*. 2001). We have reported that resveratrol, a dietary antioxidant compound present in red wine, has increased the activity of MnSOD in diabetic rat hearts (Thirunavukkarasu, Penumathsa *et al*. 2007). In addition, it has been reported to influence the levels of thioredoxin and heme oxygenase 1 enzyme activity (Kaga, Zhan *et al*. 2005). Another important enzyme involved in the defense system is gamma glutamylcysteine synthetase, which catalyzes the rate-limiting step of glutathione synthesis, and several studies with dietary compound treatment have shown to influence their activity (Ip 1984; Eaton and Hamel 1994; Huber, Scharf *et al*. 2002). Glutathione (GSH) is also by itself a potent antioxidant and scavenges predominantly O_2^-, ˙OH, RO˙ and ROO˙. Measurement of GSH and the ratio between oxidized and reduced GSH has been shown to be a parameter to determine the antioxidant effects in several dietary studies (Boyne and Ellman 1972). Hence, increased levels of antioxidant enzymes during oxidative stress are also seen in several pathological conditions (Forbes, Coughlan *et al*. 2008; Forstermann 2008; Galecka, Jacewicz *et al*. 2008; Inoguchi and Takayanagi 2008; Loscalzo 2008; Madsen-Bouterse and Kowluru 2008; Murray, Oquendo *et al*. 2008; Rebrin and Sohal 2008; Tsimikas 2008). Hence, measurement of antioxidant enzymes during the supplementation of dietary constituents can be considered a cytoprotective effect and also should be supported by other methods developed to monitor the impact of antioxidants on proteins involved in cell signaling, which includes Western blots, Northern blots, real-time PCR and ELISA, along with measurement of kinases such as protein kinase MAPKs, etc.

GENOMICS-BASED APPROACHES TO IDENTIFY THE ANTIOXIDANT EFFECT

During the last decade, several reports have shown that antioxidants are capable of affecting differential gene expression in cells and laboratory animals as well as humans by using several new techniques such as differential display, serial analysis of gene expression, DNA microarrays, and gene chips, which have been developed to simultaneously detect the large number of alterations of biological functions. Gene arrays have been used recently in analyzing the extent of changes in gene expression due to dietary antioxidant treatment, which provides more insight into the molecular functions of oxidants and antioxidants. The main application of these techniques is that they can be used in the development of new biomarkers of oxidative stress and also to study the toxicity of oxidants and antioxidants as well as for screening new antioxidants. Results of these types of studies will help people modify their diet either to avoid or to include certain nutrients/antioxidants/phytochemicals along with the better understanding of our fundamental knowledge of the interactions between life processes and our diet. Measurement of nutritionally responsive genome activity and its application depends on many important factors such as model objects, type of nutrition, diet, or antioxidants. Choice of platform, proper experimental design, selection of statistical tool for data analysis, and interpretation along with pathway analysis are explained in detail in the previous section. Reports obtained using –omic technology to study effect of antioxidants have been increasing in the past few years, and they can be classified into three major categories: *in vitro* and *in vivo* studies using animals, and human studies.

In vitro STUDIES

Several microarray experiments related to dietary antioxidants (e.g., resveratrol, epigallocate-chingalleate, curcumin, sulphoraphane, genistein, etc.) have been carried out in both *in vitro* and *in vivo* (Table 9.2), but still the *in vitro* studies predominate the *in vivo* studies. *In vitro* experiments using human umbilical vein endothelial cells (HUVEC) treated with 0.1uM of polyphenols (ferulic acid, quercetin, and resveratrol) and subjected to microarray experiments showed a significant amount of differential gene expression. The results showed significant (twofold) increase in the gene expression levels (233 genes) overall by all the three compounds, the majority found to be with resveratrol. Fewer genes were affected by quercetin or ferulic acid (Nicholson, Tucker *et al.* 2008). Diebolt *et al.* have also reported previously the ability of red wine polyphenols resveratrol to alter gene expression in rat aortas (Diebolt, Bucher *et al.* 2001). Another study by Scherf *et al.* has shown the gene expression profiles in sixty human cancer cell lines for a drug discovery screen, which involves 8000 genes tested for approximately 70,000 compounds. The data were clustered according to different parameters and are available in the National Cancer Institute (NCI) database, which helps other investigators select the gene of their interest based upon these results and synthesize a custom array for their drug of interest (Scherf, Ross *et al.* 2000). Murray *et al.* (Murray, Whitfield *et al.* 2004) have used human fibroblasts and HeLa cells to compare the effects of different types of stress (heat shock, oxidative stress, etc.). Another report by Yoneda *et al.* (Yoneda, Chang *et al.* 2003) has shown the toxic effects of H_2O_2 and tobacco smoke in bronchial epithelial cells, whereas Morgan *et al.* (Morgan, Ni *et al.* 2002) have used human liver cell line HepG2 to explore the effects of different ROS-generating chemicals. Nair *et al.* (Nair, Rao *et al.* 2004) found that quercetin treatment with human prostate cancer cell lines decreased the expression of several genes related to cell cycle and oncogene, whereas Murtaza *et al.* (Murtaza, Marra *et al.* 2006) showed downregulation of several cell cycle and apoptosis-related genes in human adenocarcinoma cells. Epigallocatechin gallate (Wolfram, Raederstorff *et al.* 2006) showed significant downregulation of genes related to fatty acid synthesis, oxidation, and activation, and in turn upregulated genes involved in glycolysis and glucose transport when treated with rat hepatoma cells. Studies with resveratrol also reported alteration of genes in relation to antioxidant enzymes (NQO1, NADPH oxireductase 1, thioredoxin reductase 1), apoptosis, etc., on treatment with human ovarian cancer cell lines (Yang, Kim *et al.* 2003).

In vivo STUDIES

Animal studies using -omics methodology to study the effects of antioxidants and their effect on ROS have been done by several groups and were published in peer-reviewed journals. Examples of some of the studies are included in this chapter. Yoshihara *et al.* (Yoshihara, Ishigaki *et al.* 2002) have studied the changes in the gene expression patterns extensively by using SOD and

Table 9.2 Examples for results obtained with dietary antioxidants in microarray experiments. (Adapted from Knasmuller *et al.*)

Treatment/aim	Cells/AS	Results
In vitro investigations		

Quercetin
Aim: study the
  impact on growth
  inhibition of
  prostate cancer
  cells
	PK-3, NLCap,
	  DU-145 human
	  prostate cancer
	  cells lines; BG-9
	  normal human
	  fibroblast cell line
	AS: number of genes
	  not specified
		In prostate cancer cell lines ↓ of the expression of cell cycle related genes (e.g., E- and D-type cycline genes) was observed as well as ↓ of tumor suppressor genes (e.g., CBP, PTEN, MSH2, TGFβR1, ALK-5) and ↓ of oncogenes (p53, CDK2) (Nair, Rao *et al.* 2004)

Quercetin
Aim: study the
  antitumor activity
  of Q
	Human CO-115 colon
	  adenocarcinoma
	  cells
	AS: whole genome
	  microarray
		5000–7000 genes were affected, specific effects on genes related to cell cycle arrest (CDKN-group) and modulation of the expression of apoptosis-related genes (e.g., p53-related genes), only one antioxidant gene (LOXL3) was altered (Murtaza, Marra, *et al.* 2006)

Genistein
Aim: study the impact
  on prostate cancer
  development
	Human prostate
	  cancer cell lines
	  LNCaP and PC-3
	AS: 557 genes related
	  to cancer
		11 genes were strongly altered; ↑glutathione peroxidase 1, aldolase A, quiescin Q6, ras homolog gene family, member D; ↓ of apoptosis inhibitor (survivin), MAPK6, fibronectin 1, topoisomerase IIa etc (Suzuki, Koike, *et al.* 2002)

Epigallocatechin
  gallate
Aim: study effects on
  cell division
	SH-SY5Y human
	  neuroblastoma cell
	  lines
	AS: 25 genes
		↓ of expression of apoptotic genes (Bax and Bcl-2) and of the cell cycle inhibitor Gadd45 as well as caspase 6 (Weinreb, Mandel, *et al.* 2003; Weinreb, Mandel, *et al.* 2003)

Epigallocatechin
  gallate
Aim: investigation of
  the protective role
  of EGCG on breast
  cancer
  development
	Dimethyl
	  benzanthracene
	  transformed breast
	  cancer cell line
	  D3-1
	AS: 7500 genes
		EGCG treatment caused changes in transformed cells that promote a more "normal" phenotype. ↓ of AhR (a transcription factor involved in the biological responses to polycyclic aromatic hydrocarbons, e.g., CYP1A1, CYP1A2 and CYP1B1). Also changes involved in nucleo-cytoplasmatic transport (SCS-1) were affected ↑ of dehydrocarbonases (AKR) (Guo, Yang, *et al.* 2005)

Table 9.2 (*Continued*)

Treatment/aim	Cells/AS	Results
Epigallocatechin gallate Aim: study impact of gene regulation of glucose metabolism	Rat H4IIE hepatoma cells AS: size not specified	↓of number of genes involved in fatty acid synthesis, oxidation, and activation as well as in triacethylglycerol synthesis ↑ of genes involved in glycolysis and glucose transport (Wolfram, Raederstorff, *et al.* 2006)
Diallyl disulfide Aim: investigation on cell division and overall tumor behaviour	HCT-15 human colon tumor cell line AS: size of the array not specified	↓ of cell cycle regulated genes (e.g., Cdk6) as well as oncogenes, tumor suppressors and extracellular matrix and communication genes ↑cell cycle proteins (cdk2, 3, 4, etc.) as well as different growth factors, microfilaments, protein turnover-related genes ↑ DNA damage-related genes (e.g., XP-related genes) (Knowles and Milner 2003)
Diallyl trisulfide Aim: investigation of lipid-lowering properties	Human-derived hepatoma cells (HepG2) AS: 452 genes	Only three genes were altered: ↑PPAR-a (peroxisome-activated receptor alfa) and HNF-4a (hepatocyte nuclear factor 4 alfa), ↓ CYP7A1 (involved in the oxidation of xenobiotics) (Zhou, Tan, *et al.* 2005)
Gallic acid Aim: investigation of antioxidant effects	Human chronic myogenous leukemia cell K562 AS: 82 genes (antioxidant enzymes and DNA repair)	Several antioxidant genes↓, i.e., GPX, thioredoxin (TXN), thioredoxin peroxidase (AOE372), several DNA polymerase genes (POLD1 and 2), X-ray repair genes (XRCC5), DNA-3-methyladenine glycosidase (Abdelwahed, Bouhlel, *et al.* 2007)
Vanillin and cinnamaldehyde Aim: investigation of prevention of formation of spontaneous mutations	Mismatch-repair deficient (MMR2) human colon cancer cell line HCT116 AS: 14,500 genes	Eight genes affected that play a role in DNA damage and oxidative and stress response, e.g., ↑HMOX1 (heme oxygenase), HSPA1B (heat shock protein) ↓ of fourteen genes involved in cell growth and differentiation: (MAPK2, FGFR2, TGFB111, etc.) (King, Shaughnessy, *et al.* 2007)

(*Continued*)

Table 9.2 (*Continued*)

Treatment/aim	Cells/AS	Results
Resveratrol Aim: general chemoprotective properties	Human ovarian cancer cell line PA-1 AS: 7448 genes	118 genes were altered ↑ of antioxidant enzymes (NQO 1, NAD(P)H quinone oxireductase 1, thioredoxin reductase 1), apoptosis-related genes (p21), investigation of time course of gene regulation (Yang, Kim, *et al.* 2003)
Lycopene Aim: investigation of the effects on breast cancer genes	Human breast cancer cell lines (MCF7, MDA-MB-231) and the fibroblastic cell line MCF-10a AS: 202 genes	Changes in genes related to apoptosis, cell cycle, and signaling. Apoptosis-related genes: p53, caspase 8, TNF (all ↓), cell cycle genes: cyclin E, etc. ↑. Also various DNA-repair genes ↑ (Chalabi, Delort, *et al.* 2006)
Vitamin E Aim: investigations on gene expression related to prevention of DNA-adduct formation by lipid peroxidation products	Differentiated and undifferentiated human colonocytes (with/without oxidative stress) AS: global gene expression	A total of 118 genes were affected that concern the cell cycle (cyclin D1, p27, p21, MAPKs, CDK-2); DNA-repair (p21, RAD54 homologous recombination) and also connective tissue-related genes (Lunec, Halligan, *et al.* 2004)
Apple flavonoids Aim: study patterns of gene expression (not specified)	HT29 human colon cancer cells AS: 96 genes of drug metabolism	↑of glutathione tranferases (GSTP-1, GSTP-2, MGST2) and cytochromes CYCP4F3, CHST5 genes ↓ of EPHX1 (Veeriah, Kautenburger, *et al.* 2006)
Resveratrol Aim: investigations on the role of macrophage inhibitory cytokine-1 (MIC-1) in resveratrol-induced growth inhibition of human pancreatic cancer cell lines	CD18 and S2-013 human pancreatic cell lines AS: global gene expression	↑ of expression of MIC-1 in both cell lines ↑of expression of growth differentiation factor 15 (GDF15), sensescence-associated epithelial membrane protein (SEMP), major histocompatibility complex, class I, F (HLA-F) and seventeen other genes in S2-013 cell line (Golkar, Ding, *et al.* 2007)

Table 9.2 (*Continued*)

Treatment/aim	Cells/AS	Results
Genistein (G) Aim: investigations on the effect of G on global gene expression patterns in androgen-responsive human prostate cancer cell line	LNCaP androgen-responsive human prostate cancer cell line AS: global gene expression	Twenty-eight genes were affected Nineteen androgen upregulated genes were ↓ by G (TNF-induced protein, prostatic kallikrein 2, prostate specific antigen kallikrein 3 and others) Four androgen downregulated genes ↑ by G (dopa decarboxylase, BRCA-1-associated RING domain 1, butyrylcholinesterase, phosphoinositide-3-kinase) five genes ↑ by both androgen and G (stearoyl- CoA desaturase, UDP-glucose dehydrogenase) (Takahashi, Lavigne, *et al.* 2004)
Phenethylisothiocyanate (PEITC) Aim: to further clarify the molecular effects of PEITC in causing death of human colon adenocarcinoma cells	HCT-116 human colon adenocarcinoma cell line AS: small cluster of apoptosis-related genes	↑ of expression of GADD45 (Powolny, Takahashi, *et al.* 2003)
Quercetin Aim: to elucidate possible mechanisms involved in inhibition of proliferation of tumor cells	Caco-2 human colon cancer cell line AS: expression of 4000 human genes	↓ expression of cell cycle genes (for example CDC6, CDK4 and cyclin D1), # cell proliferation and induced cell cycle arrest ↑of expression of several tumor suppressor genes. In addition, genes involved in signal transduction pathways like beta catenin/TCF signaling and MAPK signal transduction were influenced by quercetin (van Erk, Roepman, *et al.* 2005)
Vitamin E Aim: effects on global gene expression – time dependency	Male Fisher 344 rats fed with vitamin E-deficient diet or vitamin E-supplemented diet (60 mg/kg) AS: 7000 genes t.o. testes and liver Several time points were monitored	Five genes and seven sequence taqs were altered at least at three time points (gamma glutamylcysteinyl synthetase, GSH synthetase, 5-a-steroid reductase, factor IX, scavenger receptor CD36), furthermore, subunit of 8-GCS, coagulation factor IX, steroid reductase type 1, etc. (Barella, Muller, *et al.* 2004; Rota, Barella, *et al.* 2004)

(Continued)

Table 9.2 (*Continued*)

Treatment/aim	Cells/AS	Results
Turmeric supplements Aim: mechanism of action of turmeric in the treatment of arthritis	Female Lewis rats were treated i.p. with 25 mg rhamnose/g bw and with various doses of turmeric extract i.p. (0.5–1.0ml/g) AS: 31,000 genes t.o. arthritic joints	↑ of a variety of chemokines and cytokines (GRO/KC, MCP-1, MIP-1a, MIP3a, CXC chemokine LIX) and adhesion factors that facilitate inflammatory cell recruitment to the joint for another group of target genes. COX2 was ↓ significantly, gene expression controlled by transcriptional factors TIL-2, TNFa was ↑ (Funk, Frye, *et al.* 2006)
Epigallocatechin gallate Aim: comparison of the impact of Nrf2 on expression profiles	Male C57BL/6J (Nrft/t) and Female C57BL/6J/Nrf (-/-) knockout mice EGCG: 200 mg/kg bw oral single dose AS: 34,000 genes t.o. liver and colon	EGCG regulated 671 Nrf2 dependent genes and 256-independent genes in the liver; in the colon 228 Nrf2 regulated genes and 98 independent genes were identified. Genes fell into the functional categories: proteolysis, detoxification, transport, cell growth and apoptosis, cell adhesion and transcription factors (Shen, Xu, *et al.* 2005)
Red wine polyphenolics (RWP) Aim: investigate the effect on the prevention of carcinogenesis in the intestinal mucosa	Male Fisher 344 rats fed with high fat diet RWP were fed in the diet (50 mg/kg) AS: 5677 genes t.o. colon mucosa	20 genes ↑, 366 ↓, among the ↓ genes 41 were related to immune- and anti-inflammatory response. In addition, genes that are involved in steroid metabolism were also affected, ↓ of genes involved in energy metabolism pathways (COX6-8, etc.). ↑ of cholesterol 7a-hydroxylase (CYP7A1) (Dolara, Luceri, *et al.* 2005)
Resveratrol (R) Aim: investigation on stress response in rat liver	Male and female CD rats t.o. liver Treatment: different doses of R (0·3–3·0 gm/kg/day), oral administration, 28 days AS: 140 stress-related genes	Dose dependent effects ↑ of CYP-450 isoenzymes as well as CYP-reductase induction of AO genes such as SOD2 and thisulfate sulfurtransferase (TST) at the highest dose, but decrease of these genes at lower doses. Other genes affected were MAPK, p38, CAT, HO-1, as well as UGT-encoding genes. In female, the number of genes altered was ca. 2-fold higher than in male (Hebbar, Shen, *et al.* 2005)

Table 9.2 (*Continued*)

Treatment/aim	Cells/AS	Results
Genistein (G) Aim: study impact of the isoflavone on fatty acid metabolism	Female C57BL/6 mice fed low-fat diet, a high-fat diet or high-fat diet supplemented with genistein (2 g/kg), 12 weeks AS: 6531 transcripts t.o. liver	Ninety-seven genes were altered by high-fat diet, eighty of them were normalized by G supplementation. Many of them were associated with cholesterol biosynthesis, but additionally also genes involved in detoxification and inflammatory processes were affected, e.g., metallothionein 1, GST, kallikrein B (plasma1), serine proteinase inhibitor (clade A, member 3G), serine protease inhibitor 1–5. Also, apoptosis related genes were affected (Kim, Sohn, *et al.* 2005)
Vitamin C (Vc) Aim: impact on lifespan and antioxidant defense mechanisms	C57BL/6 mice, sex not specified, maintained either at normal temperature (Þ228C) or Þ78C (increase of metabolism) and supplemented diet (180 mg/kg), lifespan feeding AS: 163 genes t.o. live	After 6 months, no gene alterations were found, after 22 months: 3 genes ↑ (COX2, p21, UDP–glucuronosyltransferase) 8 genes ↓ : e.g. SOD, quinone NAD(P)H-dehydrogenase and different other genes not involved in AO defense (Selman, McLaren et al. 2006)
Caffeic acid phenetyl ester Aim: impact on atherosclerosis	Apolipoprotein E-deficient mice (Apoe 2 /2) and normal C57/B6, 30 mg/kg bw 12 weeks AS: 3758 genes expression t.o. aorta analyses were conducted in untreated mice and mice that were under oxidative stress	Altered genes are not clearly specified; overall expression induced by oxidative stress was reverted by CAPE-treatment Authors mention that basic transcription factors, growth factors, and cytokines as well as cell adhesion genes are ↓ in apoe -/- mice after CAPE treatment (Hishikawa, Nakaki, *et al.* 2005)

*AS-Array Size; SD-Sprague Dawley Rats; ↓- downregulation; ↑-upregulation.

Nrf2 knockout mice. Barella *et al.* performed a global gene expression profile in rat liver to understand the molecular mechanism of vitamin E when included in the diet at various different time points (Barella, Muller *et al.* 2004). They identified five genes (gamma glutamylcysteinyl synthetase, GSH synthetase, 5-α-steroid reductase, factor IX, scavenger receptor CD36, etc.) and seven sequence taqs to be altered in three various time points (Barella, Muller *et al.* 2004). Colon mucosa from red wine polyphenol-treated Fisher 344 rats showed upregulation of 20 genes and downregulation of 366 genes among which 41 genes were related to immune, anti-inflammatory response, and energy metabolism pathway (Dolara, Luceri *et al.* 2005). Another study with liver samples from resveratrol treatment of both male and female CD rats showed induction of genes such as SOD2, thiosulfate sulfurtransferase, MAPK, p38, catalase, HO-1, etc. (Hebbar, Shen *et al.* 2005). Human studies using the microarray technology have also reported that oxidative stress plays an important role in disease conditions such as rheumatoid arthritis, inflammatory bowel disease, diabetes, coronary heart disease, etc. (Heller, Schena *et al.* 1997; Kim, Lim *et al.* 2003; Jison, Munson *et al.* 2004). A study conducted with monocytes isolated from normal and apoE smokers supplemented with vitamin C showed twenty-one & seventy-one genes were altered in normal and apoE smokers respectively (Majewicz, Rimbach et al. 2005). Hoffmann *et al.* supplemented a mixture of polyphenols to healthy individuals followed by isolation of lymphocytes and showed alteration in expression of fifty-six genes, out of which mostly coded for the drug metabolism (Hofmann, Liegibel *et al.* 2006). Even though results from these studies show significant alterations in the gene expression patterns caused by ROS, the effect of antioxidant supplementation care should be given toward experimental design, selection of model (*in vitro* or *in vivo*), organ difference, etc., which might have a huge impact on the results of array studies.

NUTRIGENETICS AND NUTRIGENOMICS: TWO SIDES OF A COIN

Nutrigenomics focuses on differences among different dietary nutrients/antioxidants or conditions on changes in gene expression, whereas nutrigenetics focuses on variations between individuals with respect to the effects of dietary compounds (Ordovas and Mooser 2004; Mutch, Wahli *et al.* 2005); that is, it examines the effect of genetic variation on the interaction between diet and disease. The relationship between nutrition and health should be focused so that it can prevent the onset of diseases and disorders related to nutrition both in the case of malnutrition or imbalance of nutrients (Fig. 9.1). Gillies *et al.* have reported from clinical observations that genomics responses with specific dietary components in certain individuals might vary depending upon the presence or absence of specific genetic polymorphisms leading to alterations in the promoter region or coding/noncoding sequences of gene, which affects their expression levels directly (Simopoulos 1996; Simopoulos 1999; Gillies 2003). Interindividual genetic variations in nutrition research and phenotypes have been reported early (Simopoulos 1996; Simopoulos 1999; Murray 2001). Hence, the importance of genomic research in the nutrition field is unquestionable, but due to scattered implementation of expensive research facilities and bioinformatics capacities, only gradual progress has been achieved so far. To focus the basis of nutrition research and its link toward health will only be attained when we uncouple the direction from biomedical approach and link it toward multiple minor changes in metabolism that are the major sources of chronic nutrition-related disorders like diabetes, obesity, cardiovascular disorders, osteoporosis, etc. Hence, the major approach should be to utilize the genomic approach to maintain an inventory of all relevant parameters to unravel mechanisms and define biomarker sets. The major point to focus on is that we are dealing with more complicated cases of multiple dosing of complex mixtures of bioactive compounds rather than single exposure/single time point studies, and hence more attention should be given toward this observed real-life toxicology.

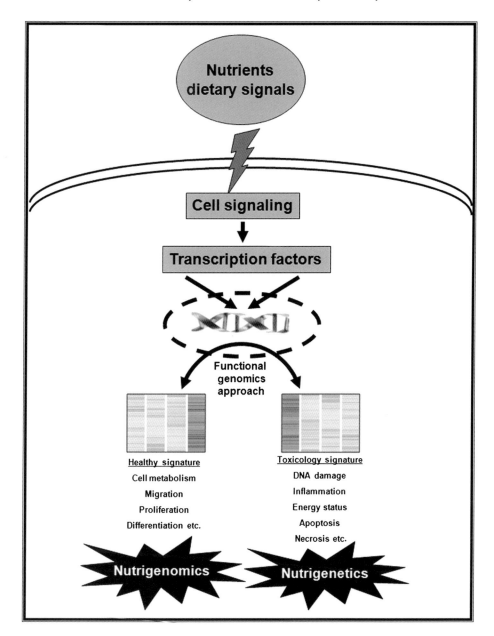

Figure 9.1 Nutrigenetics and nutrigenomics; two sides of a coin. For color detail, please refer to the color plate section.

SAFETY AND EFFICACY ASSESSMENT OF DIETARY NUTRIENTS/ANTIOXIDANTS

Risk assessment process has undergone several refinements over the past decade (Oberemm, Onyon *et al.* 2005). Assessment of human exposure and dose dependency determines whether there is a risk for human health or not. In general, conventional ways of profiling toxicology of substances entirely depends on time-consuming and expensive animal studies. Introduction of new molecular techniques along with genomics has led to the new research field called toxicogenomics, which combines genetics, transcriptomics, proteomics, metabonomics, and bioinformatics along with conventional toxicology to understand the interaction between gene environment and its relation with disease. Hence, the first step in the safety assessment of the dietary nutrients is to identify the adverse effects of this nutrient regardless of its dose and its mechanism that should be followed by quantification of these adverse effects so that the dose response relationships identified can be compared with risk characterization. This character-ization method usually starts with a detailed toxicological experiment carried out in animals to determine the no-observed adverse effect level (NOAEL) or lowest observed adverse effect (LOAEL), which then can be used to fix the acceptable daily intake for a particular drug/nutrient or a mixture of drug/nutrient. This conventional approach is often time-consuming, and only a few genes or proteins can be studied simultaneously. Toxicity of defined mixtures of nephro-toxicants, pesticides, carcinogens, fertilizers, and food additives along with dietary constituents such as quercetin, resveratrol, ferulic acid, and curcumin have been studied by this approach (Groten, Butler et al. 2000). However, based upon the results from these studies, it was difficult to understand the mechanistic-based interactions in detail because the focus mainly directed toward establishing that a no-observed-adverse-effect-level is the base of hazard assessment (Ommen and Groten 2004). Recent advances in the field of genomics and array technology have revolutionized the new area of functional genomics toward both safety and efficacy evaluation of dietary nutrients. It allows us to determine the expression of thousands of genes in a single experiment and has led to the new field of toxicogenomics, which integrates functional genomics along with classical toxicology (Stierum, Heijne et al. 2005) and hence will result in revolutionizing the mechanistic research for food mixtures and dietary antioxidants (Fig. 9.2).

TOXICOLOGICAL GENOMIC APPROACH IN THE ASSESSMENT OF DIETARY NUTRIENTS/ANTIOXIDANTS

The functional genomic approach brings both nutritionists and toxicologists together by bridging the gap between markers of effects (RDA and NOAEL) used by them. For example, in a study, a nutritionist mainly focuses on studying the effect of dietary nutrients on the intestinal health, thereby looking for the changes in the homeostasis of the intestinal cells after exposure to dietary nutrients (e.g., cell proliferation, cell differentiation, inflammatory responses, etc.) whereas toxicologists emphasize the severe and irreversible effects of dysregulation and inflammatory responses (e.g., cell death) (Fig. 9.1). But when the study is done utilizing the functional genomic approach, the results show changes in thousands of gene expressions corresponding to several pathways. Hence, both nutritionists and toxicologists can look into the results and focus on genes related to their specific pathway and field of interest. Several recent safety evaluations of novel foods and dietary nutrients have been designed and studied related to this field, and a few examples of these have been discussed below.

Olive leaf extract, which has been reported to be effective in treating fever and malaria, contains several compounds with potential antimicrobial activities (Hanbur 1984), antioxidant and anti-inflammatory activities (Visioli and Galli 1994; Petroni, Blasevich et al. 1995; Caruso, Berra et al. 1999; Coni, Di Benedetto et al. 2000; de la Puerta, Martinez-Dominguez et al. 2000). Recently, it has been used for AIDS patients to strengthen their immune system, to relieve chronic fatigue, and to boost the effects of anti-HIV medications. Lee-Huang et al. examined the effects of olive leaf preparation using DNA microarray to characterize the gene expression profiles associated with HIV-1 infection and olive leaf treatment. Several important genes such as heat shock proteins hsp27, hsp90, DNA damage inducible transcript 1 Gadd45, were upregulated whereas antiapoptotic BCL2-associated X protein Bax was downregulated during HIV-1 infection. Olive leaf extract treatment was found to reverse many of these changes due to HIV-1 infection (Lee-Huang, Zhang et al. 2003).

Gingko biloba leaf extract has been used as a dietary supplement for several years to treat symptoms associated with various brain and circulatory disorders, such as memory problems.

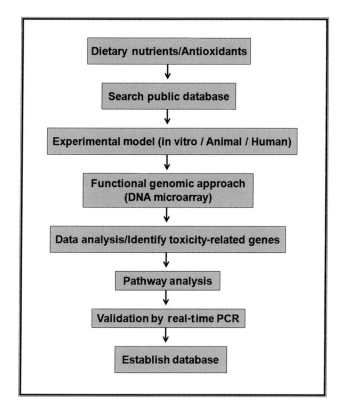

Figure 9.2 Steps involved in assessing the safety and efficacy of dietary antioxidants with the help of functional genomics.

It is one of the most popular supplements available in the market. EGb 761 (*Gingko biloba* leaf extract available commercially) was examined by Watanabe *et al.* for neuromodulatory effects on gene expression in mouse cortex and hippocampus. Mice were supplemented with this extract for 30 days followed by isolation of RNA from cortex and hippocampi separately and subjected to affymetrix gene-chip array. Forty-three genes from cortex and thirteen genes from hippocampus were differentially expressed (Gohil 2002). In 2002, Gohil *et al.* reported that almost 150 genes were differentially regulated in human bladder carcinoma cells with effect to the same leaf extract treatment, which represented a variety of functional groups, including some that might play an important role in activating the antioxidant defense system. Recent advances in DNA microarray and toxicogenomics has led to many toxicogenomics studies, and there exist some symbols of success.

TOXICOGENOMICS AS A PREDICTIVE TOOL

Toxicogenomics should not be only assumed to replace traditional toxicology, rather it should be more successfully employed to predict and understand the molecular reactions that underlie the adverse reaction along with classic parameters. The most important challenge in applying predictive toxicogenomics is to identify certain gene expression changes correlated and predictive of toxicological reaction. The first step toward achieving this is by establishing toxicogenomic databases of gene expression profiles (Yang, Blomme *et al.* 2004; Yang, Abel *et al.* 2006), that is, these databases should contain the gene expression profiles of many known dietary nutrients, pharmaceutical agents, toxicants, and control compounds at multiple doses and time points in different experimental models (Guerreiro, Staedtler *et al.* 2003; Waring, Cavet *et al.* 2003; Karpinets, Foy *et al.* 2004; Sawada, Takami *et al.* 2005). The main goal of establishing these databases is that it allows us to identify gene expression patterns, which are associated with a known toxic mechanism or pathological outcome in many cases. Several studies have reported that toxicogenomics could be a powerful tool for predicting and identifying toxicity of any dietary compound at an early time point. In spite of some limitations to this approach, the toxicogenomic approach could help pharmaceutical and nutraceutical companies save both time and resources by allowing them to either eliminate or deprioritize drug candidates with unacceptable safety margins much earlier in the drug discovery process.

CONCLUSION

Use of gene expression data in a risk assessment of dietary nutrients/antioxidants revolves around multiple and multifactorial measurements. In this regard, the future lies not only with the development of the technology but with the storage, management, and interpretation of the vast quantity of gene expression data. Differential gene expression approaches with the use of microarrays is emerging with proven success in drug safety evaluations for both predictive and mechanistic toxicology applications and will help researchers and clinicians make use of readily available information in the database along with expediting the process of drug discovery and development. Further, toxicity and efficacy measures can be conducted concurrently, as the same genes and gene products are involved in pharmacological and toxicological results, which will result in comprehensive, integrated assessments of products with intended beneficial properties such as pharmaceuticals and dietary nutrients. Hence, the future of this field depends on understanding genetics, cellular signal transduction, nutritional aspects, and toxicological responses with respect to compounds and to fuse these disciplines to provide a more holistic understanding of the effects of dietary nutrients.

ACKNOWLEDGMENT

This study was supported by grants HL-56803 and HL-69910. I would also like to thank T. Mahesh, Ph.D, for his help during chapter preparation.

REFERENCES

Aardema, M. J., MacGregor, J. T., Toxicology and genetic toxicology in the new era of "toxicogenomics": impact of "-omics" technologies. *Mutat Res* 2002, *499*(1), 13–25.

Abdelwahed, A., Bouhlel, I., *et al.,* Study of antimutagenic and antioxidant activities of gallic acid and 1,2,3,4,6-pentagalloylglucose from Pistacia lentiscus. Confirmation by microarray expression profiling. *Chem Biol Interact* 2007, *165*(1), 1–13.

Abid, M. R., Tsai, J. C., *et al.,* Vascular endothelial growth factor induces manganese-superoxide dismutase expression in endothelial cells by a Rac1-regulated NADPH oxidase-dependent mechanism. *FASEB J* 2001, *15*(13), 2548–2550.

Addya, S., Shiroto, K., *et al.,* Ischemic preconditioning-mediated cardioprotection is disrupted in heterozygous Flt-1 (VEGFR-1) knockout mice. *J Mol Cell Cardiol* 2005, *38*(2), 345–351.

Afman, L., Muller, M., Nutrigenomics: from molecular nutrition to prevention of disease. *J Am Diet Assoc* 2006, *106*(4), 569–576.

Alon, U., Barkai, N., *et al.,* Broad patterns of gene expression revealed by clustering analysis of tumor and normal colon tissues probed by oligonucleotide arrays. *Proc Natl Acad Sci USA* 1999, *96*(12), 6745–6750.

Amiri, K. I., Richmond, A., Role of nuclear factor-kappa B in melanoma. *Cancer Metastasis Rev* 2005, *24*(2), 301–313.

Anagnostopoulou, M. A., Kefalas, P., *et al.*, Analysis of antioxidant compounds in sweet orange peel by HPLC-diode array detection-electrospray ionization mass spectrometry. *Biomed Chromatogr* 2005, *19*(2), 138–148.

Angel, P., Karin, M., The role of Jun, Fos and the AP-1 complex in cell-proliferation and transformation. *Biochim Biophys Acta* 1991, *1072*(2-3), 129–157.

Baker, J. E., Felix, C. C., *et al.*, Myocardial ischemia and reperfusion: direct evidence for free radical generation by electron spin resonance spectroscopy. *Proc Natl Acad Sci U S A* 1988, *85*(8), 2786–2789.

Barella, L., Muller, P. Y., *et al.*, Identification of hepatic molecular mechanisms of action of alpha-tocopherol using global gene expression profile analysis in rats. *Biochim Biophys Acta* 2004, *1689*(1), 66–74.

Ben-Dor, A., Shamir, R., *et al.*, Clustering gene expression patterns. *J Comput Biol* 1999, *6*(3–4): 281–297.

Bolstad, B. M., Irizarry, R. A., *et al.*, A comparison of normalization methods for high density oligonucleotide array data based on variance and bias. *Bioinformatics* 2003, *19*(2), 185–193.

Boscoboinik, D., Szewczyk, A., *et al.*, Alpha-tocopherol (vitamin E) regulates vascular smooth muscle cell proliferation and protein kinase C activity. *Arch Biochem Biophys* 1991, *286*(1), 264–269.

Boyne, A. F., Ellman, G. L., A methodology for analysis of tissue sulfhydryl components. *Anal Biochem* 1972, *46*(2), 639–653.

Brazma, A., Vilo, J., Gene expression data analysis. *FEBS Lett* 2000, *480*(1), 17–24.

Canales, R. D., Luo, Y., *et al.*, Evaluation of DNA microarray results with quantitative gene expression platforms. *Nat Biotechnol* 2006, *24*(9), 1115–1122.

Cano, E., Mahadevan, L. C., Parallel signal processing among mammalian MAPKs. *Trends Biochem Sci* 1995, *20*(3), 117–122.

Carlisle, A. J., Prabhu, V. V., *et al.*, Development of a prostate cDNA microarray and statistical gene expression analysis package. *Mol Carcinog* 2000, *28*(1), 12–22.

Carluccio, M. A., Siculella, L., *et al.*, Olive oil and red wine antioxidant polyphenols inhibit endothelial activation: antiatherogenic properties of Mediterranean diet phytochemicals. *Arterioscler Thromb Vasc Biol* 2003, *23*(4), 622–629.

Carr, K. M., Bittner, M., *et al.*, Gene-expression profiling in human cutaneous melanoma. *Oncogene* 2003, *22*(20), 3076–3080.

Caruso, D., Berra, B., *et al.*, Effect of virgin olive oil phenolic compounds on in vitro oxidation of human low density lipoproteins. *Nutr Metab Cardiovasc Dis* 1999, *9*(3), 102–107.

Cash, T. P., Pan, Y., *et al.*, Reactive oxygen species and cellular oxygen sensing. *Free Radic Biol Med* 2007, *43*(9), 1219–1225.

Chalabi, N., Delort, L., *et al.*, Gene signature of breast cancer cell lines treated with lycopene. *Pharmacogenomics* 2006, *7*(5), 663–672.

Christen, Y., Olano-Martin, E., *et al.*, Egb 761 in the postgenomic era: new tools from molecular biology for the study of complex products such as Ginkgo biloba extract. *Cell Mol Biol (Noisy-le-grand)* 2002, *48*(6), 593–599.

Cobb, M. H., Goldsmith, E. J., How MAP kinases are regulated. *J Biol Chem* 1995, *270*(25), 14843–14846.

Coni, E., Di Benedetto, R., *et al.*, Protective effect of oleuropein, an olive oil biophenol, on low density lipoprotein oxidizability in rabbits. *Lipids* 2000, *35*(1), 45–54.

Coombes, K. R., Highsmith, W. E., *et al.*, Identifying and quantifying sources of variation in microarray data using high-density cDNA membrane arrays. *J Comput Biol* 2002, *9*(4), 655–669.

Davis, P. A., Polagruto, J. A., *et al.*, Effect of apple extracts on NF-kappaB activation in human umbilical vein endothelial cells. *Exp Biol Med (Maywood)* 2006, *231*(5), 594–598.

de la Puerta, R., Martinez-Dominguez, E., *et al.*, Effect of minor components of virgin olive oil on topical antiinflammatory assays. *Z Naturforsch [C]* 2000, *55*(9–10), 814–819.

de Sousa, R. R., Queiroz, K. C., *et al.*, Phosphoprotein levels, MAPK activities and NFkappaB expression are affected by fisetin. *J Enzyme Inhib Med Chem* 2007, *22*(4), 439–444.

Diebolt, M., Bucher, B., *et al.*, Wine polyphenols decrease blood pressure, improve NO vasodilatation, and induce gene expression. *Hypertension* 2001, *38*(2), 159–165.

Diplock, A. T., Charleux, J. L., *et al.*, Functional food science and defence against reactive oxidative species. *Br J Nutr* 1998, *80*(Suppl 1), S77–112.

Dizdaroglu, M., Jaruga, P., *et al.*, Free radical-induced damage to DNA: mechanisms and measurement. *Free Radic Biol Med* 2002, *32*(11), 1102–1115.

Dolara, P., Luceri, C., *et al.*, Red wine polyphenols influence carcinogenesis, intestinal microflora, oxidative damage and gene expression profiles of colonic mucosa in F344 rats. *Mutat Res* 2005, *591*(12), 237–246.

Eaton, D. L., Hamel, M., Increase in gamma-glutamylcysteine synthetase activity as a mechanism for butylated hydroxyanisole-mediated elevation of hepatic glutathione. *Toxicol Appl Pharmacol* 1994, *126*(1), 145–149.

Eisen, M. B., Spellman, P. T., *et al.*, Cluster analysis and display of genome-wide expression patterns. *Proc Natl Acad Sci U S A* 1998, *95*(25), 14863–14868.

Elliott, R., Ong, T. J., Nutritional genomics. *BMJ* 2002, *324*(7351), 1438–1442.

Forbes, J. M., Coughlan, M. T., *et al.*, Oxidative stress as a major culprit in kidney disease in diabetes. *Diabetes* 2008, *57*(6), 1446–1454.

Forstermann, U., Oxidative stress in vascular disease: causes, defense mechanisms and potential therapies. *Nat Clin Pract Cardiovasc Med* 2008, *5*(6), 338–349.

Funk, J. L., Frye, J. B., *et al.*, Efficacy and mechanism of action of turmeric supplements in the treatment of experimental arthritis. *Arthritis Rheum* 2006, *54*(11), 3452–3464.

Galecka, E., Jacewicz, R., *et al.*, [Antioxidative enzymes–structure, properties, functions]. *Pol Merkur Lekarski* 2008, *25*(147), 266–268.

Galitski, T., Molecular networks in model systems. *Annu Rev Genomics Hum Genet* 2004, *5*, 177–187.

Gatzidou, E. T., Theocharis, S. E., Toxicogenomics: *A Powerful Tool for Toxicity Assessment*. John Wiley and Sons, 2008.

Gebhart, E., Antimutagens. Data and problems. *Humangenetik* 1974, *24*(1), 1–32.

Gillies, P. J., Nutrigenomics: the Rubicon of molecular nutrition. *J Am Diet Assoc* 2003, *103*(12 Suppl 2), S50–5.

Gohil, K., Genomic responses to herbal extracts: lessons from in vitro and in vivo studies with an extract of Ginkgo biloba. *Biochem Pharmacol* 2002, *64*(56), 913–917.

Golkar, L., Ding, X. Z., *et al.*, Resveratrol inhibits pancreatic cancer cell proliferation through transcriptional induction of macrophage inhibitory cytokine-1. *J Surg Res* 2007, *138*(2), 163–169.

Gopalakrishnan, A., Tony Kong, A. N., Anticarcinogenesis by dietary phytochemicals: cytoprotection by Nrf2 in normal cells and cytotoxicity by modulation of transcription factors NF-kappa B and AP-1 in abnormal cancer cells. *Food Chem Toxicol* 2008, *46*(4), 1257–1270.

Groten, J. P., Butler, W., *et al.*,. An analysis of the possibility for health implications of joint actions and interactions between food additives. *Regul Toxicol Pharmacol* 2000, *31*(1), 77–91.

Guerreiro, N., Staedtler, F., *et al.*, Toxicogenomics in drug development. *Toxicol Pathol* 2003, *31*(5), 471–479.

Guo, S., Yang, S., *et al.*, Green tea polyphenol epigallocatechin-3 gallate (EGCG) affects gene expression of breast cancer cells transformed by the carcinogen 7,12-dimethylbenz[a]anthracene. *J Nutr* 2005, *135*(12 Suppl), 2978S–2986S.

Halliwell, B., Whiteman, M., Measuring reactive species and oxidative damage in vivo and in cell culture: how should you do it and what do the results mean? *Br J Pharmacol* 2004, *142*(2), 231–255.

Hanbur, D., On the Febrifuge properties of the Olive (Olea europa, L.). *Pharmaceut J Provincial Trans* 1984, 353–354.

Hardiman, G., Microarray platforms–comparisons and contrasts. *Pharmacogenomics* 2004, *5*(5), 487–502.

Hebbar, V., Shen, G., et al., Toxicogenomics of resveratrol in rat liver. *Life Sci* 2005, *76*(20), 2299–2314.

Heller, R. A., Schena, M., et al., Discovery and analysis of inflammatory disease-related genes using cDNA microarrays. *Proc Natl Acad Sci U S A* 1997, *94*(6), 2150–2155.

Hess, K. R., Zhang, W., et al., Microarrays: handling the deluge of data and extracting reliable information. *Trends Biotechnol* 2001, *19*(11), 463–468.

Hishikawa, K., Nakaki, T., et al., Oral flavonoid supplementation attenuates atherosclerosis development in apolipoprotein E-deficient mice. *Arterioscler Thromb Vasc Biol* 2005, *25*(2), 442–446.

Hofmann, T., Liegibel, U., et al., Intervention with polyphenol-rich fruit juices results in an elevation of glutathione S-transferase P1 (hGSTP1) protein expression in human leucocytes of healthy volunteers. *Mol Nutr Food Res* 2006, *50*(12), 1191–1200.

Huber, W. W., Scharf, G., et al., The coffee components kahweol and cafestol induce gamma-glutamylcysteine synthetase, the rate limiting enzyme of chemoprotective glutathione synthesis, in several organs of the rat. *Arch Toxicol* 2002, *75*(11–12), 685–694.

Inadera, H., Uchida, M., et al., [Advances in "omics" technologies for toxicological research]. *Nippon Eiseigaku Zasshi* 2007, *62*(1), 18–31.

Inoguchi, T., Takayanagi, R., [Role of oxidative stress in diabetic vascular complications]. *Fukuoka Igaku Zasshi* 2008, *99*(3), 47–55.

Ip, C., Comparative effects of antioxidants on enzymes involved in glutathione metabolism. *Life Sci* 1984, *34*(25), 2501–2506.

Irwin, R. D., Boorman, G. A., et al., Application of toxicogenomics to toxicology: basic concepts in the analysis of microarray data. *Toxicol Pathol* 2004, *32*(Suppl), 72–83.

Jison, M. L., Munson, P. J., et al., Blood mononuclear cell gene expression profiles characterize the oxidant, hemolytic, and inflammatory stress of sickle cell disease. *Blood* 2004, *104*(1), 270–280.

Ju, Z., Wells, M. C., et al., DNA microarray technology in toxicogenomics of aquatic models: methods and applications. *Comp Biochem Physiol C Toxicol Pharmacol* 2007, *145*(1), 5–14.

Kaga, S., Zhan, L., et al., Resveratrol enhances neovascularization in the infarcted rat myocardium through the induction of thioredoxin-1, heme oxygenase-1 and vascular endothelial growth factor. *J Mol Cell Cardiol* 2005, *39*(5), 813–822.

Kaput, J., Rodriguez, R. L., Nutritional genomics: the next frontier in the postgenomic era. *Physiol Genomics* 2004, *16*(2), 166–177.

Karpinets, T. V., Foy, B. D., et al., Tailored gene array databases: applications in mechanistic toxicology. *Bioinformatics* 2004, *20*(4), 507–517.

Kim, J. M., Chang, H. J., et al., Structure-activity relationship of neuroprotective and reactive oxygen species scavenging activities for allium organosulfur compounds. *J Agric Food Chem* 2006, *54*(18), 6547–6553.

Kim, S., Sohn, I., et al., Hepatic gene expression profiles are altered by genistein supplementation in mice with diet-induced obesity. *J Nutr* 2005, *135*(1), 33–41.

Kim, Y. H., Lim, D. S., et al., Gene expression profiling of oxidative stress on atrial fibrillation in humans. *Exp Mol Med* 2003, *35*(5), 336–349.

King, A. A., Shaughnessy, D. T., et al., Antimutagenicity of cinnamaldehyde and vanillin in human cells: Global gene expression and possible role of DNA damage and repair. *Mutat Res* 2007, *616*(1–2), 60–69.

Knasmuller, S., Nersesyan, A., et al., Use of conventional and -omics based methods for health claims of dietary antioxidants: a critical overview. *Br J Nutr* 2008, *99*(E Suppl 1), ES3–ES52.

Knoers, N. V., Monnens, L. A., Teaching molecular genetics: Chapter 1–Background principles and methods of molecular biology. *Pediatr Nephrol* 2006, *21*(2), 169–176.

Knowles, L. M., Milner, J. A., Diallyl disulfide induces ERK phosphorylation and alters gene expression profiles in human colon tumor cells. *J Nutr* 2003, *133*(9), 2901–2906.

Kundu, J. K., Surh, Y. J., Molecular basis of chemoprevention by resveratrol: NF-kappaB and AP-1 as potential targets. *Mutat Res* 2004, *555*(1–2), 65–80.

Kundu, J. K., Surh, Y. J., Breaking the relay in deregulated cellular signal transduction as a rationale for chemoprevention with anti-inflammatory phytochemicals. *Mutat Res* 2005, *591*(1–2), 123–146.

Kundu, J. K., Surh, Y. J., Epigallocatechin gallate inhibits phorbol ester-induced activation of NF-kappa B and CREB in mouse skin: role of p38 MAPK. *Ann N Y Acad Sci* 2007, *1095*, 504–512.

Kuo, W. P., Jenssen, T. K., *et al.*, Analysis of matched mRNA measurements from two different microarray technologies. *Bioinformatics* 2002, *18*(3), 405–412.

Kussmann, M., Raymond, F., *et al.*, OMICS-driven biomarker discovery in nutrition and health. *J Biotechnol* 2006, *124*(4), 758–787.

Lee, K. M., Kang, N. J., *et al.*, Myricetin down-regulates phorbol ester-induced cyclooxygenase-2 expression in mouse epidermal cells by blocking activation of nuclear factor kappa B. *J Agric Food Chem* 2007, *55*(23), 9678–9684.

Lee, M. L., Kuo, F. C., *et al.*, Importance of replication in microarray gene expression studies: statistical methods and evidence from repetitive cDNA hybridizations. *Proc Natl Acad Sci U S A* 2000, *97*(18), 9834–9839.

Lee-Huang, S., Zhang, L., *et al.,* Anti-HIV activity of olive leaf extract (OLE) and modulation of host cell gene expression by HIV-1 infection and OLE treatment. *Biochem Biophys Res Commun* 2003, *307*(4), 1029–1037.

Lennon, G. G., Lehrach, H., Hybridization analyses of arrayed cDNA libraries. *Trends Genet* 1991, *7*(10), 314–317.

Lettieri, T., Recent applications of DNA microarray technology to toxicology and ecotoxicology. *Environ Health Perspect* 2006, *114*(1), 4–9.

Levine, M., Conry-Cantilena, C., *et al.*, Vitamin C pharmacokinetics in healthy volunteers: evidence for a recommended dietary allowance. *Proc Natl Acad Sci U S A* 1996, *93*(8), 3704–3709.

Li, L., Tsao, R., *et al.*, Polyphenolic profiles and antioxidant activities of heartnut (Juglans ailanthifolia Var. cordiformis) and Persian walnut (Juglans regia L.). *J Agric Food Chem* 2006, *54*(21), 8033–8040.

Lopez-Velez, M., Martinez-Martinez, F., *et al.*, The study of phenolic compounds as natural antioxidants in wine. *Crit Rev Food Sci Nutr* 2003, *43*(3), 233–244.

Loscalzo, J., Membrane redox state and apoptosis: death by peroxide. *Cell Metab* 2008, *8*(3), 182–183.

Luhe, A., Suter, L., *et al.*, Toxicogenomics in the pharmaceutical industry: hollow promises or real benefit? *Mutat Res* 2005, *575*(1–2), 102–115.

Lunec, J., Halligan, E., *et al.*, Effect of vitamin E on gene expression changes in diet-related carcinogenesis. *Ann N Y Acad Sci* 2004, *1031*, 169–183.

Madsen-Bouterse, S. A., Kowluru, R. A., Oxidative stress and diabetic retinopathy: pathophysiological mechanisms and treatment perspectives. *Rev Endocr Metab Disord* 2008, *9*(4), 315–327.

Majewicz, J., Rimbach, G., *et al.*, Dietary vitamin C down-regulates inflammatory gene expression in apoE4 smokers. *Biochem Biophys Res Commun* 2005, *338*(2), 951–955.

Mangels, A. R., Holden, J. M., *et al.*, Carotenoid content of fruits and vegetables: an evaluation of analytic data. *J Am Diet Assoc* 1993, *93*(3), 284–296.

Masuda, Y., Kikuzaki, H., *et al.*, Antioxidant properties of gingerol related compounds from ginger. *Biofactors* 2004, *21*(1–4), 293–296.

Mathers, J., Fraser, J. A., *et al.*, Antioxidant and cytoprotective responses to redox stress. *Biochem Soc Symp* 2004, *71*, 157–176.

Maulik, N., Reactive oxygen species drives myocardial angiogenesis? *Antioxid Redox Signal* 2006, *8*(11–12), 2161–2168.

Maulik, N., Das, D. K., Redox signaling in vascular angiogenesis. *Free Radic Biol Med* 2002, *33*(8), 1047–1060.

Maulik, N., Sato, M., *et al.*, An essential role of NFkappaB in tyrosine kinase signaling of p38 MAP kinase regulation of myocardial adaptation to ischemia. *FEBS Lett* 1998, *429*(3), 365–369.

Michel, F., Bonnefont-Rousselot, D., *et al.*, [Biomarkers of lipid peroxidation: analytical aspects]. *Ann Biol Clin (Paris)* 2008, *66*(6), 605–620.

Morgan, K. T., Ni, H., *et al.*, Application of cDNA microarray technology to in vitro toxicology and the selection of genes for a real-time RT-PCR-based screen for oxidative stress in Hep-G2 cells. *Toxicol Pathol* 2002, *30*(4), 435–451.

Muller, M., Kersten, S., Nutrigenomics: goals and strategies. *Nat Rev Genet* 2003, *4*(4), 315–322.

Murray, J., Oquendo, C. E., *et al.*, Monitoring oxidative and nitrative modification of cellular proteins; a paradigm for identifying key disease related markers of oxidative stress. *Adv Drug Deliv Rev* 2008, *60*(13–14), 1497–1503.

Murray, J. I., Whitfield, M. L., *et al.*, Diverse and specific gene expression responses to stresses in cultured human cells. *Mol Biol Cell* 2004, *15*(5), 2361–2374.

Murray, R. F., Jr., Genetic variation and dietary response. *World Rev Nutr Diet* 2001, *89*, 5–11.

Murtaza, I., Marra, G., *et al.*, A preliminary investigation demonstrating the effect of quercetin on the expression of genes related to cell-cycle arrest, apoptosis and xenobiotic metabolism in human CO115 colon-adenocarcinoma cells using DNA microarray. *Biotechnol Appl Biochem* 2006, *45*(Pt 1), 29–36.

Mutch, D. M., Wahli, W., *et al.*, Nutrigenomics and nutrigenetics: the emerging faces of nutrition. *FASEB J* 2005, *19*(12), 1602–1616.

Nair, H. K., Rao, K. V., *et al.*, Inhibition of prostate cancer cell colony formation by the flavonoid quercetin correlates with modulation of specific regulatory genes. *Clin Diagn Lab Immunol* 2004, *11*(1), 63–69.

Nelson, S. K., Bose, S. K., *et al.*, The induction of human superoxide dismutase and catalase in vivo: a fundamentally new approach to antioxidant therapy. *Free Radic Biol Med* 2006, *40*(2), 341–347.

Nicholson, S. K., Tucker, G. A., *et al.*, Effects of dietary polyphenols on gene expression in human vascular endothelial cells. *Proc Nutr Soc* 2008, *67*(1), 42–47.

Nuwaysir, E. F., Bittner, M., *et al.*, Microarrays and toxicology: the advent of toxicogenomics. *Mol Carcinog* 1999, *24*(3), 153–159.

Oberemm, A., Onyon, L., *et al.*, How can toxicogenomics inform risk assessment? *Toxicol Appl Pharmacol* 2005, *207*(2 Suppl), 592–598.

Ommen, B., Groten, J. P., Nutrigenomics in efficacy and safety evaluation of food components. *World Rev Nutr Diet* 93, 134–152.

Ordovas, J. M., Mooser, V., Nutrigenomics and nutrigenetics. *Curr Opin Lipidol* 15(2), 101–108.

Ovesna, J., Slaby, O., *et al.*, High throughput 'omics' approaches to assess the effects of phytochemicals in human health studies. *Br J Nutr* 2008, *99*(E Suppl 1), ES127–ES134.

Palozza, P., Krinsky, N. I., Antioxidant effects of carotenoids in vivo and in vitro: an overview. *Methods Enzymol* 1992, *213*, 403–420.

Parker, R. S., Dietary and biochemical aspects of vitamin E. *Adv Food Nutr Res* 1989, *33*, 157–232.

Peterson, L. E., CLUSFAVOR 5.0: hierarchical cluster and principal-component analysis of microarray-based transcriptional profiles. *Genome Biol* 2002, *3*(7), SOFTWARE0002.

Petroni, A., Blasevich, M., *et al.*, Inhibition of platelet aggregation and eicosanoid production by phenolic components of olive oil. *Thromb Res* 1995, *78*(2), 151–160.

Pinkus, R., Weiner, L. M., *et al.*, Role of oxidants and antioxidants in the induction of AP-1, NF-kappaB, and glutathione S-transferase gene expression. *J Biol Chem* 1996, *271*(23), 13422–13429.

Powolny, A., Takahashi, K., *et al.*, Induction of GADD gene expression by phenethylisothiocyanate in human colon adenocarcinoma cells. *J Cell Biochem* 2003, *90*(6), 1128–1139.

Pratico, D., Reilly, M., *et al.*, Novel indices of oxidant stress in cardiovascular disease: specific analysis of F2-isoprostanes. *Agents Actions Suppl* 1997, *48*, 25–41.

Prior, R. L., Cao, G., In vivo total antioxidant capacity: comparison of different analytical methods. *Free Radic Biol Med* 1999, *27*(11–12), 1173–1181.

Quackenbush, J., Microarray data normalization and transformation. *Nat Genet* 2002, *32*(Suppl), 496–501.

Rahman, I., Biswas, S. K., Non-invasive biomarkers of oxidative stress: reproducibility and methodological issues. *Redox Rep* 2004, *9*(3), 125–143.

Rahman, I., Kelly, F., Biomarkers in breath condensate: a promising new non-invasive technique in free radical research. *Free Radic Res* 2003, *37*(12), 1253–1266.

Raychaudhuri, S., Stuart, J. M., *et al.*, Principal components analysis to summarize microarray experiments: application to sporulation time series. *Pac Symp Biocomput* 2000, 455–466.

Rebrin, I., Sohal, R. S., Pro-oxidant shift in glutathione redox state during aging. *Adv Drug Deliv Rev* 2008, *60*(13–14), 1545–1552.

Rice-Evans, C. A., Miller, N. J., Antioxidant activities of flavonoids as bioactive components of food. *Biochem Soc Trans* 1996, *24*(3), 790–795.

Rimbach, G., Minihane, A. M., *et al.*, Regulation of cell signalling by vitamin E. *Proc Nutr Soc* 2002, *61*(4), 415–425.

Ringner, M., Peterson, C., Microarray-based cancer diagnosis with artificial neural networks. *Biotechniques* 2003, Suppl, 30–35.

Ringner, M., Peterson, C., *et al.*, Analyzing array data using supervised methods. *Pharmacogenomics* 2002, *3*(3), 403–415.

Rock, C. L., Jacob, R. A., *et al.*, Update on the biological characteristics of the antioxidant micronutrients: vitamin C, vitamin E, and the carotenoids. *J Am Diet Assoc* 1996, *96*(7), 693–702; quiz 703–704.

Rota, C., Barella, L., *et al.*, Dietary alpha-tocopherol affects differential gene expression in rat testes. *IUBMB Life* 2004, *56*(5), 277–280.

Sarma, L., Kesavan, P. C., Protective effects of vitamins C and E against gamma-ray-induced chromosomal damage in mouse. *Int J Radiat Biol* 1993, *63*(6), 759–764.

Sawada, H., Takami, K., *et al.*, A toxicogenomic approach to drug-induced phospholipidosis: analysis of its induction mechanism and establishment of a novel in vitro screening system. *Toxicol Sci* 2005, *83*(2), 282–292.

Schadt, E. E., Li, C., *et al.*, Analyzing high-density oligonucleotide gene expression array data. *J Cell Biochem* 2000, *80*(2), 192–202.

Schadt, E. E., Li, C., *et al.*, Feature extraction and normalization algorithms for high-density oligonucleotide gene expression array data. *J Cell Biochem Suppl* 2001, Suppl 37, 120–125.

Schena, M., Shalon, D., *et al.*, Quantitative monitoring of gene expression patterns with a complementary DNA microarray. *Science* 1995, *270*(5235), 467–470.

Scherf, U., Ross, D. T., *et al.*, A gene expression database for the molecular pharmacology of cancer. *Nat Genet* 2000, *24*(3), 236–244.

Schreck, R., Meier, B., *et al.*, Dithiocarbamates as potent inhibitors of nuclear factor kappa B activation in intact cells. *J Exp Med* 1992, *175*(5), 1181–1194.

Schreck, R., Rieber, P., *et al.*, Reactive oxygen intermediates as apparently widely used messengers in the activation of the NF-kappa B transcription factor and HIV-1. *EMBO J* 1991, *10*(8), 2247–2258.

Selman, C., McLaren, J. S., *et al.*, Life-long vitamin C supplementation in combination with cold exposure does not affect oxidative damage or lifespan in mice, but decreases expression of antioxidant protection genes. *Mech Ageing Dev* 2006, *127*(12), 897–904.

Shen, G., Jeong, W. S., *et al.*, Regulation of Nrf2, NF-kappaB, and AP-1 signaling pathways by chemopreventive agents. *Antioxid Redox Signal* 2005, *7*(11–12), 1648–1663.

Shen, G., Xu, C., *et al.*, Comparison of (-)-epigallocatechin-3-gallate elicited liver and small intestine gene expression profiles between C57BL/6J mice and C57BL/6J/Nrf2 (-/-) mice. *Pharm Res* 2005, *22*(11), 1805–1820.

Shioda, T., Application of DNA microarray to toxicological research. *J Environ Pathol Toxicol Oncol* 2004, *23*(1), 13–31.

Simopoulos, A. P., Genetic variation and nutrition. *Biomed Environ Sci* 1996, *9*(2–3), 124–129.

Simopoulos, A. P., Genetic variation and nutrition. *World Rev Nutr Diet* 1999, *84*, 118–140.

Smith, C. V., Anderson, R. E., Methods for determination of lipid peroxidation in biological samples. *Free Radic Biol Med* 1987, *3*(5), 341–344.

Somogyi, A., Rosta, K., *et al.,* Antioxidant measurements. *Physiol Meas* 2007, *28*(4), R41–R55

Somparn, P., Phisalaphong, C., *et al.*, Comparative antioxidant activities of curcumin and its demethoxy and hydrogenated derivatives. *Biol Pharm Bull* 2007, *30*(1), 74–78.

Stierum, R., Heijne, W., *et al.*, Toxicogenomics concepts and applications to study hepatic effects of food additives and chemicals. *Toxicol Appl Pharmacol* 2005, *207*(2 Suppl), 179–188.

Surh, Y. J., NF-kappa B and Nrf2 as potential chemopreventive targets of some anti-inflammatory and antioxidative phytonutrients with anti-inflammatory and antioxidative activities. *Asia Pac J Clin Nutr* 2008, *17*(Suppl 1), 269–272.

Surh, Y. J., Chun, K. S., *et al.*, Molecular mechanisms underlying chemopreventive activities of anti-inflammatory phytochemicals: down-regulation of COX-2 and iNOS through suppression of NF-kappa B activation. *Mutat Res* 2001, *480–481*, 243–268.

Surh, Y. J., Han, S. S., *et al.*, Inhibitory effects of curcumin and capsaicin on phorbol ester-induced activation of eukaryotic transcription factors, NF-kappaB and AP-1. *Biofactors* 2000, *12*(1–4), 107–112.

Surh, Y. J., Kundu, J. K., *et al.*, Redox-sensitive transcription factors as prime targets for chemoprevention with anti-inflammatory and antioxidative phytochemicals. *J Nutr* 2005, *135*(12 Suppl), 2993S–3001S.

Suter, L., Babiss, L. E., *et al.*, Toxicogenomics in predictive toxicology in drug development. *Chem Biol* 2004, *11*(2), 161–171.

Suzuki, K., Koike, H., *et al.*, Genistein, a soy isoflavone, induces glutathione peroxidase in the human prostate cancer cell lines LNCaP and PC-3. *Int J Cancer* 2002, *99*(6), 846–852.

Swanson, K. S., Using genomic biology to study liver metabolism. *J Anim Physiol Anim Nutr (Berl)* 2008, *92*(3), 246–252.

Swartz, H. M., Using EPR to measure a critical but often unmeasured component of oxidative damage: oxygen. *Antioxid Redox Signal* 2004, *6*(3), 677–686.

Takahashi, Y., Lavigne, J. A., *et al.*, Using DNA microarray analyses to elucidate the effects of genistein in androgen-responsive prostate cancer cells: identification of novel targets. *Mol Carcinog* 2004, *41*(2), 108–119.

Tamayo, P., Slonim, D., *et al.*, Interpreting patterns of gene expression with self-organizing maps: methods and application to hematopoietic differentiation. *Proc Natl Acad Sci U S A* 1999, *96*(6), 2907–2912.

Thirunavukkarasu, M., Addya, S., *et al.*, Heterozygous disruption of Flk-1 receptor leads to myocardial ischaemia reperfusion injury in mice: application of affymetrix gene chip analysis. *J Cell Mol Med* 2008, *12*(4), 1284–1302.

Thirunavukkarasu, M., Juhasz, B., *et al.,* VEGFR1 (Flt-1+/-) gene knockout leads to the disruption of VEGF-mediated signaling through the nitric oxide/heme oxygenase pathway in ischemic preconditioned myocardium. *Free Radic Biol Med* 2007, *42*(10), 1487–1495.

Thirunavukkarasu, M., Penumathsa, S. V*., et al.*, Resveratrol alleviates cardiac dysfunction in streptozotocin-induced diabetes: Role of nitric oxide, thioredoxin, and heme oxygenase. *Free Radic Biol Med* 2007, *43*(5), 720–729.

Trayhurn, P., Proteomics and nutrition–a science for the first decade of the new millennium. *Br J Nutr* 2000, *83*(1), 1–2.

Tsimikas, S., In vivo markers of oxidative stress and therapeutic interventions. *Am J Cardiol* 2008, *101*(10A), 34D–42D.

Tusher, V. G., Tibshirani, R., *et al.*, Significance analysis of microarrays applied to the ionizing radiation response. *Proc Natl Acad Sci U S A* 2001, *98*(9), 5116–6121.

Ulrich, R., Friend, S. H., Toxicogenomics and drug discovery: will new technologies help us produce better drugs? *Nat Rev Drug Discov* 2002, *1*(1), 84–88.

van Erk, M. J., Roepman, P., *et al.*, Integrated assessment by multiple gene expression analysis of quercetin bioactivity on anticancer-related mechanisms in colon cancer cells in vitro. *Eur J Nutr* 2005, *44*(3), 143–156.

Veeriah, S., Kautenburger, T., *et al.*, Apple flavonoids inhibit growth of HT29 human colon cancer cells and modulate expression of genes involved in the biotransformation of xenobiotics. *Mol Carcinog* 2006, *45*(3), 164–174.

Villamena, F. A., Zweier, J. L., Detection of reactive oxygen and nitrogen species by EPR spin trapping. *Antioxid Redox Signal* 2004, *6*(3), 619–629.

Visioli, F., Galli, C., Oleuropein protects low density lipoprotein from oxidation. *Life Sci* 1994, *55*(24), 1965–1971.

Vrana, K. E., Freeman, W. M., *et al.*, Use of microarray technologies in toxicology research. *Neurotoxicology* 2003, *24*(3), 321–332.

Waring, J. F., Cavet, G., *et al.*, Development of a DNA microarray for toxicology based on hepatotoxin-regulated sequences. *EHP Toxicogenomics* 2003, *111*(1T), 53–60.

Weeraratna, A. T., Nagel, J. E., *et al.*, Gene expression profiling: from microarrays to medicine. *J Clin Immunol* 2004, *24*(3), 213–224.

Weinreb, O., Mandel, S., *et al.*, cDNA gene expression profile homology of antioxidants and their antiapoptotic and proapoptotic activities in human neuroblastoma cells. *FASEB J* 2003, *17*(8), 935–937.

Weinreb, O., Mandel, S., *et al.*, Gene and protein expression profiles of anti- and pro-apoptotic actions of dopamine, R-apomorphine, green tea polyphenol (-)-epigallocatechine-3-gallate, and melatonin. *Ann N Y Acad Sci* 2003, *993*, 351–361; discussion 387–393.

Wolfram, S., Raederstorff, D., *et al.*, Epigallocatechin gallate supplementation alleviates diabetes in rodents. *J Nutr* 2006, *136*(10), 2512–2518.

Yamazaki, T., Seko, Y., *et al.*, Expression of intercellular adhesion molecule-1 in rat heart with ischemia/reperfusion and limitation of infarct size by treatment with antibodies against cell adhesion molecules. *Am J Pathol* 1993, *143*(2), 410–418.

Yang, S. H., Kim, J. S., *et al.*, Genome-scale analysis of resveratrol-induced gene expression profile in human ovarian cancer cells using a cDNA microarray. *Int J Oncol* 2003, *22*(4), 741–750.

Yang, Y., Abel, S. J., *et al.*, Development of a toxicogenomics in vitro assay for the efficient characterization of compounds. *Pharmacogenomics* 2006, *7*(2), 177–186.

Yang, Y., Blomme, E. A., *et al.*, Toxicogenomics in drug discovery: from preclinical studies to clinical trials. *Chem Biol Interact* 2004, *150*(1), 71–85.

Yang, Y. H., Dudoit, S., *et al.*, Normalization for cDNA microarray data: a robust composite method addressing single and multiple slide systematic variation. *Nucleic Acids Res* 2002, *30*(4), e15.

Yoneda, K., Chang, M. M., *et al.*, Application of high-density DNA microarray to study smoke- and hydrogen peroxide-induced injury and repair in human bronchial epithelial cells. *J Am Soc Nephrol* 2003, *14*(8 Suppl 3), S284–289.

Yoon, H., Liu, R. H., Effect of selected phytochemicals and apple extracts on NF-kappaB activation in human breast cancer MCF-7 cells. *J Agric Food Chem* 2007, *55*(8), 3167–3173.

Yoshihara, T., Ishigaki, S., *et al.*, Differential expression of inflammation- and apoptosis-related genes in spinal cords of a mutant SOD1 transgenic mouse model of familial amyotrophic lateral sclerosis. *J Neurochem* 2002, *80*(1), 158–167.

Zhang, L., Wang, L., *et al.*, A new algorithm for analysis of oligonucleotide arrays: application to expression profiling in mouse brain regions. *J Mol Biol* 2002, *317*(2), 225–235.

Zhou, Y., Abagyan, R., Match-only integral distribution (MOID) algorithm for high-density oligonucleotide array analysis. *BMC Bioinformatics* 2002, *3*, 3.

Zhou, Z., Tan, H. L., *et al.*, Microarray analysis of altered gene expression in diallyl trisulfide-treated HepG2 cells. *Pharmacol Rep* 2005, *57*(6), 818–823.

Zimmermann, J. S., Kimmig, B., Pharmacological management of acute radiation morbidity. *Strahlenther Onkol* 1998, *174*(Suppl 3), 62–65.

10

Genomics Applied to Nutrients and Functional Foods in Japan: State of the Art

Yuji Nakai, Akihito Yasuoka, Hisanori Kato, and Keiko Abe

INTRODUCTION

FOOD AS A HETEROGENEOUS SYSTEM

Diet is essentially heterogeneous. Foods contain many different constituents that coexist. This means that once ingested, their metabolites could interact with one another in the body to exhibit different physiological activities at the same time. These physiological activities may work synergistically or competitively. There also exists cultural, environmental, and genetic heterogeneity among people as consumers of food.

To study the diversity of food, nutritional scientists in the past adopted reductive analytical approaches. For example, many vitamins have been discovered by following a process of using as indicators those symptoms thought to result from a shortage of certain food components, to identify components that improve the symptoms. With respect to human trials, researchers have strived to reduce the bias of sampling from the population, but they have not reached the level of discussing the bias itself or individual differences. Even so, it can be said that these approaches were highly effective in improving the health of the members of a malnourished society.

It is estimated that among the world population of six billion, 800 million face starvation and two billion suffer micronutrient deficiencies. The approaches discussed above are helpful in improving the health of these 2.8 billion people. Putting to one side the debate over the ethical issues of the geopolitical nonuniformity of nutritional supply, it is not necessarily true that the remaining half of the world's population is healthy. Excessive or imbalanced nutritional consumption gives rise to metabolic abnormalities and other new diseases. Quite a few people are in a state of partial sickness, occupying the gray area on the border of health and illness. To answer the question of how to deal with these physiological phenomena that are phenotypically unclear and that arise from multiple causes, the projects led by Japan's Ministry of Education, Science and Culture (MESC), entitled *Systematic analysis and development of food functionalities* (1984–87), *Analysis of body-modulating functions of foods* (1988–91), and *Analysis and molecular design of functional food* (1992–95), respectively, proposed a very Oriental solution and guidelines [1]. It represents a revival of the principle that medicine and diet share a common origin. These projects are primarily characterized by the fact that they sought solutions to problems occurring in the diet from the diet itself. As a result, they triggered worldwide recognition of the concept of functional food. The second feature of the projects is that, although they were inspired by a classical notion, they developed the methodology into a multifaceted one encompassing the most modern molecular biology and its sophisticated technology. Nutrigenomics is now understood as omics consisting primarily of transcriptomics, proteomics, and metabolomics usable in the field of nutritional and functional food science. Construction of a database based on

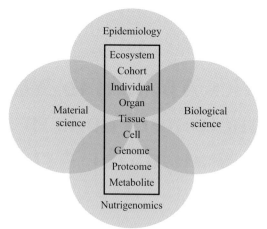

Figure 10.1 Integrated methodologies and their hieratic research targets in nutritional and functional food science.

up-to-date scientific information is essential if one is to analyze the multilateral data obtained, which includes those characteristic individuals. The project entitled *Analysis and systematization (database construction) of nonnutritive functional components in foods* (2000–05) under the initiative of Japan Science and Technology Agency (JST) has sped up the contribution of information sciences to this academic discipline. Contemporary functional food science is now in the course of transformation into a discipline based on nutrigenomics and with a focus on data sampled from the perspectives of material science, biology and epidemiology (Fig. 10.1) [2].

FOODS THAT HAVE A VARIETY OF COMPONENTS

The food functionality projects running under the aegis of the Ministry of Education, Science and Culture (MESC) proposed the concept of functional food. In response to this, many different products designated as *food for specified health use* (FOSHU) were approved by The Ministry of Health, Labour and Welfare (MHLW), which allowed some functional foods to carry labels that make health claims. There are 820 such products in total and the number is continuing to grow. They are roughly divided into eleven health claim categories including two set out in consideration of the overlapping of functions (Table 10.1) [3].

FOSHU is often explained as containing nonnutritive food functional components as opposed to traditional basic nutritional elements such as sugar, fat, minerals, vitamins, and essential amino acids. However, many of their effects are produced through the dynamic behavior of other basic nutritional factors. Food is essentially a heterogeneous system. FOSHU is the outcome of the efforts to abstract, from among the natural direct and indirect interactions between food constituents, those that promote the maintenance of good health. For example, casein phosphopeptide promotes calcium absorption [4] whereas polyphenol contained in oolong tea inhibits the digestion of fat [5]. Antioxidant substances act as scavengers for active oxygen that is generated incidentally through the metabolism of different nutritious substances. On the other hand, there are also many nutrients that exhibit unique physiological effects. Some enzymes and G protein-coupled receptors (GPCRs) have been identified as the targets for various functional peptides derived from food proteins [6]. Some plant-derived terpenoids, lignans, and polyphenols activate the metabolism of fat and sugar, exhibit estrogenic activity, and perform integrated control of metabolisms including detoxification. It is very likely that these functions are exerted through nuclear receptors [7].

Table 10.1 Eleven categories of health claims in foods for specified health use and their functional food components approved.*

Category 1: Modulation of gastrointestinal conditions

1 Oligosaccharides: lactosucrose [28]; galacto-oligo saccharides [11]; coffee bean manno-oligosaccharides including mannobiose [10]; fructo-oligosaccharides [9]; soybean oligosaccharides [6]; xylo-oligosaccharides [4]; isomalto-oligosaccharides [4]; lactulose [1]; raffinose [1]

2 Lactic acid bacteria: *Lactobacillus casei* Shirota [27]; *L. acidophilus* CK92 and *L. helveticus* CK60 [7]; *L. delbrueekii* subsp. *bulgaricus*2038 and *Streptococcus salivaricus* subsp. *thermophilus*1131 [6]; Bifidobacterium Bb-12 [6]; *B. breve* Yakurt [6]; *B. longum* BB536 [6]; *L. casei* SP and B. SP [2]; *B. lactis* LKM512 [2]; *B. lactis* FK120 [2]; *L. casei* 1 [2]; NY1301 [2]; L. GG [2]

3 Dietary fibers: indigestible dextran [225]; psyllium husks [50]; wheat bran [4]; gum guaic hydrolysate [3]; agar [3]; polydextrose [2]; low-molecular-weight sodium alginate [2]; low-molecular-weight sodium alginate and water-soluble corn fiber [1]; indigestible dextran and wheat bran [1]; dietary fiber from beer yeast [1]; indigestible dextran, reduced type [1]; and indigestible starch [1]

4 Other components: milk whey fermented by propionic acid bacterium [3]

5 Combination of chemically defined components: galacto-oligosaccharide-polydextrose [1]

Category 2: Modulation of serum cholesterol level

Chitosan [29]; soybean protein [23]; phospholipid-binding soybean peptides including Cys-Ser-Pro-His-Pro [6]; low-molecular-weight sodium alginate [6]; phytosterol [5]; phytosterol ester [3]; broccoli-cabbage peptide Ser-Met-Cys-Ser [1]; tea catechin [1]

Category 3: Modulation of serum cholesterol level and gastrointestinal conditions

Dietary fiber from psyllium husks [20]; low-molecular-weight sodium alginate [9]

Category 4: Modulation of blood pressure

Sardine peptide including Val-Tyr [39]; lactotripeptides Val-Pro-Pro and Ile-Pro-Pro [12]; dried bonito "katsuobushi" oligopeptides [9]; "wakame" seaweed peptides including Pro-Tyr-Val-Tyr and Ile-Tyr [4]; γ-aminobutyric acid [3]; casein dodecapeptide [3]; Ile-Tyr [3]; acetic acid [3]; sesame peptides including Leu-Val-Tyr [2]; "nori" seaweed oligopeptides including Ala-Lys-Tyr-Ser-Tyr [2]; "tochu" leaf glycoside [2]; royal jelly peptide including Val-Tyr, Ile-Tyr, and Ile-Val-Tyr [1]; "tochu" leaf extract [1]; and geniposide as a "tochu" leaf glycoside [1]

Category 5: Acceleration of mineral absorption

Casein phosphopeptides [3]; heme [3]; calcium citrate-malate [2]; fructo-oligosaccharide [1]

Category 6: Acceleration of mineral absorption and modulation of gastrointestinal conditions

Lactosucrose [2]; fructo-oligosaccharide [1]

Category 7: Promotion of bone health

Soybean isoflavones [13]; vitamin K as menaquinone-7 [7]; fructo-oligosaccharides [5]; menaquinone04 [1]; polyglutaminic acid [1]; milk basic protein [1]; and calcium [1]

Category 8: Maintenance of healthy teeth

Milk protein hydrolysate [25]; combined xylitol- maltitol-calcium monohydrogen phosphate-"fukuronori" seaweed furanone [22]; phosphorylated oligosaccharides, calcium salt [4]; combined xylitol-reduced palatinose-calcium monohydrogen phosphate-"fukuronori" seaweed furanone [3]; maltitol [2]; green tea fluorine [2]; combined xylitol-calcium monohydrogen phosphate-"fukuronori" seaweed furanone [1]; combined palatinose- green tea polyphenols [1]; combined maltitol-palatinose- green tea polyphenols [1]; and combined maltitol-hydrogenated palatinose- crythritol- green tea polyphenols [1]

(Continued)

Table 10.1 (*Continued*)

Category 9: Modulation of blood sugar level
Indigestible dextran [88]; wheat albumin [4]; bean extract [4]; l-arabinose [1]; and guava leaf polyphenols [1]
Category 10: Modulation of serum triacylglycerol level and blood fat percentage
Globin hydrolysate containing Val-Val-Tyr-Pro [9]; tea catechin [9]; diacylglycerol [6]; middle-chain fatty acid [5]; coffee bean manno-oligosaccharides including mannobiose [4]; oolong tea polymerized polyphenols including oolong homo-bis-flavan B [2]; combined eicosapentaenoic acid-docosahexaenoic acid [2]; and bean extract [1]; β-conglycinin [1]
Category 11: Modulation of serum triacylglycerol and cholesterol levels and of body fat percentage
Combined diacylglycerol-phytosterol (β-sitosterol) [4]

*Adapted from the Japan Health Food & Nutrition Food Association, homepage http://www.jhnfa.org/tokuho-f.html (in Japanese) by minor modification. The numbers of the products approved (19 November 2008) are parenthesized. Types: beverages (45%); fermented milk (13%); confectionery (12%); rice grains, noodles, and cereals (5%); cooking oil, vinegar, and margarine (5%); table sugar (4%); meat, fish, soybean, and their processed foods (3%); soy milk (2%); and others (7%).

Medication, under the principle that medicine and diet share a common origin, differs from the context of modern medical sciences in that the medication is to be applied as part of one's everyday diet. Food derives from other plants and animals, and it is certain that human physiological functions have evolved amid interactions with constituents that are derived from plants and animals. The physiologic activity of, for example, phytochemicals in our body is also the result of the interaction between ingested plant products and humans. FOSHU products include some prebiotic and probiotic food products, and part of their mutual interactions in human bodies have been identified. Mimicking may be a key concept in understanding functional components. It is seen that some peptides do not act as substrates for enzymes while inhibiting the activities of the enzymes, and that some flavonoids activate detoxification while exhibiting no toxicity [8]. For the future development of functional food products, it will be imperative to develop proper assay systems for the efficient detection of such substances and to customize recipes.

GENOMICS AS A MEANS OF QUALITY ASSESSMENT

Nutrigenomics refers to integrated "omics" studies that include comprehensive analyses, such as transcriptomics, proteomics, and metabolomics. Today, however, it is transcriptomics that plays the central part in nutrigenomics. There are two reasons. First, genomic study is targeted mostly at transcripts in the form of mRNA whose number is limited compared to expressed proteins and produced metabolites; the number of known transcripts is 20,000 or a little more. Moreover, RNAs have uniform chemical properties and are easy to handle. Second, as DNA microarray technologies have become widely used in genomic study, the analytical approach for DNA microarray data has been improved dramatically. The microarray analysis is therefore the mainstream of nutrigenomic studies. Although microarray has been established as a technology, the data themselves are inherently prone to noise. Thus, inappropriate choice of experimental design or data analysis method could lead to an incorrect or even erroneous conclusion [9].

In addition to the foregoing, there is a problem that derives from the point that nutrigenomics is the study of food, which is a diverse system comprising multiple components. The physiological reaction to food intake is far more complex than to a single-ingredient medication. Because food is ingested over an extended period on a daily basis, it is generally rare for the reaction

DNA microarray experiment

Preliminary experiment

Determination of assay point

Animal experiment
-Nutritional experiment
-Tissue dissection

Acquisition of raw microarray data
-RNA extraction
-Target preparation
-Hybridization to microarray, scan

DNA microarray data analysis

Raw microarray data

Quantification of expression data

Identification of differentially expressed genes

Gene-annotation enrichment analysis

Figure 10.2 A schematic representation of typical nutrigenomics study.

to take the form of a drastic change. Rather, it is often seen as a long-term accumulation of small changes. As is the case in conventional nutritional studies, model animals whose life cycles are shorter than those of humans are used, such as rats and mice. However, the study is susceptible to the environmental fluctuations, individual differences, and technical skills of researchers, as it aims to detect any minute change in reaction to complex input. Herein lies the difficulty associated with nutrigenomic studies. They require elaborate study plans, accurate experimental approaches and suitable data analysis techniques [10, 11]. Figure 10.2 represents a typical process flow for nutrigenomic studies.

Comprehensive analyses provide the advantage of leading not only conventional knowledge but also new findings that were missed out or could not be detected using conventional approaches. Although microarray analyses are now widely conducted, it is unfortunate that many studies merely follow past achievements and fail to capitalize on the advantage of comprehensive analyses. The full benefits of transcriptomics do not become evident unless the utmost effort is made both in experiment and for data analysis.

APPLICATION IN NUTRITIONAL SCIENCE

DIETARY PROTEIN

It is well known as a classic nutritional concept that both the amount and quality of any ingested protein need to be taken into consideration when its effects are discussed. Ensuring quality and quantity is essential for the maintenance of normal protein turnover, which is indispensable for homeostasis and for the net increase of protein synthesis accompanying growth, pregnancy, and lactation [12]. The quality of dietary protein is determined primarily by its composition of essential amino acid content. Meanwhile, the importance of information within dietary proteins other than their amino acid composition has been unveiled in many cases. Proteins themselves and their partially digested peptide sometimes exert specific physiological effects in the gut. Small peptides or even undigested proteins can be absorbed intact and then function to modulate a variety of target organs in the body [13].

Table 10.2 Highly up- or downregulated genes in the liver after feeding a protein-free diet (PF) or a 12% gluten diet (G) for 1 week [14].

	Number of Genes			
	PF		G	
Function of Genes	Up	Down	Up	Down
Growth factors	3	5	3	3
Receptors and signal transduction	4	25	3	3
Energy metabolism	5	10	3	6
Transport and binding proteins	6	19	2	6
Gene expression control	5	16	5	6
Stress responses	1	3	3	2
Cholesterol metabolism	0	11	15	0
Lipid metabolism	3	8	0	6
Metabolism of xenobiotics	6	5	2	2
Amino acid metabolism	7	13	3	0
Biologic oxidation	8	7	3	0
Inflammatory responses	0	2	0	0
Cell cycle	0	6	0	2
Cell structure	0	3	2	0
Ribosomal proteins	11	0	0	0
Unassigned	38	51	17	14
Total	97	184	61	50

Protein malnutrition causes a wide range of deleterious effects, although the mechanism involved has not fully been understood. To tackle this nutritionally important issue, we capitalized on DNA microarray technology to obtain a global view on the number of genes highly susceptible to protein nutrition [14]. In our first attempt, we fed rats with a protein-free diet (PF) or a wheat gluten diet (G) for 1 week, and the gene expression profiles of liver, muscle, and adipose tissue were compared with those obtained from rats fed on a casein diet (control). PF is an extreme situation of the shortage of protein supply, whereas G is a typical model of inappropriate quality because its contents of lysine and threonine are below requirement levels. A summary of the data is shown in Table 10.2, where genes that exhibited increase or decrease of twofold or more are grouped according to their functions. Observed changes included those of translation regulators and proliferation-related factors. In addition, ingestion of gluten diet caused upregulation of many enzyme genes for synthesis and catabolism of cholesterol. This diet also caused an increase in the fecal excretion of neutral and acidic sterol as well as a decrease in blood cholesterol. The results suggest that the excretion was accelerated by the function of gluten in the intestine, after which the effect of amino acid composition induced upregulation of CYP7A1 (rate-limiting enzyme of catabolism), then leading to hypocholesterolemic effect and resultant induction of synthesis pathway. In addition, the authors have recently obtained the gene expression profile of the skin of rats fed on similar diets, the result of which suggested the impaired antioxidative capacity of the skin (unpublished result). Another example of the ingestion of a plant-derived protein is provided by soy protein, which, as described later, exerts a beneficial effect on lipid metabolism when presented.

Although there are only a few transcriptome-based analyses on the effect of protein nutrition, it should be noted that Thalacker-Mercer et al. [15] have recently reported the effect of low protein intake on gene expression profile in human muscle.

UNBALANCE AND EXCESS OF AMINO ACIDS

The issue of amino acid unbalance has already been discussed in reference to the example of gluten diet. This section will be devoted to the detailed discussion on the availability of amino acids and excess intake of a single amino acid. The mechanism by which the deficiency of each essential amino acid affects the function of cells has not fully been revealed. Although the insufficient supply of amino acids as protein-building blocks naturally causes a delay in protein synthesis, other mechanisms mediating the signals of amino acid deficiency have been intensively studied. They include a pathway involving phosphorylation/dephosphorylation of translation regulators, especially of eIF4E binding protein-1 (4EBP-1). On the other hand, the 4EBP-1 gene was shown to be induced by protein malnutrition in the gluten-feeding experiment [14]. In addition, upregulation of the gene for IGFBP-1 consists of part of the mechanism of the decrease in protein synthesis and the resulting growth inhibition in inadequate amino acid supply. One of the effective measures to achieve total understanding of the amino acid signaling pathway would be realized by transcriptomic analyses using cell culture systems, as has been conducted by Peng *et al.* [16].

The revised requirement of each essential amino acid was recently released [12], and effort is being made to establish more precise values. The difficulty of reaching conclusive values stems from the lack of ideal methodology. Approaches through comprehensive gene expression analyses are likely to add precious information in this field.

Japan has a sound base of amino acid industry originating from the traditional technology of fermentation, which led to the promotion of the use of free amino acids as food ingredients. The accumulated knowledge on the functionality of each amino acid accelerated the prevalence of the purposeful intake of amino acids through supplemental products, a phenomenon also seen in many other countries. However, attention needs to be paid to the fact that a large excess of the intake of one or more amino acid causes various deleterious effects [17]. Recent efforts relating to the issue of the safety and toxicity of amino acids by the International Council on Amino Acid Science (http://www.icaas-org.com/index.php) is worth noting.

Matsuzaki et al. [18] have been using DNA microarray for the analyses of the mechanism of the toxic effects of amino acids such as leucine and cystine. In an experiment in which rats were given a 20% casein-based diet containing either 5% or 15% leucine, many enzyme genes for amino acid metabolism, carbohydrate metabolism, and lipid metabolism were upregulated in the liver. Figure 10.3 summarizes the upregulated genes associated with amino acid metabolism. It seems likely that the liver coops the toxic effects of excessive amino acids by upregulating the enzymes relating to the catabolism of amino acids.

Relatively few studies have been conducted so far for the effects of dietary proteins and amino acids on global gene expression, and much should be unveiled about these fundamental nutrients.

MINERAL DEFICIENCY

Great attention is paid to dietary minerals as well as to vitamins and related micronutrients of food origin. Magnesium (Mg), a cofactor of 300 enzymes in the body, is involved in a variety of biochemical processes and has been gaining special interest. Dietary Mg deficiency often induces physiological and even pathophysiological disorders. A number of papers have been presented reporting effects of Mg deficiency on carbohydrate, lipid, vitamin, and mineral metabolism. However, no information is available regarding a comprehensive aspect on a Mg deficiency-nutrient metabolism relationship that has been studied under the well-controlled, same experimental conditions. In the present study, we performed transcriptome analysis to comprehensively understand the effects of dietary Mg deficiency in rat liver.

Male Wistar rats aged 4 weeks were divided into two groups, and each was given a normal diet containing 0.5g Mg/kg (control group) or a similar but 0.04g Mg/kg-containing

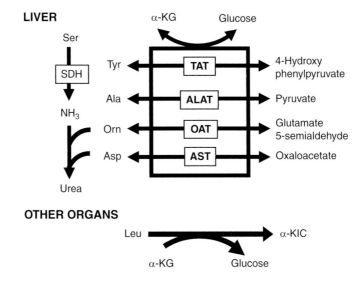

Figure 10.3 Upregulation of the genes involved in nitrogen metabolism in the liver by leucine load [18]. The framed enzymes exhibited more than a twofold increase in rats fed 5% or 15% leucine diet compared with control. Upregulation of SDH in the liver was consistent with the increase in plasma concentrations of ammonia, urea, and glutamine in rats fed excessive leucine. Enhanced expression of numbers of aminotransferases indicates a possible increase in glutamate-related transamination reactions in the liver. Symbols: SDH, serine dehydratase; TAT, tyrosine amino-transferase; ALAT, alanine aminotransferase; OAT, ornithine aminotransferase; ASAT, aspartate aminotransferase; α-KG, α-ketoglutarate; and α-KIC, α-ketoisocaproate.

diet (Mg-deficient group). The feeding was carried out under normal conditions for 28 days. During feeding, rats in both groups were pair fed to consume equal amounts of respective diets.

Clear phenomena resulting from Mg deficiency were observed for the Mg-deficient group [19]. Genomically, scatter plotting and Pearson's correlation coefficient analysis revealed a greater or lesser extent of gene expression differences between the control and Mg-deficient groups. In particular, 734 among 31,099 genes were significantly up- or downregulated by feeding with the Mg-deficient diet. These were classified into genes for the metabolism of carbohydrates, lipids, protein/amino acids, vitamins, and nucleic acids as well as those for xenobiotics detoxification, immune response, substance transport, cell proliferation, and even transcription.

In focusing on individual metabolic pathways, as for carbohydrate metabolism, genes involved in glycolysis and pentose phosphate shunt were upregulated, and those involved in gluconeogenesis and UDP-glucose metabolism were downregulated in the Mg-deficient group. Concerning lipid metabolism, genes involved in fatty acid biosynthesis and cholesterol biosynthesis were all upregulated in the Mg-deficient group, whereas those involved in fatty acid β-oxidation, acyl-CoA metabolism, and cholesterol catabolism (bile acid biosynthesis) were downregulated. Regarding protein metabolism, genes related to proteolysis and protein amino acid phosphorylation were upregulated, and those related to translation were downregulated in the Mg-deficient group. As for amino acid metabolism, genes involved in cysteine biosynthesis and glycine catabolism were upregulated, and those involved in S-adenosylmethionine biosynthesis, serine catabolism, and tRNA aminoacylation were downregulated in the Mg-deficient group. Also, changes in gene expressions for transcription factors regulating metabolic enzymes and transporters were found to take place.

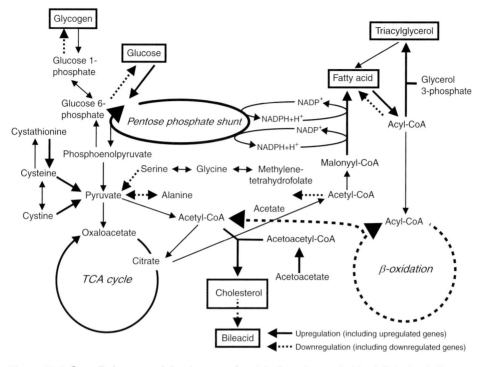

Figure 10.4 Overall view on activity changes of metabolic pathways in Mg-deficient rat's liver.

All these data suggest that dietary Mg deficiency induces a greater or lesser extent of the changes in the expression of various hepatic genes, especially in those genes involved in energy metabolism (Fig. 10.4).

FASTING

DNA microarray analysis was also made to observe the effect of fasting on rat gene expression [20]. Rats were divided into two groups — adequately fed and fasted groups. Rats in both groups were restrictedly fed for the first 7 days; they were given access to food for 6 hours a day from 10 am to 4 pm only. The two groups were separated at 10 am on day 8. The rats in the fed group were then fed in the same manner as before, whereas those in the fasted group were deprived of food. Then the white and brown adipose tissues and livers were excised from both groups at 4 pm. This approach was aimed at ensuring that the fasted group was in a suspended state of food gain for 24 hours and that the fed group was in a state of near-satiety, in order to maximize differences between the two groups. Total RNA was extracted from excised tissues, followed by the analysis of gene expression using the Affymetrix Rat Genome 230 2.0 GeneChip. Microarray analysis requires a data normalization process to make it possible to compare multiple data. There are many different methods available to normalize microarray data, but the results obtained from the analysis greatly vary depending on the normalization method. To compensate the deviation generated by the technical approach, four normalization approaches were adopted, Microarray Suite 5 (MAS5) [21], Robust Multi-array Average (RMA) [22], GLog Average (GLA) [23], and Model-Based Expression Index (MBEI) [24]. Applying these methods, genes that ranked high in all cases of the four normalization methods were selected. This approach is considered to have increased the likelihood of selecting robust gene set exhibiting expression changes.

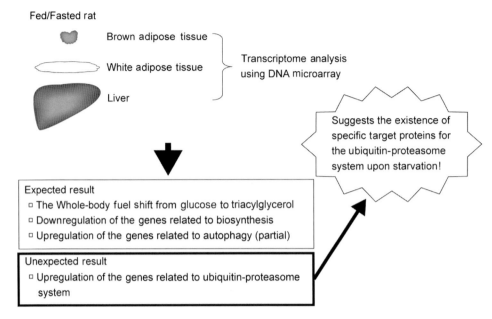

Figure 10.5 Summary of transcriptomic analysis in the brown and white adipose tissue, and the liver of fed/24-hour fasted rat.

Gene Ontology analysis conducted on the gene set confirmed that fasting leads to shifting from sugar to triacylglycerol as a source of energy. Induction of autophagy and downregulation of genes involved in biosynthesis, especially in that of proteins, were also observed. All these results were expected based on the findings of individual study in the past (Fig. 10.5).

It is known that fasting induces autophagy because intracellular organelles and proteins are used as sources of energy [25]. Degradation pathways of intracellular proteins are divided into two types — lysosomal (autophagy) pathways and ubiquitin-proteasome pathways. Generally, those in the first type are nonspecific protein degradation pathways, whereas those in the second type are specific. A recent study indicated that fasting increased the expression of TRB3 proteins in the white and brown adipose tissues of mice and accelerates the ubiquitination of acetyl-CoA carboxylase, a rate-limiting enzyme for fatty acid biosynthesis [26]. An increase in the expression of Trib3 (the rat ortholog of mouse TRB3) was actually observed in the fasted rats using the microarray analysis [20]. In addition, the expression of several types of E3 ubiquitin ligase increased in the fasting state. E3 ubiquitin ligase is responsible for substrate specificity in protein ubiquitination. Thus, there must be many different types of molecules to be ubiquitinated in the fasting state. It is interesting that protein degradation through ATP-dependent ubiquitin-proteasome pathways occurs in the time of fasting, even though energy supply is insufficient. Identification of target proteins is important for understanding the fasting response as well as energy homeostasis.

FOOD ALLERGY

Food allergy, also known as food hypersensitivity, is classified into two types. One is the immunoglobulin E (IgE)-dependent type and the other is the non-IgE-dependent type. In terms of the first type, research into clinical symptoms and the onset mechanism is relatively well advanced. Within the IgE-dependent category, immediate food allergy is the most common and may, in severe cases, develop into anaphylaxis leading to death. It is known that major allergen containing food, such as milk, eggs, wheat, seafood, soybeans, and peanuts can cause immediate-type allergy when ingested by immunologically hypersensitive people. What is currently regarded as most effective for food allergy sufferers is just to avoid the triggering foods. However, avoiding milk, eggs, wheat, and all other ingredients used in a very wide variety of food products causes considerable mental anguish and lowers quality of life (QOL). To resolve this problem, allergen-free foods have been studied as a category of functional food. Actually, hypoallergenic rice, wheat, and milk are being developed.

Many different proteins have so far been identified as allergens, but IgE epitopes for individual proteins vary from patient to patient, as does their activity with IgE. To elucidate the mechanism behind the onset of allergy and for designing hypoallergenic food products suited to individual patients, it is important to identify IgE epitopes of individual allergens. Based on this notion, numerous studies have been conducted on IgE epitopes for major allergens. To investigate the IgE epitopes, the enzyme-linked immunosorbent assay (ELISA) method using serums of food allergy patients and a method using peptide arrays spotted onto nitrocellulose membranes have been used conventionally. However, these approaches require enormous amounts of synthetic peptides and patient sera. Therefore, it was difficult to perform high-throughput analysis because of cost and labor constraints. These days, the protein or peptide microarray is emerging as a primary approach in epitope mapping. Schreffler et al.[27] prepared a peptide microarray by printing peptides on a glass slide with the epoxidized surface at intervals of 350 nanometers. The microarray was printed with a peptide set prepared by overlapping peptides consisting of 10, 15, and 20 amino acids covering all amino acid sequences of the major peanut allergen, Ara h 2, with an offset of two or three amino acid residues. A microarray immunoassay was performed using serums from forty-five patients and ten healthy subjects as samples (Fig. 10.6). The data analysis adopted hierarchical clustering similar to that for the DNA microarray. The results showed, as expected, an obvious cluster separation between the healthy subjects and patients. The patients were divided into several clusters from the signal pattern. Among the nine epitopes

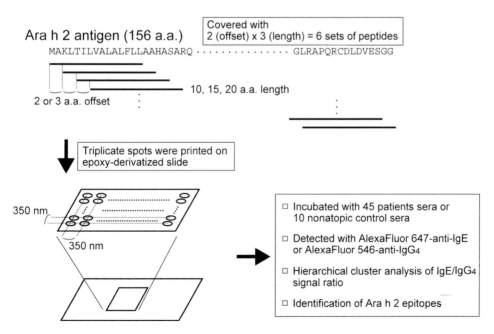

Figure 10.6 IgE and IgG4 epitope mapping of a major peanut allergen, Ara h 2, by high-resolution microarray immunoassay.

determined by the clustering of peptides, seven were identical to known epitopes obtained by the conventional method covering all sequences with nineteen peptides consisting of fifteen amino acid residues, whereas the two others have been identified as new epitopes. A similar analysis was conducted for allergens contained in milk to discover new epitopes. The peptide microarray technology is superior in throughput and in resolution to the existing methods and is expected to serve as a dominant technique for epitope mapping in the near future.

Isolation Stress

Daily exposure to social and psychological stress is associated with life style-related diseases such as hypertension, cardiovascular diseases, and cancer, as well as mental disorders [28]. It is required to clear the molecular mechanism underlying response to a social stressor for prevention of these diseases. Artificial stressors like immobilization, forced running, or electroshock were used in most animal studies [29]. However, these stress models do not mimic human psychosocial stress in our daily life. Hence, more mild and consecutive stress models are required.

Animal experiments were carried out with mice exposed to isolation stress, with the result that hypertrophy of the adrenal glands and increase of oxidative DNA damage in mouse peripheral blood were induced [30], and urinary excretion of biopyrrins as oxidative metabolites of bilirubin and serum levels of corticosterone increased [31]. It has also been shown that social isolation stress accelerated the development and growth of either transplanted or chemically induced tumors [32]. Little is known about tissue-specific gene expression in response to such social stressors. Therefore, changes in hepatic gene expression profile were evaluated for mice that were exposed to isolation stress for 30 days using a DNA microarray [33].

Male BALB/c mice (4 weeks old, Japan SLC, Shizuoka, Japan) were housed at five mice per cage. After acclimatization for 10 days, the mice were exposed to isolation stress (one mouse per cage). All cages were placed in a foamed polystyrene box to avoid social contact. To enhance the feeling of isolation, the bed volume in each cage for the isolated mice was reduced to about 2 g per cage. Hepatic RNA was extracted from each mouse and then subjected to DNA microarray analysis. The expression of 420 genes (after considering the false discovery rate) was altered in response to isolation stress for 30 days. Two hundred two genes were upregulated and 218 genes were downregulated. Gene Ontology analysis of these differentially expressed genes indicated that the lipid degradation pathway through peroxisome proliferator-activated receptor alpha-related genes were downregulated, whereas the lipid biosynthesis pathway controlled by sterol regulatory element binding factor 1, Golgi vesicle transport, and secretory pathway-related genes were significantly upregulated. These results were confirmed by real-time quantitative PCR. Nine selected genes involved in the lipid biosynthetic process (*Fasn, Elovl6,* and *Srebf1*), fatty acid metabolism (*Cyp4a10* and *Ppara*), fatty acid beta-oxidation (*Acox1* and *Ehhadh*), the pyruvate metabolic process (*Pdk4*), and the glucose metabolic process (*Igfbp1*) showed same expression as observed in the microarray experiments (Fig. 10.7). Though more increase of the body weight gain was not observed in mice exposed to isolation stress for 30 days compared with control mice, these results suggest that isolation stress may induce and/or exacerbate obesity.

We investigated whether isolation for more than 1 month with mild and consecutive stress induces the obesity in mice given high fat diet or normal diet. The body and fat tissue weight gain were observed in mice isolated for 2–3 months and high fat diet-enhanced obesity. Expression levels of hepatic genes related to lipid metabolism and plasma level of cytokines related to obesity in mice are under investigation.

APPLICATIONS IN FUNCTIONAL FOOD SCIENCE

Soy Protein Isolate

Soy protein isolate (SPI) is a common food ingredient. It is widely consumed in Japan in the form of fried or nonfried bean curd. The characteristics of SPI have been discussed not only

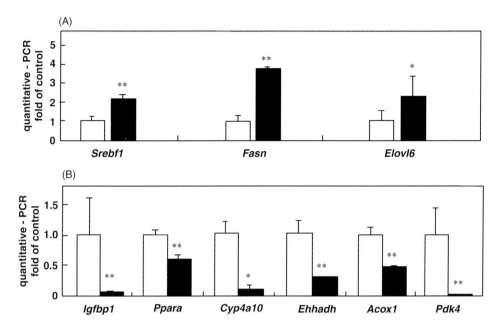

Figure 10.7 Real-time quantitative PCR analyses of social stress-related genes (*Srebf1, Fasn, Elovl6, Igfbp1, Ppara, Cyp4a10, Ehhadh, Acox1,* and *Pdk4*) in the mouse liver [33]. The expression level of each target PCR product was normalized to that of the housekeeping gene, *Gapdh*. (A) upregulated genes, (B) downregulated genes; open column: control, closed column: isolation stress for 30 days. The values represent mean ±SD ($n = 3$). $*P < 0.05$, $**P < 0.01$ compared with control mice.

in terms of nutritional functions relating to animal growth and development but also desirable physiological functions. Carroll and Hamilton [34] conducted a test on rabbits, reporting that vegetable proteins are superior to animal proteins in lowering blood cholesterol levels. Following the subsequent discovery of its antiarteriosclerosis effect, awareness of the benefits of SPI has increased. To identify the full effects of SPI intake on living bodies, especially on the liver, which plays the principal role in metabolic functions, a nutrigenomic analysis was performed using the Affymetrix DNA microarrays [35].

Sprague-Dawley rats were fed a diet containing 20% casein or SPI as crude protein for eight weeks. The SPI group showed lower blood lipid levels than was the case with the casein group. Samples of mRNA were separately prepared from the livers of three rats and examined on separate gene chips with 8740 genes. Among these, 2905 genes were confirmed as being expressed in every tested rat. A statistical analysis of these genes found significant difference from 115 genes, which made up nearly 1.3% of all genes. SPI ingestion was confirmed to considerably increase the expression of gene groups relating to steroid metabolism and antioxidation and to significantly reduce the expression in the gene group for the metabolism of fatty acids, making the gene group unique among the many different groups.

Next, a two-by-two factor analysis was performed with two types of rats whose age in weeks varied at the beginning of the test. Although the age difference produced significant differences in feed efficiency, the gene clustering analysis of exhaustive gene expression in livers has found that the major difference is caused by the difference in protein ingested rather than by the age difference. An analysis of metabolic functions of the 397 genes, the expression of which differed significantly between SPI intake and casein intake, was conducted. The results indicated that gene expression is distributed in a manner that transcends different metabolic systems. For example, the expression of a gene grouped into the PPAR-signaling pathway was significantly reduced by SPI intake, but that of Cyp7a1 alone increased significantly. Cyp7a1 is a rate-controlling enzyme for steroid synthesis. Short-term SPI intake analysis using DNA microarray has thus revealed that metabolic control takes place at the rate-determining step, although it does not occur in the case of long-term ingestion. In the future, DNA microarray analysis will further elucidate new functional peptides derived from SPI and crosstalk among different organs in the living body after SPI ingestion.

Sesamin

In 3000 B.C., sesame was already grown in the Nile basin and Hippocrates referred to sesame's tonic properties around 460 B.C. According to *The Divine Farmer's Herb-Root Classic*, an old Chinese pharmaceutical book compiled around A.D. 500, sesame falls under the category of high-level herbs that are not hazardous after one is exposed to them over the long term; it sustains five organs, increases vigor, enhances the health of the skin, and promotes the functions of the myelencephalon. Sesame oil forms an essential part of India's traditional medical treatment, known as Ayuruveda. However, it is only recently that research on its physiological activity got underway at the molecular level.

A sesame seed contains 50% fat, 20% protein, and 1.5% lignans by weight. A lignan is a constituent of lignin, a generic name for a compound resulting from two *p*-hydroxyphenylpropane molecules contained. Sesamin is one of the lignans with the largest content in sesame seeds. Like sesamolin (Fig. 10.8), it occurs at the concentration of approximately 0.5% [36]. These lignans are fat-soluble, whereas sesaminol (Fig. 10.8) exists at about 1% in sesame seeds in the form of a water-soluble glycoside [37]. Sesamin has an isomer called episesamin (Fig. 10.8), which is generated in the process of refining sesame oil. Common commercial products contain nearly the same amounts of sesamin and episesamin, although the content varies according to the degree of refinement.

To elucidate the physiologic functions of sesamin and related compounds, it is necessary to take into consideration their metabolism in the body. A test involving oral administration of sesamin by rats has detected at least four different sesamin metabolites in their bile within 24 hours [38]. They are sesamin catechol (SC1) and sesamin dicatechol (SC2), both of which are produced by the cleavage of the methylenedioxy group, as well as SC1m and SC2m, which are generated from SC1 and SC2 by methylation. Of these four substances, SC1, SC2, and SC2m exhibit a high level of antioxidant activity in 1,1-diphenyl-2-picrylhydrazyl (DPPH) assay (Fig. 10.8) similar to that of sesaminol. It is possible that an increase in antioxidant capacity following the production of these metabolites in the liver may be responsible for part of sesamin's physiological activity.

Sesamin is partially metabolized by enterobacteria into enterolactone and enterodiol (Fig. 10.8). In human serum, concentrations of both substances surge at least tenfold within 24 hours after the intake of sesame [39]. It is known that enterolactone and enterodiol have a low level of estrogenic activity. They are expected to provide the benefit of preventing and suppressing osteoporosis as well as inhibiting cancer.

The intake of sesamin promotes the metabolism of fat. This includes a decrease in the activity and gene expression of fatty acid synthase, L-type pyruvate kinase, and other fat-synthesizing enzymes and an increase in the activity and gene expression of palmitoyl-CoA oxidase carnitine, palmitoyltransferase, enoyl-CoA isomerase, and other beta-oxidation enzymes [40]. It is notable that the decline in gene expression of fat-synthesizing enzymes and the rise in gene expression of beta-oxidation enzymes, both of mitochondrial and peroxisomal type, take place at the same time. This provides a basis for considering the involvement of some integrated genetic control mechanism instead of a mere chemical effect such as an antioxidant property. Actually, it appears that peroxisome proliferator-activated receptors (PPARs) are involved. PPARs are nuclear receptors that receive fatty acids to control the metabolism of fat. Interestingly, this effect

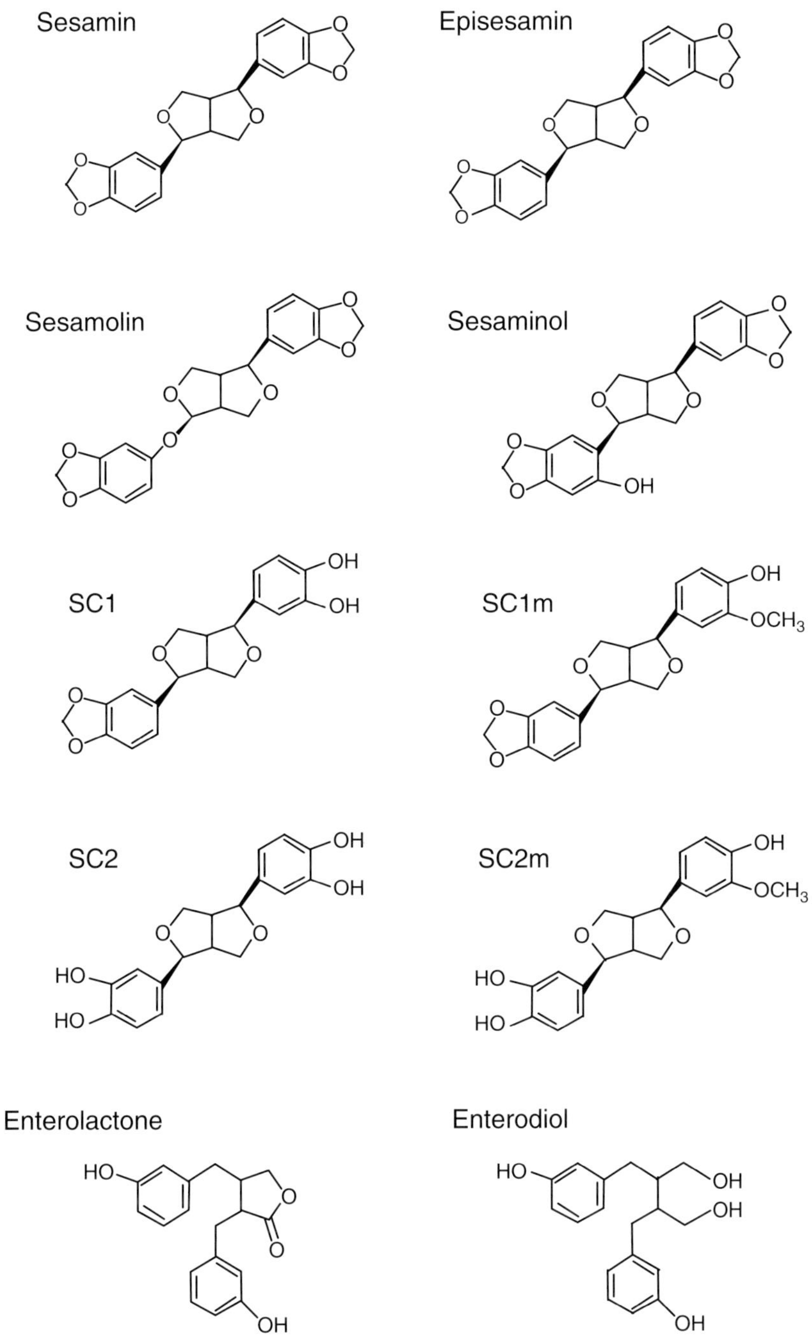

Figure 10.8 Sesamin and related compounds.

of promoting the metabolism of fat differs greatly between sesamin and its isomeric episesamin. On average, the activity of episesamin is about twice that of sesamin. The stereospecificity of the physiological effect will be a powerful tool in identifying sesamin's molecular target.

Sesamin apparently has a wide range of physiological activities, but it may be parsed as a precursor to substances that inhibit the oxidation of fats that sesame seeds contain in such large proportions. A similar reaction can take place in the human body as well. As a result, roughly two types of physiological actions may be produced as mentioned above. One is the simple physiological actions produced by its antioxidant activity. The other is the integrated physiological activities targeted at specific molecules, such as estrogenic activities and fat-consuming activities. Development of sesamin derivatives, especially those producing the second type of activities, will become important from the perspective of finding a practical application. It is also vital to pay attention to where sesamin is metabolized in the body. For instance, the deglycosylation of sesaminol glycosides by enterobacteria can be understood in the same manner as enterohepatic circulation in drug metabolism. It is a task for future research to elucidate the conversion into glycosides from the perspective of intestinal bacterial flora.

FRUCTOOLIGOSACCHARIDE

Fructooligosaccharide (FOS) is of particular interest in an irreducible saccharide. It is a combination of fructose residue of saccharose with one to three fructose molecules. Natural FOS is found in a wide variety of plants, including onions, sunchokes, and burdocks [41]. The method of industrially producing FOS at a high level of efficiency and in large quantities has been established as a technique of adding β-fructofuranosidase, an enzyme produced from *Aspergillus niger*, to the saccharose [42].

In an attempt to obtain comprehensive insight into the impact of FOS consumption as food on living bodies, we conducted an exhaustive gene expression analysis of the small intestinal mucosa tissue containing small intestinal Peyer's patches (PPs), which is believed to induce a gut immune response. The analysis observed significant changes in expression of 67 genes out of some 12,000 on the GeneChip® in the ileums of mice after FOS consumption. To estimate what biological functions are linked with the genes exhibiting expression changes in the ileums of mice after FOS consumption, the 67 genes selected were sorted by known biological functions on the basis of gene ontology. As a result, the largest group consisted of twenty-two genes related to immune responses, which made up 32.8% of all the genes, followed by nine genes associated with cell growth and maintenance (13.4%), six genes concerned with DNA/RNA processing (9.0%), five genes relating to structural formation (7.5%), another five genes linked with metabolism (7.5%), and four genes associated with signaling (6.0%) (Table 10.3). A close review of the genes related to immune responses observed a rise in expression of major histocompatibility complex (MHC) class I and II molecules as genes pertaining to antigen presentation. Among the immunity-related genes, nine genes concerned with humoral immune response formed the greatest group of genes exhibiting expression changes. The expression increased in most of them [43].

COCOA

With respect to the functionality of cocoa, it was found that mice consuming feed prepared by adding cocoa to normal diet gained less body weight than other mice consuming normal diet alone, with their fat weight reduced. Ingested cocoa is thus expected to have some effect in reducing obesity. However, little has been known about the mechanism of the effect and about what ingredients in cocoa are responsible for it. This section discusses the findings of the study conducted using DNA microarray technology. The result is summarized as: (a) in the liver where fatty acids are synthesized, the expression of fatty acid biosynthesis enzymes is inhibited; (b) the blood triacylglycerol level decreases; (c) in the white adipose tissue where fat is accumulated,

Table 10.3 Function-based classification of genes whose expressions in ileum were significantly influenced by administered FOS.

	Number of Genes	
Functions of gene	Upregulated	Downregulated
Immune response	15	7
Cell growth and/or maintenance	8	1
Structure	3	2
Metabolism	4	1
Signal transduction	3	1
DNA/RNA processing	6	0
EST or unknown	5	0
Others	10	1
Total	**54**	**13**

inhibition of expression of peroxisome proliferator-activated receptor (PPAR)-gamma results in inhibiting the expression of fatty acid transport factors; (d) as a result of the inhibited expression of sterol regulatory element-binding protein (SREBP)-1 in the white adipose tissue, expression of fatty acid biosynthesis enzymes is inhibited; and (e) expression of calorigenic factors is increased in the white adipose tissue [44].

As a result of these events, fat accumulation declines in the fatty tissue and the body weight increase is consequently inhibited as shown in Fig. 10.9.

ROYAL JELLY

Royal jelly is a milk-white viscous substance secreted from *Apis mellifera* worker bees and is a special diet given to queen bees and larvae that later grow into queen bees. It is rich in proteins, sugar, and vitamins, and has long been used as a health food ingredient. It has many different physiological functions. Given that it is especially effective in alleviating symptoms suffered by menopausal women, it is believed to contain a substance that has an effect similar to that of female hormones, and many different studies have been carried out [45–47].

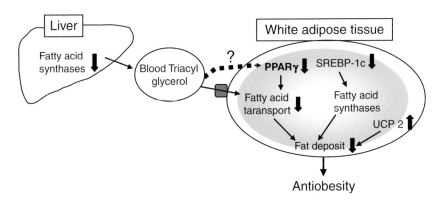

Figure 10.9 Possible antiobese effect elicited after ingestion of cocoa. PPAR-gamma: peroxisome proliferators-activated receptor-gamma, SREBP-1c: sterol regulatory element-1 binding protein-1c, UCP2: uncoupling protein-2.

Table 10.4 Numbers of genes showing more than 1.3-fold up- or downregulated genes.*

Regulation	Royal Jelly Group	E_2 Group	$E_2 \cap RJ$
Up	317	1,463	226
Down	68	519	26
Total	385	1,982	252

*A total of 19,738 genes investigated.

To examine whether royal jelly contains any substance that attaches itself to estrogen receptors, an investigation has been conducted into its binding activity to human estrogen receptors a and b. Although the binding activity of royal jelly is less than that of diethylstilbestrol, both receptor subtypes competed in inhibiting binding with 17 b estradiol (E_2). The estrogen-like substances were purified using the attachment to estrogen receptors beta as an indicator to identify the four active compounds in the ethanol soluble fraction of royal jelly. Specifically, they are 10-hydroxy-*trans*-2-decenoic acid, 10-hydroxydecanoic acid, *trans*-2-decenoic acid, and 24-methylenecholesterol. These substances accelerate the proliferation of estrogen-responsive human MCF-7 breast cancer cells, although this effect was canceled by adding tamoxifen, which is an inhibitory agent specific to estrogen receptors.

Estrogen not only acts as a female hormone to control the reproductive function but also exerts a wide range of biological regulation functions. With a focus on the relationship between estrogen and bone metabolism, the effect of royal jelly on bones was studied by performing an animal test, in which female accelerated senescence resistant (SAMR-1) mice aged 9 weeks were freely fed a diet containing 4% frozen dry royal jelly for a period of 9 weeks. This diet increased the tibial dry weight and ash weight, although the increment was smaller than in the group subject to subcutaneous administration of E_2. A transcriptome analysis of RNA prepared from the thigh bones of these mice was conducted using DNA microarrays. The number of genes exhibiting alteration was smaller in the royal jelly group than in the E_2 group. On the other hand, nearly 70% of the genes with increased expression in the royal jelly group exhibited increased expression in the E_2 group as well. This result suggests similarity between the two groups in the effect on bones (Table 10.4). These genes commonly showing change included those associated with bone formation, such as procollagen I-a1. They may be involved in the growth of bone mass caused by royal jelly. At the current stage, it is still unclear whether the estrogen-like substances mentioned above are responsible for the effect that royal jelly has on bones. However, it is possible that royal jelly as a whole produces an effect on bones similar to E_2 itself.

PROANTHOCYANIDIN

Proanthocyanidin is a condensed tannin with a flavanol skeleton. It occurs in grape seeds, astringent persimmons, pine bark, and many other plants. Having a powerful antioxidative effect, proanthocyanidin can exert powerful benefits in inhibiting arteriosclerosis, gastric ulceration, cataracts, and other diseases considered to be caused by an oxidative reaction in the body [48–52]. It also has effects of inhibiting formation of oxidized low-density lipoprotein (LDL), controlling inflammation, and improving other risk factors that form a basis for the metabolic syndrome [53,54]. The test detailed as follows was conducted with the objective of studying the impact of these diverse physiological activities that proanthocyanidin has on the pathological conditions of the metabolic syndrome.

For this test, two different types of test samples were prepared. One was an oolong tea sample with grape seed extract (GSE) containing 81.1% proanthocyanidin. The sample has a

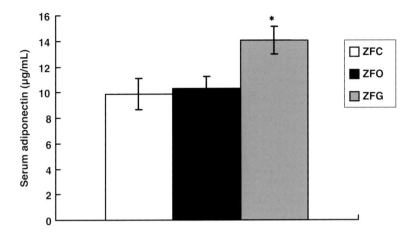

Figure 10.10 Serum adiponectin levels observed after feeding of different diets. Symbols: ZFC, fed on a solid diet and sterilized water; ZFO, fed on the same diet and GSE-free oolong tea; and ZFG, fed on the same diet and GSE-added oolong tea. $*: p = 0.025$ vs. ZFC, $p = 0.046$ vs. ZFO.

proanthocyanidin concentration of 1.07g/ml. The other is an oolong tea sample (control) with a proanthocyanidin concentration of 0.001g/ml.

Zucker fatty rats aged 5 weeks were used as obese rat models. The rats were divided into three groups. The first group was fed on a solid diet and sterilized water (ZFC), the second group on the same diet and GSE-free oolong tea (ZFO), and the third group on the same diet and GSE-added oolong tea (ZFG). Blood was extracted to prepare serums at the beginning of the test, at week 4, at week 8, at week 11, and at the end of the test in week 12.

Throughout the test period, no significant disparity was observed in body weight, diet intake, or fluid intake. A blood test found that the free fatty acid (FFA) concentration in the blood was on a strong downward trend in the ZFG group at week 8 at P value of 0.055 against the ZFC group and 0.075 against the ZFO group. Blood adiponectin level in the same group surged significantly at week 11 (P value of 0.025 against the ZFC group and 0.046 against the ZFO group). This time-dependent variation indicates that proanthocyanidin has the effect of lowering the level of FFA by inducing insulin resistance via a rise in the adiponectin level (Fig. 10.10). To closely investigate the mechanism behind this fall in the FFA level, a DNA microarray test was conducted using RNAs extracted from livers. It revealed a significant reduction in stearoyl-CoA desaturase-1 (SCD-1) mRNA expression in the ZFG group (P value of 0.043 against the ZFO group). SCD-1 is one of the genes that has in recent years been suggested to have a link with obesity. There is a report that a knockout of SCD-1 lowers the malonyl-CoA level in the liver and activates the metabolism of fatty acid [55]. In this test, the liver malonyl-CoA level was measured to observe a significant decline in the ZFG group (P value of 0.028 against the ZFC group and 0.013 against the ZFO group). All of the preceding results clearly indicate that proanthocyanidin has the effect of lowering the FFA level by reducing SCD-1 mRNA expression as well as an accompanying effect of stimulating adiponectin secretion.

APPLE POLYPHENOLS

Apple polyphenols are extracted from young apple fruit juice and fractionated by column chromatography. Procyanidin, an oligomeric or polymeric catechin, was a major constituent of the apple polyphenols [56].

The effect of the polyphenols on improving lipid metabolism was investigated by a clinical trial carried out in accordance with the objective of the Declaration of Helsinki. The trial was

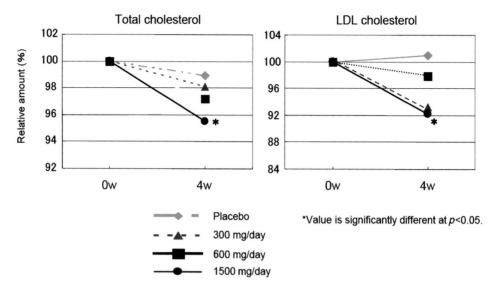

Figure 10.11 Changes in serum cholesterol levels by 4-week ingestion of apple polyphenol.

conducted using forty-eight adult volunteers with their serum total cholesterol level around the borderline ranging from 200 to 260 mg/dl. They consumed 0.3 g, 0.6 g, and 1.5 g of apple polyphenolmix in tablet form for 4 weeks, together with its placebo with no polyphenols. The trial confirmed a dose-dependent decline in the serum total cholesterol level of the apple polyphenol-consuming groups in comparison with the placebo group. The high-dose (1.5 g) group showed a significant difference (Fig. 10.11), whereas the consumption of apple polyphenols inhibits the absorption of neutral fats contained in food [57].

An animal test was performed to examine the mechanism of action and has observed that the fatty acid concentration in feces increases in a polyphenol dose-dependent manner and that the pancreatic lipase is inhibited. Thus, the test has also confirmed another mechanism in which the activity and gene expression of hepatic fatty acid synthetases diminish [58]. Apple polyphenols are observed to improve blood sugar levels and insulin levels and also to increase adiponectin and high-density lipoprotein (HDL) cholesterol.

Gene expression analysis was conducted using the nutrigenomic approach to study the effect of apple polyphenols. Interestingly, the group administered with apple polyphenol was similar in the overall gene expression pattern to the group fed on a 5% calorie restricted diet. The results suggest that the consumption of apple polyphenols leads to a vital reaction, whereas no such effect was expected from the consumption of the same calorie diet without apple polyphenols.

In addition, apple polyphenols produce antioxidant effects *in vivo*; they lower the peroxide fat level in serum and in the liver, maintain hepatic vitamin E concentration, and boost the activities of superoxide dismutase, catalase, and other bioantioxidant enzymes. Apple polyphenols are thus food material usable for preventing lifestyle-related diseases because they cause *in vivo* inhibition of the oxidation damage. With regard to safety, many different tests have been conducted, including those concerning acute toxicity, subchronic toxicity, and mutagenicity as well as a micronucleus test to verify that apple polyphenols pose no problem [59].

LYCOPENE

Among at least 750 carotenoid variants existing in nature, lycopene refers to a red carotenoid contained characteristically in tomatoes, whereas β-carotene has long been studied as a provi-tamin with a potential vitamin A activity. Lycopene was believed to be just a mere pigment.

However, it is capable of eliminating active oxygen having 100 times more antiactive oxygen activity than vitamin E, whereas β-carotene is just 50 times more, according to one report [60]. Studies on lycopene have been carried out in depth and it is now confirmed that it has a wide range of biological regulatory functions.

Because no carotenoid is biosynthesized in humans, humans cannot benefit from carotenoids without consuming the food containing them. Lycopene occurs in tomatoes and their processed products, a variety of carrot known under the name of *kintoki ninjin* in Japan, persimmons, and papayas. Among these, tomatoes and their processed derivatives make the greatest contribution to lycopene ingestion. It is reported that the absorbability of lycopene varies depending on how the food is processed or consumed and that the absorbability is increased by applying heat to the food [61]. Lycopene is distributed to serum and various organs after it is absorbed in the body; it accumulates in high concentrations in the liver and the spermaries, among other organs [62].

Many different epidemiological survey results reported indicate correlations between the ingestion of lycopene (from tomatoes) and cancer prevention, particularly with respect to prostate cancer [63]. In the United States, the Food and Drug Administration (FDA) restricts the product labeling it authorizes for processed tomato products stating that they may be effective in the prevention of prostate cancer.

In addition, it is suggested that lycopene has various other functions, including an effect of preventing arteriosclerosis by inhibiting the oxidization of low-density lipoprotein cholesterol, the effect of preventing osteoporosis, a function to improve male sterility, and the effect of inhibiting ultraviolet damage on the skin. According to a recent paper, it has been elucidated that the consumption of tomato juice containing lycopene inhibits pulmonary emphysema induced by smoking (Fig. 10.12) [64].

Lycopene is a naturally derived pigment that offers many benefits in preventing many different illnesses, but other functions are being unveiled, including effects on the respiratory system.

NEOCULIN

Neoculin is a taste-modifying protein that is isolated from the tropical fruit of *Curculigo latifolia* [65]. It is characterized by the ability not only to produce a sweet taste of its own, but also to

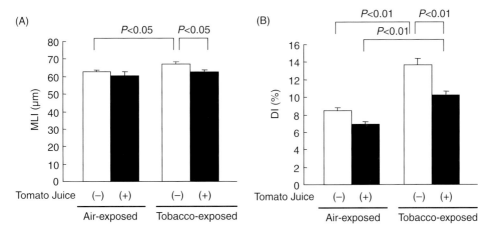

Figure 10.12 Morphometric and histological findings of the lung tissues of SAMP1 after exposure to air or tobacco smoke with or without administration of tomato juice [64]. A: MLI. Data were expressed as means $\pm$ SE($n = 6$ in each group). Open and filled bars indicate samples without or with administration of tomato juice, respectively. B: DI. Data are presented in the same manner as A ($n = 6$ in each strain).

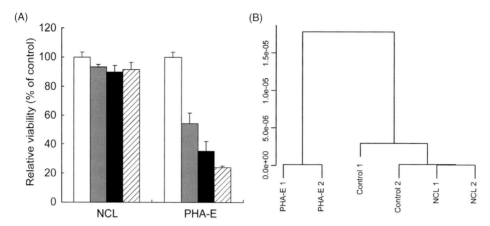

Figure 10.13 Correlation between viability and gene expression profile of Caco-2 cells. A, viability of differentiated Caco-2 cells treated with various concentrations of neoculin or PHA-E for 24 hours. B, hierarchical cluster dendrogram of microarray data obtained with lectin-treated or untreated Caco-2 cells. Differentiated Caco-2 cells were treated with neoculin (NCL) or PHA-E at 0 mg/mL (Control) or 0.03 mg/mL for 24 hours. After treatment, total RNA was extracted and subjected to DNA microarray assay. The vertical scale corresponds to the distance between clusters.

change sourness into sweetness. One expected application is to use it as a low-calorie, nonglycemic sweetener. Chemically, neoculin forms a heterodimer consisting of an N-glycosylated acidic subunit (neoculin acidic subunit, NAS) and a nonglycosylated basic subunit (neoculin basic subunit, NBS). It is similar in crystal structure to mannose-binding lectins derived from monocots, such as snowdrop and garlic lectin [66]. Without heat treatment, they exhibit deleterious physiological functions, especially toxicity to small intestine. With regard to safety to be ensured in using neoculin in a food additive, the correlations between the cytotoxicity and the gene expression profile was examined using Caco-2 cell, a cell line derived from human colonic adenocarcinoma, as a model for small intestinal epithelial cells.

First, neoculin or *Phaseolus vulgaris* agglutinin-E$_4$ (PHA-E) was added to differentiated Caco-2 cells to evaluate the cell viability by the number of living cells after trypan blue staining. As shown in Fig. 10.13A, only 54.3% of the cells survived the addition of the smallest quantity of PHA-E (at a concentration of 0.03 mg/ml) in comparison with the control group. The relative viability (% of control) was 35.3% at 0.1 mg/ml and 23.8% at 0.3 mg/ml. On the other hand, the relative viability after the addition of neoculin was no less than 89% with all concentrations, suggesting that neoculin does not exhibit cytotoxicity to Caco-2 cells, unlike PHA-E.

Next, DNA microarray analysis was performed to compare the gene expression profile in Caco-2 cells 24 hours after the addition of neoculin or PHA-E. Two independent mRNA samples were extracted from each of the three groups prepared, namely, the control group, the neoculin-treated group, and the PHA-E-treated group, and subjected to DNA microarray analysis. A hierarchical clustering analysis revealed that the control group and the neoculin-treated group formed the same cluster (Fig. 10.13B). In contrast, clusters of the PHA-E-treated group were quite different from the cluster of the neoculin-treated and control groups (Fig. 10.13B). These results indicate that the neoculin-treated and control groups are close to each other in terms of the gene expression profile whereas the PHA-E group is distinctly different from the two other groups. It implies that neoculin, highly homologous to lectin in crystal structure, has no significant effect on the Caco-2 cell viability and gene expression [67]. As a matter of course, it is impossible to discuss safety solely from the perspective of the gene expression profile.

However, it is expected that the selection of genes correlated with cytotoxicity as markers and, in the future, the prediction of toxicity or safety using these markers will help simplify the biological safety assessment.

PERSPECTIVES

FOOD-RELATED SYSTEMS BIOLOGY

There are a variety of biomarkers that can be classified into two types: markers that cause certain events and those that result from them event. However, it is difficult to define the cause-result relationship by means of metabolomics or proteomics. Indeed, both are of great practical importance, but they are insufficient as basic sciences that seek to understand the "why" and "because" of any event that occurs. Dealing with the underlying factors of biological events, transcriptomics serves as an indispensable complement to them. Although there are some exceptions, the three omics, that is, transcriptomics (T), proteomics (P), and metabolomics (M), are nearly interlinked. A new trend in nutrigenomics treats these three in a coordinated manner. It meets the needs of both basic science and applied science at the same time (Fig. 10.14).

An in-depth analysis of the T-P-M coordination will open a new basic science under the name of systems biology as a science that integrates data obtained by all approaches in advanced biology to cover the whole biological phenotypes or phenomena of living bodies.

PERSONALIZED NUTRITION FOR ANTIOBESITY AND POTENTIAL UTILITY OF MADE-TO-ORDER FOODS

Physiological effects of functional food are substantially influenced not only by qualitative factors of the harmful and beneficial ingredients they contain, but also by personal differences on the side of the consumers. The same dose of the same drug may be effective for some people but may have no effect on others, whereas side effects may occur to some people but not to others. Safe food could do harm if it is eaten in excessive quantities. Functional food must have an adverse effect in the case of excessive consumption or imbalanced dieting. Antioxidant ingredients are promoted intensively, but it is necessary to pay attention to consuming too much

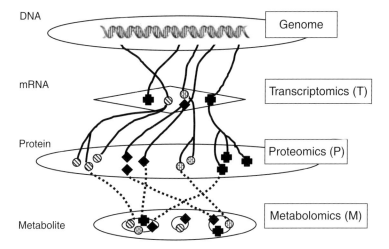

Figure 10.14 T-P-M coordination to be investigated for the second wave in nutritional and functional food genomics.

Table 10.5 Different factors elicit different mechanisms of antiobesity.

Factor	Source	Mechanism
Val-Val-Tyr-Pro	Hemoglobin	Lipase inhibition
Oolong polyphenols	Oolong tea	Ditto
1, 3-Diacylglycerol	Cooking oil	Slow lipid absorption
Middle-chain fatty acids	Cooking oil	Ditto
Mannooligosaccharides	Coffee	Lipid transport inhibition
Caffeine	Coffee, etc.	Lipid metabolism modulation
Tea catechins	Green tea	Ditto
Isoflavons	Soybean	Ditto
Cacao polyphenols	Cocoa	Ditto
β-Conglycinin	Soybean	Antiobese hormone secretion
L-Carnitine	Supplement	Fatty acid transport to mitochondria
Capsaicin	Chili	β3 Adrenergic receptor activation

of any such substance at one time as it might become a substance that accelerates oxidation while in the body.

Given these considerations, what exactly are personal differences? To summarize the latest findings, they reflect variations in the genetic makeup of individuals. Humans have 22,000 different genes, each of which differs slightly in its nucleotide sequence. Various forms of mutant genes are generated by a variation in any one of the tens of thousands of sequenced A, T, G, or C. Such a variation is called a single nucleotide polymorphism (SNP). It makes a difference in whether and how strongly the living body responds to drugs and food.

For example, β-adrenergic receptor gene has an SNP with a base change from T to C. This change leads to the W64R mutation in the proteins generated (tryptophan replaced by arginine at the 64th position from the N terminal) [68]; the variation occurs to about 33% of the humans in Japan. Normal β-adrenergenic receptor triggers the burning of fat whereas the W64R variant has no such effect. In other words, the consumption of chili has no effect on preventing obesity in one-third of the population. This is how personal differences come into play.

For these reasons, it is now considered necessary to design a diet that takes account of such personal differences. Foods designed in this manner are called "made-to-order" foods. Analogously to such clothes as tailored to suit the shape of the person who wears them. People needing an antiobesity food should select one from proper functional ingredients (Table 10.5).

In nutrition sciences, personalized nutrition is now being recognized as important. In the near future, it will be jointly used with the existing regular nutrition formula. Similar signs are emerging in the area of food. Made-to-order functional food products will come onto the market in the not-too-distant future. Nutritional and/or functional food SNPs will be practically available after in-depth insights into food-related genomics.

ACKNOWLEDGMENTS

It is noted that the International Life Science Institute (ILSI) of Japan established an endowed chair, named Functional Food Genomics Laboratory, at the University of Tokyo in 2003 under the cosponsorship of thirty-two food companies as ILSI Japan members and also that, since then, the chair has been contributing to most of the topical findings presented in this chapter. The financial support by all these companies is greatly appreciated. We would like to express our sincere thanks to a number of the manuscript contributors who are, in order of description, Dr. Tomoko Nemoto, The University of Tokyo; Dr. Kayoko Shimoi, University of Shizuoka; Dr.

Nobuhiko Tachibana, Fuji Oil Co., Ltd.; Dr. Tomoyuki Fukasawa, Meiji Seika Co.; Dr. Shozo Ohta, Api Co., Ltd.; Dr. Ikuko Masuda-Nishimura, Kikkoman Corp.; Dr. Manabu Sami, Asahi Breweries, Ltd.; and Dr. Takahiro Inakuma, Kagome Co., Ltd. Our special thanks should be given to Professor Soichi Arai, Tokyo University of Agriculture, for pertinent discussion.

REFERENCES

1. Arai, S., Functional food science born in Japan and propagated globally. *Science & Technology In Japan* 2006, *25*, 2–5.
2. Abe, K., Functional food science and nutrigenomics. *Science & Technology In Japan* 2006, *25* (99), 10–12.
3. Arai, S., Yasuoka, A., Abe, K., Functional food science and food for specified health use policy in Japan: state of the art. *Curr Opin Lipidol* 2008, *19*, 69–73.
4. Lee, Y. S., Noguchi, T., Naito, H., Phosphopeptides and soluble calcium in the small intestine of rats given a casein diet. *Br J Nutr* 1980, *43*, 457–467.
5. Nakai, M., Fukui, Y., Asami, S., Toyoda-Ono, Y., *et al.*, Inhibitory effects of oolong tea polyphenols on pancreatic lipase in vitro. *J Ag Food Chem* 2005, *53*, 4593–4598.
6. Meisel, H., FitzGerald, R. J., Opioid peptides encrypted in intact milk protein sequences. *Br J Nutr* 2000, *84*, S27–S31.
7. Moutsatsou, P., The spectrum of phytoestrogens in nature: our knowledge is expanding. *Hormones* 2007, *6*, 173–193.
8. Huang, W., Zhang, J., Moore, D. D., A traditional herbal medicine enhances bilirubin clearance by activating the nuclear receptor CAR. *J Clin Invest* 2004, *113*, 137–143.
9. Dupuy, A., Simon, R. M., Critical review of published microarray studies for cancer outcome and guidelines on statistical analysis and reporting. *J Natl Cancer Inst* 2007, *99*, 147–157.
10. Garosi, P., De Filippo, C., van Erk, M., Rocca-Serra, P., *et al.*, Defining best practice for microarray analyses in nutrigenomic studies. *Br J Nutr* 2005, *93*, 425–432.
11. Kadota, K., Nakai, Y., Shimizu, K., A weighted average difference method for detecting differentially expressed genes from microarray data. *Algorithms Mol Biol* 2008, *3*, 8.
12. World Health Organization, *Protein and Amino Acid Requirements in Human Nutrition: Report of a Joint WHO/FAO/UNU Expert Consultation (Technical Report Series)*. WHO Press, 2007.
13. Hartmanna, R., Meisela, H., Food-derived peptides with biological activity: from research to food applications. *Curr Opin Biotechnol* 2007, *18*(2), 163–169.
14. Endo, Y., Fu, Z., Abe, K., Arai, S., *et al.,* Dietary protein quantity and quality affect rat hepatic gene expression. *The Journal of Nutrition* 2002, *132*(12), 3632–3637.
15. Thalacker-Mercer, A. E., Fleet, J. C., Craig, B. A., Carnell, N. S., *et al.*, Inadequate protein intake affects skeletal muscle transcript profiles in older humans. *Amer J Clin Nutr* 2007, *85*(5), 1344–1352.
16. Peng, T., Golub, T. R., Sabatini, D. M., The immunosuppressant rapamycin mimics a starvation-like signal distinct from amino acid and glucose deprivation. *Mol Cell Biol* 2002, *22*(15), 5575–5584.
17. Harper, A. E., Benevenga, N. J., Wohlhueter, R. M., Effects of ingestion of disproportionate amounts of amino acids. *Physiol Rev* 1970, *50*(3), 428–558.
18. Matsuzaki, K., Kato, H., Sakai, R., Toue, S., Transcriptomics and metabolomics of dietary leucine excess. *J Nutr* 2005, *135*(6 Suppl), 1571S–1575S.
19. Nemoto, T., Matsuzaki, H., Uehara, M., Suzuki, K., Magnesium-deficient diet-induced reduction in protein utilization in rats is reversed by dietary magnesium supplementation. *Magnesium Research* 2006, *19*, 19–27.
20. Nakai, Y., Hashida, H., Kadota, K., Minami, M., *et al.*, Up-regulation of genes related to the ubiquitin-proteasome system in the brown adipose tissue of 24-h-fasted rats. *Biosci Biotechnol Biochem* 2008, *72*, 139–148.

21. Hubbell, E., Liu, W.-M., Mei, R., Robust estimators for expression analysis. *Bioinformatics* 2002, *18*, 1585–1592.
22. Irizarry, R. A., Hobbs, B., Collin, F., Beazer-Barclay, Y. D., *et al.*, Exploration, normalization, and summaries of high density oligonucleotide array probe level data. *Biostatistics* 2003, *4*, 249–264.
23. Zhou, L., Rocke, D. M., An expression index for Affymetrix GeneChips based on the generalized logarithm. *Bioinformatics* 2005, *21*, 3983–3989.
24. Li, F, Wong, W. H., Model-based analysis of oligonucleotide arrays: expression index computation and outlier detection. *Proc Natl Acad Sci U S A* 2001, *98*, 31–36.
25. Finn, P. F., Dice, J. F., Proteolytic and lipolytic responses to starvation. *Nutrition* 2006, *22*, 830–844.
26. Qi, L., Heredia, J. E., Altarejos, J. Y., Screaton, R., *et al.*, TRB3 links the E3 ubiquitin ligase COP1 to lipid metabolism. *Science* 2006, *312*, 1763–1766.
27. Shreffler, W. G., Lencer, D. A., Bardina, L., Sampson, H. A., IgE and IgG4 epitope mapping by microarray immunoassay reveals the diversity of immune response to the peanut allergen, Ara h 2. *J Allergy Clin Immunol* 2005, *116*, 893–899.
28. Raikkonen, K., Keltikangas-Jarvinen, L., Adlercreutz, H., Hautanen, A., Psychosocial stress and the insulin resistance syndrome. *Metabolism* 1996, *45*, 1533–1538.
29. Sato, T., Yamamoto, H., Sawada, N., Nashiki, K., *et al.*, Restraint stress alters the duodenal expression of genes important for lipid metabolism in rat. *Toxicology* 2006, *227*, 248–261.
30. Nishio, Y., Nakano, Y., Deguchi, Y., Terato, H., *et al.*, Social stress induces oxidative DNA damage in mouse peripheral blood cells. *Genes and Environment* 2007, *29*, 17–22.
31. Miyashita, T., Yamaguchi, T., Motoyama, K., Unno, K., *et al.*, Social stress increases biopyrrins, oxidative metabolites of bilirubin, in mouse urine. *Biochem Biophys Res Commun* 2006, *349*, 775–780.
32. Weinberg, J., Emerman, J. T., Effects of psychosocial stressors on mouse mammary tumor growth. *Brain Behav Immun* 1989, *3*, 234–246.
33. Motoyama, K., Nakai, Y., Miyashita, T., Fukui, Y., *et al.*, Isolation stress for 30 days alters hepatic gene expression profiles, especially with reference to lipid metabolism in mice. *Physiol Genom*, 2009, *37*, 79–87.
34. Hamilton, R. M., Carroll, K. K., Plasma cholesterol levels in rabbits fed low fat, low cholesterol diets: effects of dietary proteins, carbohydrates and fibre from different sources. *Atherosclerosis* 1976, *24*(1–2), 47–62.
35. Tachibana, N., Matsumoto, I., Fukui, K., Arai, S., *et al.*, Intake of soy protein isolate alters hepatic gene expression in rats. *J Agric Food Chem* 2005, *53*(10), 4253–4257.
36. Fukuda, Y., Osawa, T., Namiki, M., Ozaki, T., Studies on antioxidative substances in sesame seed. *Agric Biologic Chem* 1985, *49*, 301–306.
37. Katsuzaki, H., Kawakishi, S., Osawa, T., Sesaminol glucosides in sesame seeds. *Phytochemistry* 1994, *35*, 773–776.
38. Nakai, M., Harada, M. Nakahara, K., Akimoto, K., *et al.*, Novel antioxidative metabolites in rat liver with ingested sesamin. *J Agric Food Chem* 2003, *51*, 1666–1670.
39. Peñalvo, J. L., Heinonen, S. M., Aura, A. M., Adlercreutz, H., Dietary sesamin is converted to enterolactone in humans. *J Nutr* 2005, *135*, 1056–1062.
40. Tsuruoka, N., Kidokoro, A., Matsumoto, I., Abe, K., *et al.*, Modulating effect of sesamin, a functional lignan in sesame seeds, on the transcription levels of lipid- and alcohol-metabolizing enzymes in rat liver: a DNA microarray study. *Biosci Biotechnol Biochem* 2005, *69*, 179–188.
41. Mitsuoka, T., Hidaka, H., Eida, T., Effect of fructo-oligosaccharides on intestinal microflora. *Nahrung* 1987, *31*, 427–436.
42. Oku, T., Tokunaga, T., Hosoya, N., Nondigestibility of a new sweetener, "Neosugar," in the rat. *J Nutr* 1984, *114*, 1574–1581.
43. Fukasawa, T., Murashima, K., Matsumoto, I., Hosono, A., *et al.*, Identification of marker genes for intestinal immunomodulating effect of a fructooligosaccharide by DNA microarray analysis. *J Agric Food Chem* 2007, *55*, 3174–3179.

44. Matsui, N., Ito, R., Nishimura, E., Yoshikawa, M., *et al.,* Ingested cocoa and prevent high-fat diet-induced obesity by regulating the expression of genes for fatty acid metabolism. *Nutrition* 2005, *21*, 594–601.

45. Mishima, S., Suzuki, K.-M., Isohama, Y., Kuratsu, N., *et al.*, Royal jelly has estrogenic effects in vitro and in vivo. *J Ethnopharmacol* 2005, *101*, 215–220.

46. Narita, Y., Nomura, J., Ohta, S., Inoh, Y., *et al.*, Royal jelly stimulates bone formation: physiologic and nutrigenomic studies with mice and cell lines. *Biosci Biotechnol Biochem* 2006, *70*, 2508–2514.

47. Suzuki, K.-M., Isohama, Y., Maruyama, H., Yamada, Y., *et al.*, Estrogenic activities of Fatty acids and a sterol isolated from royal jelly. *Evid Based Complement Alternat Med* 2008, *5*, 295–302.

48. Yamaguchi, F., Yoshimura, Y., Nakazawa, H., Ariga, T., Free radical scavenging activity of grape seed extract and antioxidants by electron spin resonance spectrometry in an H2O2/NaOH/DMSO system. *J Agric Food Chem* 1999, *47*, 2544–2548.

49. Koga, T., Moro, K., Nakamori, K., Yamakoshi, J., *et al.*, Increase of antioxidative potential of rat plasma by oral administration of proanthocyanidin-rich extract from grape seeds. *J Agric Food Chem* 1999, *47*, 1892–1897.

50. Yamakoshi, J., Kataoka, S., Koga, T., Ariga, T., Proanthocyanidin-rich extract from grape seeds attenuates the development of aortic atherosclerosis in cholesterol-fed rabbits. *Atherosclerosis* 1999, *142*, 139–149.

51. Saito, M., Hosoyama, H., Ariga, T., Kataoka, S., *et al.,* Antiulcer activity of grape seed extract and procyanidins. *J Agric Food Chem* 1998, *46*, 1460–1464.

52. Yamakoshi, J., Saito, M., Kataoka, S., Tokutake, S. Procyanidin-rich extract from grape seeds prevents cataract formation in hereditary cataractous (ICR/f) rats. *Thromb Res* 2002, *50*, 4983–4988.

53. Sano, A., Uchida, R., Saito, M., Shioya, N., *et al.*, Beneficial effects of grape seed extract on malondialdehyde-modified LDL. *J Nutr Sci Vitaminol* 2007, *53*, 174–182.

54. Garbacki, N., Tits, M., Angenot, L., Damas, J., Inhibitory effects of proanthocyanidins from Ribes nigrum leaves on carrageenin acute inflammatory reactions induced in rats. *BMC Pharmacology* 2004, *4*, 25.

55. Miyazaki, M., Dobrzyn, A., Sampath, H., Lee, S. H., *et al.*, Reduced adiposity and liver steatosis by stearoyl-CoA desaturase deficiency are independent of peroxisome proliferator-activated receptor-alpha. *J Biol Chem* 2004, *279*, 35017–35024.

56. Shoji, T., Mutsuga, M., Nakamura, T., Kanda, T., *et al.*, Isolation and structural elucidation of some procyanidins from apple by low-temperature nuclear magnetic resonance. *J of Agric Food Chem* 2003, *51*, 3806–3813.

57. Akazome, Y., Kanda, T., Ikeda, M., Shimasaki, H., Serum cholesterol-lowering effect of apple polyphenols in healthy subjects. *J Oleo Sci* 2005, *54*(3), 143–151.

58. Ohta, Y., Sami, M., Kanda, T., Saito, K., *et al.*, Gene expression analysis of the anti-obesity effect by apple polyphenols in rats fed a high fat diet or a normal diet. *J Oleo Sci* 2006, *55*(6), 305–314.

59. Shoji, T., Akazone, Y., Kanda, T., Ikeda, M., The toxicology and safety of apple polyphenol extract. *Food Chem Toxicol* 2004, *42*, 959–967.

60. Di Mascio, P., Kaiser, S., Sies, H., Lycopene as the most efficient biological carotenoid singlet oxygen quencher. *Arch Biochem Biophys* 1989, *274*, 532–538.

61. Gartner, C., Stahl, W., Sies, H., Lycopene is more bioavailable from tomato paste than from fresh tomatoes. Am J Clin Nutr 1997, *66*, 116–122.

62. Stahl, W., Schwarz, W., Sundquist, A. R., Sies, H., cis-trans isomers of lycopene and beta-carotene in human serum and tissues. *Arch Biochem Biophys* 1992, *294*, 173–177.

63. Giovannucci, E., Tomatoes, tomato-based products, lycopene, and cancer: review of the epidemiologic literature. *J Natl Cancer Inst* 1999, *91*, 317–331.

64. Kasagi, S., *et al., Am J Physiol Lung Cell Mol Physiol* 2005, *290*, 396–404.

65. Shirasuka, Y., Nakajima, K., Asakura, T., Yamashita, H., *et al.,* Neoculin as a new

taste-modifying protein occurring in the fruit of Curculigo latifolia. *Biosci Biotechnol Biochem* 2004, *68*, 1403–1407.

66. Shimizu-Ibuka, A., Morita, Y., Terada, T., Asakura, T., *et al.*, Crystal structure of neoculin: insights into its sweetness and taste-modifying activity. *J Mol Biol* 2006, *359*, 148–158.

67. Shimizu-Ibuka, A., Nakai, Y., Nakamori, K., Morita, Y., *et al.*, Biochemical and genomic analysis of neoculin compared to monocot mannose-binding lectins. *J Agric Food Chem* 2008, *56*, 5338–5344.

68. Yanagisawa, Y., Hasegawa, K., Dever, G. J., Otto, C. T., *et al.*, Uncoupling protein 3 and peroxisome proliferator-activated receptor gamma2 contribute to obesity and diabetes in palauans. *Biochem Biophys Res Comm* 2001, *281*(3), 772–778.

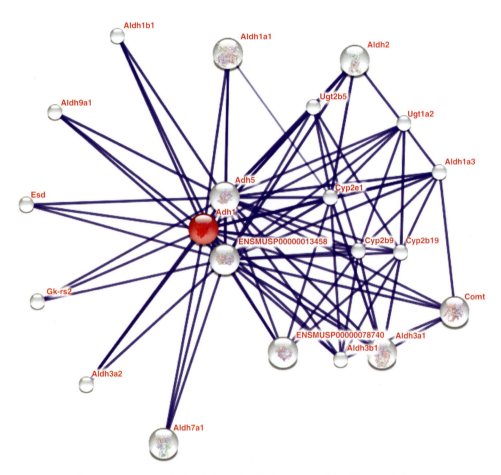

Figure 6.2 Gene networks analysis of alcohol dehydrogenase 1 by STRING 8.0.

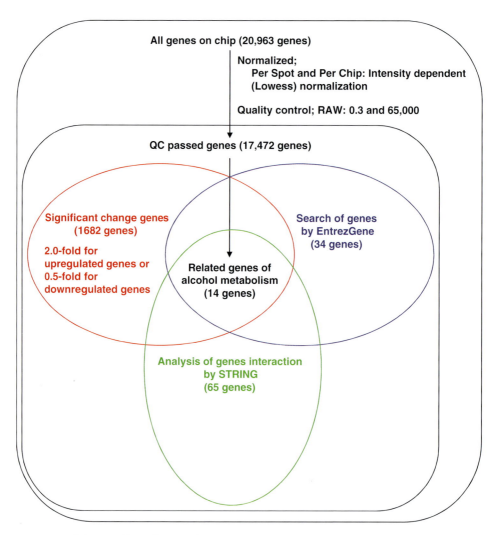

Figure 6.3 Scheme of identification of regulatory genes by *Brassica oleracea* var. *acephala* DC.

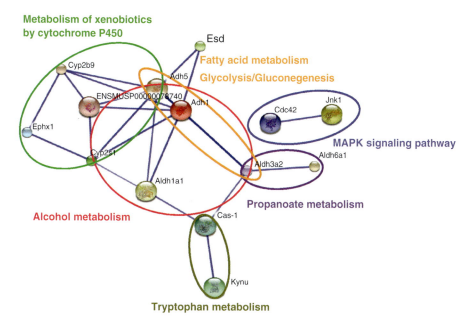

Figure 6.4 Genes network that shows remarkable change by *Brassica oleracea* var. *acephala* DC.

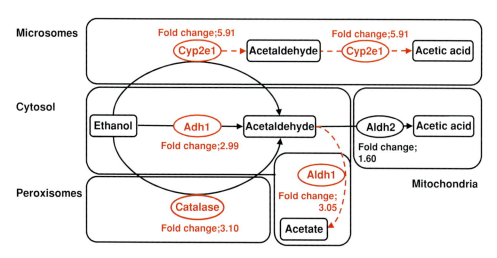

Figure 6.5 Effect of *Brassica oleracea* var. *acephala* DC on alcohol metabolism.

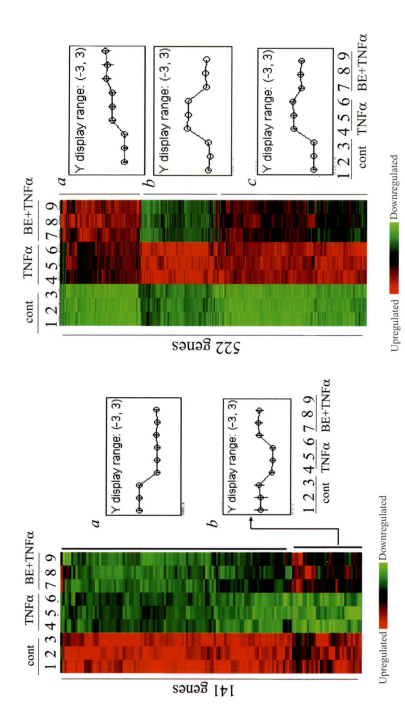

Figure 7.1 Hierarchiaal cluster images illustrating Boswellia-sensitive genes induced or downregulated by TNFα in human microvascular endothelial cells. For clear graphic display of the specific clusters of genes showing an increase (A) or decrease (B) in expression following TNFα treatment to HMEC cells, a count percentage analysis was performed. Genes that were found up- or downregulated in 100% of replicates and all comparisons (nine out of nine) following TNFα treatment were selected. These select candidate genes were subjected to hierarchical clustering to identify clusters of genes induced/downregulated by TNFα and are sensitive to Boswellia. Red to green gradation in color represents higher to lower expression signal. Reproduced with permission from Roy *et al.*, *DNA Cell Biol* 2005, *24*, 244–255.

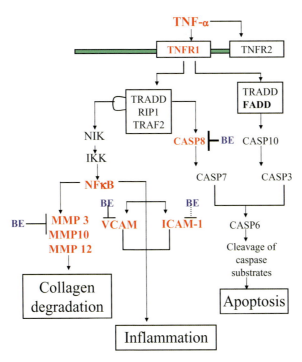

Figure 7.2 Boswellia-sensitive TNF-α induced signaling pathways in human microvascular endothelial cells. Pathway construction is based on GeneChipTM expression data and appropriate software resources (see Materials & Methods). To obtain insights on the effects of Boswellia on specific pathways induced by TNFα- in endothelial cells, the results of GeneChip analysis were mapped onto known pathways associated with inflammation, apoptosis, and collagen degradation. GenMAPP, *KEGG*, and gene ontology (GO) were used to develop the pathways. Genes shown in red are candidates identified using GeneChip assay that were upregulated following TNFα. Blunt arrow marked with BE marks the genes whose expression levels are fully (solid line) or partly (broken lines) normalized by BE pretreatment. TNFR, TNFα receptor; CASP, caspase; BE, Boswellia extract; TRADD, TNFR1-associated protein with death domain; NIK, NFκB-inducing kinase; IKK, Iκb kinase; MMP, matrix metalloproteinase; VCAM, vascular cell adhesion molecule; ICAM, intercellular adhesion molecule. Reproduced with permission from Roy *et al.*, *DNA Cell Biol* 2005, *24*, 244–255.

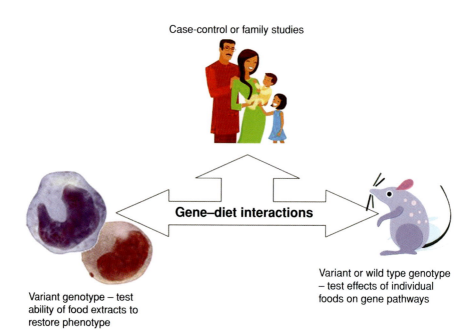

Case-control or family studies

Gene–diet interactions

Variant genotype – test
ability of food extracts to
restore phenotype

Variant or wild type genotype
– test effects of individual
foods on gene pathways

Figure 8.1 Methods of studying nutrigenomics and gastrointestinal health, as utilized by Nutrigenomics New Zealand (www.nutrigenomics.org.nz). A fundamental part of studies recognizes human variation, usually studied as either SNPs or copy number variants, and the effect that these
have on the normal phenotype. In the case of gastrointestinal health, a challenge to homeostasis may be necessary to reveal the implication of variation on the phenotype. Once key genes
are identified, then animal models may be utilized to study the effects of various foods and food
components on cellular gene expression (transcriptomics), proteins (proteomics), or metabolites
(metabolomics). It may also be appropriate to study cecal or fecal microbiota. *In vitro* or tissue
culture models provide a high-throughput approach to test whether nutrients or food extracts
can overcome the effect of the variant genotype. Evidence from all three lines of study can be
integrated to select foods for human clinical trials, which involve stratifying subjects according
to genotype and using a biomarker approach to relate genotype to the trial endpoint. Such approaches provide a pathway to develop personalized approaches in optimizing foods or diets to
ameliorate the effects of a variant genotype.

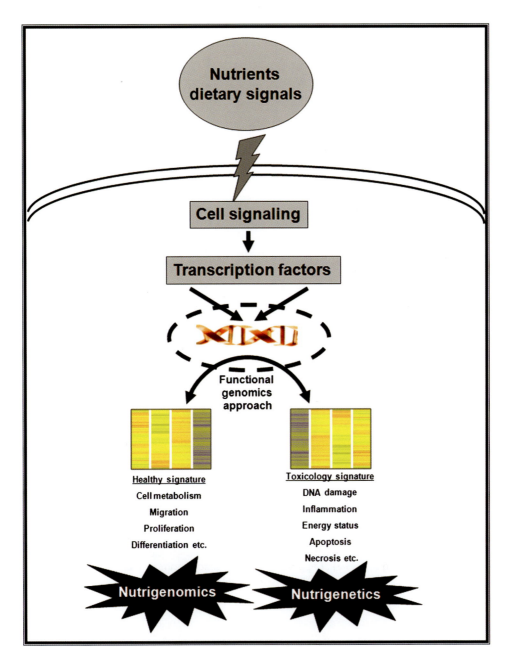

Figure 9.1 Nutrigenetics and nutrigenomics; two sides of a coin.

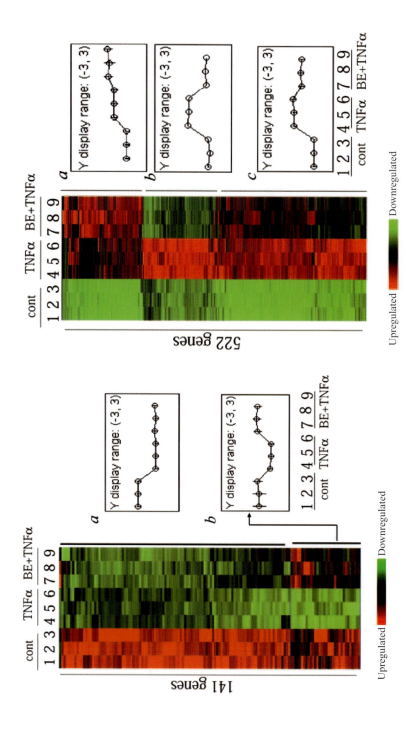

Figure 11.2 Hierarchical cluster images illustrating BE-sensitive genes induced or downregulated by TNFα in human microvascular endothelial cells. For clear graphic display of the specific clusters of genes showing a decrease (A) or an increase (B) in expression following TNFα treatment to HMEC cells, a count percentage analysis was performed. Genes that were found up- or downregulated in 100% of replicates and all comparisons (nine out of nine) following TNFα treatment were selected. These select candidate genes were subjected to hierarchical clustering to identify clusters of genes that are induced/downregulated by TNFα and are sensitive to *Boswellia*. Red to green gradation in color represents higher to lower expression signal.

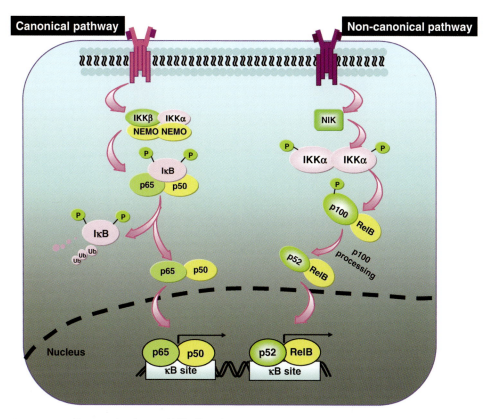

Figure 12.2 Two mechanisms of NF-κB activation.

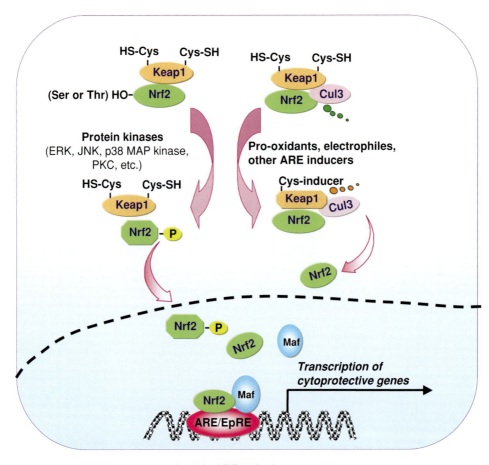

Figure 12.3 Proposed pathways for Nrf2-ARE activation.

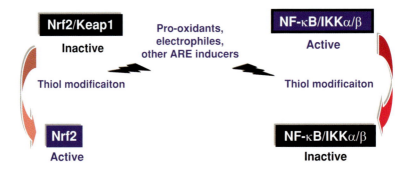

Figure 12.4 Differential effects of thiol modification on IKK-NF-κB- and Nrf2-mediated signal transduction.

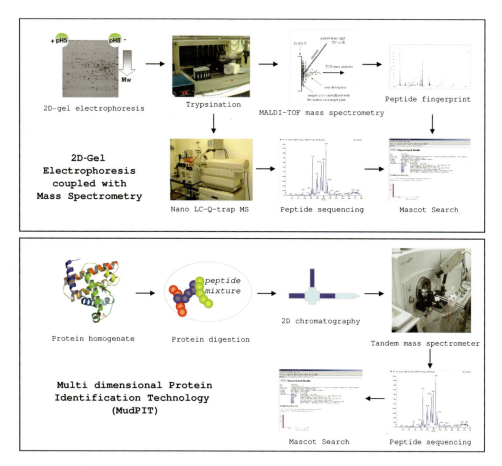

Figure 14.1 Proteomic approaches to nutrition research.

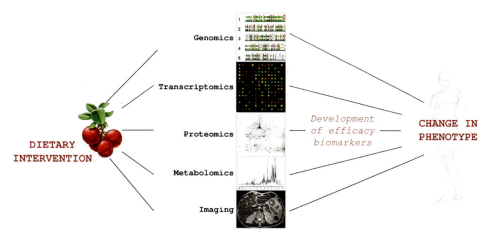

Figure 14.3 Use of OMICS technology in assessing the effects of dietary interventions on the human phenotype.

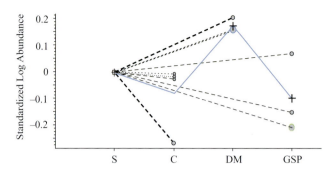

Figure 15.1 Representative line image of NADH-ubiquinone oxidoreductase from C, DM, GSP by proteomics analysis. S: standard; C: untreated control group; DM: untreated diabetic group; GSP: GSP treated diabetic group.

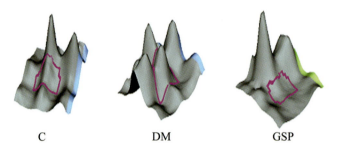

C   DM   GSP

Figure 15.2 Representative 3D images of NADH-ubiquinone oxidoreductase from C, DM, GSP by proteomics analysis. C: untreated control group; DM: untreated diabetic group; GSP: GSP treated diabetic group.

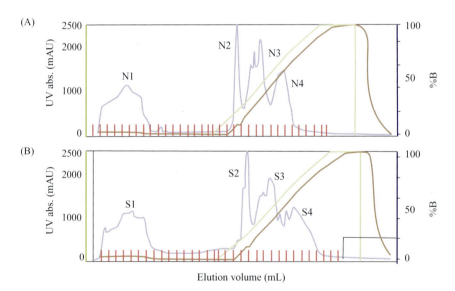

Figure 16.3 Ion exchange chromatograms of normal and regenerating pancreatic extracts. (A) Normal pancreatic extract, and (B) pancreatic extract obtained 2 days after a 60% pancreatec-tomy. Fractionation was performed using a fast protein liquid chromatography (FPLC) equipped with a MONO-Q column. The four major peaks were detected at 280 nm and each fraction was designated as N1, N2, N3, and N4 for normal pancreas and S1, S2, S3, and S4 for fractions of regenerating pancreatic extract, respectively. (From Ref. 20.)

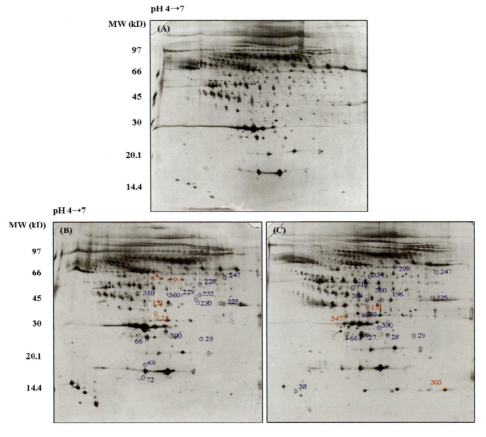

Figure 16.5 2-DE protein profiles of serum samples from the normoalbuminuric patient group (A), microalbuminuric patient group (B), and CRF patient group (C). Proteins (60 μg) were separated via IEF using 24 cm, pH 4–7 IPG strips, and 11%–16% gradient SDS-PAGE. The gels were visualized using silver staining, and their maps were analyzed with Image Master 2D Elite Software (GE Healthcare, Sweden). (From Ref. 82.)

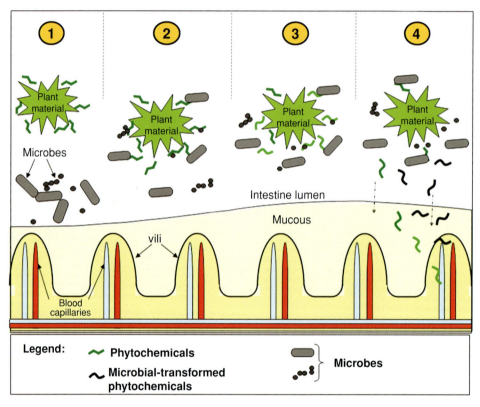

Figure 19.2 Metametabolomics approach – Phytochemical research involves the study of multiple metabolomes and their interaction. Metabolites from plant material containing phytochemicals (as part of the diet) interact with the gut microflora and with the animal metabolic processes. Microorganisms in the bowel can biotransform or degrade some phytochemical compounds enhancing their absorption or changing their biological activity. (Kindly prepared by Xavier Duportet)

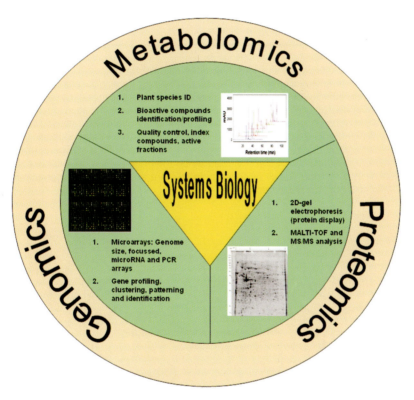

Figure 20.1 Major key aspects of the omics technologies used in phytomedicine research.

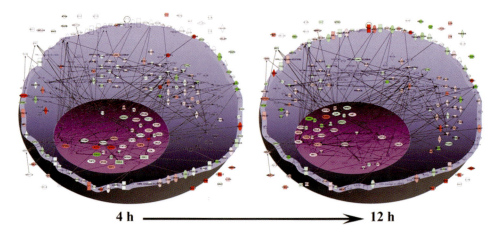

4 h ⟶ 12 h

Figure 20.2 A typical display of signaling networks, connecting cell surface, cytosolic, and nuclear compartments, for functional and differential immunity gene expression in mouse bone marrow-derived dendritic cells in response to treatment with specific medicinal phytocompounds. A prototypical cell was constructed from 293 representative genes involved in immunity of immature mouse bone marrow dendritic cells. Genes for which the expression statistically increased are colored red, those for which expression decreased are shown in green. Temporal changes in apparent expression. The response to phytocompounds extract [BF/S+L/Ep] administration in mouse immature dendritic cells can be viewed as an integrated cell-wide response, propagating and resolving over time. Result of genes networks are analysis by the Ingenuity Pathways program. (Yin *et al.*, unpublished data)

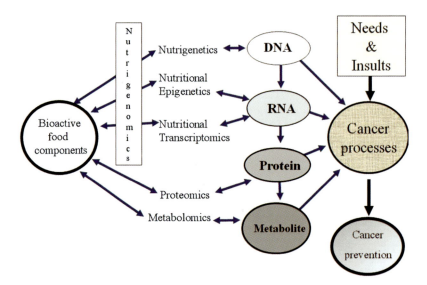

Figure 21.1 The "omic" can influence the response to bioactive food components. The needs for food components for normal growth and development along with specific insults including excess calories, nutrient–nutrient interactions, bacteria, viruses, and environmental contaminants can influence the ability of food components to influence cancer risk and tumor behavior. Because these interrelationships are not linear, effective bioinformatic tools are needed to evaluate the significance of foods and their components and to determine who will benefit and/or be placed at risk due to dietary intervention.

11
Genetic Basis of Anti-Inflammatory Properties of *Boswellia* Extract

Golakoti Trimurtulu, Chandan K. Sen, Alluri V. Krishnaraju, and Krishanu Sengupta

INTRODUCTION

Inflammation is typically a protective mechanism triggered in response to noxious stimuli, trauma, or infection to guard the body and to hasten the recovery process. As such, the complex inflammatory process per se may not constitute a serious disease condition. However, inflammation that is unchecked or fails to respond poses health risks to millions of people worldwide. Inflammation is also a common factor underlying many disease indications including wounds, burns, heart attacks, arthritis, asthma, and cancer.

The study of genetic basis and molecular mechanisms of inflammatory diseases has thus become one of the hottest areas in biomedical research these days with a particular emphasis on gene regulation. Nutrigenomics constitutes the study of the effects of bioactive compounds from food on gene expression. An increasing body of scientific evidence collected during the last few years has demonstrated that phytochemicals derived from food ingredients and plants alter the expression of genes in the human body. Detailed mechanisms of action for their active role in the prevention of inflammation have been elucidated for many natural ingredients. Employing novel genomic approaches to identify key elements of acute and chronic inflammation has thus been a key aspect of nutrigenomics to identify individual compounds, as well as extracts, derived from food ingredients to alter the expression of inflammation-related genes in the human body.

Inflammatory mediators and inflammatory cells modulate the inflammatory response. A number of inflammatory mediators, such as kinins, cytokines, eicosanoids, enzymes, and adhesion molecules act on specific targets leading to the local release of other mediators from leukocytes and also attract leukocytes to the site of inflammation. Inflammation can be controlled by inhibiting the formation of inflammatory mediators, such as eicosanoids. Eicosanoids, prostaglandins, and leukotrienes are produced primarily from arachidonic acid (Fig. 11.1) that has been released from the cell membranes. The formation of prostaglandins and leukotrienes from arachidonic acid can be suppressed by inhibiting cyclooxygenase and lipoxygenase, respectively. The two known isoforms of cyclooxygenase (COX) are COX-1 and -2. Although these two isoforms share similar structure, they differ markedly in their pattern of regulation and physiological functions. COX-1 is readily detected in many tissue types and is thought to be responsible for ''housekeeping'' activities, such as gastrointestinal cytoprotection, renal blood flow regulation, and platelet aggregation. COX-2, in contrast, is considered the inducible isoform. It is generally not detected in most tissues but can be found in large amounts in macrophages and other inflammatory cell types following exposure to cytokines, growth factors, and mitogens [1,2]. The alternative pathway (Fig. 11.1) for inflammation is mediated by 5-lipoxygenase. Lipoxygenases

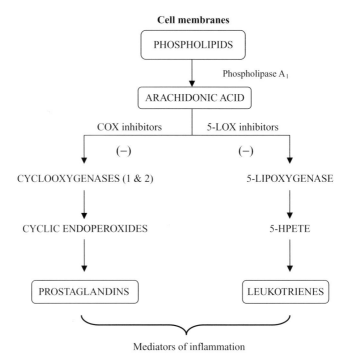

Figure 11.1 Eicosanoid biosynthetic pathways.

(LOXs) comprise a family of nonheme iron-containing dioxygenases, representing the key enzymes in the biosynthesis of leukotrienes that have been postulated to play an important role in the pathophysiology of several inflammatory and allergic diseases. The products of LOX-catalyzed oxygenation (hydroperoxyeicosatetraenoic acids (HPETE), hydroxyeicosatetraenoic acids (HETE), leukotrienes, and lipoxins) are the important inflammatory mediators. According to currently used nomenclature, the LOXs are classified with respect to their positional specificity of arachidonic acid oxygenation (5-LOX, 9-LOX, 12-LOX, 15-LOX)[3].

BOSWELLIA SERRATA

Gum resin of *Boswellia serrata*, most commonly known as Indian frankincense, has been used traditionally for the treatment of a wide range of indications including rheumatism, menstrual pain, and wrinkles [4]. It was an important anti-inflammatory agent in Ayurveda, the traditional Indian system of medicine [5]. The *in vitro* and *in vivo* experiments have supported the traditional claims on *Boswellia* products. The ethanolic extract of the gum resin, for example, inhibited neutrophils [6] *in vitro* and inhibited carrageenan-induced paw edema [7] in rats on par with phenylbutazone. Oral administration of boswellic acids to mice, after intraplueral injection of carrageenan, inhibited polymorphonuclear infiltration into the pleural cavity, a response similar to indomethacin [8].

   Boswellia gum resin and its extracts also demonstrated significant therapeutic improvements in human clinical trials [9–11], as summarized in the following, confirming the *in vivo* anti-inflammatory effects. A randomized, double-blind, placebo-controlled, crossover clinical trial with *Boswellia* extract on a group of patients with osteoarthritis of knee exhibited statistically significant mean improvements with respect to the reduction in pain, decreased swelling, and increased knee flexion [9]. In an open, nonrandomized equivalence study, thirty patients with chronic colitis were administered with either *Boswellia* gum (300 mg thrice daily) or sulfasalazine (1 g thrice a day) and the therapeutic effects shown by *Boswellia* were comparable to those exhibited by sulfasalazine [10]. In a similar study with forty-two ulcerative colitis patients, *Boswellia* gum and sulfasalazine also showed comparable improvements in abdominal pain, diarrhea, and histopathology without any statistically significant difference between the two treatment groups [11]. In another equivalency study, *Boswellia* standardized extract exhibited therapeutic improvements comparable to, or better than, mesalazine in a randomized, double-blind study on patients with active Crohn's disease [12]. *Boswellia* gum resin also showed statistically significant improvement in patients with bronchial asthma in a 6-week, double-blind, placebo-controlled study [13].

MECHANISM OF ACTION

The source of the anti-inflammatory actions of *Boswellia* gum resin has been attributed to a group of triterpene acids [14] called boswellic acids that were isolated [15] from the gum resin of *Boswellia serrata*. These compounds exert anti-inflammatory activity by inhibiting 5-lipoxygenase (5-LOX). In addition to their 5-lipoxygenase inhibitions, boswellic acids also inhibit human leukocyte elastase (HLE), an enzyme of different proinflammatory pathways [16]. Catechins, quercetin, and boswellic acids are some of the important naturally occurring 5-lipoxygenase inhibitors. Although quercetin is a potent inhibitor, it belongs to the redox type of 5-lipoxygenase inhibitor. Redox type inhibitors are not selective and thus interfere with some useful biochemical pathways causing undesirable side effects. Boswellic acids, on the other hand, are selective nonredox-type 5-lipoxygenase inhibitors.

AKBA

Drugs that target lipoxygenases and leukotrienes have recently become emerging therapies for inflammatory diseases and cancer [17]. A detailed study on the structural requirements for boswellic acids indicated that, of all the six acids, AKBA shows most pronounced inhibitory activity against 5-LOX [18, 19] with an IC_{50} of 1.5 μM [19] against 5-lipoxygenase enzyme in intact cells. AKBA acts by a unique mechanism, binding to 5-LOX in a calcium-dependent and reversible manner, and inhibits its activity as an allosteric regulator and not as a redox-type or competitive inhibitor [20]. Other boswellic acids inhibited 5-LOX only partially and incompletely, whereas the noninhibitory triterpenoid constituents in the extract such as amyrin and its derivatives antagonized the biological activity of AKBA [19,21]. AKBA is a minor constituent in the natural *Boswellia serrata* extracts, though some of the partially active boswellic acids are present in higher proportions (up to 25%). AKBA concentration even in a higher-grade commercial material (85% boswellic acids) typically varies in the range of 2% to 5%. AKBA has thus become the subject of intensive research for many groups across the globe because of its potential for the treatment of chronic inflammatory disorders.

Development of 5-LOXIN® (BE-30)

5-LOXIN® is a novel *Boswellia serrata* extract selectively enriched in AKBA concentration. 5-LOXIN® contains minimum 30% AKBA (BE-30) and is produced using a commercially viable process developed by the researchers at the Laila Impex R&D Center (Indian Patent No. 205269). Its efficacy at the enzyme and cellular level and animal level has been clearly established [22, 23]. The safety has been proven by the total spectrum of safety studies [24]. Finally, the proof of concept in humans was established by a double-blind, placebo-controlled human clinical study [25]. Keeping in perfect consonance with its higher AKBA content, BE-30 exhibited significantly better inhibitory activity against 5-lipoxygenase ($IC_{50} = 40 \mu g/ml$) when compared to other commercially available *Boswellia* extracts ($IC_{50} > 100 \mu g/ml$). It has also been found that BE-30 is more efficacious as an antibacterial and antiproliferative agent compared to the extracts containing lower concentrations of AKBA.

Genetic Basis for Efficacy of BE-30

Tumor necrosis factorα (TNFα) is a pleotropic inflammatory cytokine produced by the immune system that suppresses tumor cell proliferation. Subsequent studies established that TNFα is a key mediator of inflammation [26–28]. TNFα is primarily produced in endothelial cells (EC). EC are squamous epithelial cells that form a thin lining to the interior surface of the vasculature and help reduce turbulence to the blood flow. EC are crucial elements in the inflammatory etiology that participate through production and secretion of several proinflammatory cytokines including interleukins, M-CSF, G-CSF, GM-CSF, gro alpha, MCP, etc. EC also expresses a series of glycoproteins and cell surface proteins, which help recruit leukocytes to the site of inflammation by binding to circulating leukocytes. Inflammatory response in EC is potentially induced by TNFα [29]. Genomics has recently become the cutting-edge tool in inflammation research. With the advent of microarray technology the study of the amelioration of the genes associated with inflammation has become faster and economical. Thus, the genetic basis for the anti-inflammatory effects of the standardized extract BE-30 was tested in a system of TNFα-induced gene expression in human microvascular endothelial cells (HMEC). This was the first whole human genome screen to delineate the genetic basis of the anti-inflammatory efficacy of a medicinal plant derivative.

Gene Chip Probe Array Analysis

GeneChip™ analysis was done to identify the sets of genes in HMEC that are sensitive to TNFα. The HMEC cells were grown in MCDB-131 under standard cultured conditions using seed culture in 100 mm plates at a density of 1×10^6 cells/plate [22]. Twenty-four hours after the seeding, the HMEC cells were treated with BE-30 25μg/mL or vehicle (DMSO). Forty-eight hours after treatment, the cells were challenged with 50ng/mL of human recombinant TNF α for 6 hours. The cells were harvested and subjected to RNA extraction. The RNA fraction was purified from DNA contamination using DNA-free™ kit (Ambion, Austin, TX). After the quality assessment, the RNA samples were hybridized to probes using Affymetrix human genome arrays U 133 Plus 2.0 containing 47,000 transcripts including the entire human genome. After hybridization at 45°C for 16 hours, the arrays were washed, stained, and scanned using Affymetrix Gene Chip Scanner 3000. Affymetrix Microarray Suite 5.0 and Data Mining Tool 2.0 were used for raw data analysis. For processing additional data and to obtain mean change of each gene expression by hierarchical clustering, dChip software [30] was used. BE-30-sensitive TNF-α inducible genes were processed using comparative analysis approach.

In nine of nine pair-wise comparisons, TNFα upregulated 522 genes and downregulated 141 genes (Fig. 11.2). Of the 522 genes induced by TNFα in HMEC, 113 genes were clearly sensitive to BE-30 treatment. BE-30 sensitive TNFα-inducible genes that exhibited significant response

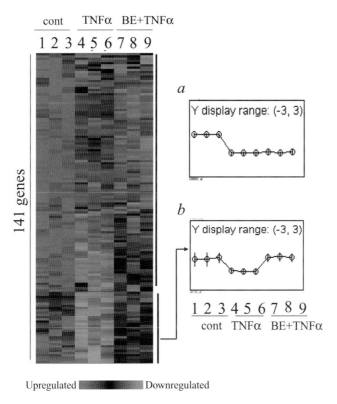

Figure 11.2 Hierarchical cluster images illustrating BE-sensitive genes induced or downregulated by TNFα in human microvascular endothelial cells. For clear graphic display of the specific clusters of genes showing a decrease (A) or an increase (B) in expression following TNFα treatment to HMEC cells, a count percentage analysis was performed. Genes that were found up- or downregulated in 100% of replicates and all comparisons (nine out of nine) following TNFα treatment were selected. These select candidate genes were subjected to hierarchial clustering to identify clusters of genes that are induced/downregulated by TNFα and are sensitive to *Boswellia*. Red to green gradation in color represents higher to lower expression signal. For color detail, please refer to the color plate section.

were subjected to further processing for identification of signaling pathways using Affymetrix gene ontology mining tools *KEGG* (*Kyoto Encyclopedia of Genes and Genomes*), Gen-MAAP, DAVID (Database for Annotation, Visualization, and Integrated Discovery Verification), GO (Gene Ontology), and Locus Link. The genes that were significantly modulated in the treatment group supplemented with both TNFα and BE-30 compared to the control group treated with TNFα alone are summarized in Table 11.1 along with their functional category. These genes were identified as directly related to inflammation, cell adhesion (ICAM-1, VCAM-1), proteolysis, peroxisome proliferation, fatty acid metabolism, and angiogenesis, etc. The downregulation of a significant number of inflammatory genes induced by TNFα in HMEC by AKBA-enriched *Boswellia* extract (BE-30) strongly supports the traditional claims of *Boswellia*.

More importantly, TNFα-inducible expression of ICAM-1 and VCAM-1 was significantly sensitive to BE-30 treatment. Cell adhesion molecules (CAMs) are proteins expressed on the cell surface that help binding with other cells or with the extracellular matrix (ECM). CAMs enable extravasation of leukocytes during physiological and pathological processes. Three families of CAMs (selectins, integrins, and immunoglobulins) play an important role in leukocyte endothelial interactions. Intercellular adhesion molecule-1 (ICAM-1 or CD 54), vascular cell adhesion

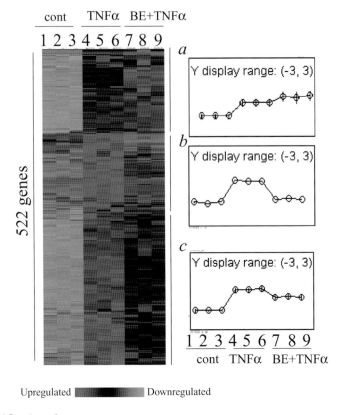

Figure 11.2 (*Continued*)

molecule-1 (VCAM-1 or CD 106) are induced by cytokines and aid in recruitment of circulating leukocytes to the inflamed site through adhesion and emigration [31, 32]. These cell adhesion molecules are directly implicated in the pathogenesis of inflammation, and their expression in endothelial cells and lymphocytes is regulated by the inflammatory cytokines. VCAM-1 plays a key role in recruiting leukocytes to the site of inflammation. Further studies to quantitatively establish the effect of BE-30 on TNFα-inducible expression of VCAM-1 manifested that the upregulated VCAM-1 expression was completely abrogated in the human microvascular en-dothelial cells pretreated with BE-30. This further supported the anti-inflammatory properties of *Boswellia* extracts in general and enriched extract BE-30 in particular.

PROTEOMICS

GeneChip study on *Boswellia* products further revealed that matrix metalloproteinases (MMPs) belong to a prominent class of TNFα-inducible genes sensitive to *Boswellia* extracts. Matrix metalloproteinases (MMPs) constitute a large family of zinc-dependent metallo-endopeptidases that degrade various components of the extracellular matrix (ECM) in both normal and diseased tissues. The first MMP discovered was a collagenase in the tail of a tadpole undergoing metamor-phosis. The MMP family consists of more than sixty-four members, of which twenty-three were identified in humans. These enzymes include collagenases and elastases, which were segregated into three groups based on their substrate preference. Collagenases degrade connective tissue collagen, whereas gelatinases degrade basement membrane collagens. However, the targets of

Table 11.1 List of TNFα-inducible genes that were sensitive to BE-30.

SC probe set	Gene[b]	TNFα mean[a]	TNFα SD	BE-30 + TNFα mean[a]	BE-30 + TNFα SD	Functional category
203868_s_at	VCAM1	172.74	5.83	22.23	4.90	Cell adhesion/ inflammation
205681_at	BCL2A1	26.18	1.80	8.14	0.36	antiapoptosis
1555759_a_at	CCL5	26.19	2.04	6.09	0.36	Chemokine
1405_I_at	CCL5	32.21	2.12	5.64	0.53	Chemokine
205890_s_at	UBD	83.34	13.26	4.45	2.16	Proteolysis
204932_at	TNFRSF11B	11.86	0.72	3.80	0.30	Inflammation
205828_at	MMP3	16.76	2.40	3.77	0.43	Proteolysis
229437_at	BIC	13.74	0.61	3.43	0.35	Carcinogenesis
221371_at	TNFSF18	8.87	0.67	2.97	0.89	Inflammation
223484_at	NMES1	15.82	1.04	2.95	0.12	Tumor suppressor
219209_at	MDA5	7.25	0.16	2.82	0.41	Differentiation
206211_at	SELE	11.21	0.60	2.62	0.76	Cell adhesion
207339_s_at	LTB	10.03	1.08	2.58	1.00	Inflammation
202023_at	EFNA1	7.82	0.31	2.39	0.15	Signal transduction
206553_at	OAS2	12.29	0.75	2.27	1.70	RNA binding
204655_at	CCL5	8.75	0.84	2.22	0.11	Chemokine
205680_at	MMP10	8.44	0.19	2.20	0.14	Proteolysis
229450_at	IFIT4	5.62	0.40	1.98	0.13	Immune/inflammation
24393_x_at	DNAH5	3.29	0.43	1.95	0.53	Microtubule
213338_at	RISI	3.68	0.07	1.94	0.14	Apoptosis
221085_at	TNFSF15	5.68	0.34	1.92	0.46	Inflammation
226847_at	FST	4.56	0.08	1.78	0.17	TGF Beta signaling
211122_s_at	CXCL11	5.26	0.59	1.64	0.76	Chemokine
204897_at	PTGER4	3.19	0.07	1.60	0.07	G-protein receptor
212448_at	NEDD4L	3.40	0.35	1.58	0.53	Proteolysis
232517_s_at	PRIC285	2.90	0.32	1.56	0.08	Peroxisome proliferation
20474_at	IFIT4	4.26	0.18	1.55	0.25	Immune/inflammation
204298_s_at	LOX	2.90	0.21	1.54	0.05	Collagen metabolism
233500_x_at	LLT1	5.71	0.60	1.51	0.70	Signal transduction
219593_at	SLC15A3	2.73	0.15	1.51	0.01	Transport
206825_at	OXTR	2.83	0.11	1.49	0.08	G-protein receptor
203595_s_at	IFIT5	2.60	0.11	1.49	0.07	Immune/inflammation
203835_at	GARP	3.17	0.21	1.47	0.15	Unknown
38037_at	DTR	2.61	0.06	1.46	0.10	Hypertrophy model
229865_at	FAD104	2.39	0.14	1.41	0.25	Protein-tyrosine phosphatase
221653_x_at	APOL2	2.16	0.26	1.41	0.20	Fatty acid metabolism
226757_at	IFIT2	2.60	0.25	1.40	0.05	Immune/inflammation
201662_s_at	ACSL3	2.88	0.30	1.40	0.16	Fatty acid metabolism
220132_s_at	LLT1	5.21	0.99	1.40	0.09	Signal transduction
225344_at	NCOA7	2.03	0.05	1.34	0.05	Receptor activity
230820_at	SMURF2	2.30	0.24	1.33	0.05	Proteolysis

(Continued)

Table 11.1 (*Continued*)

SC probe set	Gene[b]	TNFα mean[a]	TNFα SD	BE-30 + TNFα mean[a]	BE-30 + TNFα SD	Functional category
205596_s_at	SMURF2	2.05	0.06	1.32	0.05	Proteolysis
222062_at	IL27RA	2.54	0.05	1.32	0.18	Immune/inflammation
206995_x_at	SCARF1	2.18	0.20	1.31	0.10	Endothelial scavenger receptor
204533_at	CXCL10	4.63	0.38	1.31	0.04	Chemokine
227489_at	SMURF2	2.25	0.36	1.29	0.03	Proteolysis
209567_at	RRS1	2.04	0.12	1.29	0.12	Ribosome biogenesis
211588_s_at	PML	5.21	0.89	1.29	0.49	Transcription
202086_at	MX1	10.67	0.46	1.28	0.41	Immune/inflammation
208075_s_at	CCL7	4.06	0.75	1.25	0.68	Chemokine
201645_at	TNC	2.35	0.08	1.23	0.09	Adhesion molecule
202464_s_at	PFKFB3	2.04	0.16	1.23	0.08	RNA binding
224013_s_at	SOX7	2.04	0.11	1.22	0.05	Development
201661_s_at	ACSL3	2.57	0.17	1.21	0.18	Fatty acid metabolism
209277_s_at	TFPI2	2.76	0.26	1.21	0.11	Protease inhibitor
200704_at	LITAF	2.36	0.05	1.19	0.03	Inflammation
227020_at	YPEL2	2.41	0.04	1.18	0.05	Unknown
201660_at	ACSL3	2.55	0.07	1.16	0.06	Fatty acid metabolism
219522_at	FJX1	2.29	0.20	1.16	0.13	Differentiation
219279_at	DOCK10	2.75	0.12	1.16	0.15	Cell cycle
210001_s_at	SOCS1	2.37	0.33	1.15	0.23	Signal transduction
228607_at	OAS2	2.90	0.42	1.14	0.19	RNA binding
206133_at	HSXIAPAF1	3.07	0.84	1.14	0.21	Antiapoptosis
209969_s_at	STAT1	2.15	0.15	1.14	0.11	TGFβ signaling
204580_at	MMP12	16.42	0.56	1.13	0.42	Proteolysis
204702_s_at	NFE2L3	2.42	0.22	1.13	0.07	Transcription
218400_at	OAS3	2.55	0.19	1.12	0.13	RNA binding
241916_at	PLSCR1	2.27	0.37	1.12	0.09	Clearance of apoptotic cells
201150_s_at	TIMP3	2.19	0.18	1.12	0.08	Proteolysis
237169_at	TNC	3.58	0.27	1.11	0.33	Cell adhesion
226436_at	RASSF4	2.26	0.10	1.11	0.15	Signal transduction
236471_at	NFE2L3	2.46	0.09	1.11	0.06	Transcription
225163_at	FRMD4	2.15	0.03	1.11	0.08	Cytoskeletal binding
204972_at	OAS2	3.27	0.29	1.09	0.17	RNA binding
232020_at	SMURF2	2.62	0.07	1.06	0.03	Proteolysis
220104_at	ZC3HAV1	2.17	0.18	1.06	0.07	Immune
204994_at	MX2	2.58	0.12	1.05	0.20	Immune
221898_at	T1A-2	2.11	0.07	1.03	0.04	Angiogenesis
213506_at	F2RL1	2.01	0.07	1.02	0.10	G-protein receptor
208461_at	HIC1	2.21	0.19	1.01	0.14	Transcription
202446_s_at	PLSCR1	2.20	0.06	0.99	0.05	Clearance of apoptotic cells
237206_at	MYOCD	3.18	0.37	0.98	0.18	Transcription
204879_at	T1A-2	2.19	0.30	0.98	0.19	Angiogenesis

Table 11.1 (*Continued*)

SC probe set	Gene[b]	TNFα mean[a]	TNFα SD	BE-30 + TNFα mean[a]	BE-30 + TNFα SD	Functional category
222912_at	ARRB1	2.12	0.11	0.98	0.12	Signal transduction
214059_at	IFI44	2.91	0.25	0.97	0.04	Immune
206757_at	PDE5A	3.11	0.72	0.97	0.17	Purine metabolism
205483_s_at	G1P2	2.24	0.09	0.96	0.02	Immune
205660_at	OASL	2.10	0.16	0.94	0.12	RNA binding
218995_s_at	EDN1	2.12	0.15	0.93	0.19	Signal transduction
228617_at	HSXIAPAF1	4.99	0.26	0.91	0.62	Antiapoptosis
241986_at	BMPER	2.47	0.29	0.91	0.21	Differentiation
226722_at	FAM20C	2.26	0.26	0.90	0.05	Unknown
218832_x_at	ARRB1	2.28	0.39	0.89	0.07	Signal transduction
242726_at	ACSL3	2.02	0.13	0.85	0.15	Fatty acid metabolism
219211_at	USP18	2.80	0.11	0.84	0.19	Deubiquitination
210797_s_at	OASL	2.32	0.11	0.80	0.26	RNA binding
227657_at	RNF150	2.10	0.12	0.80	0.07	Unknown
201601_x_at	IFITM1	2.11	0.31	0.77	0.09	Cell cycle
222802_at	EDN1	2.21	0.09	0.77	0.07	Signal transduction
205552_s_at	OAS1	4.12	0.21	0.74	0.23	RNA binding
230746_s_at	STC1	2.21	0.19	0.74	0.12	Signal transduction
214453_s_at	IFI44	3.48	0.18	0.71	0.13	Immune/inflammation
214022_s_at	IFITM1	2.78	0.10	0.57	0.04	Cell cycle
212977_at	CMKOR1	2.60	0.13	0.52	0.06	G-protein receptor
206801_at	NPPB	2.02	0.03	0.44	0.04	Cardiovascular homeostasis
202869_at	OAS1	3.81	0.13	0.43	0.02	RNA binding
235276_at	EPSTI1	2.34	0.16	0.42	0.13	Unknown
203153_at	IFIT1	2.11	0.06	0.33	0.03	Immune/inflammation
206385_s_at	ANK3	2.62	0.33	0.27	0.15	Signal transduction
214329_x_at	TNFSF10	2.64	0.19	0.27	0.10	Apoptosis

stromelysins are ECM proteoglycans, laminin, fibronectin, and gelatin. MMPs are expressed at low levels in normal adult tissues and are upregulated during different physiological and pathological remodeling processes such as embryonic development, tissue repair, inflammation, rheumatoid arthritis, angiogenesis, tumor invasion, and metastasis. Inhibitors are predicted to have clinical benefits in arthritis and metastasis. Most MMPs are constitutively expressed *in vitro* at low levels by different cell types, such as keratinocytes, fibroblasts, macrophages, endothelial cells, mast cells, eosinophils, and neutrophils [33, 34].

Hence, the efficacy of BE-30 was further studied in human microvascular endothelial cells against TNFα-inducible MMP expression [23]. The robust BE-30 sensitive candidate genes were then subjected to further processing for the identification of BE-30-sensitive signaling pathways. The use of resources such as GenMAPP, *KEGG*, DAVID, and GO led to the recognition of the primary BE-30 sensitive TNFα-inducible pathways. Members of the MMP family of genes were subjected to hierarchical clustering. MMP3, MMP10, MMP12, and MMP19 were recognized as TNFα-inducible genes sensitive to BE-30 (Fig. 11.3). A quantitative real-time PCR approach was used in this study to follow-up on the results from the gene microarray study. TNFα

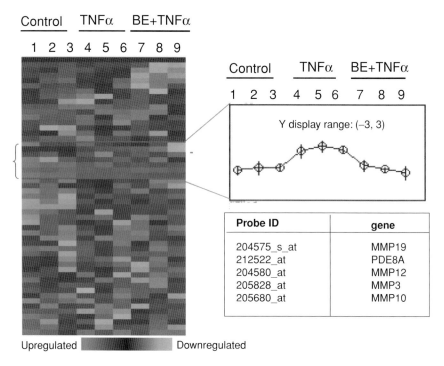

Figure 11.3 TNFα-induced *Boswellia*-sensitive genes of the MMP family in HMEC. GeneChip data were subjected to hierarchical clustering to visualize expression pattern of BE-sensitive MMP-family genes. TNFα, 50 ng/ml, 6 hours; TNFα + Boswellia extract, 48 hours pretreatment with BE-30 (25 μg/ml) followed by TNFα for 6 hours. The line graphs show average pattern of gene expression in corresponding cluster graph. Red to green gradation in color represents higher to lower expression signal. MMP19, matrix metalloproteinase 19; PDE8A, phosphodiesterase 8A; MMP12, matrix metalloproteinase 12 (macrophage elastase); MMP3, matrix metalloproteinase 3 (stromelysin 1, progelatinase), MMP10, matrix metalloproteinase 10 (stromelysin 2).

caused a dose-dependent induction of MMP3, MMP10, and MMP12 in HMEC. Pretreatment of HMEC for 2 days with BE-30 potently prevented TNFα-induced expression of MMP3 and MMP10, as well as MMP12 mRNA in HMEC (Fig. 11.4). HMEC was treated with 25 μg/ml *Boswellia* extracts (BE-3 or BE-30) for 48 hours. BE were prepared in DMSO at concentrations such that the final concentration of the solvent in cell suspension never exceeded 0.1% v/v. Controls were treated with matching volume of DMSO. After the treatment period, cells were activated with recombinant human TNFα (50 ng/ml) for 6 hours. Expression of MMP3, −10, and −12 mRNA was determined using real-time PCR. The data presented are normalized for β-actin (housekeeping gene) expression. Pretreatment of HMEC for 2 days with BE-30 potently prevented TNFα-induced expression of MMP3, MMP10, as well as MMP12 mRNA (Fig. 11.5).

The observed effects of BE-30 on TNFα-induced MMP3 gene and mRNA expression was further tested at the level of protein expression (MMP3) using ELISA. To establish qualitatively the correlation between AKBA concentration and efficacy, the observed outcome with BE-30 was compared with that of the commercial extract containing 3% AKBA (BE-3). Consistent with the quantitative gene expression findings, BE-30 potently inhibited TNFα-induced expression of MMP3 protein. The concentration of AKBA in the extracts was found to have an influence on

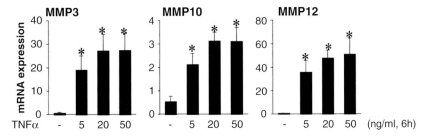

Figure 11.4 TNFα-induced MMP gene expression. HMEC were activated with various concentrations (5-50 ng/ml) of recombinant human TNFα for 6 hours. Expression levels of MMP-3, -10, and -12 mRNA were determined using real-time PCR. The data presented are normalized for β-actin. Each bar represents mean $\pm$ SD. $*p < 0.001$, significantly different compared to control (untreated) sample.

observed efficacy. BE-30 was significantly more potent than BE-3 in preventing TNFα-induced MMP3 protein expression (Fig. 11.6).

To further validate the effect of BE-30 on TNFα-induced MMP3 expression, the effect of BE-30 on TNF α-induced MMP3 activity was determined using Biotrak MMP3 activity assay system (Amersham Biosciences, USA). The HEMCs were treated with 25 μg of BE-3, BE-30, or control (DMSO). The cells were then treated with recombinant human TNFα for 24 hours. The total (pro and active) MMP3 activity was then measured. This assay system consists of pro form of a detection enzyme that can be activated by captured active MMP3 and converted into an active detection enzyme through a single proteolytic event. The results showed significant induction of MMP3 activity in HMECs after TNFα treatment. However, such inducible MMP3 activity was significantly inhibited by both *Boswellia* extracts. BE-30 showed significantly more potent MMP3 inhibitory activity when compared to BE-3 in HMECs (Fig. 11.7).

The MMP3 inhibitory potential of BE-30 was further evaluated in phorbol-12-myristate-13-acetate (PMA)-induced A2058, human melanoma cells (unpublished). The supernatants

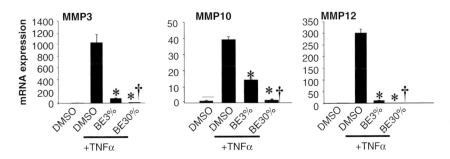

Figure 11.5 *Boswellia* significantly downregulated TNFα-induced MMP mRNA expression. HMEC were pretreated with 25 μg/ml *Boswellia* extracts (BE-3 or BE-30) for 48 hours. BE were prepared in DMSO at concentrations such that the final concentration of the solvent in cell suspension never exceeded 0.1% v/v. Controls were treated with equivalent volume of DMSO. After the treatment duration, cells were activated with recombinant human TNFα (50 ng/ml) for 6 hours. Expression of MMP-3, -10, and -12 mRNA was determined using real-time PCR. The data presented are normalized for β-actin (housekeeping gene) expression. Each bar represents mean $\pm$ SD. $*p < 0.05$, significantly lower compared to TNFα treated sample. $^{\dagger}p < 0.05$, significantly lower compared to BE-3-treated samples.

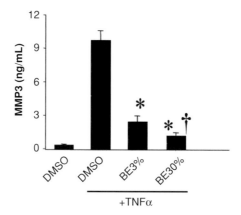

Figure 11.6 *Boswellia*-inhibited TNFα-induced MMP-3 protein expression. HMEC were treated with 25 μg/ml *Boswellia* extracts (BE-3 or BE-30) for 48 hours. After the treatment period, cells were activated with recombinant human TNFα (50 ng/ml) for 24 hours. Pro-MMP-3 levels in culture media were determined using ELISA. Data (mean ± SD) are presented as actual MMP-3 levels (ng/ml) in the culture media. *$p < 0.05$, significantly lower compared to TNF-α treated sample. †$p < 0.05$, significantly lower compared to BE-3 treated samples.

obtained from control, treated cells were used to measure MMP3 production by MMP3 ELISA Development Kit (R&D System, Minneapolis, MN, USA). Cell culture experiments show BE-30 is able to inhibit MMP3 production in PMA-induced human melanoma cells in a dose-dependent manner. Quantitative measurement of MMP3 in the culture supernatants revealed an IC$_{50}$ of BE-30 for MMP3 production is 25 μg/ml [35].

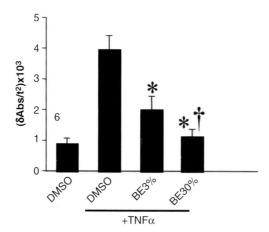

Figure 11.7 *Boswellia* downregulated TNFα-induced MMP-3 activity. HMEC were treated with 25 μg/ml *Boswellia* extracts (BE-3 or BE-30) for 48 hours. After the treatment period, cells were activated with recombinant human TNF-α (50 ng/ml) for 24 hours. Pro and active MMP-3 activity was determined using a Biotrak MMP-3 activity assay system. Data (mean ± SD) are presented as actual MMP-3 activity in the culture media. The MMP-3 activity is represented by the rate of change of absorbance at 405 nm, that is, δAbs405/t2, where t is the incubation time in hours. Data are expressed as $\{(\delta Abs405/t2) \times 103\}$. *$p < 0.05$, significantly lower compared to TNFα treated sample. †$p < 0.05$, significantly lower compared to BE-3 treated samples.

IN VIVO STUDIES

The *in vitro* evidences including gene expression and protein expression studies motivated us to test *in vivo* anti-inflammatory properties of BE-30 in rat models of inflammation and experimental arthritis. BE-30 significantly inhibited carrageenan-induced rat paw inflammation in albino Wistar rats [22]. Animals were orally supplemented with BE-30 in 1% CMC (25, 50, and 100 mg per kg) or ibuprofen (50 mg/kg body weight; positive control) for 30 days. The control group received same volume of vehicle (1% CMC) orally. Following treatment for 30 days, 1 hour after the last dose 0.1 ml of 1% solution of carrageenan was injected subcutaneously in the subplantar region of left hind paw. The paw volume was measured using a plethysmometer before and after 4 hours of the subcutaneous injection of carrageenan. The difference in paw volume between 4 hours after the injection and the initial paw volume was noted as edema volume. Percentage inhibition in edema was calculated by comparing mean edema of control group and the treated groups. BE-30 exhibited dose-dependent reduction in paw edema. It showed 18.23%, 23.65%, and 27.07% inhibition at 25, 50, and 100 mg per kg body weight respectively (Fig. 11.8).

BE-30 also exhibited significant therapeutic effects against Freund's adjuvant-induced arthritis in Wistar albino rats [23]. Animals were orally supplemented with BE-30 (25, 50, and 100 mg/kg body weight) or *Boswellia* extract containing 3% AKBA (BE-3) 100 mg/kg body weight in 1% carboxymethylcellulose (CMC) or prednisolone 10 mg/kg body weight for 30 days. Oral supplementation of BE-30 offered dose-dependent and statistically significant reduction in paw

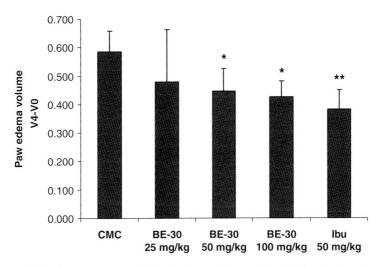

Figure 11.8 BE inhibits carrageenan-induced rat paw inflammation. Wistar albino rats were orally supplemented with BE in 1% CMC (25, 50, and 100 mg per kg) or ibuprofen (50 mg/kg body weight; positive control) for 30 days. The control group received same volume of vehicle (1% CMC). Following treatment for 30 days, 1 hour after the last dose, 0.1 ml of 1% solution of carrageenan was injected subcutaneously in the subplantar region of left hind paw. The difference in paw volume between 4 hours after the injection and the initial paw volume was noted as edema volume. Percentage inhibition in edema was calculated by comparing mean edema of control group and the treated groups. Each represent mean ± SD. $*P < 0.05$ significantly different compared to CMC treated control group. $**P < 0.005$ significantly different compared to CMC treated control group. Ibu, ibuprofen (oral 50 mg/kg body weight for 30 days); BE (oral 25, 50, and 100 mg/kg body weight for 30 days); CMC, oral 1% carboxymethylcellulose for 30 days.

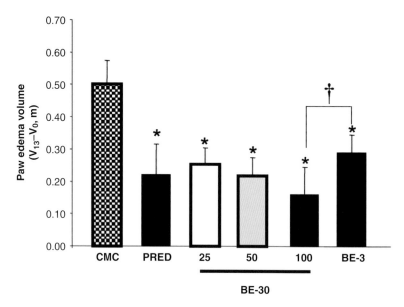

Figure 11.9 Percentage inhibition of adjuvant induced rat paw edema. *Boswellia* extract protects against Freund's adjuvant-induced arthritis in rats. Wistar albino rats ($n = 6$) were supplemented with BE-30 (25, 50, and 100 mg/kg body weight oral), BE-3 (100 mg/kg body weight oral), or prednisolone (PRED, 10 mg/kg body weight oral) in 1% carboxymethylcellulose (CMC) for 30 days. The control group (marked CMC) received the same volume of vehicle (i.e., 1% CMC) orally. On day 31, 50 μL of complete Freund's adjuvant was injected subcutaneously in the subplantar region of left hind paw. Paw edema = V13-V0. V0 and V13 represent the paw volume on days 0 and 13, respectively, after adjuvant injection.

edema and showed 49.3%, 56.7%, and 68% reduction in paw edema, respectively at 25, 50, and 100 mg per kg body weight. Prednisolone 10 mg/kg showed similar (55.7%) protection to that of BE-30 at 50 mg/kg dose. BE-30 showed significantly higher inhibition against adjuvant-induced inflammatory response compared to BE-3 (Fig. 11.9).

SAFETY OF 5-LOXIN®

A broad spectrum of safety studies of BE-30 conducted in rats and rabbits indicated its safety [24]. No acute oral toxicity was observed in either male or female Sprague-Dawley rats supplemented with BE-30 at a dose of 5 g/kg body weight, a dose level considered to be several-fold higher than the recommended daily human dose (100 mg/day). BE-30 was classified as nonirritating to skin at a dose of 2.0 g/kg when topically applied to skin in an acute dermal toxicity study conducted on New Zealand albino rabbits and showed no irritation to eye in primary eye irritation tests performed also on New Zealand albino rabbits. A dose-dependent 90-day subchronic toxicity study was conducted on male and female Sprague-Dawley rats. The animals in the treatment groups were supplemented with a feed containing 0.025%, 0.25%, or 2.5% of BE-30 corresponding to 0.2 g, 2.0 g, or 20.0 g of human equivalence dose (HED), respectively, for 90 days. Selected groups of animals were sacrificed after 30, 60, and 90 days of treatment and target organs such as adrenal glands, brain, epididymis, esophagus, eyes, heart, intestine, kidney, liver, lymph nodes, lungs, mammary glands, ovaries or testes, pancreas, pituitary, prostate, salivary glands, seminal vesicles, skin, spleen, stomach, thymus gland, thyroid gland, trachea, uterus, and urinary bladder were extracted and evaluated for toxic symptoms. Hematology, serum chemistry, and histopathological evaluations did not show any adverse effects in any of the organs tested. A comprehensive perusal of the safety data indicated that the no-observed-adverse-effect level (NOAEL) for male and female Sprague-Dawley rats supplemented with BE-30 *ad libitum* is presumed to be at least 20.0 g per day HED. Finally, the genotoxic effect of BE-30 was evaluated using the bacterial reverse mutation test (Ames test) and the results showed that BE-30 was nonmutagenic up to the highest tested concentration of 3000 mcg/plate [36]. BE-30 did not exhibit clastogenic potential to induce micronucleated reticulocytes of mouse peripheral blood in micronucleus assay in BALB/c mice. Also, BE-30 did not induce structural chromosome aberration both with and without metabolic activation in Chinese hamster cells [37]. These two studies further confirm the nongenotoxic nature of BE-30.

Safety Data

Acute Dermal Toxicity: LD_{50} 2000 mg/kg
Acute Oral Toxicity: LD_{50} 5000 mg/kg
Primary Skin Irritation: Non irritating to skin
Primary Eye Irritation: Mildly irritating to eye
Chronic Toxicity: LD_{50} 2.5 gm/kg

HUMAN CLINICAL STUDIES VALIDATE GENE CHIP STUDIES

Articular cartilage is composed of chondrocytes and a cartilage matrix. Collagen is the main component of cartilage matrix, which provides tensile strength and elasticity to cartilage. Under normal physiological conditions, degeneration and synthesis of cartilage macromolecules are under equilibrium state. At physiological concentrations, proteases help remodeling of tissues. However, overexpressed proteases during various pathological conditions, such as arthritis, degenerate cartilage and other connective tissues. Out of four classes of proteases, MMPs received major attention due to their key role in arthritic tissue degeneration. Synovial and serum MMP levels are well correlated with collagen degeneration and clinical symptoms. Out of several MMPs, MMP3 plays a central role in arthritis etiology due to its capability to degrade various macromolecules of joint cartilage, inducing other metalloproteinases and synergistically destructing cartilage with other metalloproteinases [38].

Osteoarthritis is a common, chronic, progressive, degenerative disorder, which commonly affects the knee joint and disables the elderly. Osteoarthritis is a growing health concern that has become a major challenge to the health professionals. In Western populations it is one of the most frequent causes of pain, loss of function, and disability in adults. Significant efficacy of BE-30 in downregulation of key inflammatory genes including adhesion molecules and MMPs was observed in GeneChip studies and reduction in expression of MMPs, and inhibiting the activity of MMP3 was observed in MMP expression and activity studies. These results were corroborated with *in vivo* efficacy studies in inflammation and arthritis rat models. These results prompted us to test efficacy of BE-30 in a human clinical study.

A 90-day double-blind, placebo-controlled human clinical study to evaluate the efficacy and tolerability of BE-30 in the treatment of osteoarthritis of knee [25] was conducted at Alluri Sitarama Raju Academy of Medical Sciences (ASRAM), Eluru, AP, India. The study protocol was approved by IRB of ASRAM (IRB# 06 001) and the clinical trial registration number was ISRTCN 5212803.

Mild to moderate osteoarthritis subjects were screened according to eligibility criteria of the study, which was devised according to the American College of Rheumatology, and seventy-five subjects who gave voluntary consent were recruited into the study. Selected subjects were randomly allocated in two treatment groups (BE-30 100 and 250 mg/day) and a placebo group by a randomization table generated using validated software RANCODE.

All the selected subjects underwent baseline evaluation and completed a questionnaire, providing demographics, medical history, and nutritional status. Widely acceptable and validated pain and physical function scores including Western Ontario and McMaster Universities (WOMAC) Index of Osteoarthritis, Lequesne functional index (LFI), and visual analog scale (VAS) were used to evaluate the efficacy of BE-30. These pain scores were widely used and corroborate well with clinical symptoms and disease markers [39–41]. ECGs were recorded for all the subjects. Blood and urine samples were collected from all the subjects for the evaluation of serum biochemical parameters, hematology, and urinalysis. Study supplement was issued by the study pharmacist and the patients were advised not to consume pain relieving medicines other than rescue medication. All parameters were assessed for the subjects during 7, 30, 60, and 90 days with an exception of synovial fluid, which was collected only on day 0 and day 90.

At both dose levels (100 and 250 mg/day) tested, BE-30 conferred clinically and statistically significant improvements in pain, joint stiffness, and physical function scores in osteoarthritis patients. Interestingly, significant improvement in pain scores was observed in both the treatment groups supplemented with BE-30 at as early as 7 days. BE-30 at 100 mg daily dose showed 48.83% ($P < 0.001$), 23.79% ($P < 0.036$), 39.61% ($P = 0.009$), 42.5% ($P = 0.120$), and 28.62% ($P = 0.100$) improvements, respectively, in VAS, LFI, WOMAC pain, WOMAC stiffness, and WOMAC functional ability scores. BE-30 at 250 mg showed 65.94% ($P < 0.001$), 31.34% ($P < 0.017$), 52.05% ($P < 0.001$), 62.22% ($P = 0.014$), and 49.34% ($P = 0.002$) improvements, respectively, in VAS, LFI, WOMAC pain, WOMAC stiffness, and WOMAC functional ability scores. In corroboration with the improvements in pain scores in the treatment groups, significant reduction of synovial fluid MMP3, a potent cartilage-degrading enzyme, was also documented.

MMP3, also known as stromelysin, transin, progelatinase is a proteglycanase produced predominantly by connective tissue cells. MMP3 has a broad range of substrate specificity and capability of inducing other metalloproteinases. Together with other metalloproteinases it synergistically degrades various components of cartilage and became a predominant etiological factor in various degenerative diseases including arthritis. Hence, the level of MMP3 in synovial fluid should show an inverse relation to cartilage integrity and joint health. To study the effect of BE-3 on MMP3 levels in knee joints, synovial fluid from the inflamed knee joints

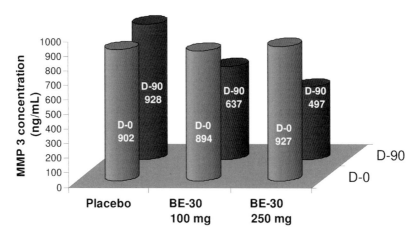

Figure 11.10 Synovial MMP-3 levels. Each bar represents mean concentration of MMP-3 (ng/ml synovial fluid). Presented are the matrix metalloproteinase (MMP)-3 levels in synovial fluid collected from BE-30 treated and placebo patients of osteoarthritis. At day 90, there was no significant change in MMP-3 concentration in the placebo group compared with baseline. In comparison with the placebo group, at the end of the study the groups receiving 100 mg/day and 250 mg/day BE-30 showed 31.37% ($P = 0.002$), and 46.4% ($P < 0.001$) reductions in MMP-3 concentration, respectively.

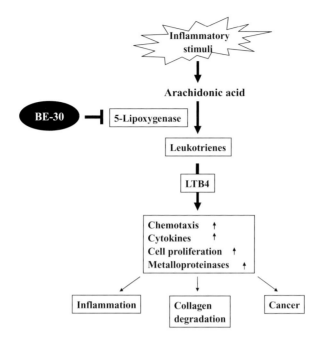

Figure 11.11 Overview. Flow diagram represents the overview of pathways induced by inflamma-tory stimuli that lead to various pathological conditions. BE-30 blocks the leukotriene biosynthesis by inhibiting 5-lipoxygenase activity and regulates pathological factors responsible for a variety of diseases. Small up-arrows (↑) indicate up regulatory effect.

of subjects were aseptically collected for MMP3 evaluation. The MMP3 levels were measured quantitatively by the ultrasensitive ELISA kit (R & D Systems, USA). Synovial fluid samples were incubated on capture antibody coated 96-well microplates. Specifically bound MMP3 was detected by appropriate biotinylated detection antibody and was probed with horseradish per-oxidase enzyme. The specific immune reaction was detected by substrate solution and the color development was read using a microplate reader (Bio-Rad, Hercules, CA, USA). The sensitivity of MMP3 detection ELISA kit is 9 pg/ml. The MMP3 data for the placebo group and the two treatment groups are summarized in Fig. 11.10.

Compared to the baseline data, both the treatment groups showed highly significant reductions in synovial fluid MMP3 concentrations at the end of 90-day study period. The MMP3 level in the placebo group was virtually unchanged at the end of the 90-day period when compared to the baseline value. When compared to the placebo group the low-dose (100 mg) and high-dose (250 mg) groups showed 31.37% ($p = 0.002$) and 46.4% ($p < 0.001$) reductions in MMP3 concentrations, respectively. The high-dose group also showed significant reduction ($P < 0.0001$) in synovial MMP3 concentration when compared to the low-dose group [25]. These clinical findings further substantiated the data obtained from GeneChip studies, which showed downregulation of genes having anti-inflammatory and proteolytic functions. The safety parameters were virtually unchanged in the treatment groups when compared with those in the placebo group. This further confirms safety and tolerability of BE-30.

CONCLUSION

BE-30 may thus represent a first plant-derived product that underwent whole genome screening and had its molecular basis of anti-inflammatory properties established through detailed gene expression studies. The sensitivity of the inducible expression of many inflammatory genes including TNFR1, ICAM-1, and VCAM-1 to BE-30 offers a mechanistic rationale for the observed efficacy of *Boswellia* extracts. The inflammatory properties exhibited by BE-30 in GeneChip study were supported by potent efficacy shown by BE-30 in many *in vitro* and *in vivo* models of inflammation. The different true validation of GeneChip study findings, however, came from human clinical study. This placebo-controlled, double-blind, human clinical study not only improved the joint function significantly but also significantly reduced the concentration of cartilage-degrading enzyme MMP3 in synovial fluid consistent with the observed sensitivity of the inducible expression of MMP3, MMP10, and MMP12 to BE-30 in human microvascular endothelial cells. This chapter reviews the detailed studies carried out by us and other previous workers relating the molecular basis of anti-inflammatory properties of AKBA and its enriched product BE-30. Based on the previous information, BE-30 can be used as a potential therapeutic candidate against various types of inflammatory manifestations (Fig. 11.11).

REFERENCES

1. Simmons, D. L., Xie, W., Chipman, J. G., Evett, G. E., Multiple cyclooxygenases: cloning of a mitogen inducible form. In: J. M. Bailey (ed.), *Prostaglandins, Leukotrienes, Lipoxins, and PAF*. New York: Plenum, pp. 67–78, 1992.
2. Smith, W. L., Dewitt, D. L., Prostaglandin endoperoxide H synthases-1 and -2. *Adv Immunol* 1996, *62*, 167–215.
3. Yamamoto, S., Mammalian lipoxygenases: molecular structures and functions. *Biochim Biophys Acta* 1992, *1128*, 117–131.
4. Ethan, B., Heather, B., Theresa, D. H., Ivo, F., *et al., Boswellia*: An evidence-based systematic review by the natural standard research collaboration. *J Herb Pharmacother* 2004, *4*, 63–83.
5. Chatterjee, G. K., Pal, S. D. *Indian Drugs* 1984, *21*, 431.
6. Ammon, H. P., Mack, T., Singh, G. B., Safayhi, H., Inhibition of leukotriene B4 formation in rat peritoneal neutrophils by an ethanolic extract of the gum resin exudate of *Boswellia serrata*. *Planta Med* 1991, *57*, 203–207.
7. Singh, G. B., Atal, C. K., Pharmacology of an extract of salai guggal ex-*Boswellia serrata*, a new non-steroidal anti-inflammatory agent. *Agents Actions* 1986, *18*, 407–412.
8. Ammon, H. P. T., Boswellic acids (components of frankincense) as the active principle in treatment of chronic inflammatory diseases. *Wien Med Wochenschr* 2002, *152*, 373–378.
9. Kimmatkar, N., Thawani, V., Hingorani, L., Khiyani, R., Efficacy and tolerability of *Boswellia serrata* in treatment of osteoartritis of knee-A randomized double blind placebo controlled trial. *Phytomedicine* 2003, *10*, 3–7.
10. Gupta, I., Parihar, A., Malhotra, P., Gupta, S., *et al.,* Effects of gum resin of *Boswellia serrata* in patients with chronic colitis. *Planta Med* 2001, *67*, 391–395.
11. Gupta, I., Parihar, A., Malhotra, P., Singh, G. B., *et al.,* Effects of *Boswellia serrata* gum resin in patients with ulcerative colitis. *Eur J Med Res* 1997, *2*, 37–43.
12. Gerhardt, H., Seifert, F., Buvari, P., Vogelsang, H., *et al.,* Therapy of active Crohn disease with *Boswellia serrata* extract H15. *Z Gastroenterol* 2001, *39*, 11–17.
13. Gupta, I., Gupta, V., Parihar, A., Gupta, S., *et al.,* Effects of *Boswellia serrata* gum resin in patients with bronchial asthma: results of a double-blind, placebo-controlled, 6-week clinical study. *Eur J Med Res* 1998, *3*, 511–514.
14. Safayhi, H., Sailer, E. R., Anti-inflammatory actions of pentacyclic triterpenes. *Planta Med* 1997, *63*, 487–493.
15. Pardhy, R. S., Bhattacharya, S. C., B-boswellic acid, acetyl-b-boswellic acid, acetyl-11-keto-b-boswellic acid & 11-keto-b-boswellic acid and four pentacyclic triterpene acids from the resin of *s* Roxb. *Indian J Chem* 1978, *16B*, 176–178.
16. Safayhi, H., Rall, B., Sailer, E. R., Ammon, H. P. T., Inhibition by boswellic acids of human leukocyte elastase. *J Pharmacol Exp Ther* 1997, *281*, 460–463.

17. Poff, C. D., Balazy, M., Drugs that target lipoxygenases and leukotrienes as emerging therapies for asthma and cancer. *Curr Drug Targets Inflamm Allergy* 2004, *3*, 19–33.
18. Safayhi, H., Mack, T., Sabieraj, J., Anazodo, M. I., *et al.,* Boswellic acids: novel, specific, nonredox inhibitors of 5-lipoxygenase. *J Pharmacol Exp Ther* 1992, *26*, 1143–1146.
19. Sailer, E. R., Subramanian, L. R., Rall, B., Hoernlein, R. F., *et al.,* Acetyl-11-keto-beta-boswellic acid (AKBA): structure requirements for binding and 5-lipoxygenase inhibitory activity. *Br J Pharmacol* 1996, *117*, 615–618.
20. Sailer, E. R., Schweizer, S., Boden, S. E., Ammon, H. P. T., *et al.,* Characterization of an acetyl-11-keto-beta-boswellic acid and arachidonate-binding regulatory site of 5-lipoxygenase using photoaffinity labeling. *Eur J Biochem* 1998, *256*, 364–368.
21. Safayhi, H., Sailer, E. R., Ammon, H. P. T., Mechanism of 5-lipoxygenase inhibition by acetyl-11-keto-beta-boswellic acid. *Mol Pharmacol* 1995, *47*, 1212–1216.
22. Roy, S., Khanna, S., Shah, H., Rink, C., *et al.,* Human genome screen to identify the genetic basis of the anti-inflammatory effects of Boswellia in microvascular endothelial cells. *DNA Cell Biol* 2005, *24*(4), 244–255.
23. Roy, S., Khanna, S., Krishnaraju, A.V., Subbaraju, G. V., *et al.,* Regulation of vascular response to inflammation: inducible matrix metalloprotenase-3 expression in human microvascular endothelial cells is sensitive to anti-inflammatory Boswellia. *Antioxid Redox Signal* 2006, *8* (3&4), 653–660.
24. Lalithakumari, K., Krishnaraju, A. V., Subbaraju, G. V., Yasmin, T., *et al.,* Safety and toxicological evaluation of a novel, standardized 3-acetyl-11-keto-β-boswellic acid (AKBA)-enriched *Boswellia serrata* extract (5-Loxin). *Toxicol Mech Methods* 2006, *16*, 199–226.
25. Sengupta, K., Alluri, K. V., Satish, A. R., Mishra, S., *et al.,* Double blind, randomized, placebo controlled study of the efficacy and safety of 5-Loxin® for treatment of osteoarthritis of the knee. *Arthritis Res Ther* 2008, *10*(4), R85–R96.
26. Warren, J. S., Ward, P. A., Johnson, K. J., Tumor necrosis factor: A plurifunctional mediator of acute inflammation. *Modern Pathol* 1988, *1*, 242–247
27. Strieter, R. M., Kunkel, S. L., Bone, R. C., Role of tumor necrosis factor-alpha in disease states and inflammation. *Crit Care Med* 1993, *21*, S447–S463.
28. Sullivan, K. E., Regulation of inflammation. *Immunol Res* 2003, *27*, 529–538.
29. Pober, J. S., Endothelial activation: Untracellular signaling pathways. *Arthritis Res* 2002, *4*(3), S109–S116.
30. Li, C., Wong, W. H., Model-based analysis of oligonucleotide arrays: Expression index computation and outlier detection. *Proc Natl Acad Sci* 2001, *98*, 31–36.
31. Albelda, S. M., Smith, C. W., Ward, P. A., Adhesion molecules and inflammatory injury. *FASEB* 1994, *8*, 504–512.
32. Barreiro, O., Yáñez-Mó, M., Serrador, J. M., Montoya, M. C., *et al.,* Dynamic interaction of VCAM-1 and ICAM-1 with moesin and ezrin in a novel endothelial docking structure for adherent leukocytes. *J Cell Biol* 2002, *157*(7), 1233–1245.
33. Massova, I., Kotra, L. P., Fridman, R., Mobashery, S., Matrix metalloproteinases: Structures, evolution, and diversification. *FASEB Journal* 1998, *12*, 1075–1095.
34. Herouy, Y., The role of matrix metalloproteinases (MMPs) and their inhibitors in venous leg ulcer healing. *Phlebolymphology* 2004, *44*, 231–243.
35. Sengupta, K., Tummala, T., Golakoti, T., 30% 3-O-acetyl-11-keto-β-boswellic acid inhibits phorbol-myristate acetate stimulated matrix metalloproteinase-3 in A2058 human melanoma cells. Study no. LI/CMB/MMP-3/02/08. Submitted to the archival files, Laila Impex R&D Center, 2008.
36. Indrani, B. K., Bacterial reverse mutation test with 5-Loxin. Study No. 4477/05, Toxicology Department, Advinus Therapeutics Private Limited, Post Box No. 5813, Plot Nos 21 & 22, Peenya Industrial Area, Phase II, Bangalore, India 560 058, 2006.

37. Chang, J-T., Micronucleus assay in mice 5-Loxin. Study numbers MN00075 and CA00094, Center of Toxicology and Preclinical Sciences, Development Center for Biotechnology, 101, Lane 169, Kangning St, Xizhi City, Taipei County 221, Taiwan, 2007.
38. Takahashi, M., Naito, K., Abe, M., Sawada, T., *et al.,* Relationship between radiographic grading of osteoarthritis and the biochemical markers for arthritis in knee osteoarthritis. *Arthritis Res Ther* 2004, *6*(3), R208–R212.
39. Bellamy, N., Buchnan, W. W., Goldsmith, C. H., Campbell, J., *et al.,* Validation study of WOMAC: a health status instrument for measuring clinically important patient relevant outcomes to anti-rheumatic drug therapy in patients with osteoarthritis of the hip or knee. *J Rheumatol* 1988, *15*, 1833–1840.
40. Lequesne, M. G., Mery, C., Samson, M., Gerard, P., Indexes of severity for osteoarthritis of the hip and knee validation-value in comparison with other assessment tests. *Scand J Rheumatol* 1987, *65*, 85–89.
41. Chapman, C. R., Casey, K. L., Dubner, R., Foley, K. M., *et al.,* Pain measurement: an overview. *Pain* 1985, *22*, 1–31.

12

Nutrigenomic Perspectives on Cancer Chemoprevention with Anti-inflammatory and Antioxidant Phytochemicals: NF-κB and Nrf2 Signaling Pathways as Potential Targets

Hye-Kyung Na and Young-Joon Surh

INTRODUCTION

Cancer is a global health problem – particularly with the steady rise in life expectancy, increasing urbanization, and subsequent changes in lifestyle. Over the past 2 or 3 decades, we have witnessed enormous progress in the development of a vast variety of anticancer drugs and strategies. Nonetheless, we do not have a magic bullet that can completely and selectively destroy malignant cells. In this context, more attention should be paid to the prevention of cancer. Increased consumption of fruit and vegetables has been recommended for prevention of cancer. A vast variety of chemical substances derived from plants, collectively called "phytochemicals" ("*phyto*"-in Greek means plants), possess substantial anticarcinogenic activities. Because neoplastic transformation, in general, is a multistage process that may take more than decades, there are ample opportunities to intervene in the pathogenesis of cancer, especially at early phases of oncogenesis. Chemoprevention, an attempt to use nontoxic chemical substances or their mixtures to interfere with neoplastic development, is considered to be an effective and reliable strategy to reduce the risk of human malignancies. Over the past few decades, there has been a growing body of interest in identifying naturally occurring chemopreventive agents, particularly those present in fruits, vegetables, and spices. Figure 12.1 illustrates the chemical structures of representative edible phytochemicals known to possess chemopreventive potential. Plant-based products are, in general, inexpensive and relatively safe compared to synthetic agents. Therefore, a series of human intervention trials are being considered with individual phytochemicals or their combination with known synthetic chemopreventive agents. Based on preclinical results, selected phytochemicals have been evaluated in clinical interventions for various cancers [1]. However, precise assessment of underlying mechanisms of individual components is necessary before undertaking large-scale human trials. This chapter deals with two representative redox-sensitive transcription factors, NF-κB and Nrf2, as critical targets of some chemopreventive phytochemicals.

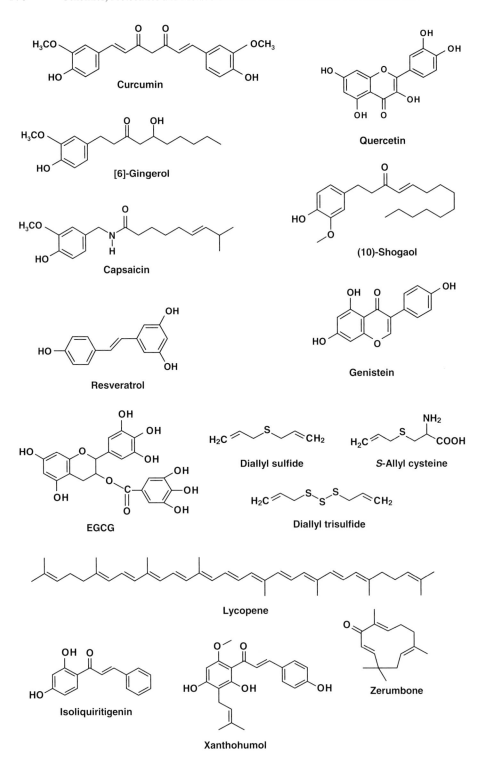

Figure 12.1 Chemical structures of representative chemopreventive phytochemicals with anti-inflammatory and/or antioxidative activities.

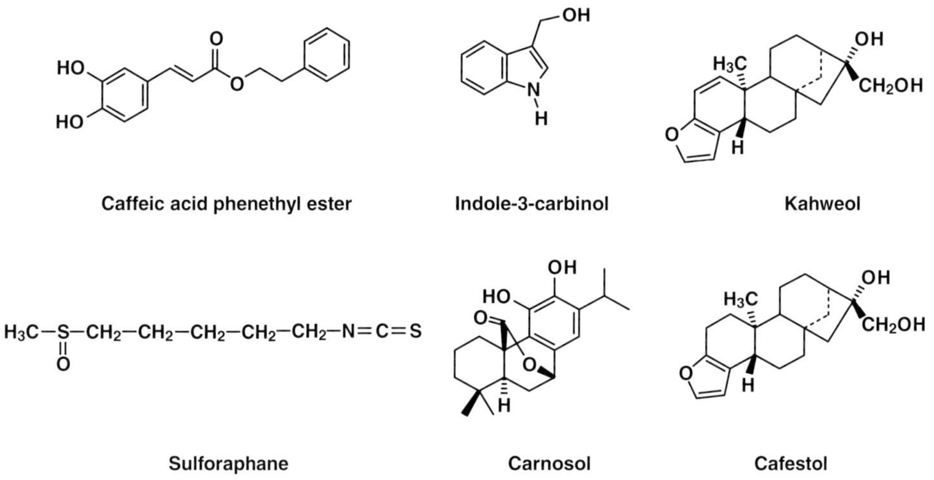

| Caffeic acid phenethyl ester | Indole-3-carbinol | Kahweol |

| Sulforaphane | Carnosol | Cafestol |

Figure 12.1 (*Continued*)

MOLECULAR-BASED CANCER CHEMOPREVENTION

Carcinogenesis is a complex process that involves a series of individual steps, each of which accompanies distinct molecular and cellular alterations. Initiation is a rapid and irreversible process that involves a chain of extracellular and intracellular events. These include the initial uptake or exposure to a carcinogenic agent, its distribution and transport to organs and tissues where metabolic activation and detoxification can occur, and the covalent interaction of ultimate electrophilic species with target cell DNA, leading to genotoxic damage. In contrast to initiation, tumor promotion is considered a relatively lengthy and reversible process in which actively proliferating preneoplastic cells accumulate. Progression, the final stage of neoplastic transformation, involves the growth of a tumor with increased invasiveness and metastatic potential.

The chemopreventive effects that most edible phytochemicals exert are likely to be the sum of several distinct mechanisms. These include blockade of metabolic activation and subsequent DNA binding of carcinogens, stimulation of detoxification, repair of DNA damage, suppression of cell proliferation and metastasis or angiogenesis, induction of differentiation or apoptosis of precancerous or maliganant cells, etc. [2]. Because the cellular signaling network often goes awry in carcinogenesis, it is fairly rational to target intracellular signaling cascades for chemoprevention. Components of signaling networks include protein kinases, such as the family of proline-directed serine/threonine kinases named mitogen-activated protein (MAP) kinases, protein kinase C (PKC), phosphatidylinositol-3-kinase (PI3K), protein kinase B/Akt, glycogen synthase kinase (GSK), and their downstream targets including redox-sensitive transcription factors. Many chemopreventive phytochemicals can work as regulators of specific signal transduction pathways to elicit beneficial effects [2].

NUCLEAR FACTOR-KAPPA B (NF-κB)

Although inflammation acts as an adaptive host defense against infection or injury, inadequate resolution of inflammatory responses often ends up with various chronic diseases including cancer. Mounting evidence from preclinical and clinical studies suggests that chronic inflammation plays a multifaceted role in carcinogenesis. It is estimated that 15% to 20% of human malignancies are associated with chronic inflammation as a consequence of persistent pathogen infections [3]. In response to proinflammatory stimuli, activated inflammatory/immune cells generate reactive oxygen species (ROS) and reactive nitrogen species (RNS), which can function as chemical effectors in inflammation-associated carcinogenesis. Thus, one of the plausible mechanisms by which chronic inflammation causes malignant transformation include oxidative/nitrosative DNA damage leading to activation of oncogenes and/or inactivation of tumor suppressor genes. Furthermore, a wide array of DNA-binding proteins are aberrantly activated in response to inflammatory stimuli, which can cause inappropriate induction of various proinflammatory genes.

NF-κB, a ubiquitously expressed eukaryotic transcription factor, regulates the expression of genes encoding those proteins that are essential for inflammatory responses, cell survival, immune reactions, and cell proliferation. Activation of NF-κB occurs in response to diverse stimuli including ligation of innate immune receptors, antigen-receptor engagement, and proinflammatory cytokines [4–6]. NF-κB activation upon exposure to external stimuli is typically rapid and transient for timely regulation of target gene expression. However, constitutively overactivated NF-κB signaling often occurs in many types of cancers, making this transcription factor an attractive target for the development of anticancer as well as anti-inflammatory drugs.

The NF-κB family contains at least five structurally related members – p50, p52 (the N-terminal fragments of the longer NF-κB1/p105 and NF-κB2/p100 proteins, respectively), p65 (RelA), c-Rel, and RelB [7]. These proteins form dimers and are normally sequestered in the cytoplasm of resting cells by the inhibitory family of IκB proteins. The IκB family comprises IκBα, IκBβ, IκBε, and the C-terminal of p105 and p100. A prototypic NF-κB-IκB complex expressed in the majority of cell types is a heterodimer of p50 and p65 associated with IκBα as an inactive complex in cytoplasm. Upon stimulation with mitogens, proinflammatory cytokines, UV radiation, viral infection, bacterial toxins, etc., IκBα becomes phosphorylated by activated IκB kinases (IKKs). Phosphorylated IκBα, upon ubiquitination, is directed to proteasomes for degradation. The degradation of IκBα allows NF-κB to translocate to the nucleus for binding to a -κB element located in the promoter regions of various proinflammatory genes including cox-2, thereby controlling their expression.

A pivotal regulator of all inducible NF-κB signaling pathways is IKK complex that consists of two kinases (IKKα and IKKβ) and a regulatory subunit named NF-κB essential modulator (NEMO)/IKKγ. The IKKs function as a bridge between inflammation and cancer [8]. IKKβ plays a role in tumor promotion whereas IKKα is mainly involved in metastatogenesis [8]. Two major pathways, the classical (or canonical) and noncanonical mechanisms of NF-κB activation have been identified. The "canonical" mechanism involves NEMO/IKKγ– and IKKβ-dependent IκB phosphorylation and degradation, liberating NF-κB complexes typified by the p50–p65 heterodimer (Fig. 12.2). Canonical NF-κB signaling is induced by tumor necrosis factor (TNF), interleukin-1 (IL-1), lipopolysaccharide (LPS), and antigen-receptor engagement. The canonical pathway regulates the vast majority of genes activated by NF-κB, including those encoding proinflammatory and immunomodulatory cytokines (e.g., IL-1, IL-2, IL-6, and TNF), chemokines (e.g., CXCL8, CCL2, and CCL3), leukocyte adhesion molecules (e.g., E-selectin, intercellular adhesion molecule-1, and vascular cell adhesion molecule-1), and various prosurvival and antiapoptotic genes (e.g., Bcl-2, Bcl-X$_L$, and XIAP) [5,9]. The noncanonical pathway is absolutely dependent upon an upstream kinase named NF-κB inducing kinase (NIK) that phosphorylates and activates IKKα, thereby directly phosphorylating NF-κB2/p100 at the C-terminal domain. This results in the generation of p52 bound to RelB, and p52-RelB dimer translocates to the nucleus. NIK is rapidly turned over in resting cells through the ubiquitin ligase activity of the TRAF3 adaptor protein that ubiquitinates NIK, triggering its proteasomal degradation [10]. Ligation of noncanonical pathway-inducing receptors sequesters TRAF3, leading to the accumulation of newly synthesized NIK and activation of IKKα [10]. Noncanonical NF-κB signaling is induced by ligation of a subset of TNF-receptor family members including the lymphotoxin-β receptor, CD40, receptor activator of NF-κB, and B cell-activating factor receptor. Most stimuli that activate the canonical pathway do not stimulate the noncanonical pathway. However, noncanonical stimuli can activate both NF-κB pathways.

Functional loss of NF-κB and its regulator is associated with reduced susceptibility to carcinogenesis. Knockout of IKKβ in liver and hematopoietic cells substantially reduced

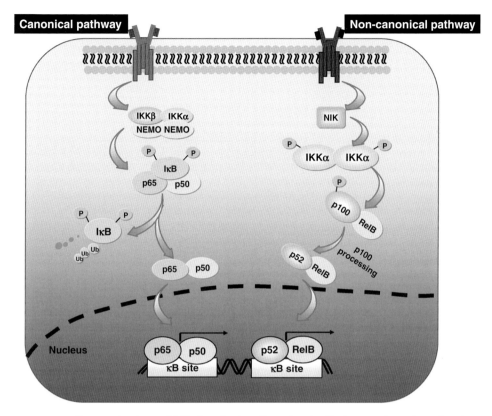

Figure 12.2 Two mechanisms of NF-κB activation. For color detail, please refer to the color plate section.

diethylnitrosamine-induced elevation of TNF-α and IL-6 and suppressed tumorigenesis in mice [11]. Inactivation of NF-κB in multidrug resistance-2 (mrd2)-null mice by overexpressing a superrepressor of IκBα enhanced apoptosis of transformed hepatocytes and attenuated tumorigenesis [12]. In addition, LPS-induced colon adenocarcinoma progression was regressed after deletion of NF-κB [13]. Thus, the NF-κB represents an important target for cancer prevention.

Several dietary phytochemicals, such as curcumin, [6]-gingerol, capsaicin, resveratrol, quercetin, sulforaphane, genistein, epigallocatechin gallate (EGCG), myricetin, guggulsterone, zerumbone, indole-3-carbinol, ellagic acid, lycopene, caffeic acid phenethyl ester, emodin, and S-allyl cysteine, are natural chemopreventive agents that have been found to be potent inhibitors of NF-κB [2, 14]. These phytochemicals may block one or more events in the NF-κB signaling pathway, such as stimulation of IκBα phosphorylation, translocation of NF-κB into the nucleus, DNA binding of the dimers, or interactions with the basal transcriptional machinery.

CURCUMIN

Curcumin, a yellow coloring agent contained in turmeric (*Curcuma longa* L., Zingiberaceae), has been reported to possess strong antitumor promotional as well as anti-inflammatory and antioxidant activities. Curcumin and structurally related curcuminoids from plants of the ginger family exert their therapeutic effects by suppressing the activation of NF-κB and expression of NF-κB-regulated gene products, such as cyclooxygenase-2 (COX-2), cyclin D1, adhesion molecules, metalloproteinases (MMPs), inducible nitric oxide synthase (iNOS), Bcl-2, and Bcl-X_L [15]. Topically applied curcumin inhibited activation of NF-κB and expression of COX-2 in mouse skin stimulated with the tumor promoter 12-O-tetradecanoylphorbol-13-acetate (TPA) [16]. Curcumin blocked TPA, TNF-α, or fecapentaene-mediated NF-κB transactivation by inhibiting the NIK/IKK signaling in colon epithelial cells [17]. Curcumin induced apoptosis in mouse melanoma cells [18] and human neuroblastoma cells [19] via inactivation of NF-κB. In addition, curcumin blunted radiation-induced NF-κB activation in human neuroblastoma cells [20] and colorectal cancer xenografts [21]. Moreover, curcumin treated to mouse leukemia cells suppressed the expression of P-glycoprotein associated with multidrug resistance by targeting NF-κB [22]. A nonrandomized, open-label, phase II clinical trial with curcumin was conducted for twenty-five pancreatic cancer patients. The patients took capsules containing curcumin for 8 weeks without any apparent toxicity [23]. Oral administration of curcumin downregulated the expression of COX-2 and phosphorylation of signal transducer and activator of transcription 3 (STAT 3) in peripheral blood mononuclear cells from the majority of patients, but the compound showed poor bioavailability [23]. A synthetic monoketone analog of curcumin termed 3,5-bis(2-flurobenzylidene) piperidin-4-one (EF24) exhibited potent anticancer activity through direct suppression of IKK activity [24].

[6]-Gingerol

[6]-gingerol, a pungent phenolic substance derived from the root of ginger (*Zingiber officinale* Roscoe, Zingiberaceae), inhibited TPA-induced TNF-α production, ornithine decarboxylase activity, and skin tumor promotion in female ICR mice [25]. Topically applied [6]-gingerol inhibited TPA-induced phosphorylation of p65 at Ser 536 and its interaction with the coactivator cAMP response element binding protein-binding protein (CBP/p300) in mouse skin, thereby rendering NF-κB transcriptionally inactive [26]. In addition, [6]-gingerol has been shown to inhibit UVB-induced activation of NF-κB and COX-2 expression in hairless mouse skin and also in an immortalized human keratinocytes cell line [27]. [6]-gingerol induced apoptosis in the gastric cancer cells by blocking TRAIL-induced NF-κB activation [28].

Capsaicin

Capsaicin, a major pungent principle of hot chili pepper (*Capsicum annuum* L., Solanaceae) with potential anti-inflammatory and antitumor promoting properties, also suppressed TPA-induced activation of NF-κB in mouse skin *in vivo* [29] as well as in cultured human promyelocytic leukemia HL-60 [30] and human myeloid ML-1a cells [31]. Capsaicin, by inhibiting TNF-α stimulated proteasome activity, abrogated the degradation of IκBα, thereby preventing the activation of NF-κB in prostate cancer cells [32]. In another study, capsaicin inhibited the degradation of IκBα, resulting in the decrease of NF-κB/p65 DNA binding activity and Bcl-2 expression in T-cell leukemia cells [33].

Resveratrol

Resveratrol, a phytoalexin present in grapes and red wine, inhibited TPA-induced phosphorylation of IκBα and subsequent p65 nuclear translocation in mouse skin by blocking IKKα and IKKβ [34]. Resveratrol attenuated the TPA-induced NF-κB activation and NF-κB-dependent luciferase activity in human fibrosarcoma HT1080 cells [35]. Resveratrol inhibited hyaluronan/CD44-mediated β-catenin/NF-κB p65 acetylation via suppression of NF-κB activation, which was associated with a significant decrease in both P-glycoprotein and Bcl-xL gene expression and enhancement of capsase-3 activity and chemosensitivity in breast cancer MCF-7 cells [36]. Resveratrol enhanced the radiosensitivity of human nonsmall cell lung cancer [37] and TRAIL-resistance of melanoma cells [38] via NF-κB inactivation. Resveratrol induced apoptosis of human multiple myeloma cells and sensitized these cells to chemotherapeutic agents. These effects were attributed to its inhibition of constitutively active NF-κB through suppression of IKK and the phosphorylation of IκBα and of p65 [39]. Piceatannol (3,4,3',5'-tetrahdroxy-*trans*-stilbene), a hydroxylated analog of resveratrol, has been reported to have antiproliferative, anti-inflammatory, and antioxidant properties. Piceatannol suppressed TNF-induced NF-κB activation, IκBα phosphorylation, IKK activation, and p65 phosphorylation, and p65 nuclear translocation, but had no effect on degradation of IκBα in myeloid cells [40]. Piceatannol abrogated the expression of TNF-induced NF-κB dependent genes, such as MMP-9, COX-2, and cyclin D1 [40].

QUERCETIN

Quercetin is a flavonoid derived from various foods of plant origin and has been widely used as a major constituent of nutritional supplements. Quercetin inhibited IκB degradation and NF-κB activity in pulmonary epithelial cells treated with IL-1 [41]. Quercetin decreased the gene expression and production of TNF-α, IL-1β, IL-6, and IL-8 and these effects appear to be associated with downregulation of NF-κB in TPA-stimulated human mast cells [42].

SULFORAPHANE

Sulforaphane, [1-isothiocyanato-(4R,S)-(methylsulfinyl)butane], a representative isothiocyanate present in broccoli and other cruciferous vegetables, exerts its chemopreventive effects in various tumor models. Sulforaphane inhibited LPS-induced activation of NF-κB and COX-2 expression in cultured mouse macrophages [43]. Interestingly, sulforaphane-induced suppression of NF-κB was associated with neither degradation of IκB nor nuclear translocation of NF-κB, but rather attributable to its direct binding to essential thiol groups of p50 subunit of NF-κB. Contrary to these findings, treatment of human mammary epithelial cells with sulforaphane inhibited TPA-induced COX-2 expression by blocking IKK activities and subsequent phosphorylation and degradation of IκBα, leading to suppression of NF-κB activation [44]. Alternatively, sulforaphane may interact with glutathione (GSH) or other redox regulators like thioredoxin and Ref-1, resulting in perturbation of an intracellular reducing milieu required for the proper DNA binding of NF-κB [45].

GENISTEIN

Genistein, an isoflavonoid present in soybeans, exhibits anticarcinogenic effects. Several studies have shown that genistein inhibits cancer cell proliferation and triggers apoptosis in various human cancers. Exposure of breast cancer MDA-231 cells to genistein caused a concentration-dependent decrease in NF-κB/p65 protein levels and DNA-binding activity of NF-κB, which may account for its inhibition of cell growth and induction of apoptosis [46]. Genistein inhibited the radiation-induced activation of NF-κB in prostate cancer cells [47,48]. In addition, suppression of NF-κB activity by genistein abolished drug resistance induced by chemotherapeutic agents such as erlotinib [49], cisplatin [50,51], and docetaxel [52].

NUCLEAR FACTOR ERYTHROID 2 (NF-E2)-RELATED FACTOR (NRF2)

Environmental-related factors (e.g., tobacco smoking, diet, alcohol, ionizing radiations, biocides, pesticides, and viral infections) and other health-related factors (e.g., obesity or the aging process) capable of generating ROS provoke oxidative stress [53]. Oxidative stress is one of the major causes of carcinogenesis [54]. Oxidative stress occurs as a consequence of an imbalance between ROS production and cellular capability to inactivate or detoxify these reactive species.

Therefore, the protection against oxidative cell damage confers the first line of defense against carcinogenic insults. This can be achieved either by blocking the formation of ROS/RNS or stimulating their elimination. Likewise, neutralization or elimination of carcinogens is often catalyzed by a series of phase II detoxifying enzymes that catalyze the conjugation of electrophilic carcinogenic species with various endogenous moieties, such as GSH, glucuronide, and sulfate, which renders the carcinogens water-soluble, thereby facilitating their excretion out of the body. The induction of phase II detoxifying or antioxidant enzymes is one of the most important components of cellular cytoprotection whereby a diverse array of electrophilic and oxidative toxicants can be eliminated from the cell before they damage the target cell DNA [55].

The redox-sensitive transcription factor, Nrf2, plays a key role in regulating induction of phase II detoxifying or antioxidant enzymes via interaction with a *cis*-acting DNA element called antioxidant-response element (ARE) or the electrophile-responsive element (EpRE) [55]. These include NAD(P)H:quinone oxidoreductase 1 (NQO1), glutathione *S*-transferase (GST) Ya subunit, glutamate cystein ligase (GCL), and heme oxygenase-1 (HO-1). The involvement of this redox-sensitive transcription factor in regulating the induction of antioxidant and carcinogen detoxifying enzymes has been directly evidenced by disruption of the *nrf2* gene in mice [56,57]. Nrf2 knockout mice are predisposed to developing the chemically induced DNA damage and exhibited a higher gene mutation frequency in the lung than did the wild-type mice [58]. Likewise, *nrf2*-null mice developed the much larger number of tumors in the forestomach [56], liver [59], urinary bladder [60], skin [57], and colon [61] than did the wild-type mice after treatment with a carcinogen. Therefore, Nrf2 is considered another important molecular target for cancer prevention [62]

Kelch-like erythorid CNC homologue (ECH)-associated protein 1 (Keap1) is a member of a large family of proteins containing an N-terminal region (two cysteines), Broad complex, Tramtrack, and Bric-a-brac (BTB) region (three cysteines), linker region (eight cysteines), and Kelch repeat region (DGR, nine cysteines, and six Kelch motifs), and C-terminal region (three cysteines) [63]. Keap1 functions as a substrate adaptor protein for a Cul3-dependent E3 ubiquitin ligase complex [64]. The cytoplasmic protein Keap1 interacts with Nrf2 and represses its function [65].

Although the molecular mechanisms involved in the Nrf2-driven transcriptional activation of antioxidant enzymes are not fully understood, the dissociation of Nrf2 from Keap1 as a consequence of Keap1 cysteine thiol modification used to be considered a plausible mechanism underlying Nrf2 activation by electrophiles and pro-oxidants [66]. According to the new paradigm, the direct interaction of the highly reactive cysteine residues of Keap1 with phase II enzyme inducers as well as electrophiles causes conformational changes of this repressor protein, which abrogate the capability of Keap1 to aid proteasomal degradation of Nrf2. In addition to activation of Nrf2 through oxidation or chemical modification of Keap1 cysteine thiols, phosphorylation of specific serine [67] or threonine [68] residues of Nrf2 may facilitate the release of Nrf2 from the Keap1 repression and its subsequent nuclear translocation. As a result, Nrf2 forms a heterodimer with a small Maf protein (e.g., Marf K, Marf G, or Maf F) and binds to ARE sequences in the promoter region of the genes encoding many antioxidant enzymes (Fig. 12.3). Activation of several upstream kinases, such as MAP kinases [69], PI3K/Akt [70], PKC [67,71], and casein kinase-2 (CK-2) [72] have been considered to facilitate nuclear translocation and transcriptional activation of Nrf2. On the other hand, GSK3β negatively regulates Nrf2 signaling via phosphorylation of Nrf2 at tyrosine [73] or serine [74] residues. Nrf2 activation by dietary chemopreventives through modulation of one or more of the upstream kinases or thiol modification has recently been reviewed ([75] and *vide infra*).

SULFORAPHANE

Sulforaphane has been known to induce genes encoding phase 2 detoxifying and antioxidant enzymes through activation of Nrf2. Sulforaphane upregulated the expression of detoxifying enzymes including NQO1, GST, and GCL in the small intestine [76] and liver [77] of *nrf2*-wild-type mice, whereas the *nrf2*-null mice displayed lower levels of these enzymes. Sulforaphane induced

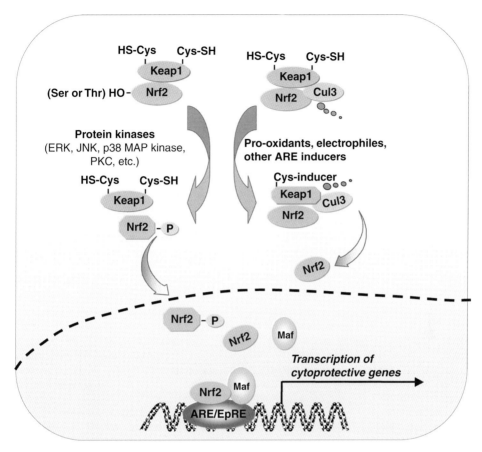

Figure 12.3 Proposed pathways for Nrf2-ARE activation. For color detail, please refer to the color plate section.

thioredoxin in murine retina [78] and human adenocarcinoma Caco-2 cells [79], which appeared to be mediated via Nrf2-ARE binding. Sulforaphane induced Nrf2-driven phase II enzyme expression by modulating the activation of MAP kinases [55, 80–83]. Sulforaphane and its sulfide analogue, erucin, elevated the mRNA expression of NQO1, UDP-glucuronosyltransferases 1A1, and multidrug transporter (MRP) 2 in Caco-2 cells by activating PI3K/Akt- or MAP kinase kinase (MEK)/extracellular signal-regulated kinase (ERK)-mediated signaling [81]. Involvement of ERK1/2 and c-Jun-N-terminal kinase (JNK) in sulforaphane-induced ARE-transcription activities was also observed in murine keratinocytes [82]. Sulforaphane induced HO-1 expression in HepG2 cells by downregulating p38 MAP kinase, thereby activating the Nrf2-ARE signaling [83].

In addition to the modulation of upstream kinases, the mechanism of Nrf2 activation by sulforaphane involves a direct modification of cysteine residue(s) present in Keap1, facilitating dissociation of Nrf2 from Keap1. Zhang et al. [84] have suggested that sulforaphane inhibits the Keap1-dependent ubiquitination of Nrf2, which increases steady-state levels of Nrf2, leading to enhanced nuclear localization and transcriptional activity of this transcription factor. Particularly, modification of Cys151 of Keap1 by sulforaphane is required for its suppression of Keap1-dependent Nrf2 ubiquitination [84]. Sulforaphane may alter the redox state of Cys 151 and reduces the ability of Nrf2-bound Keap1 proteins to associate with the Cul3-Rbx1 core complex, thereby increasing the stability and accumulation of Nrf2. In addition, sulforaphane is an electrophile that can react with cysteine thiols of Keap1 to form thionoacyl adducts, thereby inducing stability of Nrf2 [85].

Chemopreventive effects of sulforaphane are mediated, at least in part, through Nrf2 activation. The gastrointestinal GPx (GI-GPx) is a selenium-dependent protein and involved in the control of inflammation and malignant growth [86]. It has been reported that gastrointestinal tumor formation was enhanced in *gpx2/gpx1* double knockout mice, corroborating the anticarcinogenic activity of GI-GPx [87]. Sulforaphane activated the ARE binding site located in the GI-GPx promoter through overexpression of Nrf2 in human hepatoma (HepG2) or human colon cancer (Caco-2) cells [86]. Topical application of sulforaphane (100 nmol) for 14 consecutive days inhibited skin carcinogenesis induced by 7,12-dimethylbenz[*a*]anthracene plus TPA in wild-type C57BL/6 mice, whereas no such chemopreventive effects of sulforaphane were elicited in the *nrf2*-deficient mice [57].

A sulforaphane analog, 6-methylsulphinylhexylisothiocyanate isolated from Japanese wasabi was also found to induce cytoprotective gene expression via the Nrf2-ARE signaling [88]. Likewise, phenethyl isothiocyanate induced HO-1 expression and ARE activity in human prostate cancer (PC3) cells via ERK- and JNK-mediated phosphorylation of Nrf2 and subsequent Nrf2 nuclear translocation [69].

CURCUMINOIDS

The chemopreventive effects of curcuminoids have been extensively investigated and well defined [89], [90]. As part of its chemopreventive mechanism, curcumin targets the Nrf2-ARE signaling pathway to induce phase II detoxifying and antioxidant enzymes. Given orally *ad libitum* for 16 days, dietary curcumin (0.05%) enhanced not only the Nrf2 levels but also its nuclear translocation and the ARE binding in liver and lung of mice [91]. Dietary curcumin enhanced the expression of carcinogen detoxifying enzymes such as GST isoforms and NQO1 in parallel with the activation of Nrf2, leading to increased detoxification of benzo[*a*]pyrene [91]. Oral administration of curcumin at 200 mg/kg for 4 consecutive days resulted in enhanced nuclear accumulation and the ARE-binding of Nrf2 and HO-1 upregulation in rat liver, and these effects account for the cytoprotective effect against liver toxicity [92].

Curcumin induced nuclear localization of Nrf2 and HO-1 expression effectively in wild-type mouse embryonic fibroblasts, but not in those from *nrf2*-deficient mice [93]. Many cytoprotective genes were induced in liver and small intestine of wild-type C57BL/6J, but not in C57BL/6J/*nrf2*(−/−) mice, given a single oral dose of curcumin (1000 mg/kg) [94]. Curcumin also induced expression of GSTP1, GCL, and HO-1 via Nrf2 activation in various cells [93, 95–97]. The curcumin-induced expression of antioxidant enzymes via Nrf2-ARE signaling was mediated by activation of PKC delta, PI3K, and p38 MAPK [98, 99]. In addition, curcumin-induced HO-1 expression and Nrf2 activation were ROS-dependent [100].

Structurally, curcumin has two α, β unsaturated carbonyl groups and can hence act as a Michael reaction acceptor, thereby causing thiol modification of Keap1. Consistent with this notion, tetrahydrocurcumin, which lacks an electrophilic α,β-unsaturated carbonyl functional moiety, failed to induce Nrf2-ARE binding as well as HO-1 induction when given orally to rats [92]. Demethoxycurcumin and *bis*-demethoxycurcumin induced expression of HO-1, GCL, and NQO-1 mRNA and *HO-1* promoter activity, and activated Nrf2 more effectively than curcumin in mouse pancreatic-β (MIN6) cells [101]. However, demothoxycurcumin and *bis*-demethoxycurcumin exhibited almost similar potency to induce QR activity as compared to that observed with curcumin [102].

EGCG

EGCG, the major active catechin component of green tea, has been known to possess antioxidant, anti-inflammatory, and chemopreventive properties [103, 104]. EGCG was found to be the most potent Nrf2 activator among the green tea polyphenols, as evidenced by its pronounced ability to induce ARE-luciferase reporter gene transactivation [105]. EGCG has been reported to activate Nrf2 and induce expression of HO-1 in endothelial cells [106] and B-lymphoblasts [93]. Although EGCG-induced HO-1 expression was attributed to activation of Akt and ERK1/2 in endothelial cells [106], p38 MAP kinase as well as Akt is involved in HO-1 induction and Nrf2 nuclear translocation in B lympoblasts treated with EGCG [93]. Activation of ERK1/2 and Akt induced by EGCG was involved in HO-1 expression in human mammary epithelial MCF-10A cells [107]. Similarly, a nontoxic dose of EGCG increased the ARE-luciferase activity and the expression of ARE-regulated genes by activating MAP kinases in HepG2 cells [105]. EGCG inhibited the growth and liver/pulmonary metastasis of colon tumor implanted orthotopically in the cecum of nude mice, and this anticancer effect was proposed to be partly mediated by activating the Nrf2-UGT1A signal pathway [108]. Oral administration of EGCG at 200 mg/kg induced the Nrf2-dependent gene expression in the liver and small intestine of Nrf2 wild-type mice but not in Nrf2-deficent mice [109].

ALLYL SULFIDES

Garlic oil contains several organosulfur compounds, such as diallyl sulfide (DAS), diallyl disulfide (DADS), and diallyl trisulfide (DATS), capable of inducing carcinogen-detoxifying enzymes. Chen et al. examined Nrf2-driven ARE activity and antioxidant gene expression by garlic organosulfur compounds in HepG2 cells [110]. Of the three allyl sulfides derived from garlic, DATS was most potent in terms of inducing ARE activation and expression of cytoprotective enzymes, such as HO-1 and NQO1. In addition, DAS induced NQO1 5-fold in wild-type mice, whereas induction was completely absent in nrf2 (−/−) mice, indicating that DAS also activates Nrf2 [111]. DAS induced HO-1 expression, Nrf2 protein expression, nuclear translocation, and DNA binding activity in HepG2 cells [112]. Both ERK and p38 pathways appeared to be involved in DAS-induced Nrf2 nuclear translocation and HO-1 gene expression [112], whereas MAP kinases induced by DATS did not affect the ARE activity [110]. DAS- and DATS-induced Nrf2 activation is presumably mediated by generation of ROS. The thiol antioxidant N-acetyl-L-cysteine (NAC) and catalase blocked not only DAS-induced ROS production but also ERK activation as well as nuclear translocation of Nrf2, and also HO-1 expression. Moreover, cotreatment with thiol antioxidants NAC and GSH inhibited the ARE activity and the Nrf2 accumulation induced by DATS [110]. DAS-treatment rendered the HepG2 cells resistant to oxidative stress caused by hydrogen peroxide or arachidonic acid, and this was attributable to its induction of HO-1 as pharmacologic inhibition of HO-1 activity blunted the cytoprotective effects of DAS [112]. It is noteworthy that the prooxidant activity of DATS contributes to Nrf2-driven antioxidant enzyme induction, which conferred the protection against oxidative cell death. Three major MAP kinases, that is, ERK, JNK, and p38, were activated by DATS treatment. However, the inhibition of these enzymes did not affect DATS-induced ARE activity. Likewise, the PKC pathway was not directly involved in DATS-induced ARE activity, instead the calcium-dependent signaling pathway might be responsible for the DATS-induced cytoprotective effect.

RESVERATROL

Resveratrol exerts antioxidant, anti-inflammatory, and chemopreventive activities by modulating various events in cellular signaling [113]. Resveratrol prevented chemically induced tumorigenesis in many experimental models [114–117]. As a mechanism of carcinogen detoxification and cellular antioxidant defense, resveratrol induced NQO1 activity in murine hepatoma cells [118] and human K562 cells [119]. The stimulation of NQO1 gene expression by resveratrol involved the stimulation of ARE signaling, which was accompanied by an increase in the state of phosphorylation of Nrf2 and its translocation to the nucleus [119]. The compound was found to induce HO-1 expression and activity in human aortic smooth muscle [120] and rat pheochoromocytoma (PC12) cells [121] via activation of NF-κB and Nrf2, respectively.

Treatment of human primary small airway epithelial and human alveolar epithelial (A549) cells with cigarette smoke extract (CSE) dose dependently decreased GSH levels and GCL activity, effects that were associated with enhanced production of ROS [122]. Resveratrol restored CSE-depleted GSH levels by upregulation of GCL via activation of Nrf2 and also quenched CSE-induced release of ROS.

Pungent Vanilloids

Capsaicin induced expression of HO-1 by activating PI3K/Akt-mediated activation of Nrf2 signaling in a ROS-dependent manner in HepG2 cells [123]. It was suggested that a quinone metabolite or other reactive forms of capsaicin would covalently modify NQO-1, and inhibit its activity, leading to production of ROS. The resulting overproduction of ROS is speculated to stimulate PI3K/Akt-mediated activation of Nrf2 [123]. (10)-Shogaol, a pungent ingredient of ginger (*Zingiber officianale* Roscoe, Zingiberaceae) has been reported to interact with the cysteine 151 residue of human Keap1 to form an akylated adduct [124]. The alkylation of Keap1 by this electrophilic natural product may contribute to its antioxidant, anti-inflammatory, and chemopreventive properties.

Lycopene

Lycopene, a natural antioxidant present predominantly in tomato products, has been reported to exert chemopreventive activity, especially against prostate and mammary carcinogenesis. The antioxidant properties of lycopene are thought to be primarily responsible for its chemopreventive effects [125]. Lycopene activated GCL promoter activity in MCF-7 cells [126]. GCL-ARE and NQO1-ARE reporter activities were induced in HepG2 cells as well [126]. Lycopene elevated the mRNA and/or protein levels of GCL and NQO1, enhanced the cellular GSH level, and reduced ROS generation in MCF-7 and HepG2 cells. In addition, the induction of NQO1 and GCL by lycopene was diminished in HepG2 cells ectopically expressing a dominant negative mutant Nrf2, suggesting that lycopene induced the previously mentioned antioxidant enzymes via Nrf2 activation [126]. Treatment of both cells with lycopene also lowered the intracellular ROS level.

Coffee-Derived Diterpenes

Epidemiological studies have revealed an inverse relationship between coffee consumption and the risk of certain types of cancer [127]. Dietary administration of coffee (3% or 6%) for 5 days showed significantly elevated expression of mRNA transcripts of NQO1 and GSTA-1 in liver and small intestine, and that of UGTA-6 and GCLC in small intestine of *nrf2*$^{+/+}$ mice as compared to *nrf2*$^{-/-}$ animals [128]. The coffee-derived diterpenes, cafestol and kahweol, when treated to embryonic fibroblasts isolated from *nrf2*$^{+/+}$ mice, increased NQO1 mRNA expression to a greater extent than that achieved with embryonic fibroblasts from *nrf2*$^{-/-}$ mice.

Carnosol

Carnosol, an orthophenolic diterpene present in rosemary (*Rosmarinus officinalis,* Lamiaceae), induced HO-1 expression at both protein and mRNA levels by increasing the binding of Nrf2 to ARE and induced Nrf2-dependent activation of *HO-1* promoter in PC12 cells via upregulation of MAP kinases [129]. Cinnamaldehyde, another dietary diterpene present in dried stem bark of *Cinnamomum cassia* Presl. (Lauraceae), induced HO-1 protein expression, increased Nrf2 nuclear translocation and the ARE-luciferase reporter activity in human endothelial cells [130].

XANTHOHUMOL

Xanthohumol, a chemopreventive sesquiterpene derived from hops (*Humulus lupulus* L.), exhibited capability to induce expression of antioxidant enzymes. Pretreatment of hepa1c1c7 cells with xanthohumol diminished menadione-induced DNA damage via upregulation of NQO1 [131]. Xanthohumol induced NQO1 expression in an ARE-dependent manner, partly through alkylation of a cysteine residue in the Keap1 [131]. Ben-dor and colleagues also suggested that xanthohumol alkylated the cysteine 151 residue (located in BTB domain), and C319 (located in central linker domain) and C613 (located in Kelch repeat domain), thereby contributing to the Nrf2-dependent ARE activation [124]. Likewise, isoliquiritigenin derived from licorice has also been shown to alkylate cysteine 151 residue of Keap-1, thereby inducing the ARE activity [124]. Cumulatively, the data suggest that xanthohumol upregulates the transcription of ARE-mediated detoxifying genes by directly binding to Keap1 protein.

ZERUMBONE

Zerumbone, a sesquiterpene occurring in zingiberaceous plants in Southeast Asian countries, has been shown to have anti-inflammatory effects in several independent experimental studies [132]. Zerumbone enhanced the cellular GSH level and induced a battery of antioxidant enzymes, such as GSTP-1, GCL, GPx, and HO-1 in normal rat liver epithelial (RL34) cells [133]. Treatment of RL34 cells with zerumbone (25 μM) showed increased nuclear accumulation of Nrf2, whereas its reduced analogues such as α-humulene or 8-hydroxy-α-humulene failed to activate Nrf2 and induce aforementioned antioxidant enzymes, suggesting that the α,β-carbonyl moiety at the 8 position is crucial for Nrf2 activation and antioxidant enzyme induction by zerumbone.

CHALCONES

Chalcone, an α,β-unsaturated flavonoid, possesses anti-inflammatory properties. Chalcone upregulated the nuclear levels of Nrf2 and increased the ARE-luciferase activity and also the thioredoxin reductase promoter activity in bovine aortic endothelial cells [134]. It also induced expression of thioredoxin reductase as well as HO-1 in the same cells. Some synthetic chalcone derivatives, such as 2',4',6'-tris(methoxymethoxy)chalcone and 3',4',5',3,4,5-hexamethoxychalcone diminished NF-κB activation, whereas they induced HO-1 expression [135], [136].

CROSSTALK BETWEEN NRF2 AND NF-κB SIGNALING PATHWAYS

Compelling evidence supports that the Nrf2-ARE signaling is associated with the cellular defense against inflammation. Thus, the Nrf2-ARE pathway has been proposed to be a promising target for anti-inflammation and cancer prevention [137]. According to a recent report, the aggravation of DSS-induced colitis in *nrf2*$^{-/-}$ mice was associated with decreased expression of HO-1, NQO-1, UGT1A1, and GSTμ-1 [61]. Levels of proinflammatory mediators, such as COX-2, iNOS, IL-1β, IL-6, and TNF-α, were significantly elevated in the colonic tissues of *nrf2*$^{-/-}$ mice as compared to their wild-type counterparts [61]. In a recent study, Khor *et al.* also reported that the tumor incidence, the multiplicity, the size, and the stage of progression were increased in *nrf2*-deficient mice in an azoxymethane plus DSS-induced colon carcinogenesis [138]. Inflammatory cytokines, through induction of oxidative stress, suppressed NQO1 activity in both cholangiocarcinoma cells and normal HeLa Chang liver cells [139]. Moreover, activation of the NF-κB pathway by tumor promoter could be attenuated by diverse Nrf2 activators, such as sulforaphane, curcumin, and phenethyl isothiocyanate (PEITC). These results indicate that dysfunction of Nrf2 seems to accelerate NF-κB-mediated proinflammatory reactions, directly or indirectly [140]. Proper activation of Nrf2 signaling hence represents a promising strategy for prevention of inflammation-associated cancer. Therefore, Nrf2-mediated antitumorigenic effects are likely to be achieved not only by potentiation of antioxidant machinery but also through suppression of proinflammatory pathways mediated by NF-κB signaling.

The molecular mechanisms underlying suppression of NF-κB and activation of Nrf2-ARE are totally independent of each other or may involve a crosstalk between two signaling pathways. Recently, Liu *et al.* [141] have suggested that overexpression of p65 liberates CBP from Nrf2 by competitive interaction with the CH1-KIX domain of CBP, which resulted in inactivation of Nrf2. The status of p65 phosphorylation was not associated with obstruction of MafK-mediated recruitment of CBP to ARE. Overexpression of p65 facilitated the recruitment of HDAC3, the repressor of ARE by promoting the interaction of HDAC3 with either CBP or Mafk, leading to local histone hypoacetylation, thus serving as a negative regulator of Nrf2-ARE signaling. Recent studies have revealed that cysteine thiols present in various transcription factors, such as NF-κB, AP-1, and p53 function as redox sensors in fine-tuning of transcriptional regulation of many genes essential for maintaining cellular homeostasis. It is noticeable that NF-κB binds to DNA preferentially when these cysteines are reduced and that redox-related modifications of cysteine thiols, such as the formation of disulfides [142] or nitrosothiol [143], disrupts DNA binding ability of NF-κB. Liu *et al.* reported that chalcone abrogated the activation of NF-κB on IL-6- and LPS-treated endothelial cells [134]. Furthermore, chalcone upregulated the levels of Nrf2 in nuclear extracts and increased the ARE activity, whereas it suppresseed both IL-6 and LPS-induced signaling pathways by disrupting the thiol-dependent intracellular redox state. In contrast, Nrf2 signaling is activated by a compound possessing an α,β-unsaturated carbonyl moiety capable of modifying the cysteine thiolds present in Keap1. Therefore, thiol modification by electrophilic phase II inducers may result in opposite effects in the NF-κB- and Nrf2-mediated signal transducing pathways (Fig. 12.4).

Nair *et al.* [144] performed multiple sequence alignment of Nrf2 and NF-κB1 genes in five mammalian species, including human, chimpanzee, dog, mouse, and rat to explore conserved biological features. The comparative analyses of transcription factor-binding sites in these two gene promoters revealed that many matrix families were conserved between Nrf2 and NF-κB promoter regions in both humans and mice. In addition, a conserved transcription factor-binding site for NF-κB itself was found to be present in murine promoter regions of Nrf2 and NF-κB, lending further support to the possible crosstalk between these two transcription factors. Curcumin-induced aldose reductase via activation of Nrf2 was abolished by a pharmacological NF-κB inhibitor Bay11-7082 or siRNA knock down of *p65* gene, suggesting the involvement of NF-κB in the cellular response to oxidative stress and toxic aldehydes [145].

Recently, a double-edged sword nature of the Nrf2-HO-1 axis was confirmed in the various cancer tissues. Nrf2 enhances resistance of cancer cells to chemotherapeutic drugs including

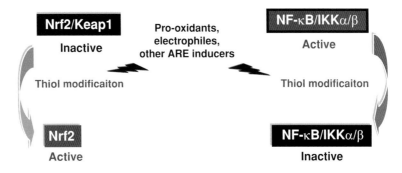

Figure 12.4 Differential effects of thiol modification on IKK-NF-κB- and Nrf2-mediated signal transduction. For color detail, please refer to the color plate section.

tamoxifen [146, 147]. It has been shown that genetic alteration or mutation of Keap 1 confers constitutive activation of Nrf2 in breast and gallbladder cancer. Based on the previous findings, Nrf2 has different roles in cancer chemoprevention and carcinogenesis. The intercellular network between Nrf2 and NF-κB signaling may also be differentially regulated in normal and cancer cells.

Sulforaphane has been reported to inhibit NF-κB activation in human colon cancer HT-29 [148] and human prostate cancer PC-3 [149] cells. It selectively reduced DNA binding of NF-κB without blocking LPS-induced degradation of IκB or nuclear translocation of NF-κB [43]. The sulforaphane-mediated attenuation of NF-κB DNA binding activity was prevented by the sulfhydryl-reducing agent mercaptoethanol, suggesting that sulforaphane could either directly inactivate NF-κB subunits by binding to essential cysteine residues or indirectly interact with GSH or other redox regulators like thioredoxin and Ref-1 relevant for NF-κB function. In contrast, increased nuclear accumulation of NF-κB as well as Nrf2 was observed in human adenocarcinoma Caco-2 cells after sulforaphane treatment [79].

CONCLUSION

Oxidative stress and inflammatory tissue injuries are two of the most critical factors implicated in multistage carcinogenesis. Therefore, suppression of abnormally amplified inflammatory signaling and restoration/potentiation of antioxidant machinery that is improperly working or repressed represent important strategies for chemoprevention. NF-κB and Nrf2 are major redox-sensitive transcription factors that are involved in regulating proinflammatory and antioxidant genes, respectively. Many chemopreventive phytochemicals with anti-inflammatory activities have been shown to inhibit NF-κB activation via multiple mechanisms. Some antioxidative phytochemicals not only scavenge ROS but also induce Nrf2-driven synthesis of antioxidant or phase II detoxification enzymes, thereby fortifying inherent cellular defense capacity against oxidative or electrophilic insults. Interestingly, a majority of chemopreventive phytochemicals possess both anti-inflammatory and antioxidant properties. In consideration of close association between anti-inflammatory and antioxidant properties effects mediated by the same phytochemicals, it is worthwhile identifying molecules that mediate the crosstalk between NF-κB and Nrf2 signaling. Cysteine thiols present in redox-sensitive transcription factors and their regulators (e.g., IKK and Keap1) are recognized to function as redox sensors involved in fine-tuning of transcriptional regulation of many genes essential for maintaining cellular homeostasis. Thus, oxidation or covalent modification of thiol groups present in redox-sensitive transcription factors and their regulating molecules can constitute an essential component of molecular target-based chemoprevention and cytoprotection with anti-inflammatory and antioxidant phytochemicals [137, 150].

ACKNOWLEDGMENTS

This work was supported by the research grants from BioGreen 21 Program (No.2007301-034-027) and also by Basic Science Research Program.

REFERENCES

1. Greenwald, P., McDonald, S. S., Anderson, D. E., An evidence-based approach to cancer prevention clinical trials. *Eur J Cancer Prev* 2002, *11* (Suppl 2), S43–S47.
2. Surh, Y. J., Cancer chemoprevention with dietary phytochemicals. *Nat Rev Cancer* 2003, *3*, 768–780.
3. Kuper, H., Adami, H. O., Trichopoulos, D., Infections as a major preventable cause of human cancer. *J Intern Med* 2000, *248*, 171–183.

4. Gilmore, T. D., Introduction to NF-κB: players, pathways, perspectives. *Oncogene* 2006, *25*, 6680–6684.
5. Hayden, M. S., Ghosh, S., Shared principles in NF-κB signaling. *Cell* 2008, *132*, 344–362.
6. Hayden, M. S., West, A. P., Ghosh, S., NF-κB and the immune response. *Oncogene* 2006, *25*, 6758–6780.
7. Solt, L. A., May, M. J., The IκB kinase complex: master regulator of NF-κB signaling. *Immunol Res* 2008, *42*, 3–18.
8. Karin, M., The IκB kinase - a bridge between inflammation and cancer. *Cell Res* 2008, *18*, 334–342.
9. Bonizzi, G., Karin, M., The two NF-κB activation pathways and their role in innate and adaptive immunity. *Trends Immunol* 2004, *25*, 280–288.
10. Liao, G., Zhang, M., Harhaj, E.W., Sun, S. C., Regulation of the NF-κB-inducing kinase by tumor necrosis factor receptor-associated factor 3-induced degradation. *J Biol Chem* 2004, *279*, 26243–26250.
11. Maeda, S., Kamata, H., Luo, J. L., Leffert, H., *et al.*, IKKβ couples hepatocyte death to cytokine-driven compensatory proliferation that promotes chemical hepatocarcinogenesis. *Cell* 2005, *121*, 977–990.
12. Pikarsky, E., Porat, R. M., Stein, I., Abramovitch, R., *et al.*, NF-κB functions as a tumour promoter in inflammation-associated cancer. *Nature* 2004, *431*, 461–466.
13. Luo, J.L., Maeda, S., Hsu, L.C., Yagita, H., *et al.*, Inhibition of NF-κB in cancer cells converts inflammation- induced tumor growth mediated by TNF to TRAIL-mediated tumor regression. *Cancer Cell* 2004, *6*, 297–305.
14. Aggarwal, B. B., Shishodia, S., Molecular targets of dietary agents for prevention and therapy of cancer. *Biochem Pharmacol* 2006, *71*, 1397–1421.
15. Shishodia, S., Sethi, G., Aggarwal, B. B., Curcumin: getting back to the roots. *Ann N Y Acad Sci* 2005, *1056*, 206–217.
16. Chun, K. S., Keum, Y. S., Han, S. S., Song, Y. S., *et al.*, Curcumin inhibits phorbol ester-induced expression of cyclooxygenase-2 in mouse skin through suppression of extracellular signal-regulated kinase activity and NF-κB activation. *Carcinogenesis* 2003, *24*, 1515–1524.
17. Plummer, S. M., Holloway, K. A., Manson, M. M., Munks, R. J., *et al.*, Inhibition of cyclo-oxygenase 2 expression in colon cells by the chemopreventive agent curcumin involves inhibition of NF-κB activation via the NIK/IKK signalling complex. *Oncogene* 1999, *18*, 6013–6020.
18. Marin, Y. E., Wall, B. A., Wang, S., Namkoong, J., *et al.*, Curcumin downregulates the constitutive activity of NF-κB and induces apoptosis in novel mouse melanoma cells. *Melanoma Res* 2007, *17*, 274–283.
19. Freudlsperger, C., Greten, J., Schumacher, U., Curcumin induces apoptosis in human neuroblastoma cells via inhibition of NFκB. *Anticancer Res* 2008, *28*, 209–214.
20. Aravindan, N., Madhusoodhanan, R., Ahmad, S., Johnson, D., *et al.*, Curcumin inhibits NFκB mediated radioprotection and modulate apoptosis related genes in human neuroblastoma cells. *Cancer Biol Ther* 2008, *7*, 569–576.
21. Kunnumakkara, A. B., Diagaradjane, P., Guha, S., Deorukhkar, A., *et al.*, Curcumin sensitizes human colorectal cancer xenografts in nude mice to gamma-radiation by targeting nuclear factor-κB-regulated gene products. *Clin Cancer Res* 2008, *14*, 2128–2136.
22. Choi, B. H., Kim, C. G., Lim, Y., Shin, S. Y., *et al.*, Curcumin down-regulates the multidrug-resistance mdr1b gene by inhibiting the PI3K/Akt/NF κ B pathway. *Cancer Lett* 2008, *259*, 111–118.
23. Dhillon, N., Aggarwal, B. B., Newman, R. A., Wolff, R. A., *et al.,* Phase II trial of curcumin in patients with advanced pancreatic cancer. *Clin Cancer Res* 2008, *14*, 4491–4499.
24. Kasinski, A. L., Du, Y., Thomas, S. L., Zhao, J., *et al.,* Inhibition of IκB kinase-nuclear factor-κB signaling pathway by 3,5-bis(2-flurobenzylidene)piperidin-4-one (EF24), a novel monoketone analog of curcumin. *Mol Pharmacol* 2008, *74*, 654–661.

25. Park, K. K., Chun, K. S., Lee, J. M., Lee, S. S., *et al.*, Inhibitory effects of [6]-gingerol, a major pungent principle of ginger, on phorbol ester-induced inflammation, epidermal ornithine decarboxylase activity and skin tumor promotion in ICR mice. *Cancer Lett* 1998, *129*, 139–144.

26. Kim, S. O., Kundu, J. K., Shin, Y. K., Park, J. H., *et al.*, [6]-Gingerol inhibits COX-2 expression by blocking the activation of p38 MAP kinase and NF-κB in phorbol ester-stimulated mouse skin. *Oncogene* 2005, *24*, 2558–2567.

27. Kim, J. K., Kim, Y., Na, K. M., Surh, Y. J., *et al.,* [6]-Gingerol prevents UVB-induced ROS production and COX-2 expression in vitro and in vivo. *Free Radic Res* 2007, *41*, 603–614.

28. Ishiguro, K., Ando, T., Maeda, O., Ohmiya, N., *et al.,* Ginger ingredients reduce viability of gastric cancer cells via distinct mechanisms. *Biochem Biophys Res Commun* 2007, *362*, 218–223.

29. Han, S. S., Keum, Y. S., Seo, H. J., Chun, K. S., *et al.,* Capsaicin suppresses phorbol ester-induced activation of NF-κB/Rel and AP-1 transcription factors in mouse epidermis. *Cancer Lett* 2001, *164*, 119–126.

30. Han, S. S., Keum, Y. S., Chun, K. S., Surh, Y. J., Suppression of phorbol ester-induced NF-κB activation by capsaicin in cultured human promyelocytic leukemia cells. *Arch Pharm Res* 2002, *25*, 475–479.

31. Singh, S., Natarajan, K., Aggarwal, B. B., Capsaicin (8-methyl-N-vanillyl-6-nonenamide) is a potent inhibitor of nuclear transcription factor-κ B activation by diverse agents. *J Immunol* 1996, *157*, 4412–4420.

32. Mori, A., Lehmann, S., O'Kelly, J., Kumagai, T., *et al.*, Capsaicin, a component of red peppers, inhibits the growth of androgen-independent, p53 mutant prostate cancer cells. *Cancer Res* 2006, *66*, 3222–3229.

33. Zhang, J., Nagasaki, M., Tanaka, Y., Morikawa, S., Capsaicin inhibits growth of adult T-cell leukemia cells. *Leuk Res* 2003, *27*, 275–283.

34. Kundu, J. K., Shin, Y. K., Kim, S. H., Surh, Y. J., Resveratrol inhibits phorbol ester-induced expression of COX-2 and activation of NF-κB in mouse skin by blocking IκB kinase activity. *Carcinogenesis* 2006, *27*, 1465–1474.

35. Park, J. S., Kim, K. M., Kim, M. H., Chang, H. J., *et al.,* Resveratrol inhibits tumor cell adhesion to endothelial cells by blocking ICAM-1 expression. *Anticancer Res* 2009, *29*, 355–362.

36. Bourguignon, L. Y., Xia, W., Wong, G., Hyaluronan-mediated CD44 interaction with p300 and SIRT1 regulates beta-catenin signaling and NFκB-specific transcription activity leading to MDR1 and Bcl-xL gene expression and chemoresistance in breast tumor cells. *J Biol Chem* 2009, *284*, 2657–2671.

37. Liao, H. F., Kuo, C. D., Yang, Y. C., Lin, C. P., *et al.*, Resveratrol enhances radiosensitivity of human non-small cell lung cancer NCI-H838 cells accompanied by inhibition of nuclear factor-κ B activation. *J Radiat Res (Tokyo)* 2005, *46*, 387–393.

38. Ivanov, V. N., Partridge, M. A., Johnson, G. E., Huang, S. X., *et al.*, Resveratrol sensitizes melanomas to TRAIL through modulation of antiapoptotic gene expression. *Exp Cell Res* 2008, *314*, 1163–1176.

39. Bhardwaj, A., Sethi, G., Vadhan-Raj, S., Bueso-Ramos, C., *et al.*, Resveratrol inhibits proliferation, induces apoptosis, and overcomes chemoresistance through down-regulation of STAT3 and nuclear factor-κB-regulated antiapoptotic and cell survival gene products in human multiple myeloma cells. *Blood* 2007, *109*, 2293–2302.

40. Ashikawa, K., Majumdar, S., Banerjee, S., Bharti, A. C., *et al.*, Piceatannol inhibits TNF-induced NF-κB activation and NF-κB-mediated gene expression through suppression of IκB kinase and p65 phosphorylation. *J Immunol* 2002, *169*, 6490–6497.

41. Ying, B., Yang, T., Song, X., Hu, X., *et al.*, Quercetin inhibits IL-1 beta-induced ICAM-1 expression in pulmonary epithelial cell line A549 through the MAPK pathways. *Mol Biol Rep* 2009, *36*, 1825–1832.

42. Min, Y. D., Choi, C. H., Bark, H., Son, H. Y., *et al.,* Quercetin inhibits expression of inflammatory cytokines through attenuation of NF-κB and p38 MAPK in HMC-1 human mast cell line. *Inflamm Res* 2007, *56*, 210–215.

43. Heiss, E., Herhaus, C., Klimo, K., Bartsch, H., *et al.,* Nuclear factor κ B is a molecular target for sulforaphane-mediated anti-inflammatory mechanisms. *J Biol Chem* 2001, *276*, 32008–32015.

44. Kim, H.-N., Na, H.-K., Kim, E.-H., Surh, Y.-J., *Proc Am Assoc Cancer Res.* 2007, Los Angeles Convention Center, Los Angeles, CA.

45. Heiss, E., Gerhauser, C., Time-dependent modulation of thioredoxin reductase activity might contribute to sulforaphane-mediated inhibition of NF-κB binding to DNA. *Antioxid Redox Signal* 2005, *7*, 1601–1611.

46. Li, Z., Li, J., Mo, B., Hu, C., *et al.*, Genistein induces cell apoptosis in MDA-MB-231 breast cancer cells via the mitogen-activated protein kinase pathway. *Toxicol In Vitro* 2008, *22*, 1749–1753.

47. Raffoul, J. J., Wang, Y., Kucuk, O., Forman, J. D., *et al.,* Genistein inhibits radiation-induced activation of NF-κB in prostate cancer cells promoting apoptosis and G2/M cell cycle arrest. *BMC Cancer* 2006, *6*, 107–116.

48. Singh-Gupta, V., Zhang, H., Banerjee, S., Kong, D., *et al.,* Radiation-induced HIF-1α cell survival pathway is inhibited by soy isoflavones in prostate cancer cells. *Int J Cancer* 2009, *124*, 1675–1684.

49. El-Rayes, B. F., Ali, S., Ali, I. F., Philip, P. A., *et al.*, Potentiation of the effect of erlotinib by genistein in pancreatic cancer: the role of Akt and nuclear factor-κB. *Cancer Res* 2006, *66*, 10553–10559.

50. Ali, S., Varghese, L., Pereira, L., Tulunay-Ugur, O. E., *et al.*, Sensitization of squamous cell carcinoma to cisplatin induced killing by natural agents. *Cancer Lett* 2009, *278*, 201–209.

51. Banerjee, S., Zhang, Y., Wang, Z., Che, M., *et al.*, In vitro and in vivo molecular evidence of genistein action in augmenting the efficacy of cisplatin in pancreatic cancer. *Int J Cancer* 2007, *120*, 906–917.

52. Li, Y., Kucuk, O., Hussain, M., Abrams, J., *et al.*, Antitumor and antimetastatic activities of docetaxel are enhanced by genistein through regulation of osteoprotegerin/receptor activator of nuclear factor-κB (RANK)/RANK ligand/MMP-9 signaling in prostate cancer. *Cancer Res* 2006, *66*, 4816–4825.

53. Mena, S., Ortega, A., Estrela, J. M., Oxidative stress in environmental-induced carcinogenesis. *Mutat Res 2009*, *674*, 36–44.

54. Toyokuni, S., Molecular mechanisms of oxidative stress-induced carcinogenesis: from epidemiology to oxygenomics. *IUBMB Life* 2008, *60*, 441–447.

55. Kong, A. N., Owuor, E., Yu, R., Hebbar, V., *et al.*, Induction of xenobiotic enzymes by the MAP kinase pathway and the antioxidant or electrophile response element (ARE/EpRE). *Drug Metab Rev* 2001, *33*, 255–271.

56. Ramos-Gomez, M., Kwak, M. K., Dolan, P. M., Itoh, K., *et al.*, Sensitivity to carcinogenesis is increased and chemoprotective efficacy of enzyme inducers is lost in nrf2 transcription factor-deficient mice. *Proc Natl Acad Sci U S A* 2001, *98*, 3410–3415.

57. Xu, C., Huang, M. T., Shen, G., Yuan, X., *et al.*, Inhibition of 7,12-dimethylbenz(a) anthracene-induced skin tumorigenesis in C57BL/6 mice by sulforaphane is mediated by nuclear factor E2-related factor 2. *Cancer Res* 2006, *66*, 8293–8296.

58. Aoki, Y., Hashimoto, A. H., Amanuma, K., Matsumoto, M., *et al.,* Enhanced spontaneous and benzo(a)pyrene-induced mutations in the lung of Nrf2-deficient *gpt* delta mice. *Cancer Res* 2007, *67*, 5643–5648.

59. Kitamura, Y., Umemura, T., Kanki, K., Kodama, Y., *et al.*, Increased susceptibility to hepatocarcinogenicity of Nrf2-deficient mice exposed to 2-amino-3-methylimidazo [4,5-f]quinoline. *Cancer Sci* 2007, *98*, 19–24.

60. Iida, K., Itoh, K., Maher, J. M., Kumagai, Y., *et al.*, Nrf2 and p53 cooperatively protect against BBN-induced urinary bladder carcinogenesis. *Carcinogenesis* 2007, *28*, 2398–2403.

61. Khor, T. O., Huang, M. T., Kwon, K. H., Chan, J. Y., *et al.*, Nrf2-deficient mice have an increased susceptibility to dextran sulfate sodium-induced colitis. *Cancer Res* 2006, *66*, 11580–11584.

62. Lee, J. S., Surh, Y. J., Nrf2 as a novel molecular target for chemoprevention. *Cancer Lett* 2005, *224*, 171–184.

63. Adams, J. R., Ali, S., Bennett, C. L., Pricing, profits and pharmacoeconomics–for whose benefit? *Expert Opin Pharmacother* 2001, *2*, 377–383.

64. Zhang, D. D., Lo, S. C., Cross, J. V., Templeton, D. J., *et al.,* Keap1 is a redox-regulated substrate adaptor protein for a Cul3-dependent ubiquitin ligase complex. *Mol Cell Biol* 2004, *24*, 10941–10953.

65. Motohashi, H., Yamamoto, M., Nrf2-Keap1 defines a physiologically important stress response mechanism. *Trends Mol Med* 2004, *10*, 549–557.

66. Eggler, A. L., Liu, G., Pezzuto, J. M., van Breemen, R. B., *et al.,* Modifying specific cysteines of the electrophile-sensing human Keap1 protein is insufficient to disrupt binding to the Nrf2 domain Neh2. *Proc Natl Acad Sci U S A* 2005, *102*, 10070–10075.

67. Huang, H. C., Nguyen, T., Pickett, C. B., Regulation of the antioxidant response element by protein kinase C-mediated phosphorylation of NF-E2-related factor 2. *Proc Natl Acad Sci U S A* 2000, *97*, 12475–12480.

68. Lo, S. C., Li, X., Henzl, M. T., Beamer, L. J., *et al.,* Structure of the Keap1:Nrf2 interface provides mechanistic insight into Nrf2 signaling. *EMBO J* 2006, *25*, 3605–3617.

69. Xu, C., Yuan, X., Pan, Z., Shen, G., *et al.*, Mechanism of action of isothiocyanates: the induction of ARE-regulated genes is associated with activation of ERK and JNK and the phosphorylation and nuclear translocation of Nrf2. *Mol Cancer Ther* 2006, *5*, 1918–1926.

70. Rojo, A. I., Sagarra, M. R., Cuadrado, A., GSK-3beta down-regulates the transcription factor Nrf2 after oxidant damage: relevance to exposure of neuronal cells to oxidative stress. *J Neurochem* 2008, *105*, 192–202.

71. Huang, H. C., Nguyen, T., Pickett, C. B., Phosphorylation of Nrf2 at Ser-40 by protein kinase C regulates antioxidant response element-mediated transcription. *J Biol Chem* 2002, *277*, 42769–42774.

72. Apopa, P. L., He, X., Ma, Q., Phosphorylation of Nrf2 in the transcription activation domain by casein kinase 2 (CK2) is critical for the nuclear translocation and transcription activation function of Nrf2 in IMR-32 neuroblastoma cells. *J Biochem Mol Toxicol* 2008, *22*, 63–76.

73. Jain, A. K., Jaiswal, A. K., GSK-3β acts upstream of Fyn kinase in regulation of nuclear export and degradation of NF-E2 related factor 2. *J Biol Chem* 2007, *282*, 16502–16510.

74. Salazar, M., Rojo, A. I., Velasco, D., de Sagarra, R. M., *et al.,* Glycogen synthase kinase-3β inhibits the xenobiotic and antioxidant cell response by direct phosphorylation and nuclear exclusion of the transcription factor Nrf2. *J Biol Chem* 2006, *281*, 14841–14851.

75. Chen, C., Kong, A. N., Dietary chemopreventive compounds and ARE/EpRE signaling. *Free Radic Biol Med* 2004, *36*, 1505–1516.

76. Thimmulappa, R. K., Mai, K. H., Srisuma, S., Kensler, T. W., *et al.*, Identification of Nrf2-regulated genes induced by the chemopreventive agent sulforaphane by oligonucleotide microarray. *Cancer Res* 2002, *62*, 5196–5203.

77. Hu, R., Xu, C., Shen, G., Jain, M. R., *et al.*, Gene expression profiles induced by cancer chemopreventive isothiocyanate sulforaphane in the liver of C57BL/6J mice and C57BL/6J/Nrf2 (−/−) mice. *Cancer Lett* 2006, *243*, 170–192.

78. Tanito, M., Masutani, H., Kim, Y. C., Nishikawa, M., *et al.*, Sulforaphane induces thioredoxin through the antioxidant-responsive element and attenuates retinal light damage in mice. *Invest Ophthalmol Vis Sci* 2005, *46*, 979–987.

79. Jakubikova, J., Sedlak, J., Bod'o, J., Bao, Y., Effect of isothiocyanates on nuclear accumulation of NF-κB, Nrf2, and thioredoxin in caco-2 cells. *J Agric Food Chem* 2006, *54*, 1656–1662.

80. Juge, N., Mithen, R. F., Traka, M., Molecular basis for chemoprevention by sulforaphane: a comprehensive review. *Cell Mol Life Sci* 2007, *64*, 1105–1127.

81. Jakubikova, J., Sedlak, J., Mithen, R., Bao, Y., Role of PI3K/Akt and MEK/ERK signaling pathways in sulforaphane- and erucin-induced phase II enzymes and MRP2 transcription, G2/M arrest and cell death in Caco-2 cells. *Biochem Pharmacol* 2005, *69*, 1543–1552.

82. Manandhar, S., Cho, J. M., Kim, J. A., Kensler, T. W., *et al.*, Induction of Nrf2-regulated genes by 3H-1, 2-dithiole-3-thione through the ERK signaling pathway in murine keratinocytes. *Eur J Pharmacol* 2007, *577*, 17–27.

83. Keum, Y. S., Yu, S., Chang, P. P., Yuan, X., *et al.*, Mechanism of action of sulforaphane: Inhibition of p38 mitogen-activated protein kinase isoforms contributing to the induction of antioxidant response element-mediated heme oxygenase-1 in human hepatoma HepG2 cells. *Cancer Res* 2006, *66*, 8804–8813.

84. Zhang, D.D., Hannink, M., Distinct cysteine residues in Keap1 are required for Keap1-dependent ubiquitination of Nrf2 and for stabilization of Nrf2 by chemopreventive agents and oxidative stress. *Mol Cell Biol* 2003, *23*, 8137–8151.

85. Hong, F., Freeman, M. L., Liebler, D. C., Identification of sensor cysteines in human Keap1 modified by the cancer chemopreventive agent sulforaphane. *Chem Res Toxicol* 2005, *18*, 1917–1926.

86. Banning, A., Deubel, S., Kluth, D., Zhou, Z., *et al.*, The GI-GPx gene is a target for Nrf2. *Mol Cell Biol* 2005, *25*, 4914–4923.

87. Brigelius-Flohe, R., Banning, A., Part of the series: from dietary antioxidants to regulators in cellular signaling and gene regulation. Sulforaphane and selenium, partners in adaptive response and prevention of cancer. *Free Radic Res* 2006, *40*, 775–787.

88. Morimitsu, Y., Nakagawa, Y., Hayashi, K., Fujii, H., *et al.*, A sulforaphane analogue that potently activates the Nrf2-dependent detoxification pathway. *J Biol Chem* 2002, *277*, 3456–3463.

89. Surh, Y. J., Chun, K. S., Cancer chemopreventive effects of curcumin. *Adv Exp Med Biol* 2007, *595*, 149–172.

90. Thangapazham, R. L., Sharma, A., Maheshwari, R. K., Multiple molecular targets in cancer chemoprevention by curcumin. *AAPS J* 2006, *8*, E443–E449.

91. Garg, R., Gupta, S., Maru, G., Dietary curcumin modulates transcriptional regulator(s) of phase I and phase II enzymes in benzo(a)pyrene-treated mice: mechanism of its anti-initiating action. *Carcinogenesis* 2008.

92. Farombi, E. O., Shrotriya, S., Na, H. K., Kim, S. H., *et al.*, Curcumin attenuates dimethylnitrosamine-induced liver injury in rats through Nrf2-mediated induction of heme oxygenase-1. *Food Chem Toxicol* 2008, *46*, 1279–1287.

93. Andreadi, C. K., Howells, L. M., Atherfold, P. A., Manson, M. M., Involvement of Nrf2, p38, B-Raf, and nuclear factor-κB, but not phosphatidylinositol 3-kinase, in induction of hemeoxygenase-1 by dietary polyphenols. *Mol Pharmacol* 2006, *69*, 1033–1040.

94. Shen, G., Xu, C., Hu, R., Jain, M. R., *et al.*, Modulation of nuclear factor E2-related factor 2-mediated gene expression in mice liver and small intestine by cancer chemopreventive agent curcumin. *Mol Cancer Ther* 2006, *5*, 39–51.

95. Nishinaka, T., Ichijo, Y., Ito, M., Kimura, M., *et al.*, Curcumin activates human glutathione S-transferase P1 expression through antioxidant response element. *Toxicol Lett* 2007, *170*, 238–247.

96. Dickinson, D. A., Iles, K. E., Zhang, H., Blank, V., *et al.*, Curcumin alters EpRE and AP-1 binding complexes and elevates glutamate-cysteine ligase gene expression. *FASEB J* 2003, *17*, 473–475.

97. Balogun, E., Hoque, M., Gong, P., Killeen, E., *et al.*, Curcumin activates the haem oxygenase-1 gene via regulation of Nrf2 and the antioxidant-responsive element. *Biochem J* 2003, *371*, 887–895.

 98. Rushworth, S. A., Ogborne, R. M., Charalambos, C. A., O'Connell, M. A., Role of protein kinase C delta in curcumin-induced antioxidant response element-mediated gene expression in human monocytes. *Biochem Biophys Res Commun* 2006, *341*, 1007–1016.

 99. Kang, E. S., Woo, I. S., Kim, H. J., Eun, S. Y., *et al.*, Up-regulation of aldose reductase expression mediated by phosphatidylinositol 3-kinase/Akt and Nrf2 is involved in the protective effect of curcumin against oxidative damage. *Free Radic Biol Med* 2007, *43*, 535–545.

100. McNally, S. J., Harrison, E. M., Ross, J. A., Garden, O. J., *et al.*, Curcumin induces heme oxygenase 1 through generation of reactive oxygen species, p38 activation and phosphatase inhibition. *Int J Mol Med* 2007, *19*, 165–172.

101. Pugazhenthi, S., Akhov, L., Selvaraj, G., Wang, M., *et al.*, Regulation of heme oxygenase-1 expression by demethoxy curcuminoids through Nrf2 by a PI3-kinase/Akt-mediated pathway in mouse beta-cells. *Am J Physiol Endocrinol Metab* 2007, *293*, E645–E655.

102. Dinkova-Kostova, A. T., Talalay, P., Relation of structure of curcumin analogs to their potencies as inducers of Phase 2 detoxification enzymes. *Carcinogenesis* 1999, *20*, 911–914.

103. Na, H. K., Surh, Y. J., Intracellular signaling network as a prime chemopreventive target of (−)-epigallocatechin gallate. *Mol Nutr Food Res* 2006, *50*, 152–159.

104. Park, O. J., Surh, Y. J., Chemopreventive potential of epigallocatechin gallate and genistein: evidence from epidemiological and laboratory studies. *Toxicol Lett* 2004, *150*, 43–56.

105. Chen, C., Yu, R., Owuor, E. D., Kong, A. N., Activation of antioxidant-response element (ARE), mitogen-activated protein kinases (MAPKs) and caspases by major green tea polyphenol components during cell survival and death. *Arch Pharm Res* 2000, *23*, 605–612.

106. Wu, C. C., Hsu, M. C., Hsieh, C. W., Lin, J. B., *et al.*, Upregulation of heme oxygenase-1 by Epigallocatechin-3-gallate via the phosphatidylinositol 3-kinase/Akt and ERK pathways. *Life Sci* 2006, *78*, 2889–2897.

107. Na, H. K., Kim, E. H., Jung, J. H., Lee, H. H., *et al.*, (−)-Epigallocatechin gallate induces Nrf2-mediated-antioxidant enzyme expression via activation of PI3K and ERK in human mammary epithelial cells. *Arch Biochem Biophys* 2008, *476*, 171–177.

108. Yuan, J. H., Li, Y. Q., Yang, X. Y., Inhibition of epigallocatechin gallate on orthotopic colon cancer by upregulating the Nrf2-UGT1A signal pathway in nude mice. *Pharmacology* 2007, *80*, 269–278.

109. Shen, G., Xu, C., Hu, R., Jain, M. R., *et al.*, Comparison of (−)-epigallocatechin-3-gallate elicited liver and small intestine gene expression profiles between C57BL/6J mice and C57BL/6J/Nrf2 (−/−) mice. *Pharm Res* 2005, *22*, 1805–1820.

110. Chen, C., Pung, D., Leong, V., Hebbar, V., *et al.*, Induction of detoxifying enzymes by garlic organosulfur compounds through transcription factor Nrf2: effect of chemical structure and stress signals. *Free Radic Biol Med* 2004, *37*, 1578–1590.

111. Fisher, C. D., Augustine, L. M., Maher, J. M., Nelson, D. M., *et al.*, Induction of drug-metabolizing enzymes by garlic and allyl sulfide compounds via activation of constitutive androstane receptor and nuclear factor E2-related factor 2. *Drug Metab Dispos* 2007, *35*, 995–1000.

112. Gong, P., Hu, B., Cederbaum, A. I., Diallyl sulfide induces heme oxygenase-1 through MAPK pathway. *Arch Biochem Biophys* 2004, *432*, 252–260.

113. Kundu, J. K., Surh, Y. J., Molecular basis of chemoprevention by resveratrol: NF-κB and AP-1 as potential targets. *Mutat Res* 2004, *555*, 65–80.

114. Jang, M., Cai, L., Udeani, G. O., Slowing, K. V., *et al.*, Cancer chemopreventive activity of resveratrol, a natural product derived from grapes. *Science* 1997, *275*, 218–220.

115. Aziz, M. H., Reagan-Shaw, S., Wu, J., Longley, B. J., *et al.*, Chemoprevention of skin cancer by grape constituent resveratrol: relevance to human disease? *FASEB J* 2005, *19*, 1193–1195.

116. Banerjee, S., Bueso-Ramos, C., Aggarwal, B. B., Suppression of 7,12-dimethylbenz(a) anthracene-induced mammary carcinogenesis in rats by resveratrol: role of nuclear factor-κB, cyclooxygenase 2, and matrix metalloprotease 9. *Cancer Res* 2002, *62*, 4945–4954.

117. Li, Z. G., Hong, T., Shimada, Y., Komoto, I., *et al.*, Suppression of N-nitrosomethy lbenzylamine (NMBA)-induced esophageal tumorigenesis in F344 rats by resveratrol. *Carcinogenesis* 2002, *23*, 1531–1536.

118. Gerhauser, C., Klimo, K., Heiss, E., Neumann, I., *et al.*, Mechanism-based in vitro screening of potential cancer chemopreventive agents. *Mutat Res* 2003, 163–172, 523–524.

119. Hsieh, T. C., Lu, X., Wang, Z., Wu, J. M., Induction of quinone reductase NQO1 by resveratrol in human K562 cells involves the antioxidant response element ARE and is accompanied by nuclear translocation of transcription factor Nrf2. *Med Chem* 2006, *2*, 275–285.

120. Juan, S. H., Cheng, T. H., Lin, H. C., Chu, Y. L., *et al.*, Mechanism of concentration-dependent induction of heme oxygenase-1 by resveratrol in human aortic smooth muscle cells. *Biochem Pharmacol* 2005, *69*, 41–48.

121. Chen, C. Y., Jang, J. H., Li, M. H., Surh, Y. J., Resveratrol upregulates heme oxygenase-1 expression via activation of NF-E2-related factor 2 in PC12 cells. *Biochem Biophys Res Commun* 2005, *331*, 993–1000.

122. Kode, A., Rajendrasozhan, S., Caito, S., Yang, S. R., *et al.*, Resveratrol induces glutathione synthesis by activation of Nrf2 and protects against cigarette smoke-mediated oxidative stress in human lung epithelial cells. *Am J Physiol Lung Cell Mol Physiol* 2008, *294*, L478–L488.

123. Joung, E. J., Li, M. H., Lee, H. G., Somparn, N., *et al.*, Capsaicin induces heme oxygenase-1 expression in HepG2 cells via activation of PI3K-Nrf2 signaling: NAD(P)H:quinone oxidoreductase as a potential target. *Antioxid Redox Signal* 2007, *9*, 2087–2098.

124. Luo, Y., Eggler, A. L., Liu, D., Liu, G., *et al.*, Sites of alkylation of human Keap1 by natural chemoprevention agents. *J Am Soc Mass Spectrom* 2007, *18*, 2226–2232.

125. Hwang, E. S., Bowen, P. E., Can the consumption of tomatoes or lycopene reduce cancer risk? *Integr Cancer Ther* 2002, *1*, 121–132; discussion 132.

126. Ben-Dor, A., Steiner, M., Gheber, L., Danilenko, M., *et al.*, Carotenoids activate the antioxidant response element transcription system. *Mol Cancer Ther* 2005, *4*, 177–186.

127. Cavin, C., Holzhaeuser, D., Scharf, G., Constable, A., *et al.*, Cafestol and kahweol, two coffee specific diterpenes with anticarcinogenic activity. *Food Chem Toxicol* 2002, *40*, 1155–1163.

128. Higgins, L. G., Cavin, C., Itoh, K., Yamamoto, M., *et al.*, Induction of cancer chemopreventive enzymes by coffee is mediated by transcription factor Nrf2. Evidence that the coffee-specific diterpenes cafestol and kahweol confer protection against acrolein. *Toxicol Appl Pharmacol* 2008, *226*, 328–337.

129. Martin, D., Rojo, A. I., Salinas, M., Diaz, R., *et al.*, Regulation of heme oxygenase-1 expression through the phosphatidylinositol 3-kinase/Akt pathway and the Nrf2 transcription factor in response to the antioxidant phytochemical carnosol. *J Biol Chem* 2004, *279*, 8919–8929.

130. Liao, B. C., Hsieh, C. W., Liu, Y. C., Tzeng, T. T., *et al.*, Cinnamaldehyde inhibits the tumor necrosis factor-α-induced expression of cell adhesion molecules in endothelial cells by suppressing NF-κB activation: Effects upon IκB and Nrf2. *Toxicol Appl Pharmacol* 2008, *229*, 161–171.

131. Dietz, B. M., Kang, Y. H., Liu, G., Eggler, A. L., *et al.*, Xanthohumol isolated from Humulus lupulus Inhibits menadione-induced DNA damage through induction of quinone reductase. *Chem Res Toxicol* 2005, *18*, 1296–1305.

132. Murakami, A., Tanaka, T., Lee, J. Y., Surh, Y. J., *et al.*, Zerumbone, a sesquiterpene in subtropical ginger, suppresses skin tumor initiation and promotion stages in ICR mice. *Int J Cancer* 2004, *110*, 481–490.

133. Nakamura, Y., Yoshida, C., Murakami, A., Ohigashi, H., *et al.*, Zerumbone, a tropical ginger sesquiterpene, activates phase II drug metabolizing enzymes. *FEBS Lett* 2004, *572*, 245–250.

134. Liu, Y. C., Hsieh, C. W., Wu, C. C., Wung, B. S., Chalcone inhibits the activation of NF-κB and STAT3 in endothelial cells via endogenous electrophile. *Life Sci* 2007, *80*, 1420–1430.

135. Lee, S. H., Sohn, D. H., Jin, X. Y., Kim, S. W., *et al.*, 2',4',6'-tris(methoxymethoxy) chalcone protects against trinitrobenzene sulfonic acid-induced colitis and blocks tumor necrosis

factor-α-induced intestinal epithelial inflammation via heme oxygenase 1-dependent and independent pathways. *Biochem Pharmacol* 2007, *74*, 870–880.

136. Alcaraz, M. J., Vicente, A. M., Araico, A., Dominguez, J. N., *et al.*, Role of nuclear factor-κB and heme oxygenase-1 in the mechanism of action of an anti-inflammatory chalcone derivative in RAW 264.7 cells. *Br J Pharmacol* 2004, *142*, 1191–1199.

137. Surh, Y. J., Kundu, J. K., Na, H. K., Lee, J. S., Redox-sensitive transcription factors as prime targets for chemoprevention with anti-inflammatory and antioxidative phytochemicals. *J Nutr* 2005, *135*, 2993S–3001S.

138. Khor, T. O., Huang, M. T., Prawan, A., Liu, Y., *et al.*, Increased susceptibility of Nrf2 knockout mice to colitis-associated colorectal cancer. *Cancer Prev Res (Phila Pa)* 2008, *1*, 187–191.

139. Prawan, A., Buranrat, B., Kukongviriyapan, U., Sripa, B., *et al.*, Inflammatory cytokines suppress NAD(P)H:quinone oxidoreductase-1 and induce oxidative stress in cholangiocarcinoma cells. *J Cancer Res Clin Oncol* 2009, *135*, 515–522.

140. Li, W., Khor, T. O., Xu, C., Shen, G., *et al.*, Activation of Nrf2-antioxidant signaling attenuates NFκB-inflammatory response and elicits apoptosis. *Biochem Pharmacol* 2008, *76*, 1485–1489.

141. Liu, G. H., Qu, J., Shen, X., NF-κB/p65 antagonizes Nrf2-ARE pathway by depriving CBP from Nrf2 and facilitating recruitment of HDAC3 to MafK. *Biochim Biophys Acta* 2008, *1783*, 713–727.

142. Matthews, J. R., Wakasugi, N., Virelizier, J. L., Yodoi, J., *et al.*, Thioredoxin regulates the DNA binding activity of NF-κ B by reduction of a disulphide bond involving cysteine 62. *Nucleic Acids Res* 1992, *20*, 3821–3830.

143. Matthews, J. R., Botting, C. H., Panico, M., Morris, H. R., et al., (1996) Inhibition of NF-κB DNA binding by nitric oxide. *Nucleic Acids Res* 1996, *24*, 2236–2242.

144. Nair, S., Doh, S. T., Chan, J. Y., Kong, A. N., *et al.*, Regulatory potential for concerted modulation of Nrf2- and NFKB1-mediated gene expression in inflammation and carcinogenesis. *Br J Cancer* 2008, *99*, 2070–2082.

145. Kang, E. S., Kim, G. H., Kim, H. J., Woo, I. S., *et al.*, Nrf2 regulates curcumin-induced aldose reductase expression indirectly via nuclear factor κB. *Pharmacol Res* 2008, *58*, 15–21.

146. Kim, S. K., Yang, J. W., Kim, M. R., Roh, S. H., *et al.*, Increased expression of Nrf2/ARE-dependent anti-oxidant proteins in tamoxifen-resistant breast cancer cells. *Free Radic Biol Med* 2008, *45*, 537–546.

147. Wang, X. J., Sun, Z., Villeneuve, N. F., Zhang, S., *et al.*, Nrf2 enhances resistance of cancer cells to chemotherapeutic drugs, the dark side of Nrf2. *Carcinogenesis* 2008, *29*, 1235–1243.

148. Jeong, W. S., Kim, I. W., Hu, R., Kong, A. N., Modulatory properties of various natural chemopreventive agents on the activation of NF-κB signaling pathway. *Pharm Res* 2004, *21*, 661–670.

149. Xu, C., Shen, G., Chen, C., Gelinas, C., *et al.*, Suppression of NF-κB and NF-κB-regulated gene expression by sulforaphane and PEITC through IκBα, IKK pathway in human prostate cancer PC-3 cells. *Oncogene* 2005, *24*, 4486–4495.

150. Na, H. K., Surh, Y. J., Transcriptional regulation via cysteine thiol modification: a novel molecular strategy for chemoprevention and cytoprotection. *Mol Carcinog* 2006, *45*, 368–380.

Section 3

Proteomics

13

Proteomics Analysis of the Functionality of *Toona sinensis* by 2D-Gel Electrophoresis

Sue-Joan Chang and Chun-Yung Huang

INTRODUCTION

Proteomics is the study of the "proteome," which is recognized as the comprehensive analysis of a protein complement in a cell, tissue, or biological fluid at a given time. The word "proteome" was coined in late 1994 at the Siena 2D electrophoresis meeting [1] and described the entire collection of proteins of an organism, including products arising from events such as the processing of mRNA transcripts (i.e., alternative splicing) and posttranslational modifications (e.g., phosphorylation, glycosylation, and oxidation) [2, 3]. Proteomics-based studies are focused on the interactions of multiple proteins and their role as part of a biological system rather than the structure and function of one single component. Proteins are directly involved in all cellular activities, cell phenotype, and hence the tissue or organ. This phenotype varies under normal physiological conditions (e.g., cell-cycle stage, differentiation, function, and age) or in response to pathophysiological stresses. Environmental changes may alter protein structure and functions, ultimately leading to the progression of disease. The concept of comparing the global protein profile from tissues between two biological states has given rise to a new era in biological science. Therefore, proteomic studies can be applied to a range of biological systems to answer the desired research question, including human studies, animal models, and/or cell culture systems [4]. Furthermore, more studies in food-related products have applied proteomic analysis to elucidate the functionalities of phytoactive and/or bioactive components. In this article, certain functionalities of *Toona sinensis* explored by our laboratory are discussed and used as examples to elucidate the functionality of *Toona sinensis* by 2D-gel electrophoresis-based proteomics.

TOONA SINENSIS AS A NOVEL ANTIOXIDANT

ANTIOXIDANT ACTIVITY IN LEAF EXTRACTS OF *Toona sinensis*

Toona sinensis Roem (TS), a widely distributed arbor in Asia, is a nutritious food in Chinese society and a popular vegetarian cuisine in Taiwan. Leaves of TS (TSL) have been used as an Oriental medicine for treatment of enteritis, dysentery, metabolic disease, infection, and itch. Leaf extracts of TS provide novel functions including (1) antioxidant properties and effective protection from atherogenesis [5]; (2) strong 1,1-diphenyl-2-picrylhydrazyl (DPPH) radicalscavenging activities and inhibitory effects on lipid peroxidation [6]; (3) protection against hydrogen peroxide-induced oxidative stress and DNA damage in Madin-Darby canine kidney (MDCK) cells [7]; (4) antilow-density lipoprotein (LDL) glycative activity [8]; and (5) improvement of learning and memory through the reduction of lipid peroxidation and s-amyloid plaques in the brains of senescence-accelerated mice [9]. In Chinese ancient book, TS was recognized as a therapeutic herbal medicine for treatment of wet dreams. However, the scientific evidence for TSL in prevention or treatment of sperm dysfunctions is scarce.

POSSIBLE ACTIVE COMPOUNDS IN TSL EXTRACTS

Certain phytochemical components were identified from TS leaves, including quercetin, rutin, methyl gallate, gallic acid, and catechin [7,10], as widely distributed representatives of flavonoids that possessed antioxidative properties [11,12,13]. In our laboratory, leaves of *Toona sinensis* were boiled in reverse osmosis (RO) filtered water for 30 minutes and extracted by serious different concentrations of ethanol. The filtered concentrates were lyophilized to obtain a crude extract (TSL-CE), TSL-1, TSL-2 (supernatant), TSL-2P (pellet), TSL-3 (ethanol, supernatant), TSL-3P (ethanol, pellet), TSL-4 (ethanol, supernatant), TSL-4P (ethanol, pellet), TSL-5 (H_2O, supernatant), TSL-5P (H_2O, pellet), TSL-6 (H2O, supernatant), and TSL-7 (H_2O, pellet). Among these, TSL-2, TSL-2P, and TSL-6 were found to be the most active extracts. Therefore, they were utilized for the 2D-gel electrophoresis-based proteomic analysis of the functionalities of the *Toona sinensis* leaves (TSL). In addition, TSL-2, TSL-2P, and TSL-6 were dissolved in methanol and subjected to HPLC for analysis of possible active components. It was found that 1 mg/ml TSL-2 contained approximately 50.3% gallic acid, 1.4% quercetin, 2.49% rutin, 1.3% catechin, 0.19% vitamin C, and 44.32% unknown compounds. One mg/ml TSL-2P contained approximately 22.53% gallic acid, 1.45% quercetin, 2.2% rutin, 0.66% ethyl gallate, 0.2% vitamin C, and 72.96% unknown compounds. Moreover, 1 mg/ml TSL-6 contained approximately 7.85% gallic acid, 0.92% quercetin, 1.17% rutin, 0.36% vitamin C, and 89.7% unknown compounds. Therefore, gallic acid may be the active component contributed to antioxidant in TSL extracts.

FUNCTIONALITY OF *Toona sinensis* BY 2D-GEL ELECTROPHORESIS

2D-gel electrophoresis-based proteomic approach was carried out to investigate the proteins modulated by TSL extracts in rat testes under oxidative stress. Sprague-Dawley (S-D) rats were IP injected with H_2O_2 every other day and fed with normal diet (control), vitamin C, or TSL extracts (TSL-2; 0.053 g/kgBW/day, TSL-2P; 0.94 g/kgBW/day or TSL-6; 0.013 g/kgBW/day) daily for 8 weeks. Testicular proteins were extracted for 2-DE analysis. Protein spot determination and comparative analysis of 2-DE among control, vitamin C, and TSL groups showed that ten protein spots were modulated by TSL-2 at least more than 1.5-fold in the rat testes under oxidative stress, whereas four proteins including glutathione S-transferase (GST), phospholipids hydroperoxide glutathione peroxidase (PHGPx), fatty acid binding protein 9 (FABP9), and thioredoxin were further identified by MS MALDI-TOF. Three protein spots were modulated by TSL-2P more than 1.5-fold in the rat testes under oxidative stress, whereas one protein (phospholipids hydroperoxide glutathione peroxidase, PHGPx) was further identified by MS MALDI-TOF. Moreover, eleven protein spots were modulated by TSL-6 more than two-fold in the rat testes under oxidative stress, whereas nine proteins including tumor rejection antigen gp96, 3-hydroxy-3-methylglutaryl-Coenzyme A synthase 2 (HMG CoA synthase 2), glutathione transferase Mu 6 (GST Mu6), cofilin 2, pancreatic trypsin 1, heat shock protein 1-β (HSP 1-β), peptidylprolyl isomerase A, type II keratin Kb1, and heat shock 90kDa protein 1-β were identified by MS MALDI-TOF.

FUNCTIONS OF TSL-2

Four proteins modulated by TSL-2, including GST, PHGPx, FABP9, and thioredoxin, are involved in antistress and sperm functions. GSTs are a family of phase II detoxification enzymes that catalyze the conjugation of glutathione (GSH) to a wide variety of endogenous and exogenous electrophilic toxic compounds. GSTs are divided into two distinct super-family members:

the membrane-bound microsomal and cytosolic family members. Microsomal GSTs play a key role in the endogenous metabolism of leukotrienes and prostaglandins. Cytosolic GSTs play a regulatory role in signaling transduction pathway that participates in cellular survival and death signals. Aside from a pivotal role in detoxification and defense against oxidative stress, GSTs may play a role in the etiology of diseases, including tumor, neurodegenerative diseases, multiple sclerosis, and asthma [14, 15]. Previous studies have demonstrated that a mix of glucosinolate breakdown products (isothiocyanates, nitriles, and thiocyanates) in cruciferous vegetables synergistically upregulate phase II detoxification enzymes (quinone reductase and GST) [16]. Using stannous chloride ($SnCl_2$) to generate reactive oxygen species (ROS), Yousef and colleagues found $SnCl_2$ significantly decreased activity of GST in rabbit blood plasma. On the contrary, ascorbic acid alone increased activity of GST. Ascorbic acid also alleviated $SnCl_2$-induced harmful effects [17]. We found that vitamin C upregulated GST and TSL-2 downregulated GST. Consistent with Yousef *et al.*'s results, vitamin C, via induction of GST, alleviated oxidative damage in rat testes. Previous studies from our laboratory revealed that TSL-2 exhibited prooxidant effect and elevated levels of ROS in human spermatozoa. Consistent with our previous findings, TSL-2 was found to possess prooxidant effect and reduce GST expression by using proteomic analysis. Glutathione peroxidase (GPx) converts hydrogen peroxide to H_2O at the expense of oxidizing GSH to its disulfide form (GSSG). GSSG is returned to the GSH form by glutathione reductase (GR) using NADPH. In general, intracellular glutathione peroxidases comprise two distinct proteins, classical GPx (cGPx) and phospholipid hydroperoxide glutathione peroxidase (PHGPx), which exist in the nucleus, mitochondria, and cytosol. PHGPx belongs to the selenoprotein family, whose members contain selenocysteine at the active site and regulatory domain [18]. PHGPx and vitamin E synergistically scavenged lipid hydroperoxyl radicals and inhibited lipid peroxidation [19]. PHGPx was also found to suppress apoptosis in RBL2H3 cells [20]. Selenium is essential for male fertility and has also been implicated in the fertilization capacity of spermatozoa in both livestock and humans. Selenium deficiency is associated with impaired sperm motility, structural alterations of the midpiece, and loss of flagellum. The selenoprotein PHGPx is abundantly expressed in spermatids and displays high activity in postpubertal testis. Therefore, PHGPx plays a pivotal role in male fertility through its dual functions (a soluble active enzyme and enzymatically inactive structural protein) during sperm maturation [21]. We found that PHGPx was upregulated by control and downregulated by TSL-2. TSL-2 was known to exhibit prooxidant property, and thus reduced PHGPx expression. Fatty acid binding protein (FABP) is a cytosolic protein that binds unsaturated long-chain fatty acids and acyl-CoA esters [22]. So far, nine different FABPs, with tissue-specific distribution, have been identified. The primary role of all the FABPs is to regulate fatty acid uptake and intracellular transport. Among these FABPs, FABP9 is predominantly expressed in testis [23] and involved in spermatogenesis, testicular germ cell apoptosis, and sperm quality preservation [24, 25]. Here, we found that FABP9 was upregulated by control and followed by vitamin C and then TSL-2. The possible reason was that, in control, elevated H_2O_2-induced testicular germ cell apoptosis led to the upregulation of FABP9. TSL-2 and vitamin C protected testicular germ cell from apoptosis and gave rise to the decreased FABP9 expression as compared to control. Thioredoxin is reported as one of the most important thiol-based systems being involved in many physiological as well as pathophysiological processes. Thioredoxins function as general protein disulphide reductases. Mammalian male germ cells possess a set of three testis-specific thioredoxins (named Sptrx-1, -2, and -3, respectively) that are expressed either in different structures within the sperm cell or at different stages of sperm development. Sptrx-1 is located in the developing tail of elongating spermatids, transiently associated with the longitudinal columns of the fibrous sheath. Sptrx-2 is also associated with the sperm fibrous sheath, but unlike Sptrx-1, Sptrx-2 becomes a structural part of the mature sperm tail and can be detected in ejaculated spermatozoa. Sptrx-3 can be found in Golgi-derived vesicles associated with the developing spermatid acrosome and its function might be related to the posttranslational modification of proteins required for acrosomal biogenesis [26]. We found that thioredoxin was upregulated by vitamin C and control and downregulated by TSL-2. However, function of TSL extracts in thioredoxins of rat testes under oxidative stress needs to be further investigated due to the distinct role of Sptrx-1, Sptrx-2, and Sptrx-3.

FUNCTIONS OF TSL-2P

One protein, PHGPx, was modulated by TSL-2P using 2D-gel electrophoresis-based proteomics. PHGPx is abundantly expressed in spermatids and displays high activity in postpubertal testis. PHGPx, a soluble active enzyme and enzymatically inactive structural protein, plays a crucial role in male fertility during sperm maturation [21]. We found that TSL-2P downregulated PHGPx expression using proteomic analysis and exhibited prooxidant property at high concentration. Therefore, TSL-2P impairs sperm functions via downregulation of PHGPx.

FUNCTIONS OF TSL-6

Nine proteins modulated by TSL-6 and identified by MS MALDI-TOF are gp96, HMG CoA synthase 2, GST Mu6, cofilin 2, pancreatic trypsin 1, HSP 1-β, peptidylprolyl isomerase A, type II keratin Kb1, and heat shock 90kDa protein 1-β. Gp96, a member of the heat shock protein 90 family, has been implicated in the formation of a functional zona-receptor complex on the surface of mammalian spermatozoa. Gp96 was subsequently found to become coexpressed on the surface of live mouse spermatozoa following capacitation *in vitro* and was lost once these cells had undergone the acrosome reaction, as would be expected of cell surface molecules involved in sperm–egg interaction. These findings indicated that GP96 was intimately involved in the mechanisms by which mammalian spermatozoa both acquire and express their ability to recognize the zona pellucida (ZP) [27]. Studies also suggested that elevated ROS resulted in decline of sperm–egg interaction [28]. We found that gp96 was highly upregulated by TSL-6 and downregulated by vitamin C and control. The ROS elevated by H_2O_2 in control leading to decreased gp96 expression implicated a decline of sperm–egg interaction. TSL-6, compared to the control, highly induced gp96 expression, suggesting that TSL-6 may facilitate capacitation and recognition of zona pellucida in sperm under oxidative stress [27]. HMG-CoA synthase catalyzes the condensation of acetoacetyl-CoA and acetyl-CoA to form HMG-CoA plus free CoA. HMG-CoA synthases are located in two different compartments: the cytosol and the mitochondria. The HMG-CoA produced by the cytosolic HMG-CoA synthase is transformed into mevalonate (a precursor of cholesterol) by the action of HMG-CoA reductase. The HMG-CoA produced inside the mitochondria by the mitochondrial HMG-CoA synthase is transformed into acetoacetate by the action of HMG-CoA lyase. Acetoacetate is transformed into hydroxybutyrate and acetone; all of these are known as ketone bodies [29, 30]. Testis and ovary express the gene for the ketogenic mitochondrial HMG-CoA synthase, which is involved in *de novo* cholesterogenesis in gonads [31]. In our study, the HMG-CoA synthase identified by MALDI-TOF was mitochondrial HMG-CoA synthase, which was upregulated by TSL-6 and control and downregulated by vitamin C. Therefore, TSL-6 induced HMG-CoA synthase expression to regulate hormone production in gonads under oxidative stress. GST Mu6 is a member of the GST family. GST Mu6 is mainly expressed in brain, testis, and lung and functions as a detoxicative enzyme [32]. In our study, the highest amount of GST Mu6 was upregulated by control and followed by TSL-6 and downregulated by vitamin C. Studies have shown that methyl gallate (MG) from TS prevents intracellular GSH from being depleted following an exposure of H_2O_2 in MDCK cells [7]. Certain phytochemical components in TSL have been isolated and identified, including quercetin, rutin, methyl gallate, gallate, and catechin [7, 10]. We found that approximately 7.85% gallic acid, which is the most abundant and active antioxidant of TSLs, is comprised in TSL-6. The gallic acid in TSL-6 prevents GSH from being depleted under oxidative stress; therefore, activity of GST Mu6 was not pronouncedly needed and its expression was declined as compared to the control. Vitamin C, being an antioxidant possessing the similar mechanism as TSL-6, declined GST Mu6 expression as compared to control. Cofilin is an actin-modulating protein of 20 kDa, which is widely distributed throughout muscle and nonmuscle cells. Two cofilin subtypes, muscle type (M-type) and nonmuscle type (NM-type), were found to exist in mammals [33]. Nonmuscle cofilin is a component of tubulobulbar complexes with finger-like structures that form at the interface between maturing spermatids and Sertoli cells prior to sperm release and at the interface between two Sertoli cells near the base of the seminiferous epithelium [34]. We found that cofilin 2 was upregulated by vitamin C and downregulated by control and TSL-6. We suggest that vitamin C increases cofilin 2 expression and maintains the contact interface between spermatids and Sertoli cells. Expression of cofilin is also related to survival and differentiation of spermatogenic cells. Studies revealed that, in stress condition, cofilin phosphorylation was severely impaired, and cofilin frequently accumulated in the nucleus and led to apoptosis of germ cells [35]. Previous studies from our laboratory have shown that TSL-6 alleviated H_2O_2-induced apoptosis in human spermatozoa. It is thus interesting to further study whether TSL-6 reversed apoptosis through impairment of cofilin phosphorylation. Pancreatic trypsin 1 is a kind of testicular serine proteinase that participates in the later stages of male germ cell maturation and fertilization [36]. Acrosin, a serine protease with trypsin-like cleavage specificity, is one of the proteins present in the sperm acrosome. Using acrosin-deficient ($Acr^{-/-}$) mutant mice conclusively showed that sperms do not require acrosin to penetrate the zona pellucida. Further experiments of the $Acr^{-/-}$ mouse sperm have provided evidence that the major role of acrosin is to accelerate the dispersal of acrosomal components during the acrosome reaction of sperms. Acrosomal trypsin-like protease (named pancreatic trypsin 1) other than acrosin is probably essential for the sperm penetration of ZP in mouse [37]. In our study, pancreatic trypsin 1 was upregulated by vitamin C, followed by TSL-6 and control. Vitamin C and TSL-6 reversed H_2O_2-induced decreased levels of pancreatic trypsin 1 expression and promoted acrosome reaction of sperm. Exposure of cells to environmental stress including heat shock, oxidative stress, heavy metals, or pathologic conditions results in the inducible expression of heat shock proteins (HSPs) that function as molecular chaperones or proteases. HSPs have been classified into six major families according to their molecular size: HSP100, HSP90, HSP70, HSP60, HSP40, and small heat shock proteins. Heat shock protein 1 belongs to HSP90 family. Roles of HSP90 were characterized in signal transduction (e.g., interaction with steroid hormone receptors, tyrosine kinases, serine/threonine kinases), refolding and maintaining of proteins *in vitro*, autoregulation of the heat shock response, cell cycle, and proliferation [38]. In response to environmental stress such as oxidative stress and heat shock, HSP90 is intensively expressed. HSP90 appears to interact with intermediately folded proteins and to prevent their aggregation but lacks the ability of HSP70 to refold denatured proteins [39]. Accordingly, HSP90 and HSP70 generally coexpress and mutually assist in protein refolding. In our study, HSP90 was upregulated by vitamin C, followed by control and then TSL-6. Hassen *et al*. showed that zearalenone (ZEN), a fusarial mycotoxin, was cytotoxic to Hep G2 cells through induction of oxidative DNA damage, depletion of GSH, and induction of HSP70 and HSP90 expression. However, significant reduction of the oxidative DNA damage as well as heat shock protein induction occurred when cells were pretreated with vitamin E prior to exposure to ZEN [40]. Consistent with Hassen *et al*.'s results, we found that TSL-6, with antioxidant property, protected rat testes from oxidative damage and concomitantly declined HSP90 expression. In addition, vitamin C recovered oxidative damage in rat testes via refolding of denatured proteins by induction of HSP90 expression as compared to control [39]. Our findings suggest that TSL-6 and vitamin C mediate distinct regulatory mechanism in HSP90 of rat testes under oxidative stress. Peptidylprolyl isomerase A is highly expressed in spermatogonia and Sertoli cells in adult mouse testes. Using peptidylprolyl isomerase A-depleted mouse model, it has been shown that germ cells in postnatal peptidylprolyl isomerase $A^{-/-}$ testis were able to initiate and complete spermatogenesis and production of mature spermatozoa. However, there was a progressive and age-dependent degeneration of the spermatogenic cells in peptidylprolyl isomerase $A^{-/-}$ testis that led to the germ cell loss by 14 months of age [41]. Moreover, peptidylprolyl isomerase A was reported to possess nuclease activity. In the event of 2-methoxyethanol-induced apoptosis of spermatocyte, peptidylprolyl isomerase A was highly expressed and able to cleave substrate DNA into a pattern of DNA fragmentation consisting of ~180–200 base pairs [42]. We found that peptidylprolyl isomerase A was upregulated by vitamin C, followed by control and then TSL-6. Vitamin C upregulated peptidylprolyl isomerase A expression to maintain proper cell cycle progression in testes under oxidative stress. Under oxidative stress, peptidylprolyl isomerase A was induced to cleave substrate DNA. TSL-6, with antioxidant effect, protected DNA from oxidative damage via reducing peptidylprolyl isomerase A expression.

The 2D-gel electrophoresis-based proteomic results showed that TSL extracts differentially regulate the expression profiles of proteins involved in sperm-oocyte interaction, detoxification, spermatogenesis, hormone production, and protein misfolding/refolding in rat testes under oxidative stress. TSL-2P is the upstream extract of TSL-6. TSL-2P exhibits spermatic regulatory functions similar to TSL-6; however, the proteins modulated by TSL-2P are less than that of TSL-6. We speculate that TSL-6, the downstream extract of TSL-2P, retains more optimal active component than that of TSL-2P. The most important finding, according to the functions of expressed proteins, TSL-2 and TSL-6 are first discovered to inhibit and promote sperm functions, respectively, which may be beneficial for development of functional products for male contraceptive purpose and treatment of male infertility, respectively.

TSL EXTRACTS AND HUMAN SPERM FUNCTIONS

INTRACELLULAR ROS LEVELS

ROS are free radicals derived from the metabolism of oxygen. The production of ROS, such as superoxide ($O2^-$), hydrogen peroxide (H_2O_2), and the hydroxyl radical (OH^-), normally occurs in cells. ROS play an important role in multiple cellular physiologic processes, such as phagocytosis, and in many signaling processes. ROS previously have been found to have dual effects on human spermatozoa. Physiologically, ROS are required for capacitation and acrosome reaction. At high concentrations, on the other hand, ROS induce motility loss and lead to sperm dysfunctions, including poor spermzona pellucida binding and sperm-oocyte fusion [43, 44]. ROS induce membrane lipid peroxidation in sperm, and the toxicities of generated fatty acid peroxides are important causes of impaired sperm functions and male infertility [45, 46, 47]. We found that TSL exhibited anti/prooxidant effects as indicated by intracellular ROS levels. Low concentrations of various TSL extracts reduced intracellular ROS levels of human spermatozoa as compared to the control at different time intervals, suggesting an antioxidant effect of TSL extracts. On the contrary, high concentrations of various TSL extracts except for TSL-6 exhibited elevated intracellular ROS levels of human spermatozoa as compared to the control at different time intervals, indicating a prooxidant effect of TSL extracts at high concentrations. Treatment of UVA radiation-induced oxidative damaged HaCaT (human keratinocyte line) cells with 10 μM EGCG (epigallocatechin-3-gallate) reduced ROS levels. However, in the absence of UVA, a prooxidant effect was detected upon 50 μM EGCG treatment [48]. Certain phytochemical components of TSL had been isolated and identified, including quercetin, rutin, methyl gallate, gallate, and catechin [7, 10], as widely distributed representatives of flavonoids. Labuta and colleagues [11] have reported that flavonoids, such as quercetin, rutin, catechin, and EGCG possessed anti/prooxidant properties. Consistent with Labuta et al.'s results, we also found that TSL extracts possessed anti/prooxidant properties.

SPERM MOTILITY, MITOCHONDRIAL MEMBRANE POTENTIAL (MMP), AND ADENOSINE TRIPHOSPHATE (ATP) LEVELS

A representative sperm computer-assisted sperm analysis (CASA) shows three velocity indicators (VAP, VSL, and VCL). VAP and VSL represent the "progression" of sperm motion, and VCL represents the "vigor" of sperm motion. Sperm motility parameters provide useful information for assessing sperm quality and predicting fertilization rates in *in vitro* fertilization (IVF) [49, 50]. ATP levels and MMP are responsible for sperm motility [51]. Using the hypoxanthine/xanthine oxidase system to generate ROS, Armstrong and colleagues [52] determined that spermatozoa ATP levels and sperm forward progression, after ROS treatment, were reduced compared to control. Neither superoxide dismutase (SOD) nor dimethyl sulfoxide (DMSO) reversed these effects; however, protection was observed with catalase (CAT). We found that low concentrations of TSL extracts augmented VAP, VSL, and VCL. However, high concentrations of TSL extracts except TSL-6 decreased VAP, VSL, and VCL. Low concentrations of TSL extracts had no effect on percentage of high MMP cells. However, high concentrations of TSL extracts, most TSL extracts but not TSL-6, decreased the percentage of high MMP in human spermatozoa. When we examined ATP levels (energy source for spermatozoa) of sperm, we found that all TSL extracts, but not TSL-6, diminished ATP levels in a dose-dependent manner. These findings noted that low concentrations of all TSL extracts promoted sperm motility and might attribute to the antioxidant properties of TSL extract at low concentration. On the contrary, high concentrations of TSL extracts but not TSL-6 revealed prooxidant effects verified by intracellular ROS levels, CASA, percentage of high MMP cells as well as ATP levels, contributing to the impairment of sperm motility.

SPERM CHROMATIN STRUCTURE ASSAY (SCSA)

Measurement of oxidative DNA damage, which was characterized as sperm chromatin structure assay (SCSA), provided profitable information for assessment of sperm quality and prediction of pregnancy rate in IVF [53, 54]. SCSA was analyzed using flow cytometry [55]. We found that low concentrations of TSL extracts had no effect on DNA integrity. High concentrations of TSL extracts but not TSL-6 damaged double-strand DNA dose dependently.

CYTOTOXICITY

Prooxidant effect leads to cell death [56, 57]. TSL extracts induced apoptosis in human ovarian cancer cells [58] as well as human premyelocytic leukemia cells [59]. We found that high concentrations of TSL extracts but not TSL-6 elevated levels of apoptotic sperms. However, percentage of necrotic cells showed no significant differences among all TSL extracts. Therefore, TSL extracts but not TSL-6 induced apoptosis in human spermatozoa.

HUMAN SPERM FUNCTIONS

Certain antioxidant agents such as vitamin C, vitamin E, CAT, and GPx/GR effectively decreased ROS levels and improved sperm motility, sperm-oocyte binding and penetration, and fertilization rate of IVF [60, 61, 62, 63]. Propolis was also reported effective on protecting human spermatozoa from DNA damage caused by exogenous reactive oxygen species [64]. The antioxidant functions of crude TSL extracts were documented as effective antioxidants against various oxidative systems *in vitro*, prevention of oxidative modification of human LDL, and other antioxidant functions [5, 7]. Methyl gallate from TSL extracts was reported to exhibit strong antioxidative activity when MDCK cells were challenging with H_2O_2 [7]. TSL-6 revealed antioxidant activity without prooxidant effect at both low and high concentrations. We further investigated its functions on human sperm, including intracellular ROS levels, sperm motility, MMP, ATP production, cell death, and chromatin integrity. TSL-6 increased sperm motility, percentage of high MMP cells, and ATP levels, maintained chromatin integrity and decreased intracellular ROS levels, percentage of denatured cells, and percentage of cell death under oxidative stress. TSL extracts showed antioxidant effect and did not impair sperm functions at low concentration. However, high concentration of TSL extracts except TSL-6 exhibited prooxidant properties (elevated ROS levels), reduced sperm motility, MMP, and ATP levels, impaired sperm chromatin structure, and enhanced sperm apoptosis. TSL-6, at both low and high concentrations, exhibited antioxidant effect and improved sperm functions via decreased ROS levels, increased sperm motility, MMP, and ATP levels, reduced percentage cell death, and the maintenance of chromatin integrity under oxidative stress. These results supported the antioxidant effect of TSL-6, reflecting on the induction of detoxification enzyme, spermatogenesis, and sperm-oocyte penetration-related proteins by 2D-gel electrophoresis-based proteomics.

ANTIOXIDANT ENZYMES

UNDER OXIDATIVE STRESS

SOD, CAT, GPx, GR, and GST are well recognized as antioxidant enzymes in living organisms [65, 66]. Certain antioxidant enzymes such as CAT and GPx/GR effectively decreased ROS levels and improved sperm functions [62, 63]. Liao and colleagues [9] found that SOD, CAT, and GPx activities in the brain and liver were significantly higher in the TSL extracts-treated mice. In an *in vitro* study, using α,α-diphenyl-β-pricryl-hydrazyl radical-scavenging test, the scavenging activities of TSL extracts were over 80% at a concentration of 0.625 mg/ml [9]. We found that TSL-2 and TSL-2P significantly reduced activity of GPx as well as GST in testes of rats under oxidative stress. However, TSL-6 did not affect activities of antioxidant enzymes in testes of rats under oxidative stress.

UNDER NORMAL PHYSIOLOGICAL CONDITION

Certain antioxidant enzymes were modulated by TSL extracts under oxidative stress. We further investigated whether these antioxidant enzymes were also modulated by TSL extracts under normal physiological condition. TSL-2 decreased SOD, CAT, and GR activities and increased GST activity in testes of mice under normal physiological conditions. TSL-2P reduced activity of GR while TSL-6 enhanced GPx activity and declined GR activity in testes of mice under normal physiological condition. Accordingly, TSL extracts exhibited specific regulatory effects on antioxidant enzyme activities in normal mouse testes.

Proteomic analysis showed that certain antioxidant enzymes, including GST (downregulated by TSL-2), PHGPx (downregulated by TSL-2 and TSL-2P), and GST Mu6 (upregulated by TSL-6) were modulated by TSL extracts. Biochemical analysis also provided evidence that TSL-6 increased activity of GPx. On the contrary, TSL-2, which declines the activities of GPx and GST under oxidative stress and that of SOD, CAT, and GR under normal physiological conditions, are potentially used for contraceptive applications.

CONCLUSION

The 2D-gel electrophoresis-based proteomic analysis reveals the protein profile modulated by TSL extracts being involved in sperm-oocyte interaction, detoxification, spermatogenesis, hormone production, and protein misfolding/refolding in testes under oxidative stress. Biochemical analysis provides evidence supporting the proteomic analysis that TSL extracts exhibit regulatory functions in testes under oxidative stress and normal physiological conditions. TSL-2P, an upstream extract of TSL-6 and possessing less optimal active component than that of TSL-6, modulates fewer proteins than that of TSL-6 by 2D-gel electrophoresis. In addition, according to the functions of expressed proteins, TSL-2 is discovered to inhibit sperm functions and TSL-6 is discovered to promote sperm functions, indicating TSL-2 and TSL-6 may be beneficial for development of functional products for male contraceptive purpose and treatment of male infertility, respectively.

REFERENCES

1. Wilkins, M. R., Sanchez, J. C., Gooley, A. A., Appel, R. D., Humphery-Smith, I., Hochstrasser, D. F. and Williams, K. L. 1996. Progress with proteome projects: why all proteins expressed by a genome should be identified and how to do it. *Biotechnol Genet Eng Rev 13*:19–50.

2. Mann, M., Hendrickson, R. C. and Pandey, A. 2001. Analysis of proteins and proteomes by mass spectrometry. *Annu Rev Biochem 70*:437–473.

3. Figeys, D. 2003. Proteomics in 2002: a year of technical development and wide-ranging applications. *Anal Chem 75*(12):2891–2905.

4. Lam, L., Lind, J. and Semsarian, C. 2006. Application of proteomics in cardiovascular medicine. *Int J Cardiol 108*(1):12–19.

5. Hseu, Y. C., Chang, W. H., Chen, C. S., Liao, J. W., Huang, C. J., Lu, F. J., Chia, Y. C., Hsu, H. K., Wu, J. J. and Yang, H. L. 2008. Antioxidant activities of Toona Sinensis leaves extracts using different antioxidant models. *Food Chem Toxicol 46*(1):105–114.

6. Cho, E. J., Yokozawa, T., Rhyu, D. Y., Kim, H. Y., Shibahara, N. and Park, J. C. 2003. The inhibitory effects of 12 medicinal plants and their component compounds on lipid peroxidation. *Am J Chin Med 31*(6):907–917.

7. Hsieh, T. J., Liu, T. Z., Chia, Y. C., Chern, C. L., Lu, F. J., Chuang, M. C., Mau, S. Y., Chen, S. H., Syu, Y. H. and Chen, C. H. 2004. Protective effect of methyl gallate from Toona sinensis (Meliaceae) against hydrogen peroxide-induced oxidative stress and DNA damage in MDCK cells. *Food Chem Toxicol 42*(5):843–850.

8. Hsieh, C. L., Lin, Y. C., Ko, W. S., Peng, C. H., Huang, C. N. and Peng, R. Y. 2005. Inhibitory effect of some selected nutraceutic herbs on LDL glycation induced by glucose and glyoxal. *J Ethnopharmacol 102*(3):357–363.

9. Liao, J. W., Hsu, C. K., Wang, M. F., Hsu, W. M. and Chan, Y. C. 2006. Beneficial effect of Toona sinensis Roemor on improving cognitive performance and brain degeneration in senescence-accelerated mice. *Br J Nutr 96*(2):400–407.

10. Liao, J. W., Chung, Y. C., Yeh, J. Y., Lin, Y. C., Lin, Y. G., Wu, S. M. and Chan, Y. C. 2007. Safety evaluation of water extracts of Toona sinensis Roemor leaf. *Food Chem Toxicol 45*(8):1393–1399.

11. Labuda, J., Buckova, M., Heilerova, L., Silhar, S. and Stepanek, I. 2003. Evaluation of the redox properties and anti/pro-oxidant effects of selected flavonoids by means of a DNA-based electrochemical biosensor. *Anal Bioanal Chem 376*(2):168–173.

12. Alia, M., Ramos, S., Mateos, R., Granado-Serrano, A. B., Bravo, L. and Goya, L. 2006. Quercetin protects human hepatoma HepG2 against oxidative stress induced by tert-butyl hydroperoxide. *Toxicol Appl Pharmacol 212*(2):110–118.

13. Kim, Y. J. 2007. Antimelanogenic and antioxidant properties of gallic acid. *Biol Pharm Bull 30*(6):1052–1055.

14. Keppler, D. 1999. Export pumps for glutathione S-conjugates. *Free Radic Biol Med 27*(9–10):985–991.

15. Townsend, D. M. and Tew, K. D. 2003. The role of glutathione-S-transferase in anti-cancer drug resistance. *Oncogene 22*(47):7369–7375.

16. Nho, C. W. and Jeffery, E. 2001. The synergistic upregulation of phase II detoxification enzymes by glucosinolate breakdown products in cruciferous vegetables. *Toxicol Appl Pharmacol 174*(2):146–152.

17. Yousef, M. I., Awad, T. I., Elhag, F. A. and Khaled, F. A. 2007. Study of the protective effect of ascorbic acid against the toxicity of stannous chloride on oxidative damage, antioxidant enzymes and biochemical parameters in rabbits. *Toxicology 235*(3):194–202.

18. Imai, H. and Nakagawa, Y. 2003. Biological significance of phospholipid hydroperoxide glutathione peroxidase (PHGPx, GPx4) in mammalian cells. *Free Radic Biol Med 34*(2):145–169.

19. Maiorino, M., Coassin, M., Roveri, A. and Ursini, F. 1989. Microsomal lipid peroxidation: effect of vitamin E and its functional interaction with phospholipid hydroperoxide glutathione peroxidase. *Lipids 24*(8):721–726.

20. Nomura, K., Imai, H., Koumura, T., Arai, M. and Nakagawa, Y. 1999. Mitochondrial phospholipid hydroperoxide glutathione peroxidase suppresses apoptosis mediated by a mitochondrial death pathway. *J Biol Chem 274*(41):29294–29302.

21. Ursini, F., Heim, S., Kiess, M., Maiorino, M., Roveri, A., Wissing, J. and Flohe, L. 1999. Dual function of the selenoprotein PHGPx during sperm maturation. *Science 285*(5432):1393–1396.

22. Watanabe, M., Ono, T. and Kondo, H. 1991. Immunohistochemical studies on the localisation and ontogeny of heart fatty acid binding protein in the rat. *J Anat 174*:81–95.

23. Chmurzynska, A. 2006. The multigene family of fatty acid-binding proteins (FABPs): function, structure and polymorphism. *J Appl Genet 47*(1):39–48.

24. Kido, T. and Namiki, H. 2000. Expression of testicular fatty acid-binding protein PERF 15 during germ cell apoptosis. *Dev Growth Differ 42*(4):359–366.

25. Kido, T., Arata, S., Suzuki, R., Hosono, T., Nakanishi, Y., Miyazaki, J., Saito, I., Kuroki, T. and Shioda, S. 2005. The testicular fatty acid binding protein PERF15 regulates the fate of germ cells in PERF15 transgenic mice. *Dev Growth Differ 47*(1):15–24.

26. Jimenez, A., Prieto-Alamo, M. J., Fuentes-Almagro, C. A., Jurado, J., Gustafsson, J. A., Pueyo, C. and Miranda-Vizuete, A. 2005. Absolute mRNA levels and transcriptional regulation of the mouse testis-specific thioredoxins. *Biochem Biophys Res Commun 330*(1):65–74.

27. Asquith, K. L., Harman, A. J., McLaughlin, E. A., Nixon, B. and Aitken, R. J. 2005. Localization and significance of molecular chaperones, heat shock protein 1, and tumor rejection antigen gp96 in the male reproductive tract and during capacitation and acrosome reaction. *Biol Reprod 72*(2):328–337.

28. Kodama, H., Yamaguchi, R., Fukuda, J., Kasai, H. and Tanaka, T. 1997. Increased oxidative deoxyribonucleic acid damage in the spermatozoa of infertile male patients. *Fertil Steril 68*(3):519–524.

29. Hegardt, F. G. 1999. Mitochondrial 3-hydroxy-3-methylglutaryl-CoA synthase: a control enzyme in ketogenesis. *Biochem J 338*(Pt 3):569–582.

30. Aledo, R., Zschocke, J., Pie, J., Mir, C., Fiesel, S., Mayatepek, E., Hoffmann, G. F., Casals, N. and Hegardt, F. G. 2001. Genetic basis of mitochondrial HMG-CoA synthase deficiency. *Hum Genet 109*(1):19–23.

31. Royo, T., Pedragosa, M. J., Ayte, J., Gil-Gomez, G., Vilaro, S. and Hegardt, F. G. 1993. Testis and ovary express the gene for the ketogenic mitochondrial 3-hydroxy-3-methylglutaryl-CoA synthase. *J Lipid Res 34*(6):867–874.

32. Eaton, D. L. and Bammler, T. K. 1999. Concise review of the glutathione S-transferases and their significance to toxicology. *Toxicol Sci 49*(2):156–164.

33. Ono, S., Minami, N., Abe, H. and Obinata, T. 1994. Characterization of a novel cofilin isoform that is predominantly expressed in mammalian skeletal muscle. *J Biol Chem 269*(21):15280–15286.

34. Guttman, J. A., Obinata, T., Shima, J., Griswold, M. and Vogl, A. W. 2004. Non-muscle cofilin is a component of tubulobulbar complexes in the testis. *Biol Reprod 70*(3):805–812.

35. Takahashi, H., Koshimizu, U., Miyazaki, J. and Nakamura, T. 2002. Impaired spermatogenic ability of testicular germ cells in mice deficient in the LIM-kinase 2 gene. *Dev Biol 241*(2):259–272.

36. Hooper, J. D., Nicol, D. L., Dickinson, J. L., Eyre, H. J., Scarman, A. L., Normyle, J. F., Stuttgen, M. A., Douglas, M. L., Loveland, K. A., Sutherland, G. R. and Antalis, T. M. 1999. Testisin, a new human serine proteinase expressed by premeiotic testicular germ cells and lost in testicular germ cell tumors. *Cancer Res 59*(13):3199–3205.

37. Ohmura, K., Kohno, N., Kobayashi, Y., Yamagata, K., Sato, S., Kashiwabara, S. and Baba, T. 1999. A homologue of pancreatic trypsin is localized in the acrosome of mammalian sperm and is released during acrosome reaction. *J Biol Chem 274*(41):29426–29432.

38. Jolly, C. and Morimoto, R. I. 2000. Role of the heat shock response and molecular chaperones in oncogenesis and cell death. *J Natl Cancer Inst 92*(19):1564–1572.

39. Nollen, E. A. and Morimoto, R. I. 2002. Chaperoning signaling pathways: molecular chaperones as stress-sensing 'heat shock' proteins. *J Cell Sci 115*(Pt 14):2809–2816.

40. Hassen, W., Ayed-Boussema, I., Oscoz, A. A., Lopez Ade, C. and Bacha, H. 2007. The role of oxidative stress in zearalenone-mediated toxicity in Hep G2 cells: oxidative DNA damage, gluthatione depletion and stress proteins induction. *Toxicology 232*(3):294–302.

41. Atchison, F. W. and Means, A. R. 2003. Spermatogonial depletion in adult Pin1-deficient mice. *Biol Reprod 69*(6):1989–1997.

42. Wine, R. N., Ku, W. W., Li, L. H. and Chapin, R. E. 1997. Cyclophilin A is present in rat germ cells and is associated with spermatocyte apoptosis. Reproductive Toxicology Group. *Biol Reprod 56*(2):439–446.

43. Aitken, J. and Fisher, H. 1994. Reactive oxygen species generation and human spermatozoa: the balance of benefit and risk. *Bioessays 16*(4):259–267.

44. de Lamirande, E. and Gagnon, C. 1995. Impact of reactive oxygen species on spermatozoa: a balancing act between beneficial and detrimental effects. *Hum Reprod 10*Suppl 1: 15–21.

45. Hull, M. G., Glazener, C. M., Kelly, N. J., Conway, D. I., Foster, P. A., Hinton, R. A., Coulson, C., Lambert, P. A., Watt, E. M. and Desai, K. M. 1985. Population study of causes, treatment, and outcome of infertility. *Br Med J (Clin Res Ed) 291*(6510):1693–1697.

46. Aitken, R. J. and Clarkson, J. S. 1987. Cellular basis of defective sperm function and its association with the genesis of reactive oxygen species by human spermatozoa. *J Reprod Fertil 81*(2):459–469.

47. Alvarez, J. G., Touchstone, J. C., Blasco, L. and Storey, B. T. 1987. Spontaneous lipid peroxidation and production of hydrogen peroxide and superoxide in human spermatozoa. Superoxide dismutase as major enzyme protectant against oxygen toxicity. *J Androl 8*(5):338–348.

48. Tobi, S. E., Gilbert, M., Paul, N. and McMillan, T. J. 2002. The green tea polyphenol, epigallocatechin-3-gallate, protects against the oxidative cellular and genotoxic damage of UVA radiation. *Int J Cancer 102*(5):439–444.

49. Donnelly, E. T., Lewis, S. E., McNally, J. A. and Thompson, W. 1998. In vitro fertilization and pregnancy rates: the influence of sperm motility and morphology on IVF outcome. *Fertil Steril 70*(2):305–314.

50. Hirano, Y., Shibahara, H., Obara, H., Suzuki, T., Takamizawa, S., Yamaguchi, C., Tsunoda, H. and Sato, I. 2001. Relationships between sperm motility characteristics assessed by the computer-aided sperm analysis (CASA) and fertilization rates in vitro. *J Assist Reprod Genet 18*(4):213–218.

51. Ericsson, S. A., Garner, D. L., Thomas, C. A., Downing, T. W. and Marshall, C. E. 1993. Interrelationships among fluorometric analyses of spermatozoal function, classical semen quality parameters and the fertility of frozen-thawed bovine spermatozoa. *Theriogenology 39*(5):1009–1024.

52. Armstrong, J. S., Rajasekaran, M., Chamulitrat, W., Gatti, P., Hellstrom, W. J. and Sikka, S. C. 1999. Characterization of reactive oxygen species induced effects on human spermatozoa movement and energy metabolism. *Free Radic Biol Med 26*(7–8):869–880.

53. Virro, M. R., Larson-Cook, K. L. and Evenson, D. P. 2004. Sperm chromatin structure assay (SCSA) parameters are related to fertilization, blastocyst development, and ongoing pregnancy in in vitro fertilization and intracytoplasmic sperm injection cycles. *Fertil Steril 81*(5):1289–1295.

54. Larson-Cook, K. L., Brannian, J. D., Hansen, K. A., Kasperson, K. M., Aamold, E. T. and Evenson, D. P. 2003. Relationship between the outcomes of assisted reproductive techniques and sperm DNA fragmentation as measured by the sperm chromatin structure assay. *Fertil Steril 80*(4):895–902.

55. Evenson, D. P., Jost, L. K., Marshall, D., Zinaman, M. J., Clegg, E., Purvis, K., de Angelis, P. and Claussen, O. P. 1999. Utility of the sperm chromatin structure assay as a diagnostic and prognostic tool in the human fertility clinic. *Hum Reprod 14*(4):1039–1049.
56. Yoshino, M., Haneda, M., Naruse, M., Htay, H. H., Tsubouchi, R., Qiao, S. L., Li, W. H., Murakami, K. and Yokochi, T. 2004. Prooxidant activity of curcumin: copper-dependent formation of 8-hydroxy-2'-deoxyguanosine in DNA and induction of apoptotic cell death. *Toxicol In Vitro 18*(6):783–789.
57. Wolfler, A., Caluba, H. C., Abuja, P. M., Dohr, G., Schauenstein, K. and Liebmann, P. M. 2001. Prooxidant activity of melatonin promotes fas-induced cell death in human leukemic Jurkat cells. *FEBS Lett 502*(3):127–131.
58. Chang, H. L., Hsu, H. K., Su, J. H., Wang, P. H., Chung, Y. F., Chia, Y. C., Tsai, L. Y., Wu, Y. C. and Yuan, S. S. 2006. The fractionated Toona sinensis leaf extract induces apoptosis of human ovarian cancer cells and inhibits tumor growth in a murine xenograft model. *Gynecol Oncol 102*(2):309–314.
59. Yang, H. L., Chang, W. H., Chia, Y. C., Huang, C. J., Lu, F. J., Hsu, H. K. and Hseu, Y. C. 2006. Toona sinensis extract induces apoptosis via reactive oxygen species in human premyelocytic leukemia cells. *Food Chem Toxicol 44*(12):1978–1988.
60. Hsu, P. C., Hsu, C. C. and Guo, Y. L. 1999. Hydrogen peroxide induces premature acrosome reaction in rat sperm and reduces their penetration of the zona pellucida. *Toxicology 139*(1–2):93–101.
61. Griveau, J. F. and Le Lannou, D. 1997. Reactive oxygen species and human spermatozoa: physiology and pathology. *Int J Androl 20*(2):61–69.
62. Geva, E., Bartoov, B., Zabludovsky, N., Lessing, J. B., Lerner-Geva, L. and Amit, A. 1996. The effect of antioxidant treatment on human spermatozoa and fertilization rate in an in vitro fertilization program. *Fertil Steril 66*(3):430–434.
63. Patel, S. R. and Sigman, M. 2008. Antioxidant therapy in male infertility. *Urol Clin North Am 35*(2):319–330, x.
64. Russo, A., Troncoso, N., Sanchez, F., Garbarino, J. A. and Vanella, A. 2006. Propolis protects human spermatozoa from DNA damage caused by benzo[a]pyrene and exogenous reactive oxygen species. *Life Sci 78*(13):1401–1406.
65. Wei, Y. H. and Lee, H. C. 2002. Oxidative stress, mitochondrial DNA mutation, and impairment of antioxidant enzymes in aging. *Exp Biol Med (Maywood) 227*(9):671–682.
66. Mates, J. M. 2000. Effects of antioxidant enzymes in the molecular control of reactive oxygen species toxicology. *Toxicology 153*(1–3):83–104.

14

Application of Proteomics in Nutrition Research

Baukje de Roos

INTRODUCTION

Nutrition research has traditionally dealt with the discovery of essential nutrients that prevent malnutrition and promote optimal health in populations. However, in the last decades, with obesity rates increasing and overnutrition becoming more prevalent, investigations have started to focus on improving health and performance of individuals through functional ingredients in the diet. Many bioactive food ingredients claim to either decrease risk factors indicative for development of disease or to improve life quality by optimizing and maintaining body functions. New technologies such as genomics, transcriptomics, proteomics, metabolomics, and imaging are now slowly being introduced in nutrition research. Such technologies aim to demonstrate effectiveness of food ingredients as well as aiding discovery of novel biomarkers to measure changes in efficacy. This chapter will discuss the application of proteomics in nutrition research.

PROTEOMICS TECHNOLOGY IN NUTRITION SCIENCES

The potential value of proteomics for nutrition research has been recognized for the last decade, and many authors have written comprehensive reviews (de Roos, 2008; de Roos and McArdle, 2008; Fuchs *et al.*, 2005; Kim et al., 2004; Kussmann et al., 2006; Kussmann and Affolter, 2006; Milner, 2007) – with numbers of proteomics reviews actually exceeding the number of research papers until recently (de Roos and McArdle, 2008). The main advantage of proteomics is that it measures the functional product (protein) of gene expression and allows the identification of modifications that may relate to the activation or inactivation of proteins by dietary interventions. Indeed, proteome analysis in nutritional studies can identify proteins that provide new biomarkers for health and disease diagnosis and enable discovery of the mechanisms whereby food components influence health (de Roos, 2008; de Roos and McArdle, 2008). However, the actual use of this technique in intervention trials is still limited. This is due largely to the challenges that the proteomics technique are currently facing. High relative levels of biological and analytical variability may mask subtle effects due to dietary interventions. Furthermore, most proteomics techniques have difficulty detecting differential regulation of low abundant but often clinically relevant proteins. A recent increase in quality and sensitivity of a new generation of mass spectrometers has hugely accelerated developments in detection and identification in the proteomics field. However, we are still dealing with mass spectrometry-related problems, (Service, 2008a). Future advances in mass spectrometry should make it possible to accurately

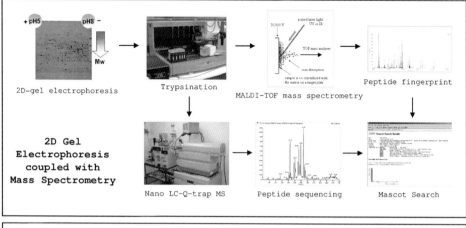

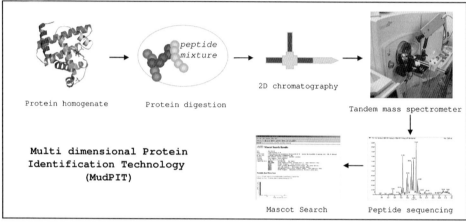

Figure 14.1 Proteomic approaches to nutrition research. For color detail, please refer to the color plate section.

measure and quantitate proteins from complex biological samples, increasing our understanding of the dynamic changes in the proteome induced by dietary components (Moresco et al., 2008).

Currently, proteomic technologies use either specific digestion of proteins or direct analysis of intact proteins after their chromatographic separation (Fig. 14.1). The multiple components of the proteome can be analyzed by mass spectrometry (MS) after an evaporation of peptides and protein by electrospray ionization (ESI) or matrix-assisted laser desorption ionization (MALDI) technologies (Kim et al., 2004). Classical two-dimensional (2D) gel electrophoresis, where proteins are separated according to charge and molecular weight, coupled with protein spot analysis by mass spectrometry, is still the most widely used technical approach in proteomics to identify changes in individual proteins of tissues, cells, and biofluids upon nutritional intervention (de Roos and McArdle, 2008; Fuchs et al., 2005; Kim et al., 2004). Using such techniques, most frequently identified proteins are those that relate to glucose and fatty acid metabolism as well as pathways involving oxidative stress, antioxidant defense mechanisms, and redox status. Although this method is one of the most labor-intensive of several types of 2D separation methods available, it actually yields a physical separation of intact polypeptides, providing information about molecular weight and iso-electric point, parameters that can be used to narrow the identification of the protein. Unfortunately, it is still difficult to visualize and detect differential regulation of low abundant, very hydrophobic, acidic, or basic proteins. For example, proteins involved in inflammatory pathways, like cytokines, are secreted in the ng/mL range, and mass spectrometry has not been sensitive enough to detect relevant changes with such low concentrations (Service, 2008a). Therefore, 2D-gel electrophoresis coupled with mass spectrometry may not represent the most sensitive tool to reveal effects of nutritional interventions on inflammatory pathways. A recent advance is the introduction of difference gel electrophoresis (DIGE) technology, allowing direct quantitative comparison of differentially labeled samples using cyanine fluorescent dyes prior to electrophoresis. When absolute protein variation between two or three samples is the primary target, this method is more reproducible and accurate and not limited by distortion due to gel-to-gel variation (Issaq and Veenstra, 2007), but it is costly and automation is difficult (Kussmann, 2007). Shotgun proteomics (i.e., digestion of the protein mixture and multidimensional chromatographic separation of peptides followed by online mass spectrometric peptide detection and sequencing) (Kussmann and Affolter, 2006) can now be combined with stable isotope labeling to allow for the quantification of changes in expression levels of hundreds to thousands of proteins in a single experiment. The most commonly adopted approaches include isotope-coded affinity tags (ICAT) (Smolka et al., 2001) and isobaric tag relative absolute protein quantitation (iTRAQ) (Streckfus et al., 2008). Quantification is based on relative changes in the levels of labeled peptides that may be common to a family of proteins with differential regulation/abundance, and thus quantification experiments can sometimes lead to ambiguous or conflicting results (Duncan and Hunsucker, 2005).

THE SEARCH FOR PLASMA BIOMARKERS

Biomarkers may reflect different stages in a biological process ranging from healthy functioning to a small deviation from a "healthy" equilibrium to disease. In clinical practice, biomarkers of interest are diagnostic biomarkers that measure the incidence and progression of a disease process (like hemoglobin A1C) or prognostic biomarkers that can predict whether a subject will be susceptible to a disorder (like plasma cholesterol levels). Such a biomarker should have a high sensitivity and specificity for the outcome it is expected to identify, explain a reasonable proportion of the outcome, independent of other established predictors, and be mechanistically involved in the disease process (Vasan, 2006). Biomarker proteins may not only play a major physiological role in target organs but could also reflect changes in mechanisms initiated by dietary intervention when circulating in the blood. The most direct approach to take in biomarker discovery is the identification of human plasma protein markers of systemic changes that might be affected by diet.

    For example, we assessed the effects of daily fish oil supplements for 6 weeks on the serum proteome using 2-DE. Serum levels of apolipoprotein A1, apolipoprotein L1, zinc-α-2-glycoprotein, haptoglobin precursor, α-1-antitrypsin precursor, antithrombin III-like protein, serum amyloid P component, and hemopexin were all significantly downregulated by fish oil compared with high oleic sunflower oil supplementation. In addition, the decrease in serum apolipoprotein A1 was associated with a significant shift toward the larger, more cholesterol-rich HDL$_2$ particles. The alterations in serum proteins and HDL size imply that fish oil activates anti-inflammatory and lipid-modulating mechanisms believed to impede the early onset of coronary heart disease (CHD). These proteins are potential diagnostic biomarkers to assess mechanisms whereby fish oils protect against CHD in humans (de Roos et al., 2008).

    A proteomics approach was also used to identify alterations in peripheral blood mononuclear cell (PBMC) proteins of healthy males who were consuming flaxseed for a week. PBMC from the same study subjects were also exposed to physiological concentrations of enterolactone (a metabolite produced from dietary lignans by colonic microflora) *ex vivo* to assess whether similar effects on the proteome could be observed as those caused by dietary flaxseed. Intervention with flaxseed in healthy males resulted in enhanced plasma enterolactone levels, although both baseline levels and maximal plasma concentrations varied considerably between subjects. In addition, a change in sixteen PBMC proteins occurred upon consumption of flaxseed. Four out of these sixteen proteins were altered in a similar manner when blood mononuclear cells were exposed *ex vivo* to enterolactone. There were enhanced levels of peroxiredoxin, decreased levels of the long-chain fatty acid β-oxidation multienzyme complex proteins, and levels of glycoprotein IIIa/II (Fuchs et al., 2007a).

PBMC proteomics was also applied to identify response biomarkers to a dietary supplementation with an isoflavone extract for 8 weeks in postmenopausal women. Twenty-nine proteins, including several involved in the anti-inflammatory response, showed significantly altered expression in the mononuclear blood cells following the soy-isoflavone intervention. Heat shock protein 70, a lymphocyte-specific protein phosphatase and proteins that promote increased fibrinolysis, such as α-enolase, were all increased, whereas those that mediate adhesion, migration, and proliferation of vascular smooth muscle cells, such as galectin-1, showed decreased expression. As no overall anti-inflammatory activity of the soy intervention was observed when clinically relevant inflammation markers were measured in plasma, it was suggested that the PBMC proteome may represent a more sensitive approach to detect inhibition of inflammatory processes and that may also respond earlier than those plasma markers classically used (Fuchs et al., 2007b).

A problem with plasma proteomics, however, is often the variability in plasma protein levels within and between subjects, and the variation in abundance between these proteins. Such sources of variation may easily obscure the biological changes under investigation, and this has led some to argue that proteomic profiling of blood is a relatively immature technology in need of further refinement (Coombes et al., 2005). The issue of variability is now being addressed in a "proof of principle" initiative funded by the European Nutrigenomics Organization (NuGO), a Network of Excellence funded by the European Commission (Baccini et al., 2008). The project started in 2007 and aims to determine biological variability in plasma, platelet, PBMC, urine and saliva proteomics, in plasma and urine metabolomics and in PBMC transcriptomics, at baseline and the spectrum and scale of changes brought about after a 36-hour fasting challenge. This study will provide a highly comprehensive human dataset and insights into both intra- and interperson variability at baseline measurements and after metabolic challenges. Biological variation originating from environmental or genetic factors, as well as differences in sample collection in human intervention studies, cannot be easily controlled. Therefore, it is important to run enough replicates to ensure sufficient statistical power to detect physiological changes in protein levels. However, power analysis, which can be used to infer the number of samples that should be analyzed to discover statistically significant results, has so far been used very little in proteomic studies (Wilkins et al., 2006).

BIOMARKER PROTEINS FROM BLOOD CELLS

Elucidation of changes in platelet and PBMC proteomes upon dietary intervention may provide a more sensitive and less variable approach to detect regulation of, for example, inflammatory processes. For many years, 2D-gel electrophoresis has been used to study platelet biology as the absence of a nucleus prevents platelets from being studied using a classic molecular approach (Gravel et al., 1995; Jenkins et al., 1976). Beside their physiological role in hemostasis, platelets play a significant role in common diseases, especially atherothrombosis and coronary artery disease. The cells are mainly involved in maintaining vascular integrity by sensing and responding to endothelial damage. They also function in wound healing as well as in activation of inflammatory and immune responses (Macaulay et al., 2005). Proteomic analysis of platelets has, thus far, used many different approaches. Initial studies focused on the global cataloguing of proteins present in resting platelets, highlighting the abundance of signaling and cytoskeletal proteins (Garcia et al., 2004; O'Neill et al., 2002). Changes in levels of proteins involved in signaling cascades regulating platelet activation and aggregation on cytoskeletal reorganization are well described (Hartwig, 2006); however, it is less clear whether, for example, dietary intervention can affect levels of cytoskeletal proteins and whether such changes are functionally relevant for signaling cascades (Garcia et al., 2005). We recently found that fish oil supplementation for 3 weeks caused significant upregulation of several structural proteins detected in platelets of healthy volunteers using 2D-gel electrophoresis. Such changes could possibly promote the cytoskeletal stability of the resting platelet and decrease the incidence of shape change in response to external stimuli (B. de Roos, unpublished results). Nutritional intervention will also affect the transcriptome (Bouwens et al., 2007) and proteome of PBMC, which consist of a mixture of lymphocytes and monocytes/macrophages. These cells can be used to distinguish certain metabolic or disease states based on lymphocyte numbers (Bell and O'Keefe, 2007; Lau et al., 2003; Ruggiero et al., 2007; Yu et al., 2007), or based on their gene expression profiles (Burczynski and Dorner, 2006; Maas et al., 2002). Intervention with dietary flaxseed differentially regulated sixteen PBMC proteins in healthy men (Fuchs et al., 2007a), whereas intervention with an isoflavone differentially regulated twenty-nine PBMC proteins in postmenopausal women (Fuchs et al., 2007b). The latter study suggests that the PBMC proteome represents a more sensitive approach to detect inhibition of inflammatory processes, which responded earlier than classical plasma markers of inflammation (Fuchs et al., 2007b).

MECHANISTIC BIOMARKERS FROM ANIMAL STUDIES

Potential biomarkers can also be identified from the use of animal models. Proteomics of target organ tissues has already provided valuable insight in mechanisms of dietary interventions identifying proteins involved in the regulation of glucose and fatty acid metabolism, oxidative stress, and the redox system (Fig. 14.2).

We demonstrated, for example, that two structurally, almost identical dietary conjugated linoleic acid (CLA) isomers had divergent effects on atherosclerosis development and insulin resistance in $Apoe^{-/-}$ mice. Consumption of the $trans10,cis12$ CLA isomer, a fatty acid present in supplements currently freely available, was proatherogenic and induced pathways involved in

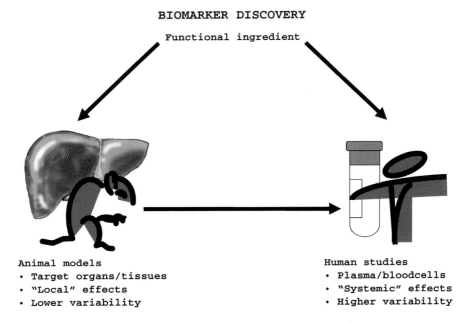

Figure 14.2 Biomarker discovery in human and animal intervention studies.

the development of insulin resistance, as evidenced by simultaneous changes in various enzymes involved in gluconeogenesis, glycolysis, and ketogenesis (de Roos et al., 2005b). Consumption of the *cis*9,*trans*11 isomer of CLA, in contrast, was antiatherogenic and specifically induced HSP 70 kD protein levels (de Roos et al., 2005b). Although this may be an indication of increased oxidative stress, we have postulated that upregulation of HSP 70kD protein upon exposure to a relatively mild stress stimulus (like *cis*9,*trans*11 CLA) could protect against exposure to a subsequent, more severe stress. This could occur through anti-inflammatory pathways, possibly mediated through NF-κB (Shimizu et al., 2002). Indeed, increased levels of human serum HSP 70 have been associated with a low coronary artery disease risk, independent of traditional risk factors (Zhu et al., 2003). The consumption of dietary fish oil and *trans*10,*cis*12 CLA induced differential expression of long-chain acyl-CoA thioester hydrolase protein as an indicator of β-oxidation of fatty acids in the liver, and the consumption of dietary fish oil, olive oil, and *trans*10,*cis*12 CLA induced differential expression of adipophilin protein as an indicator of selective hepatic lipid accumulation and triglyceride secretion (de Roos et al., 2005a, 2005b).

A systems biology approach was used to gain a better understanding of the mechanisms by which the fatty acids of olive oil, or its minor antioxidant constituents, may affect hepatic metabolic pathways, oxidative stress, and atherogenesis. Extra virgin olive oil decreased atherosclerotic plaque size after 10 weeks of intervention in *Apoe*$^{-/-}$ mice but also induced the accumulation of triglycerides in the liver, resulting in an increased hepatic fat content. These results imply that olive oils can, on the one hand, decrease the risk of atherosclerotic plaque formation, but can also increase the risk of hepatic steatosis. Proteomics was extremely helpful in unraveling the complex interactions. This detected significant upregulation of a large array of antioxidant enzymes upon consumption of extra virgin olive oils that may diminish oxidative stress instigated by hepatic steatosis and in addition, may slow down the development of atherosclerosis. Therefore, compounds in extra virgin olive oils (such as the polyphenols) may well be able to delay the onset of atherosclerotic lesions by combating oxidative stress. Indeed, the accumulation of triglycerides may not pose a major challenge to the liver and represent a relatively safe way to store triglycerides, as long as the antioxidant capacity is adequate to prevent lipotoxicity (Arbones-Mainar et al., 2007).

A combination of transcriptomics and proteomics was used to determine the influence of supraphysiological concentrations of dietary quercetin in the healthy distal colon of inbred F344 rats. This revealed that quercetin changed the cellular energy metabolism from glycolysis in the cytoplasm toward fatty acid degradation in the mitochondria. Cancer cells generally exhibit increased glycolysis and mitochondrial dysfunction and in human colon tumors an increase in glycolysis combined with decreased fatty acid degradation as an energy source. Therefore, the quercetin-mediated glycolysis-to-fatty-acid degradation shift in energy pathways could represent an additional underlying mechanism involved in the ability of quercetin to inhibit induced colorectal carcinogenesis (Dihal et al., 2008).

A combination of transcriptomics and proteomics was also applied to explain mechanisms underlying changes in hepatic lipid metabolism during zinc deficiency (Tom et al., 2005). Zinc is involved in many aspects of normal physiology, including intermediary metabolism, hormone secretion pathways, and immune defense. Deficiency has severe consequences but the mechanisms involved are largely unproven. The results of this study provided evidence for major changes in the expression of gene groups functionally linked to hepatic lipogenesis and lipolysis, with an inverse regulation that fostered the accumulation of triglycerides in liver, changing the hepatic fatty acid pattern and reducing fatty acid oxidation. The experimental findings provided support for an unbalanced gene transcription control via PPARα, thyroid hormone, and sterol regulatory element binding protein (SREBP)-dependent pathways, which may explain most of the apparently pleiotropic effects of zinc deficiency on hepatic fat metabolism (Tom et al., 2005). A second study found that zinc deficiency also affected proteins involved in fatty acid and carbohydrate metabolism in the rat aorta. In addition, zinc deficiency differentially regulated structural proteins, including zyxin and more than nine transgelin 1 proteins. Such changes may be disadvantageous for maintaining vascular health, and optimal zinc levels may thus play a protective role in the development of atherosclerosis (Beattie et al., 2008).

Proteomics was used as a tool to explore biochemical alterations in the adult brain after developmental vitamin D deficiency. For this, vitamin D-deficient female rats were mated with vitamin D normal males. Pregnant females were kept vitamin D-deficient until birth of their offspring whereupon they were returned to a control diet. At week 10, protein expression in the progeny's prefrontal cortex and hippocampus was compared with control. Developmental vitamin D deficiency led to long-lasting alterations in expression of brain proteins involved in mitochondrial, cytoskeletal, and neuronal function. Specifically, thirty-six proteins were identified that were involved in, for example, oxidative phosphorylation, redox balance, cytoskeleton maintenance, calcium homeostasis, chaperoning, and neurotransmission. An *in silico* survey of vitamin D target genes identified a putative vitamin D responsive element in the promoter region of six of these thirty-six proteins. Interestingly, many of the proteins found to be dysregulated in this study have also been identified in postmortem and genetic studies in patients with schizophrenia or multiple sclerosis, the most prominent similarity being the dysregulation of mitochondrial proteins (Almeras et al., 2007).

A rat model of oxidative stress was used to investigate the mechanism(s) whereby salicylic acid modulates potentially procancerous activity in the colon using a proteomic approach. Supplementation of salicylic acid (1 mg/kg diet) resulted in a significant change in expression of fifty-five cytosolic proteins extracted from the distal colon milled in liquid nitrogen. The functions of these proteins related to protein folding, protein transport, redox balance, energy metabolism, and cytoskeletal regulation. Partial least squares analyses was performed to identify associations between significantly altered proteins and biochemical indices of oxidative stress (plasma and colon TBARS), prostaglandin levels, and cyGPx activity data. Seven of the proteins were involved in the two major redox pathways of thioredoxin and glutathione, which implies that salicylic acid can regulate redox signaling via similar pathways to vitamin E, but may do this indirectly through modulation of interactive components of the redox signaling system (Drew et al., 2006).

A proteomics approach was used to investigate the underlying mechanisms of fructose-induced fatty livers in Syrian gold hamsters. Studies indicate that high fructose consumption is linked to the development of metabolic syndrome. Hamsters fed a high fructose diet for 8 weeks developed impaired glucose tolerance and hyperlipidemia, as well as regulation of thirty-three proteins relating to, among others, glucose metabolism (upregulation of glycerol kinase and fructose 1,6-bisphosphatase), lipid metabolism (upregulation of fatty acid binding protein and apolipoprotein A1), oxidative stress (upregulation of peroxiredoxin 2) and protein folding, and stress (upregulation of heat shock protein 70) (Zhang et al., 2008).

A hepatic proteomics approach was also used to examine the effects of dietary betaine supplementation on ethanol-induced changes in methionine and other metabolic pathways. Betaine-supplemented diets for 4 weeks caused normalization of S-adenosylmethionine (SAM) to S-adenosylhomocysteine (SAH) ratio induced by ethanol. Proteomic profiling of the liver also revealed ethanol-induced downregulation of CA-III protein, which was not restored by betaine treatment. CA-III protein has two reactive sulfhydryl groups that form disulfide linkages with glutathione under conditions of stress, but also can scavenge radicals to protect the cells from oxidative damage (Kharbanda et al., 2009).

A last example of a proteomics approach illustrates the technique's ability to measure oxidative stress and metabolic changes (Opii et al., 2008). This study investigated a 2.8-year treatment with an antioxidant-fortified diet (containing vitamin E, L-carnitine, DL-alpha-lipoic acid, vitamin C, and spinach flakes, tomato pomace, grape pomace, carrot granules, and citrus pulp) and/or a program of behavioral enrichment on oxidative damage in the parietal cortex. They used old beagles as a model for human aging as these animals naturally develop cognitive deficits and accumulate brain pathology similar to aging humans. The combined treatment caused the greatest decrease in the oxidative stress biomarkers protein carbonyls, 3-nitrotyrosine (3-NT), and the lipid perox-idation product 4-hydroxynonenal (HNE). Protein carbonyl levels of glutamate dehydrogenase [NAD (P)], glyceraldehyde-3-phosphate dehydrogenase (GAPDH), α-enolase, neurofilament triplet L protein, glutathione-S-transferase (GST), and fascin actin bundling protein were signif-icantly decreased, and the protein levels of Cu/Zn superoxide dismutase, fructose-bisphosphate aldolase C, creatine kinase, glutamate dehydrogenase, and glyceraldehyde-3-phosphate dehy-drogenase were increased in brain of the combined treatment group. The increased expression of a number of antioxidant proteins, in particular Cu/Zn SOD, correlated with improved cogni-tive function. These findings suggest that the antioxidant plus behavioral enrichment treatment decreases the levels of oxidative damage and improves the antioxidant reserve systems in the aging canine brain, and may contribute to improvements in learning and memory (Opii et al., 2008).

Although proteomics has provided great opportunities to elucidate novel mechanisms of fatty acid function in, for example, the liver of animal models, ideally such biomarkers should be validated in humans. Often, such changes are reflected in the regulation of specific (but sometimes low abundant) plasma, platelet, or PBMC proteins. Proteins circulating in human plasma or present in blood cells can then serve as valid markers of health or disease. More selective, quantitative, and sensitive methods, like ELISA or the quantitative analysis of certain peptides specific for unique proteins in complex biological mixtures (such as human plasma) in the multiple reaction monitoring (MRM) mode can then be used to evaluate and validate newly discovered candidate biomarkers in human plasma and blood cells.

CONCLUSION

To substantiate health claims of novel functional foods in a scientific way, new technologies in nutrition research, such as proteomics of human plasma and blood cells combined with other nutrigenomics technologies such as genomics, transcriptomics, metabolomics, and imaging, could provide valuable tools for extensive phenotyping and give access to holistic discovery of efficacy biomarkers (Fig. 14.3). For this, the nutrition community, as well as regulatory agencies that assess health claims, would clearly benefit from the introduction of a "dictionary of validated protein biomarkers." We are hopeful that improving proteomics technologies will generate novel biomarkers, but perhaps not as quickly as people would like (Service, 2008b). Recently, new

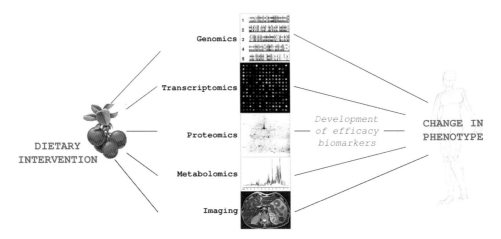

Figure 14.3 Use of OMICS technology in assessing the effects of dietary interventions on the human phenotype. For color detail, please refer to the color plate section.

approaches have been developed that provide a significantly faster and less expensive tool to quantitatively measure proteins in blood (present at low nanogram/milliliter concentrations) that can be highly "multiplexed," permitting the simultaneous measurement of multiple proteins. Also, the increasing use of triple quadrupole-derived technologies combined with linear ion trap in proteomics now enables quantitative analysis of specific peptides, including those that are posttranslationally modified, in complex biological mixtures such as human plasma and serum with high sensitivity and selectivity in the multiple reaction monitoring (MRM) mode (de Roos, 2008; de Roos and McArdle, 2008). This will eventually help us fully understand the impact of exposure to nutritional intervention in humans.

ACKNOWLEDGMENTS

The laboratory of Baukje de Roos is funded by the Scottish Government Rural and Environment Research and Analysis Directorate (RERAD).

REFERENCES

Almeras L, Eyles D, Benech P, Laffite D, Villard C, Patatian A, Boucraut J, Mackay-Sim A, McGrath J, and Feron F. 2007. Developmental vitamin D deficiency alters brain protein expression in the adult rat: implications for neuropsychiatric disorders. *Proteomics.* 7:769–780.

Arbones-Mainar JM, Ross K, Rucklidge GJ, Reid M, Duncan G, Arthur JR, Horgan GW, Navarro MA, Carnicer R, Arnal C, Osada J, and Roos BD. 2007. Extra virgin olive oils increase hepatic fat accumulation and hepatic antioxidant protein levels in APOE(-/-)() mice. *J. Proteome. Res.* 6:4041–4054.

Baccini M, Bachmaier EM, Biggeri A, Boekschoten MV, Bouwman FG, Brennan L, Caesar R, Cinti S, Coort SL, Crosley K, Daniel H, Drevon CA, Duthie S, Eijssen L, Elliott RM, van Erk M, Evelo C, Gibney M, Heim C, Horgan GW, Johnson IT, Kelder T, Kleemann R, Kooistra T, van Iersel MP, Mariman EC, Mayer C, McLoughlin G, Muller M, Mulholland F, van Ommen B, Polley AC, Pujos-Guillot E, Rubio-Aliaga I, Roche HM, de Roos B, Sailer M, Tonini G, Williams LM, and de Wit N. 2008. The NuGO proof of principle study package: a collaborative research effort of the European Nutrigenomics Organisation. *Genes Nutr.* 3:147–151.

Beattie JH, Gordon MJ, Rucklidge GJ, Reid MD, Duncan GJ, Horgan GW, Cho YE, and Kwun IS. 2008. Aorta protein networks in marginal and acute zinc deficiency. *Proteomics.* 8:2126–2135.

Bell DS and O'Keefe JH. 2007. White cell count, mortality, and metabolic syndrome in the Baltimore longitudinal study of aging. *J. Am. Coll. Cardiol.* 50:1810–1811.

Bouwens M, Afman LA, and Muller M. 2007. Fasting induces changes in peripheral blood mononuclear cell gene expression profiles related to increases in fatty acid beta-oxidation: functional role of peroxisome proliferator activated receptor alpha in human peripheral blood mononuclear cells. *Am. J. Clin. Nutr.* 86:1515–1523.

Burczynski ME and Dorner AJ. 2006. Transcriptional profiling of peripheral blood cells in clinical pharmacogenomic studies. *Pharmacogenomics.* 7:187–202.

Coombes KR, Morris JS, Hu J, Edmonson SR, and Baggerly KA. 2005. Serum proteomics profiling—a young technology begins to mature. *Nat. Biotechnol.* 23:291–292.

de Roos B. 2008. Proteomic analysis of human plasma and blood cells in nutritional studies: development of biomarkers to aid disease prevention. *Expert. Rev. Proteomics.* 5:819–826.

de Roos B, Duivenvoorden I, Rucklidge G, Reid M, Ross K, Lamers RJ, Voshol PJ, Havekes LM, and Teusink B. 2005a. Response of apolipoprotein E∗3-Leiden transgenic mice to dietary fatty acids: combining liver proteomics with physiological data. *FASEB J.* 19:813–815.

de Roos B, Geelen A, Ross K, Rucklidge G, Reid M, Duncan G, Caslake M, Horgan G, and Brouwer IA. 2008. Identification of potential serum biomarkers of inflammation and lipid modulation that are altered by fish oil supplementation in healthy volunteers. *Proteomics.* 8:1965–1974.

de Roos B and McArdle HJ. 2008. Proteomics as a tool for the modelling of biological processes and biomarker development in nutrition research. *Br. J. Nutr. 99*:S66–S71.

de Roos B, Rucklidge G, Reid M, Ross K, Duncan G, Navarro MA, Arbones-Mainar JM, Guzman-Garcia MA, Osada J, Browne J, Loscher CE, and Roche HM. 2005b. Divergent mechanisms of *cis*9, *trans*11-and *trans*10, *cis*12-conjugated linoleic acid affecting insulin resistance and inflammation in apolipoprotein E knockout mice: *a proteomics approach. FASEB J. 19*:1746–1748.

Dihal AA, van der Woude H, Hendriksen PJ, Charif H, Dekker LJ, Ijsselstijn L, de Boer VC, Alink GM, Burgers PC, Rietjens IM, Woutersen RA, and Stierum RH. 2008. Transcriptome and proteome profiling of colon mucosa from quercetin fed F344 rats point to tumor preventive mechanisms, increased mitochondrial fatty acid degradation and decreased glycolysis. *Proteomics. 8*:45–61.

Drew JE, Padidar S, Horgan G, Duthie GG, Russell WR, Reid M, Duncan G, and Rucklidge GJ. 2006. Salicylate modulates oxidative stress in the rat colon: a proteomic approach. *Biochem. Pharmacol. 72*:204–216.

Duncan MW and Hunsucker SW. 2005. Proteomics as a tool for clinically relevant biomarker discovery and validation. *Exp. Biol. Med. (Maywood) 230*:808–817.

Fuchs D, Piller R, Linseisen J, Daniel H, and Wenzel U. 2007a. The human peripheral blood mononuclear cell proteome responds to a dietary flaxseed-intervention and proteins identified suggest a protective effect in atherosclerosis. *Proteomics. 7*:3278–3288.

Fuchs D, Vafeiadou K, Hall WL, Daniel H, Williams CM, Schroot JH, and Wenzel U. 2007b. Proteomic biomarkers of peripheral blood mononuclear cells obtained from postmenopausal women undergoing an intervention with soy isoflavones. *Am. J. Clin. Nutr. 86*:1369–1375.

Fuchs D, Winkelmann I, Johnson IT, Mariman E, Wenzel U, and Daniel H. 2005. Proteomics in nutrition research: principles, technologies and applications. *Br. J. Nutr. 94*:302–314.

Garcia A, Prabhakar S, Brock CJ, Pearce AC, Dwek RA, Watson SP, Hebestreit HF, and Zitzmann N. 2004. Extensive analysis of the human platelet proteome by two-dimensional gel electrophoresis and mass spectrometry. *Proteomics. 4*:656–668.

Garcia A, Watson SP, Dwek RA, and Zitzmann N. 2005. Applying proteomics technology to platelet research. *Mass Spectrom. Rev. 24*:918–930.

Gravel P, Sanchez JC, Walzer C, Golaz O, Hochstrasser DF, Balant LP, Hughes GJ, Garcia-Sevilla J, and Guimon J. 1995. Human blood platelet protein map established by two-dimensional polyacrylamide gel electrophoresis. *Electrophoresis. 16*:1152–1159.

Hartwig JH. 2006. The platelet: form and function. *Semin. Hematol. 43*:S94–100.

Issaq HJ and Veenstra TD. 2007. The role of electrophoresis in disease biomarker discovery. *Electrophoresis. 28*:1980–1988.

Jenkins CS, Phillips DR, Clemetson KJ, Meyer D, Larrieu MJ, and Luscher EF. 1976. Platelet membrane glycoproteins implicated in ristocetin-induced aggregation. Studies of the proteins on platelets from patients with Bernard-Soulier syndrome and von Willebrand's disease. *J. Clin. Invest. 57*:112–124.

Kharbanda KK, Vigneswara V, McVicker BL, Newlaczyl AU, Bailey K, Tuma D, Ray DE, and Carter WG. 2009. Proteomics reveal a concerted upregulation of methionine metabolic pathway enzymes, and downregulation of carbonic anhydrase-III, in betaine supplemented ethanol-fed rats. *Biochem. Biophys. Res. Commun. 381*:523–527.

Kim H, Page GP, and Barnes S. 2004. Proteomics and mass spectrometry in nutrition research. *Nutrition. 20*:155–165.

Kussmann M. 2007. How to comprehensively analyse proteins and how this influences nutritional research. *Clin. Chem. Lab. Med. 45*:288–300.

Kussmann M and Affolter M. 2006. Proteomic methods in nutrition. *Curr. Opin. Clin. Nutr. Metab. Care. 9*:575–583.

Kussmann M, Raymond F, and Affolter M. 2006. OMICS-driven biomarker discovery in nutrition and health. *J. Biotechnol. 124*:758–787.

Lau B, Gange SJ, Phair JP, Riddler SA, Detels R, and Margolick JB. 2003. Rapid declines in total lymphocyte counts and hemoglobin concentration prior to AIDS among HIV-1-infected men. *AIDS. 17*:2035–2044.

Maas K, Chan S, Parker J, Slater A, Moore J, Olsen N, and Aune TM. 2002. Cutting edge: molecular portrait of human autoimmune disease. *J. Immunol. 169*:5–9.

Macaulay IC, Carr P, Gusnanto A, Ouwehand WH, Fitzgerald D, and Watkins NA. 2005. Platelet genomics and proteomics in human health and disease. *J. Clin. Invest. 115*:3370–3377.

Milner JA. 2007. Nutrition in the 'omics' era. *Forum Nutr. 60*:1–24.

Moresco JJ, Dong MQ, and Yates JR, III. 2008. Quantitative mass spectrometry as a tool for nutritional proteomics. *Am. J. Clin. Nutr. 88*:597–604.

O'Neill EE, Brock CJ, von Kriegsheim AF, Pearce AC, Dwek RA, Watson SP, and Hebestreit HF. 2002. Towards complete analysis of the platelet proteome. *Proteomics. 2*:288–305.

Opii WO, Joshi G, Head E, Milgram NW, Muggenburg BA, Klein JB, Pierce WM, Cotman CW, and Butterfield DA. 2008. Proteomic identification of brain proteins in the canine model of human aging following a long-term treatment with antioxidants and a program of behavioral enrichment: relevance to Alzheimer's disease. *Neurobiol. Aging. 29*:51–70.

Ruggiero C, Metter EJ, Cherubini A, Maggio M, Sen R, Najjar SS, Windham GB, Ble A, Senin U, and Ferrucci L. 2007. White blood cell count and mortality in the Baltimore Longitudinal Study of Aging. *J. Am. Coll. Cardiol. 49*:1841–1850.

Service RF. 2008a. Proteomics. Proteomics ponders prime time. *Science. 321*:1758–1761.

Service RF. 2008b. Proteomics. Will biomarkers take off at last? *Science. 321*:1760.

Shimizu M, Tamamori-Adachi M, Arai H, Tabuchi N, Tanaka H, and Sunamori M. 2002. Lipopolysaccharide pretreatment attenuates myocardial infarct size: A possible mechanism involving heat shock protein 70-inhibitory kappaBalpha complex and attenuation of nuclear factor kappaB. *J. Thorac. Cardiovasc. Surg. 124*:933–941.

Smolka MB, Zhou H, Purkayastha S, and Aebersold R. 2001. Optimization of the isotope-coded affinity tag-labeling procedure for quantitative proteome analysis. *Anal. Biochem. 297*:25–31.

Streckfus CF, Mayorga-Wark O, Arreola D, Edwards C, Bigler L, and Dubinsky WP. 2008. Breast cancer related proteins are present in saliva and are modulated secondary to ductal carcinoma in situ of the breast. *Cancer Invest. 26*:159–167.

Tom DH, Doring F, Fuchs D, Roth HP, and Daniel H. 2005. Transcriptome and proteome analysis identifies the pathways that increase hepatic lipid accumulation in zinc-deficient rats. *J. Nutr. 135*:199–205.

Vasan RS. 2006. Biomarkers of cardiovascular disease: molecular basis and practical considerations. *Circulation. 113*:2335–2362.

Wilkins MR, Appel RD, Van Eyk JE, Chung MC, Gorg A, Hecker M, Huber LA, Langen H, Link AJ, Paik YK, Patterson SD, Pennington SR, Rabilloud T, Simpson RJ, Weiss W, and Dunn MJ. 2006. Guidelines for the next 10 years of proteomics. *Proteomics. 6*:4–8.

Yu HH, Wang LC, Lee JH, Lee CC, Yang YH, and Chiang BL. 2007. Lymphopenia is associated with neuropsychiatric manifestations and disease activity in paediatric systemic lupus erythematosus patients. *Rheumatology. (Oxford) 46*:1492–1494.

Zhang L, Perdomo G, Kim DH, Qu S, Ringquist S, Trucco M, and Dong HH. 2008. Proteomic analysis of fructose-induced fatty liver in hamsters. *Metabolism. 57*:1115–1124.

Zhu J, Quyyumi AA, Wu H, Csako G, Rott D, Zalles-Ganley A, Ogunmakinwa J, Halcox J, and Epstein SE. 2003. Increased serum levels of heat shock protein 70 are associated with low risk of coronary artery disease. *Arterioscler. Thromb. Vasc. Biol. 23*:1055–1059.

15

Proteomics Approach to Assess the Potency of Dietary Grape Seed Proanthocyanidins

Hai-qing Gao

ABSTRACT

This is an exciting time for biological scientists as the "omics" era continues to evolve and shape the way science is understood and conducted. Understanding the interrelationship among genes, gene products, and diet is fundamental to identifying those genes that will benefit most from or be placed at risks by intervention strategies. Unraveling the multitude of nutrigenomic, proteomic, and metabolomic patterns that arise from the ingestion of foods or their bioactive food components will likely provide insight into a tailored approach to diet and health. Nutritional proteomics or nutriproteomics is the application of proteomics modality to nutrition-related research [1, 2]. It also represents the interaction of bioactive food ingredients with proteins, whereby the interaction with proteins occurs in two basically specific ways.

We are aware that diet may not only provide an adequate amount of nutrients to meet the metabolic requirements but could also contribute to improving human health status. As a consequence, plant extracts or single compounds that benefit human health need to be identified and developed for the food market to complement a balanced diet. Grape is one of the most widely consumed fruits in the world. Moderate consumption of wine is associated with a reduced risk of cancer. Grape seeds are rich in dimers, trimers, and other oligomers of flavan-3-ols, named procyanidins. Grape seed extracts are commonly available dietary supplements taken for the antioxidant activity that is attributed to its proanthocyanidin (oligomers of monomeric polyphenols) content. Grape seed proanthocyanidins (GSP) have been reported to possess a variety of potent properties including antioxidant, antinonenzymatic glycation, anti-inflammation, and antitumor, and so on [3]. Moreover, GSP have cardiovascular, renal, and cerebellar protective effects [4, 5]. GSP exhibit a novel spectrum of biological, pharmacological, therapeutic, as well as chemoprotective properties against oxygen-free radicals.

CHEMOPROTECTIVE PROPERTIES OF GSP

COMPONENTS AND MOLECULES

GSP exist as oligomers or polymers of flavan-3-ol such as (+)-catechin and (−)-epicatechin that are predicted to form helices in their global minimum-energy conformation. GSP are formed from the association of several of these monomeric units: 2-5 units for catechin oligomers, more than 5 units for catechin polymers. The oligomeric proanthocyanidin (OPC) has the strongest biological activity. Procyanidins differ in the position and configuration of their monomeric linkages. The structures of procyanidin dimers B1, B2, B3, and B4 are the best known. Among various types of procyanidins, dimers distribute the most widely in nature and have been given much attention by scientists. Trimers are named C1, C2, and so on, because of different compositions of monomers and different locations where carbon atoms are connected. People are now becoming more and more interested in GSP because of their beneficial effects on human health.

ANTIOXIDANT EFFECTS

Various environmental pollutants have been demonstrated to produce an enormous amount of free radicals, resulting in oxidative deterioration of lipids, proteins and DNA, activation of procarcinogens, inhibition of antioxidant defense systems and changes in gene expression, and significant contributions to human diseases. The free radical scavenging abilities (RSA) of GSP, vitamin E, and vitamin C against biochemically generated superoxide anion and hydroxyl radical were assessed *in vitro* at varying concentrations via cytochrome C reduction and chemiluminescence response. Chemiluminescence is a general assay for the production of reactive oxygen species, whereas cytochrome C reduction is a specific assay for superoxide anion. At 50 mg/L, GSP demonstrated 84% and 98% greater RSA against superoxide anion and hydroxyl radical, respectively, as compared to natural vitamin E, and at 100 mg/L, GSP demonstrated 439% and 575% greater RSA against superoxide anion and hydroxyl radical, respectively, as compared to vitamin C [6].

GSP could be useful in the attenuation of ultraviolet (UV) radiation-induced oxidative stress-mediated skin diseases in human skin. Treatment of normal human epidermal keratinocytes (NHEK) with GSP inhibits UV-induced hydrogen peroxide (H_2O_2), lipid peroxidation, protein oxidation, and DNA damage in NHEK. It can scavenge hydroxyl radicals and superoxide anions in a cell-free system. GSP also inhibit UVB-induced depletion of antioxidant defense components, such as glutathione peroxidase, catalase, superoxide dismutase, and glutathione. As UV-induced oxidative stress mediates activation of mitogen-activated protein kinase (MAPK) and nuclear factor-κB (NF-κB) signaling pathways, GSP inhibit UV-induced phosphorylation of ERK1/2, JNK, and p38 proteins. GSP also inhibit UV-induced activation of NF-κB, which is mediated through inhibition of degradation and activation of IκB-α and IKK-α, respectively [7].

GSP treatment can attenuate tissue malondialdehyde (MDA) level, myeloperoxidase activity, and collagen content in bile duct ligated biliary obstruction rats, suggesting its antifibrosis effect, which is possibly because of its inhibition of neutrophil infiltration and lipid peroxidation; thus, restoration of oxidant and antioxidant status in the tissue.

ANTINONENZYMATIC GLYCATION AND ANTI-INFLAMMATION EFFECTS

Using *in vitro* protein glycosylation system, glucose and bovine serum albumin were coincubated at 37°C and 50°C, respectively, as control while the experimental groups were given additional different doses of GSP or aminoguanidine. Then, the amount of advanced glycation end products (AGEs) were determined using a fluorescence photometer. Results show that GSP in the range of 1.0–2.0 g/L can inhibit the production of AGEs efficiently and the effect of GSP at the concentration of 2.0 g/L was similar to that of aminoguanidine of the same dose. Studies indicate that proanthocyanidins can effectively scavenge the reactive carbonyl species and thus inhibit the formation of AGEs.

Our previous experiments showed that GSP had antinonenzymatic glycation and anti-inflammation effects by reducing receptor of advanced glycation end products (RAGE) protein expression, subsequently leading to decreased expression of high-level vascular cell adhesion molecule 1 (VCAM-1) induced by AGEs in human umbilical vein endothelial cells (HUVEC). AGEs activated the expression of RAGE and inhibited peroxisome proliferator-activated receptors (PPAR) gamma expression in HUVEC, whereas GSP inhibited the expression of RAGE through activation of PPAR gamma in HUVEC simultaneously [3, 8, 9]. These findings indicated that GSP inhibited the cell inflammatory factor expression and protected the function of endothelial cell through activation of PPAR gamma expression and inhibition of RAGE expression.

Although several studies have postulated NF-κB as the molecular site where redox-active substances act to regulate agonist-induced ICAM-1 and VCAM-1 gene expression, inhibition of inducible VCAM-1 gene expression by GSP was not through NF-κB dependent pathway. GSP modulated inflammatory response in activated macrophages by the inhibition of nitric oxide (NO) and prostaglandin E_2 (PGE_2) production, suppression of inducible nitric oxide synthase (iNOS) expression, and NF-κB translocation. These results demonstrate an immunomodulatory role of GSP and thus a potential health benefit in inflammatory conditions that exert an overproduction of NO and PGE_2.

PROTECTIVE EFFECTS ON CARDIOVASCULAR SYSTEM

A large body of evidence has shown that GSP can improve the elasticity of blood vessels, inhibit the activity of angiotensin II, and lower blood pressure. And it is also proven to decrease plasma low-density lipoprotein (LDL) and total cholesterol (TC), decrease capillary permeability, and prevent thrombosis. In a word, GSP have been suggested to have a cardioprotective effect.

Antiatherosclerosis effects In our previous experiments, GSP can prevent atherosclerosis, and this effect may be due to modulation of lipids and antioxidant properties. Male New Zealand rabbits were divided into three groups and were fed a standard diet, standard + 1% cholesterol diet, and standard +1% cholesterol +1% grains feedstuff diet, respectively. At 12 weeks, the animals were sacrificed and the aortas were dissected for analysis. Results showed that GSP can reduce plasma TC, low-density lipoprotein cholesterol (LDL-C), triglyceride (TG), TG/high-density lipoprotein cholesterol (HDL-C), malondialdehyde (MDA), ox-LDL significantly and elevate HDL-C significantly. Immunohistochemical analysis demonstrated less thickening of aorta blood vessel wall and less foam cells comparing GSP group with high-cholesterol group.

Postprandial hyperlipemia is a well-defined risk factor for atherosclerosis. A reasonable contributing mechanism could involve the postprandial increase of plasma lipid hydroperoxides (LPO) affecting the oxidant–antioxidant balance and increasing the susceptibility of LDL to oxidation. The supplementation of a meal with GSP minimizes the postprandial oxidative stress by decreasing the oxidants and increasing the antioxidant levels in plasma, and, as a consequence, enhancing the resistance to oxidative modification of LDL. It is proved that GSP supplement can increase the plasma antioxidant capacity in the postprandial phase, and the content of LPO in chylomicrons was 1.5-fold higher after the control meal than after the GSP-supplemented meal [10].

Anti-ischemic reperfusion injury of myocardium GSP have cardioprotective effects against ischemic reperfusion-induced injury via their ability to reduce or remove, directly or indirectly, free radicals in myocardium that undergoes reperfusion after ischemia. GSP reduce the incidence of reperfusion-induced ventricular fibrillation and ventricular tachycardia. In rats treated with 100 mg/kg proanthocyanidins, the recovery of coronary flow, aortic flow, and developed pressure after 60 minutes of reperfusion was improved [11]. Electron spin resonance studies indicated that GSP remove peroxyl radical generated by 2, 2'-azobis (2-amidinopropane) dihydrochloride and reduce 7-OH.-coumarin-3-carboxylic acid attributing to its hydroxyl radical-scavenging activity.

GSP can protect the heart against myocardial injury induced by isoproterenol (ISO) in a rat model. The prior administration of GSP for 6 days a week for 5 weeks maintained the levels of the marker enzymes in all the treatment groups (GSP 50 mg, 100 mg, and 150 mg) when compared to ISO-injected rats. In the ISO-injected group there was a significant rise in thiobarbituric acid reactive substances (TBARS) and a significant decrease in glutathione,

glutathione peroxidase, glutathione S transferase, superoxide dismutase (SOD), and catalase. The administration of GSP maintained the activities of these enzymes close to normal levels. It also significantly increases the activities of mitochondrial enzymes (isocitrate dehydrogenase, succinate dehydrogenase, malate dehydrogenase, and alpha-ketoglutarate dehydrogenase) and respiratory chain enzymes (NADH dehydrogenase and cytochrome c oxidase) and significantly decreased the activities of lysosomal enzymes in the heart tissues of ISO-induced rats [12]. GSP pretreatment significantly maintains the cholesterol, phospholipids, triglycerides, and free fatty acids levels in serum and heart tissue of the ISO-induced myocardial injury in rats, too. All these findings prove the stress-stabilizing effect of GSP.

Antithrombosis effects GSP can inhibit platelet aggregation, improve endothelial function, and reduce oxidative stress. Studies show GSP, significantly decreased adenosine diphosphate-stimulated platelet reactivity and epinephrine-stimulated platelet reactivity. Rein et al. found that adding GSP to *in vitro* whole blood increased the expression of PAC-1 and P-selectin and reduced platelet reaction to agonist and did not inhibit platelet activation induced by epinephrine. Vitseva proved that incubation of platelets with GSP could lead to a decrease in platelet aggregation from 50% to 70% and lead to a dramatic decrease in superoxide release as well as a significant increase in radical-scavenging activity, decrease in reactive oxygen species release, as well as enhance platelet NO release [13]. These findings suggest potentially beneficial platelet-dependent antithrombotic and anti-inflammatory properties of GSP.

Effects of GSP on blood pressure Hypertension is a confirmed risk factor in cardiovascular diseases and one of the mechanisms of hypertension is about the dysfunction of ACE. GSP can inhibit the activity of angiotensin converting enzyme (ACE) *in vitro*. Rabbit experiments showed infusion of GSP (5 mg/kg) lessened responses to Ang I and II, which suggested a therapeutic role of GSP in hypertension [14]. However, clinic experiment of GSP in treating individuals with hypertension showed contradictory results, too.

Antiarrhythmic effects GSP prevented the incidence of reperfusion arrhythmias and significantly shortened the duration of the episodes of ventricular fibrillation. Simultaneously, the percentage of duration of normal sinus rhythm increased. These results demonstrate antiarrhythmic and cytoprotective effects of oral administration of oligomer procyanidins [15].

PROTECTIVE EFFECTS ON DIABETES AND ITS COMPLICATIONS

It has been documented that impaired homeostasis in diabetes mellitus is associated with increased production of reactive oxygen species and depletion of the antioxidant defense systems. Increased oxidative stress is a widely accepted participant in the development and progression of diabetes and its complications. Application of antioxidants, especially natural antioxidants such as GSP in treatment of diabetes, attracts more and more attention.

Protective effect on pancreas in diabetes In alloxan-induced diabetic rats treated with GSP there is a significant increase in glutathione levels and reduction in lipid peroxidation and total nitrate/nitrite content in pancreatic tissue. The protection effect of GSP on pancrea contributes to control hyperglycemia [16].

Protective effect on kidney in diabetes According to recent studies, GSP can also protect target organs like kidneys in diabetes. Diabetic rats induced by alloxan were given GSP intragastrically for 6 weeks; compared with the diabetic group, the 24-hour urinary protein, levels of blood urea nitrogen and serum creatinine, creatinine clearance rate, and ratio of kidney weight/body weight were decreased, the SOD activity in kidney rose while MDA content, NO content, and NOS activity decreased, indicating its nephroprotective effect on diabetic rats [17].

Protective effects on heart in diabetes In patients with diabetes mellitus, increased oxidative stress also accelerates the accumulation of AGEs. AGEs are thought to act through receptor-independent and -dependent mechanisms to promote myocardial damage; fibrosis and inflammation are associated with accelerated diabetic cardiomyopathy. For the cardioprotective effects on streptozocin-induced diabetic rats, GSP significantly reduced the plasma AGEs, RAGE, NF-κB, and transforming growth factor-β1 (TGF-β1) gene expression in myocardial tissue of diabetic rats. The structure of the myocardium was improved, too [18]. Therefore, GSP can ameliorate glycemia-associated cardiac damage.

Protective effects on hippocampus in diabetes Long-term chronic hyperglycemia caused the overexpression of AGEs/RAGE and NF-κB in the CA region of hippocampus in streptozocin (STZ)-induced diabetic rats. GSP decreased the expression of RAGE and NF-kappaBP65 at a daily oral dosage of 250 mg/kg. This study provides evidence that GSP can prevent structural changes of the rat brain with diabetes and it suggests that GSP might be a useful remedy in the treatment of diabetic encephalopathy [5].

Therefore, these studies may provide some recognition of GSP for the treatment of diabetes and its complications.

ANTIAGING EFFECTS

Aging is the accumulation of diverse deleterious changes in the cells and tissues leading to increased risk of different diseases. Oxidative stress is considered a major risk factor and contributes to age-related increase in DNA oxidation and DNA protein cross-links in central nervous system during aging. GSP had opposite effects on the accumulation of age-related oxidative DNA damages in spinal cord and in various brain regions such as cerebral cortex, striatum, and hippocampus [19]. Aging is also characterized by impairment of physiochemical and biological aspects of cellular functions, for example, devastation of normal cell function and membrane integrity. GSP significantly decrease cell surface charge level and concomitantly increase protein carbonyls and decrease glycoprotein, antioxidants status in erythrocytes of aged rats. Long-term supplementation of GSP increased erythrocyte surface charge density to near normal level in aged rats.

ANTIONCOGENESIS EFFECTS

Several studies proved that the growth and metastasis of tumors, such as breast cancer, lung cancer, gastric adenocarcinoma, and epidermoid carcinoma A431 could be inhibited by GSP, indicating its antioncogenesis effect, which is probably due to the upregulation of tumor suppression genes and/or downregulation of some oncogenes [20].

GSP exhibited cytotoxicity toward some cancer cells such as MCF-7 human breast cancer cells, A-427 human lung cancer cells, CRL-1739 human gastric adenocarcinoma cells, and K562 chronic myelogenous leukemic cells, while enhancing the growth and viability of normal cell lines such as normal human gastric mucosal cells and J774A.1 murine macrophage cells. GSP have been proved to have the ability to upregulate antiapoptotic gene Bcl-2 and downregulate cell cycle associated and proapoptotic genes such as c-myc and p53 [21].

In vitro treatment of breast cancer cells, 4T1, MCF-7, and MDA-MB-468 with GSP resulted in significant inhibition of cellular proliferation and viability and induction of apoptosis in 4T1 cells in a time- and dose-dependent manner. Further analysis indicated an alteration in the ratio of Bax/Bcl-2 proteins in favor of apoptosis, and the knockdown of Bax using Bax siRNA transfection of 4T1 cells resulted in blocking of GSP-induced apoptosis. Induction of apoptosis was associated with the release of cytochrome c, increasing expression of Apaf-1 and activation of caspase 3 and poly (ADP-ribose) polymerase. Treatment with the pan-caspase inhibitor (Z-VAD-FMK) resulted in partial but significant inhibition of apoptosis in 4T1 cells suggesting the involvement of both caspase activation dependent and independent pathways in the apoptosis of 4T1 cells induced by GSP. 4T1 cells were implanted subcutaneously in Bal b/c mice. Dietary GSP (0.2% and 0.5%, w/w) significantly inhibited the growth of the implanted 4T1 tumor cells and increased the ratio of Bax: Bcl-2 proteins, cytochrome c release, induction of Apaf-1, and activation of caspase 3 in the tumor microenvironment. These data suggest that GSP possess chemotherapeutic effects against breast cancer including inhibition of metastasis [22].

Cell cycles of cancer cells also get changed when being cocultured with GSP. *In vitro* treatment of human epidermoid carcinoma A431 cells with GSP inhibited cellular proliferation (13%–89%) and induced cell death (1%–48%) in a dose- and time-dependent manner. GSP-induced inhibition of cell proliferation was associated with an increase in G1-phase arrest at 24 hours, which was mediated through the inhibition of cyclin-dependent kinases (Cdk) Cdk2, Cdk4, Cdk6, and cyclins D1, D2, and E, and simultaneous increase in protein expression of cyclin-dependent kinase inhibitors (Cdki), Cip1/p21, and Kip1/p27, and enhanced binding of Cdki-Cdk. The treatment of A431 cells with GSP (20–80 mug/ml) resulted in a dose-dependent increase in apoptotic cell death (26%–58%), which was associated with an increasing protein expression of proapoptotic Bax, decreasing expression of antiapoptotic Bcl-2 and Bcl-xl, loss of mitochondrial membrane potential, and cleavage of caspase-9, caspase-3, and poly (ADP-ribose) polymerase (PARP). Pretreatment with the pan-caspase inhibitor (z-VAD-fmk) blocked the GSP-induced apoptosis in A431 cells suggesting that GSP-induced apoptosis is associated primarily with the caspase-3-dependent pathway [23]. GSP decreased the phosphatidylinositol 3-kinase (PI3K) and the phosphorylation level of Akt at ser473, and the constitutive activation of NF-κB. Treatment with GSP inhibits the expression of cyclooxygenase 2, iNOS, proliferating cell nuclear antigen, cyclin D1 and matrix metalloproteinase-9 in A431 cells compared with non-GSP-treated controls. Treatment of athymic nude mice with GSP by oral gavage (50 or 100 mg/kg body weight) reduces the growth of A431-xenografts in mice, which is associated with the inhibition of tumor cell proliferation in xenografts as indicated by the inhibition of mRNA expression of PCNA and cyclin D1, and of NF-κB activity [24].

In conclusion, GSP are a potential antioncogenesis agent that could be supplemented in cancer treatment in the future.

EFFECT IN WOUND HEALING

Angiogenesis plays a central role in wound healing. The wound site is rich in oxidants, such as hydrogen peroxide, mostly contributed by neutrophils and macrophages. GSP influence dermal wound healing *in vivo*. The potential effect of GSP on inducing vascular endothelial growth factor expression is at the transcriptional level. GSP accelerate wound contraction, and closure associated with a well-defined hyperproliferative epithelial region and higher cell density enhances deposition of connective tissue and improves histological architecture. GSP treatment also increased tenascin expression in the wound edge tissue. Tissue glutathione oxidation and 4-hydroxynonenal immunostaining results supported that GSP application enhanced the oxidizing environment at the wound site [25].

ANTIOSTEOPOROSIS

It has been proved that a high-calcium diet combined with GSP supplement is more effective in reversing mandibular condyle bone debility in rats than is a low-calcium diet, standard diet, or high-calcium diet alone. GSP included in a diet mixture with calcium has a beneficial effect on bone formation and bone strength for the treatment of bone debility caused by a low level of calcium [26].

Moreover, GSP treatment can protect brains from reperfusion injuries, too. All these studies demonstrate GSP as a safe, novel, highly potent, and bioavailable free radical scavenger and antioxidant exhibiting a broad spectrum of pharmacological, therapeutic, and chemoprotective properties. GSP, functioning at the genetic level and promoting therapeutic efficacy, is a potential natural agent in promoting human health.

PROTEOMIC PLATFORM

Proteomics can be divided into expression proteomics and cell mapping proteomics. Expression proteomics focus on differential display or "discovery" proteomics, global analysis of changes in protein expression during a biological process or in disease. Cell mapping proteomics studies protein interactions, which include: protein–protein interactions in order to identify components of functional protein complexes, pathway analysis and the analysis of protein networks; and protein–small molecule interactions such as protein-drug binding.

The extraction, display, and analysis of the individual protein of tissues and cells together comprise a complex multistage process, involving a variety of biochemical and biophysical principles. This is a rapidly developing field, and new or modified techniques continue to emerge. However, two-dimensional (2D) gel electrophoresis, coupling with spot analysis by mass spectrometry (MS), is still the most widely used technical approach.

BASED ON TWO-DIMENSIONAL GEL ELECTROPHORESIS (2-DE) PROTEOMICS

2.1.1 2-DE 2-DE is currently the core technique that can be routinely applied for parallel quantitative expression profiling of large sets of complex protein mixtures such as whole cell and tissue lists [27]. Proteins are separated on the basis of charge in the first dimension and molecular mass in the second. Typically, 1000–3000 proteins per gel can be visualized, for example, by staining with silver. Multiple forms of individual proteins can be readily visualized, and the particular subset of proteins from the proteome is determined by factors such as initial choice of sample soluble conditions and pH range of the gel strip used for the first dimension.

Analysis of gel with specialized software allows comparisons of multiple gels both within a laboratory and to comprehensive proteome databases on the Internet. Proteins of interest can then be identified on the basis of knowledge of the isoelectric point and apparent molecular size determined from the two-dimensional gels, supplemented by a combination of methods, generally applied hierarchically in the past years, and mass spectrometry technique is now applied widely for high-throughput analysis, which requires smaller amounts of material and has a higher throughput than conventional sequencing methods. Proteins or peptides are ionized by electrospray ionization from liquid state or matrix-assisted laser desorption ionization from solid state, and the mass of the ions is measured precisely by various coupled analyzers [28,29]. For example, the time-of-flight analyzer measures the time for ions to travel from the source to the detector (MALDI-TOF). If the excised protein spot is first digested by trypsin, which cleaves proteins at specific amino acid (if present), the protein can be broken into a mixture of peptides. The masses of the peptides can then be measured by mass spectrometry to produce a peptide mass fingerprint. This discriminating signature is compared with the peptide masses predicted from theoretical digestion of protein sequences currently contained within databases, and the protein can be identified. If necessary, actual sequence can be obtained by tandem mass spectrometry, in which discrete peptide ions can be selected and fragmented, and complex algorithms are used to correlate the experimental data with data derived from peptide sequence in protein databases. If the protein or peptide of interest cannot match any known sequence, generally long enough sequence can be interpreted to design suitable nucleotide-based probes and subsequently to identify new proteins and genes.

Fluorescence 2-D difference gel electrophoresis (DIGE) Despite increased reproducibility of the 2-DE process through standardization and automation, alignment of gel images is still an issue. No matter how good the separation, however, some proteins will always conjugate, thereby confounding accurate quantitative analysis. DIGE is a new technology built upon the classical gel approach for quantitative analysis (gel densitometry); it is a difference gel electrophoresis, in which separate samples are treated with unique fluorophore tags (binding covalently with lysine ε-amino groups); samples are combined and run on the same 2D-gel (MW of proteins is negligible); and quantitative analysis is based on relative intensities of fluorescing labels at specific spots (relative quantification) or to labeled standard (absolute quantification) [30]. The advantages are obvious and are described in the following: up to three individual samples can be separated on the same 2D gel; Ettan DIGE enables the incorporation of the same internal standard on every 2-D gel thereby negating the problem of gel-to-gel variation; the Cy dyes afford great sensitivity with detection of 125 pg of a single protein, and a linear response in protein concentration over at least five orders of magnitude (10^5). Meanwhile, the disadvantages are gel based-, time-, labor-, and money-intensive; also, not all proteins can be run on gels.

"GEL-FREE" PROTEOMICS

Based on mass spectrometry technology, "gel-free" proteomics can be divided into "shotgun" LC-ESI-MS/MS of total tryptic digest of proteins and nongel-based quantitative proteomics methods that can be broadly categorized into isotopic and isotope-free methods.

"Shotgun"proteomics 2-DE is inadequate for analysis of high molecular weight, hydrophobic, or highly acidic/basic proteins. So-called shotgun proteomics is an emerging concept developed to cope with this problem [31]. Protein samples are digested into a large array of small peptide fragments. The protein composition of immunoprecipitates, organelles, cultured cells, and clinical samples has been thoroughly identified by the combination of multidimensional LC and MS/MS.

LC-MS is a proteomic approach coupling of high-performance liquid chromatography and mass spectrometry to separate and to identify proteins or peptides [32]. In brief, as for sample sources if 1D LC-MS, the sample should be peptide mixture in most time, however, as for 2D LC-MS, the protein or the peptide samples could be loaded onto the first phase of LC, and only peptides are allowed to the second phase of LC; for protein separation if 1D LC-MS, the reverse phase column is adopted in most time, however, as for 2D LC-MS, the first phase column could have multiple choices, ionic exchange resin as priority, and the second phase column is usually reverse phase one; for protein identification ESI-ion trap, ESI-quatrupole TOF or ESI-FT mass spectrometry is often employed for peptide identification, but MALDI-TOF MS is also used for this process; and for annotation, "Sequest" is a major search engine of proteins adopted by Thermi-Finnigan ESI mass spectrometry, and "MasCot" is a popular search engine of proteins for number of mass spectrometries.

Electronic spray ionization (ESI) or matrix-assisted laser desorption ionization (MALDI) enables vaporization of nonvolatile biological samples from a liquid or a solid-state phase directly into the pseudo-gas phase so that the masses of biological macromolecules can be measured [33].

Quantitation by stable isotope labeling Effective and economical methods for quantitative analysis of high-throughput mass spectrometry data are essential to meet the goals of directly identifying, characterizing, and quantifying proteins from a particular cell state. Several types of *in vivo* metabolic and *in vitro* chemical and enzymatic isotope labeling methods including ICAT (isotope-coded affinity tag), SILAC (stable isotope labeling by amino acids in cell culture), and iTRAQ (isobaric tagging for relative and absolute quantification) have been developed to add a quantitative dimension to MS/MS.

SILAC was developed as a simple and accurate approach for mass spectrometric (MS)-based quantitative proteomics; this method relies on the incorporation of amino acids with substituted stable isotopic nuclei (in this case, deuterium 2H, 13C, 15N) [34]. In SILAC, two groups of cells are grown in culture media that are identical except in one respect: the first media contains the "light" and the other a "heavy" form of a particular amino acid. ICAT compares the relative abundance of two tagged proteins, each sample is covalently labeled at cysteine residues with either the heavy tag or the light tag (MW 8Da); cICAT (cleavable isotope-coded affinity tags) samples are mixed and then digested; the labeled tags are purified by a biotin affinity column [35]. iTRAQ uses up to four tag reagents that bind covalently to the N-terminus of the peptide and any lysine side-chains at the anime group (global tagging); each sample set is digested separately and then mixed with the specific iTRAQ tag. In brief, tags are designed to produce fragments in a "quiet" spectral region; the samples are then combined and subjected to LC-MS/MS [36].

Label-free strategies Recent advances in proteomics approaches developed an integrated platform called label-free quantitative proteomics. This method consists of machinery and software modules that can apply vast amounts of data generated by nanoflow LC-MS to differential protein expression analyses. Label-free methods are based on the relationship between protein abundance and sampling statistics such as peptide count, spectral count, probabilistic peptide identification scores, and sum of peptide Sequest XCorr scores (ΣXCorr).

PROTEIN CHIPS

This platform includes techniques such as surface-enhanced laser desorption ionization time of flight (SELDI-TOF)[37] and surface microarray-based systems (protein, tissue, and antibody arrays) [38].

PROTEOMICS ANALYSIS OF ACTIONS OF GSP

Nutritional proteomics or nutriproteomics is not only the application of proteomics methodology to nutrition-related research but also the representation of interaction between bioactive food ingredients and the innate proteins, whereby the interaction between bioactive food ingredients and the innate proteins occurs in two basic ways: the influence of bioactive food ingredients on protein synthesis via gene expression and the interaction between these ingredients and proteins via posttranslational modifications or small-molecule protein interactions [1,39,40]. The possible target structures, on a cellular and molecular level, as well as the enormous number of bioactive compounds to be tested, represent a very heterogeneous group of molecular structures.

The method of current nutrition study is to use sophisticated technologies to identify the molecular basis for the activity of various dietary chemicals [41]. Proteomics technologies will benefit a better understanding of the interplay between GSP and diet-related diseases such as cancer, diabetes mellitus, or neurodegenerative diseases. At present, Kim et al. utilized proteomics technology to assess the effects of GSP on proteins in the brains of normal rats that had ingested a defined diet supplemented with 5% GSP over a 6-week period [42,43]. Moreover, Gao et al. studied the mechanisms of action of GSP on kidney in streptozotocin-induced diabetic rats [44]. These studies indicate the interaction between GSP and diseases will cross all scientific disciplines.

PROTEOMICS ANALYSIS OF THE ACTIONS OF GSP IN THE BRAIN OF NORMAL RATS

Experimental procedures [42, 45] Five-week-old female Sprague-Dawley rats were segregated into two groups. One group ($n = 5$) received the AIN-76A (Tekland Industries, Madison, WI, USA) diet supplemented with 5% GSP (Kikkoman Corp., Chiba, Japan), and a second group ($n = 5$) received unsupplemented AIN-76A diet for 6 weeks. At the end of the intervention, all rats were sacrificed. Whole brains above the brain stem were dissected out. To perform a differential proteome analysis, the protein samples were extracted in the right hemisphere of brain and separated using 2-DE. The gel spots of interest were identified by MALDI-TOF MS and LC-MS/MS.

Localization analysis of differently expressed proteins The expression of thirteen proteins was found either upregulated or downregulated in the brain following ingestion of GSP. The increased proteins were the heat shock protein 60 (HSP 60), heat shock cognate protein 70 (HSC 70), HSC 71, creatine kinase brain β chain (CK-BB), and neurofilament proteins (NF-L and NF-M). Seven proteins were significantly lower in amount in the brains of animals that ingested GSP the ε isoform of 14-3-3 protein, glial fibrillary acidic protein (GFAP), actin, vimentin, α and γ subunits of enolase, and polypeptides homologous with the polypeptide sequence predicted by the RIKEN cDNA (NM025994)[42,43,45]. Using the proteomics tool of ExPASy, the subcellar localization of all different expressed proteins consists of cytoplasm, nucleus, and mitochondrion. Seventy-seven percent proteins are localized within cytoplasm.

Function analysis of differently expressed proteins Using the dynamically controlled vocabulary gene ontology and a lot of data, the function of all differently expressed proteins consists of response to stress, energy metabolism, cell anchorage, motility, and polarity. Moreover, most proteins are cytoskeleton. The cytoskeleton has the functions of anchorage, mobility, information, and polarity.

As chaperones, HSP-60, HSC-70, and HSC-71 have roles in protein folding, assemble, and secretion [46]. CK-BB plays an important role in the brain in regenerating ATP from phosphocreatine at discrete cellular sites of high ATP turnover [47]. NF-L and NF-M are cytoskeleton components of neurons that are important in brain development as well as in neuronal maintenance [48]. These proteins are found at higher levels in the brains of the animals that ingested GSP. Moreover, enolase is the enzyme that converts 2-phosphoglycerate to phosphoenolpyruvate in the glycolytic pathway [49]. It was found downregulated in the brain of normal rats with ingestion of GSP.

PROTEOMICS ANALYSIS OF THE ACTIONS OF GSP IN THE RATS DIABETIC NEPHROPATHY

Experimental procedures [44] Ten-week-old male Wistar rats received a single dose of STZ (55 mg/kg, injected into tail veins) freshly dissolved in 0.1 mol/L sodium citrate buffer (pH 4.5) after a 12-hour overnight fasting. Only rats with blood glucose higher than 16.7 mmol/L after 5 days were considered as diabetic in the fasting state.

All rats were divided into three groups: group 1, control rats (C); group 2, untreated diabetic rats (DM); group 3, treated diabetic rats (GSP, 250 mg/kg body weight/day). GSP (Lot No: G050412) were provided by Jianfeng Inc. (Tianjin, China). GSP were given in normal saline solution by intragastric administration for 24 weeks. To perform a differential proteome analysis, left kidneys from group C ($n = 3$), group DM ($n = 3$), and group GSP ($n = 3$) were dissected and rinsed thoroughly with ice-cold phosphate-buffered saline to remove blood components. The kidney lysate was labeled with Cy2, Cy3, and Cy5, and separated using 2-DE. The gel spots of interest were identified by MALDI-TOF MS and LTQ-ESI-MS/MS. Line image and 3D images of NADH- ubiquinone oxidoreductase are shown in Fig. 15.1 and Fig. 15.2.

Localization analysis of differently expressed proteins In the kidney of diabetic rats, twenty-five proteins were found to be significantly changed compared with normal. Nine proteins were found in normal level among the rats that received GSP treatment. The proteins that

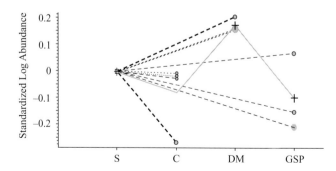

Figure 15.1 Representative line image of NADH-ubiquinone oxidoreductase from C, DM, GSP by proteomics analysis. S: standard; C: untreated control group; DM: untreated diabetic group; GSP: GSP treated diabetic group. For color detail, please refer to the color plate section.

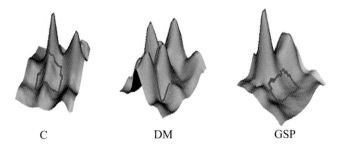

C   DM   GSP

Figure 15.2 Representative 3D images of NADH-ubiquinone oxidoreductase from C, DM, GSP by proteomics analysis. C: untreated control group; DM: untreated diabetic group; GSP: GSP treated diabetic group. For color detail, please refer to the color plate section.

were increased in expression were aflatoxin B1 aldehyde reductase (AFAR), phosphotriesterase-related protein, and beta actin. Moreover, seven proteins were significantly lower in amount in the kidneys of diabetic rats treated by GSP, namely, NADH-ubiquinone oxidoreductase (complex I), glutathione S-transferase mu (GSTM), selenium-binding protein 2 (SBP2), F1-ATPase beta subunit, glutamate carboxypeptidase (GCP), and LOC500183 protein [44]. Using the proteomics tool of ExPASy, the subcellar localization of all differently expressed proteins are cytoplasm, mitochondrion, nucleus, and cytolemma. Forty-four percent proteins are localized within cytoplasm. Thirty-three percent proteins are localized within mitochondrion.

Function analysis of differently expressed proteins Using the dynamically controlled vocabulary gene ontology and a lot of data, the function of all differently expressed proteins consists of oxidative stress, metabolism, signal transduction, cell proliferation, and apoptosis. Most proteins are oxidative stress-related proteins.

In the present study, complex I was found upregulated in the kidney of diabetic rats and downregulated to normal after GSP therapy. It is well known that increased oxidative stress has been proposed to be implicated in the pathogenesis of diabetes mellitus (DM) and diabetic nephropathy (DN) [50]. By this, it was shown that several proteins were involved in the oxidation and reduction changes in DN. Complex I is the point of entry for the major fraction of electrons that traverse the respiratory chain, eventually resulting in the reduction of oxygen. Recently, it has been reported that the inhibition of complex I leads to a blockage of more than 95% of activation-induced ROS production [51]. Moreover, complex I is not only the source of mitochondria-derived ROS but also its activity seems to be a prerequisite for subsequent ROS production via the NADPH oxidase. GSP are a commonly available dietary supplement taken for the antioxidant activity attributed to its proanthocyanidin (oligomers of monomeric polyphenols) content. It has also been reported that GSP provide significantly greater protection against ROS and free radical-induced lipid peroxidation and DNA damage than vitamins E, C, and b-carotene, as well as a combination of vitamins E plus C [6]. The results suggest that the inhibition effect of GSP on complex I at least partly contributed to the improvement of DN.

GSTM is a biotransformation enzyme and has functions in the elimination of free radicals, peroxides, electrophilic reagents, and heavy metals, the participation of cell protection and the regulation of cell growth. The expression of GSTM of diabetic rats was significantly higher than those of control rats by Western blot and 2-D DIGE. This might be the mechanism of compensation in diabetic rats. However, it was reported that the presence of the GSTM1 gene was associated with a susceptibility to 1 type DM [52]. Therefore, we need to do further research on the function of GSTM in diabetes. GCP was found upregulated in the kidney of diabetic rats and backregulated to normal after GSP therapy. GCP is a membrane peptidase expressed in the prostate, central and peripheral nervous system, kidney, small intestine, and tumor-associated neovasculature. It has been reported that the inhibition of GCP has beneficial effects on hyperalgesia, nerve function, and structural degenerative changes in diabetic polyneuropathy [53].

Moreover, aflatoxin B1 aldehyde reductase (AFAR) and beta actin were found downregulated in the kidney of diabetic rats and backregulated to normal after GSP therapy. It was reported that AFAR may act as local scavengers of osones and methylglyoxal and may protect cells and tissues against sugar-mediated damage [54]. The ability of AFAR to reduce the lipid peroxidation product 4-hydroxynonenal suggests that the reductase serves as a protection mechanism against oxidative stress. Previous studies showed that beta actin is a member of the actins family, which maintains the functions of cell structure, intracellular movement, and cell division. Beta actin disassembly is a prominent feature of diabetic nephropathy [55]. It suggested that the activation effect of GSP on AFAR at least partly contributed to increasing the activity of scavengers of AGEs in the kidney tissue. Meanwhile, we got the variation of glutathione peroxidase 1, SBP2, and phosphotriesterase-related protein. The proteins were related to the functions of deintoxication and antioxidant.

FUNCTIONAL CONFIRMATION OF PROTEINS

The proteins of interest that demonstrated statistic difference would be confirmed with different approaches alternatively. Consistently, our findings with 2-DE and 2-D DIGE were validated by immunocytochemistry and immunohistochemistry, Western blot analysis, as well as RT-PCR, respectively. In order to further study the functions of differently expressed proteins after treatment with GSP, we will apply the technologies of gene transfection and gene silencing, or gene knockout. Moreover, the interactions between proteins are important for many biological functions. We may study these associations from the perspective of biochemistry, signal transduction, and networks through protein–protein interactions.

FUTURE PERSPECTIVE

We are just entering a new era of postgenomic research and there is no doubt that nutrition science is going to be of central interest as nutrients and other food components. A wealth of protein information and novel techniques with high-throughput capabilities provide fantastic new tools for nutrition research. As those tools generate overwhelming datasets, there is clearly an emerging need for bioinformatics in nutritional sciences but also for nutritionists with a good background in cell biology and biochemistry of metabolism. The proteome analysis will be a useful technology with which to solve major nutrition-associated problems in humans and animals (including obesity, diabetes, cardiovascular disease, cancer, aging, and intrauterine fetal retardation).

Moreover, proteomics technologies are powerful tools that will be applied to dietary GSP in the future to optimize nutritional and health status. This will be accomplished by determining differently expressed proteins and identified molecular targets after treatment with GSP by which "functional" ingredients act to prevent and treat diseases. These methods can improve the velocity and efficiency of drug research and decrease the cost of new drug research.

Proteomic technologies, including 2-DE, 2-DE DIGE, and MS analyses combined with gene expression determination formed a basic research platform for investigating the actions of the GSP. In summary, our recent proteomics analyses and previous studies indicate that GSP affect mitochondrial function, cell morphology, cell proliferation and apoptosis, oxidative stress, substance metabolism, heat shock, inflammatory pathways, and neuronal survival pathways, associated with its antioxidant and anti-inflammatory activities. The diverse molecular mechanisms and cell-signaling pathways participating in the protection of brain, kidney, retina, and aorta treated with GSP make this multifunctional compound a potential agent for reducing risks of various neurodegenerative diseases and chronic complications of diabetes.

Integrated with other advanced technologies (genomics, transcriptomics, metabolomics, and bioinformatics) and systems biology, proteomics will greatly facilitate the discovery of key proteins that function to regulate metabolic pathways and whose synthesis, degradation, and modifications are affected by GSP. This will aid in enhancing our knowledge of the complex mechanisms responsible for GSP utilization and designing a contemporary paradigm for dietary prevention and intervention of disease.

Moreover, molecular imaging is a powerful technique that has the potential to analyze these factors noninvasively, repetitively, and quantitatively in an intact, living whole-body system. In addition, it can be used to track cell survival, monitor endogenous transcriptional regulation, screen transgenic animal phenotypes, study protein–protein interactions, and expedite novel drug discovery. Those who are able to incorporate these technologies to assess the potential of dietary GSP will make meaningful contributions over the next decade.

ACKNOWLEDGMENTS

We wish to thank the personnel of the Department of Geriatrics of the Qi-Lu Hospital of Shandong University. We also thank professor Qian Wang for her assistance.

REFERENCES

1. Schweigert FJ. 2007. Nutritional proteomics: methods and concepts for research in nutritional science. *Ann Nutr Metab.* *51*(2):99–107.
2. Trujillo E, Davis C, Milner J. 2006. Nutrigenomics, proteomics, metabolomics, and the practice of dietetics. *J Am Diet Assoc.* *106*(3):403–413.
3. Zhang FL, Gao HQ, Wu JM, Ma YB, You BA, Li BY, Xuan JH. 2006. Selective inhibition by grape seed proanthocyanidin extracts of cell adhesion molecule expression induced by advanced glycation end products in endothelial cells. *J Cardiovasc Pharmacol.* *48*(2):47–53.
4. Bagchi D, Sen CK, Ray SD, Das DK, Bagchi M, Preuss HG, Vinson JA. 2003. Molecular mechanisms of cardioprotection by a novel grape seed proanthocyanidin extract. *Mutat Res.* *523–524*:87–97.
5. Xu L, Li B, Cheng M, Zhang W, Pan J, Zhang C, Gao H. 2008. Oral administration of grape seed proanthocyanidin extracts downregulate RAGE dependant nuclear factor- kappa BP65 expression in the hippocampus of streptozotocin induced diabetic rats. *Exp Clin Endocrinol Diabetes.* *116*(4):215–224.
6. Bagchi D, Garg A, Krohn RL, Bagchi M, Tran MX, Stohs SJ. 1997. Oxygen free radical scavenging abilities of vitamins C and E, and a grape seed proanthocyanidin extract in vitro. *Res Comm Mol Path Pharmacol.* *95*(2):179–189.
7. Mantena SK, Katiyar SK. 2006. Grape seed proanthocyanidins inhibit UV-radiation-induced oxidative stress and activation of MAPK and NF-kappaB signaling in human epidermal keratinocytes. *Free Radic Biol Med.* *40*(9):1603–1614.
8. Ma L, Gao HQ, Li BY, Ma YB, You BA, Zhang FL. 2007. Grape seed proanthocyanidin extracts inhibit vascular cell adhesion molecule expression induced by advanced glycation end products through activation of peroxisome proliferators-activated receptor gamma. *J Cardiovasc Pharmacol.* *49*(5):293–298.
9. Zhang FL, Gao HQ, Shen L. 2007. Inhibitory effect of GSPE on RAGE expression induced by advanced glycation end products in endothelial cells. *J Cardiovasc Pharmacol.* *50*(4):434–440.
10. Natella F, Belelli F, Gentili V, Ursini F, Scaccini C. 2002. Grape seed proanthocyanidins prevent plasma postprandial oxidative stress in humans. *J Agric Food Chem.* *50*(26):7720–7725.
11. Pataki T, Bak I, Kovacs P, Bagchi D, Das DK, Tosaki A. 2002. Grape seed proanthocyanidins improved cardiac recovery during reperfusion after ischemia in isolated rat hearts. *Am J Clin Nutr.* *75*(5):894–899.
12. Karthikeyan K, Sarala Bai BR, Niranjali Devaraj S. 2007. Grape seed proanthocyanidins ameliorates isoproterenol-induced myocardial injury in rats by stabilizing mitochondrial and lysosomal enzymes:an in vivo study. *Life Sci.* *81*(23–24):1615–1621.

13. Vitseva O, Varghese S, Chakrabarti S, Folts JD, Freedman JE. 2005. Grape seed and skin extracts inhibit platelet function and release of reactive oxygen intermediates. *J Cardiovasc Pharmacol. 46*(4):445–451.

14. Peng N, Clark JT, Prasain J, Kim H, White CR, Wyss JM. 2005. Antihypertensive and cognitive effects of grape polyphenols in estrogen-depleted, female, spontaneously hypertensive rats. *Am J Physiol Regul Integr Comp Physiol. 289*(3):R771–R775.

15. Al-Makdessi S, Sweidan H, Jacob R. 2006. Effect of oligomer procyanidins on reperfusion arrhythmias and lactate dehydrogenase release in the isolated rat heart. *Arzneimittelforschung. 56*(5):317–321.

16. El-Alfy AT, Ahmed AA, Fatani AJ. 2005. Protective effect of red grape seeds proanthocyanidins against induction of diabetes by alloxan in rats. *Pharmacol Res. 52*(3):264–270.

17. Liu YN, Shen XN, Yao GY. 2006. Effects of grape seed proanthocyanidins extracts on experimental diabetic nephropathy in rats. *Wei Sheng Yan Jiu. 35*(6):703–705.

18. Cheng M, Gao HQ, Xu L, Li BY, Zhang H, Li XH. 2007. Cardioprotective effects of grape seed proanthocyanidins extracts in streptozocin induced diabetic rats. *J Cardiovasc Pharmacol. 50*(5):503–509.

19. Balu M, Sangeetha P, Murali G, Panneerselvam C. 2006. Modulatory role of grape seed extract on age-related oxidative DNA damage in central nervous system of rats. *Brain Res Bull. 68*(6):469–473.

20. Singletary KW, Meline B. 2001. Effect of grape seed proanthocyanidins on colon aberrant crypts and breast tumors in a rat dual-organ tumor model. *Nutr Cancer. 39*(2):252–258.

21. Joshi SS, Kuszynski CA, Bagchi D. 2001. The cellular and molecular basis of health benefits of grape seed proanthocyanidin extract. *Curr Pharm Biotechnol. 2*(2):187–200.

22. Mantena SK, Baliga MS, Katiyar SK. 2006. Grape seed proanthocyanidins induce apoptosis and inhibit metastasis of highly metastatic breast carcinoma cells. *Carcinogenesis. 27*(8): 1682–1691.

23. Meeran SM, Katiyar SK. 2007. Grape seed proanthocyanidins promote apoptosis in human epidermoid carcinoma A431 cells through alterations in Cdki-Cdk-cyclin cascade, and caspase-3 activation via loss of mitochondrial membrane potential. *Exp Dermatol. 16*(5):405–415.

24. Meeran SM, Katiyar SK. 2008. Proanthocyanidins inhibit mitogenic and survival-signaling in vitro and tumor growth in vivo. *Front Biosci. 13*:887–897.

25. Khanna S, Venojarvi M, Roy S, Sharma N, Trikha P, Bagchi D, Bagchi M, Sen CK. 2002. Dermal wound healing properties of redox-active grape seed proanthocyanidins. *Free Radic Biol Med. 33*(8):1089–1096.

26. Yahara N, Tofani I, Maki K, Kojima K, Kojima Y, Kimura M. 2005. Mechanical assessment of effects of grape seed proanthocyanidins extract on tibial bone diaphysis in rats. *J Musculoskelet Neuronal Interact. 5*(2):162–169.

27. Gorg A, Obermaier C, Boguth G, Harder A, Scheibe B, Wildgruber R, Weiss W. 2000. The current state of two dimensional electrophoresis with immobilized pH gradients. *Electrophoresis. 21*(6):1037–1053.

28. Hillenkamp F, Karas M. 1990. Mass spectrometry of peptides and proteins by matrix-assisted ultraviolet laser desorption/ionization. *Methods Enzymol. 193*:280–295.

29. Baldwin MA, Medzihradszky KF, Lock CM, Fisher B, Settineri TA, Burlingame AL. 2001. Matrix-assisted laser desorption/ionization coupled with quadrupole/orthogonal acceleration time-of-flight mass spectrometry for protein discovery, identification, and structural analysis. *Anal Chem. 73*(8):1707–1720.

30. Han MJ, Herlyn M, Fisher AB, Speicher DW. 2008. Microscale solution IEF combined with 2-D DIGE substantially enhances analysis depth of complex proteomes such as mammalian cell and tissue extracts. *Electrophoresis. 29*(3):695–705.

31. Swanson SK, Washburn MP. 2005. The continuing evolution of shotgun proteomics. *Drug Discov Today. 10*(10):719–725.

32. Elias JE, Haas W, Faherty BK, Gygi SP. 2005. Comparative evaluation of mass spectrometry platforms used in large-scale proteomics investigations. *Nat Methods. 2*(9):667–675.
33. Vissers JP, Blackburn RK, Moseley MA. 2002. A novel interface for variable flow nanoscale LC/MS/MS for improved proteome coverage. *J Am Soc Mass Spectrom. 13*(7):760–771.
34. Schmidt F, Strozynski M, Salus SS, Nilsen H, Thiede B. 2007. Rapid determination of amino acid incorporation by stable isotope labeling with amino acids in cell culture (SILAC). *Rapid Commun Mass Spectrom. 21*(23):3919–3926.
35. Gygi SP, Rist B, Gerber SA, Turecek F, Gelb MH, Aebersold R. 1999. Quantitative analysis of complex protein mixtures using isotope- coded affinity tags. *Nat Biotechnol. 17*(10):994–999.
36. Ross PL, Huang YN, Marchese JN, Williamson B, Parker K, Hattan S, Khainovski N, Pillai S, Dey S, Daniels S, Purkayastha S, Juhasz P, Martin S, Bartlet-Jones M, He F, Jacobson A, Pappin DJ. 2004. Multiplexed protein quantitation in Saccharomyces cerevisiae using amine-reactive isobaric tagging reagents. *Mol Cell Proteomics. 3*(12):1154–1169.
37. Tang N, Tornatore P, Weinberger SR. 2004. Current developments in SELDI affinity technology. *Mass Spectrom Rev. 23*(1):34–44.
38. Olle EW, Sreekumar A, Warner RL, McClintock SD, Chinnaiyan AM, Bleavins MR, Anderson TD, Johnson KJ. 2005. Development of an internally controlled antibody microarray. *Mol Cell Proteomics. 4*(11):1664–1672.
39. Barnes S, Kim H. 2004. Nutriproteomics: identifying the molecular targets of nutritive and non-nutritive components of the diet. *J Biochem Mol Biol. 37*(1):59–74.
40. Weinreb O, Amit T, Youdim MB. 2007. A novel approach of proteomics and transcriptomics to study the mechanism of action of the antioxidant-iron chelator green tea polyphenol (-)-epigallocatechin-3-gallate. *Free Radic Biol Med. 43*(4):546–556.
41. Fuchs D, Winkelmann I, Johnson IT, Mariman E, Wenzel U, Daniel H. 2005. Proteomics in nutrition research: principles, technologies and applications. *Br J Nutr. 94*(3):302–314.
42. Deshane J, Chaves L, Sarikonda KV, Isbell S, Wilson L, Kirk M, Grubbs C, Barnes S, Meleth S, Kim H. 2004. Proteomics analysis of rat brain protein modulations by grape seed extract. *J Agric Food Chem. 52*(26):7872–7883.
43. Kim H, Deshane J, Barnes S, Meleth S. 2006. Proteomics analysis of the actions of grape seed extract in rat brain: technological and biological implications for the study of the actions of psychoactive compounds. *Life Sci. 78*(18):2060–2065.
44. Li BY, Cheng M, Gao HQ, Ma YB, Xu L, Li XH, Li XL, You BA. 2008. Back-regulation of six oxidative stress proteins with grape seed proanthocyanidin extracts in rat diabetic nephropathy. *J Cell Biochem. 104*(2):668–679.
45. Kim H. 2005. New nutrition, proteomics, and how both can enhance studies in cancer prevention and therapy. *J Nutr. 135*(11):2715–2718.
46. Reddy RK, Lu J, Lee AS. 1999. The endoplasmic reticulum chaperone glycoprotein GRP94 with Ca(2+)-binding and antiapoptotic properties is a novel proteolytic target of calpain during etoposide-induced apoptosis. *J Biol Chem. 274*(40):28476–28483.
47. Friedman DL, Roberts R. 1994. Compartmentation of brain-type creatine kinase and ubiquitous mitochondrial creatine kinase in neurons: evidence for a creatine phosphate energy shuttle in adult rat brain. *J Comp Neurol. 343*(3):500–511.
48. Bajo M, Yoo BC, Cairns N, Gratzer M, Lubec G. 2001. Neurofilament proteins NF-L, NF-M and NF-H in brain of patients with Down syndrome and Alzheimer's disease. *Amino Acids. 21*(3):293–301.
49. Messier C, Gagnon M. 1996. Glucose regulation and cognitive functions: relation to Alzheimer's disease and diabetes. *Behav Brain Res. 75*(1–2):1–11.
50. Brownlee M. 2001. Biochemistry and molecular cell biology of diabetic complications. *Nature. 414*(6865):813–820.
51. Kaminski M, Kiessling M, Süss D, Krammer PH, Gülow K. 2007. Novel role for mitochondria: Protein kinase C{theta}-dependent oxidative signaling organelles in activation-induced T-cell death. *Mol Cell Biol. 27*(10):3625–3639.

52. Bekris LM, Shephard C, Peterson M, Hoehna J, van Yserloo B, Rutledge E, Farin F, Kavanagh TJ, Lernmark A. 2005. Glutathione-s- transferase M1 and T1 polymorphisms and associations with type 1 diabetes age-at-onset. *Autoimmunity. 38*(8):567–575.
53. Zhang W, Slusher B, Murakawa Y, Wozniak KM, Tsukamoto T, Jackson PF, Sima AA. 2002. GCPII (NAALADase) inhibition prevents long-term diabetic neuropathy in type 1 diabetic BB/Wor rats. *J Neurol Sci. 194*(1):21–28.
54. O'Connor T, Ireland LS, Harrison DJ, Hayes JD. 1999. Major differences exist in the function and tissue-specific expression of human aflatoxin B1 aldehyde reductase and the principal human aldo-keto reductase AKR1 family members. *Biochem J. 343* Pt2:487–504.
55. Clarkson MR, Murphy M, Gupta S, Lambe T, Mackenzie HS, Godson C, Martin F, Brady HR. 2002. High glucose-altered gene expression in mesangial cells. Actin-regulatory protein gene expression is triggered by oxidative stress and cytoskeletal disassembly. *J Biol Chem. 277*(12):9707–9712.

16
Proteomics and Its Application for Elucidating Insulin Dysregulation in Diabetes

Hyun-Jung Kim and Chan-Wha Kim

INTRODUCTION

Diabetes is essentially a disease of insulin dysregulation. Insulin is an animal hormone whose presence informs the body's cells that the animal is well fed, causing liver and muscle cells to take in glucose and store it in the form of glycogen and fat cells to take in blood lipids and turn them into triglycerides. There are two primary types of diabetes, type 1 and type 2. Individuals diagnosed with type 1 diabetes are incapable of producing pancreatic insulin and must rely on insulin medication for survival. Patients diagnosed with type 2 diabetes have insulin resistance, relatively low insulin production, or both. Some patients with type 2 diabetes may eventually require insulin when other medications become insufficient in controlling blood glucose levels. An impaired early insulin secretory response of the pancreatic β-cell and decreased insulin sensitivity of the liver, skeletal muscle, and fat cells contribute to the state of hyperglycemia [1–3]. Diabetes in poorly treated patients is associated with significant morbidity and mortality due to long-term microvascular and macrovascular complications such as nephropathy, retinopathy, neuropathy, myocardial infarction, and stroke [4].

Proteomics can be used to identify the entire protein complement of a cell, tissue, or microorganism at different stages of development and to examine the integrated response to a particular challenge such as hormonal, environmental, or nutritional [5]. In addition, proteomics has led to the identification of many potential new drug targets for the treatment of a variety of diseases. Therefore, proteomics will be a key area of biology during the first couple of decades in the new millennium and it will have a major impact, both directly and indirectly, on nutritional science.

The principal proteomic methodology for separating, comparing, and detecting quantitative changes in biomarker expression patterns associated with disease has classically been two-dimensional gel electrophoresis (2-DE) followed by protein identification using mass spectrometry (MS) (Fig. 16.1), including matrix-assisted laser desorption-ionization time-of-flight (MALDI-TOF), electrospray ionization (ESI)-MS, liquid chromatography (LC)-MS, surface-enhanced laser desorption/ionization (SELDI)-MS, capillary electrophoresis (CE)-MS, or hybrid instrumentation [6–12].

The purpose of this chapter is to elucidate insulin dysregulation in diabetes via proteomic approaches. Therefore, we provide some examples where proteomic applications have been used to examine insulin signaling in pancreatic islets and β-cells and the serum from target organs or cells in patients with diabetes. In addition, we will provide an overview of studies that have used proteomic approaches to examine diabetic complications.

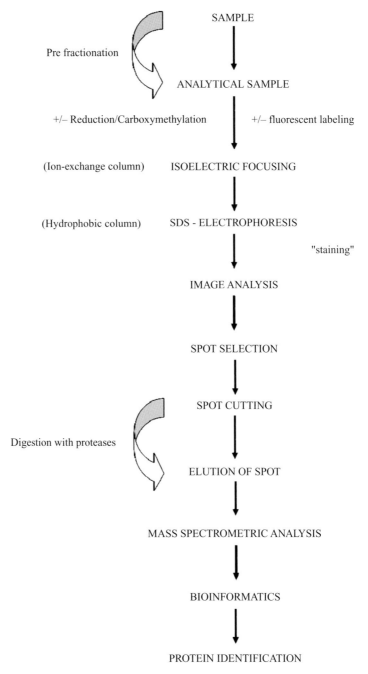

Figure 16.1 An overview of proteome analysis using two-dimensional gel electrophoresis and mass spectrometry.

DIABETES MELLITUS

Diabetes mellitus is a homogeneous clinical syndrome characterized by hyperglycemia that occurs as a consequence of absolute or relative insulin deficiency and/or defective insulin action. Although diabetes mellitus is recognized by its characteristic hyperglycemia, metabolic derangements are more pervasive and involve altered metabolism of carbohydrates, fats, and proteins. There are two types of diabetes, type 1 (or juvenile onset) and type 2 (or adult onset). Type 1 diabetes mellitus is characterized by the destruction of β-cells in the pancreatic islets, which is caused by an autoimmune process that usually leads to an absolute insulin deficiency [13, 14]. The cause of the autoimmune disorder is currently being widely investigated. At least two major components are thought to contribute to the manifestation of the disease. The first is a genetic component, where certain individuals that possess defects in certain genes will have an increased susceptibility. So far, only two genes have been identified that appear to put an individual at risk; however, there are certainly other undiscovered genes involved. The genetic component is not, in itself, sufficient to cause the autoimmunity. Unidentified environmental components are also required to produce the disease in these susceptible individuals. Type 2 diabetes mellitus is characterized by insulin resistance in peripheral tissue (liver, skeletal muscles, and adipocytes) and an insulin secretory defect of the β-cells [13, 14]. This is the most common form of diabetes mellitus and is highly associated with a family history of diabetes, age, obesity, and lack of exercise. The cause of this insulin resistance is not fully known, although it has been linked to defects in the action of insulin after it is bound to the cell surface receptor (i.e., the cascade of reactions that were mentioned earlier). Obesity is found in approximately 55% of patients diagnosed with type 2 diabetes [15]. Obesity has been shown to predispose individuals to insulin resistance. Abdominal fat is especially hormonally active and secretes a group of hormones called adipokines that may impair glucose tolerance. The most important among these is the tumor necrosis factor α (TNFα), which acts to induce a state of insulin resistance by downregulating the glucose transporter 4 (GLUT4) and increasing free fatty acid release [16, 17]. The potential roles of resistin and leptin in insulin resistance in humans remain to be conclusively delineated. In contrast, adiponectin is a recently discovered adipocytokine that enhances insulin sensitivity [18], and its blood serum levels are lower in type 2 diabetes and obese individuals, implying a causal relationship.

PROTEOMICS IN DIABETES

Despite attempts at controlling blood glucose levels with currently available drugs, diabetes remains a serious problem in clinical practice. The complexity of diabetes requires a broad-based, unbiased, scientific approach, such as proteomics, to elucidate the changes imposed by various immunological mediators and lack of insulin. Proteomic approaches have been extensively used to explore the complexity of biological problems and to identify disease biomarkers. This section summarizes the recent proteomics-based approaches that have been used to study diabetes and provides insight into the future use of this technique to better understand the complexity of diabetes.

PROTEOMICS IN PANCREAS, PANCREATIC ISLETS, AND β-CELLS

In an attempt to search for novel biomarkers that could monitor the prognosis of diabetes, Kim *et al.* [19] examined the influence of hypoglycemic fungal extracellular polysaccharides (EPS) on the differential change in the pancreatic proteome and transcriptome in streptozotocin (STZ)-induced diabetic rats using 2-DE followed by peptide mass fingerprinting (PMF) analysis. The 2-DE system separated more than 2000 individual spots, where 34 proteins out of about 500 matched spots were found to be differentially expressed. A total of 22 overexpressed and 12 underexpressed proteins in the 2-DE map were observed (*p*, 0.05) between healthy and diabetic rats, of which 26 spots were identified by PMF analysis. In this study, they found that fungal EPS provided direct signals to regulate the expression of many metabolically important genes in rat pancreatic cells, suggesting that EPS may function through complex gene regulation events. Although the proteins and genes that were altered in response to STZ and EPS treatment were not well matched, the individual results of the proteome and transcriptome analysis did help explain the differential changes in the levels of proteins and genes involved in oxidative stress, insulin biosynthesis and release, and cell growth and differentiation in diabetic pancreas. This report addresses the value of using proteomics and transcriptomics technologies in assessing the cause of diabetes.

Understanding the importance and process of regenerating new pancreas cells may be highly valuable for developing novel therapies for diabetes that involve stimulating the regeneration of β-cells in humans. To characterize the complex pattern of protein expression in regenerating pancreatic extract (RPE), Shin et al. [20] and Yang et al. [21] monitored the altered expression levels of proteins on the second day after a 60% pancreatectomy or on the third day after a 90% pancreatectomy by 2-DE (Fig. 16.2). In the study by Shin et al. [20], sixty-one of seventy-six

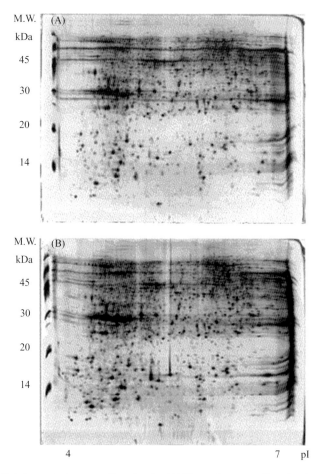

Figure 16.2 2-D gel maps of pancreatic extracts. (A) Normal pancreatic extract, and (B) regenerating pancreatic extract obtained 2 days after a 60% pancreatectomy. One hundred micrograms of protein were loaded and separated on a pH 4–7 linear IPG strip (24 cm). Fifteen percent acrylamide gels were used. (From Ref. 20.)

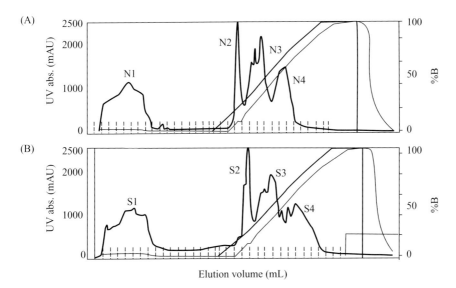

Figure 16.3 Ion exchange chromatograms of normal and regenerating pancreatic extracts. (A) Normal pancreatic extract, and (B) pancreatic extract obtained 2 days after a 60% pancreatectomy. Fractionation was performed using a fast protein liquid chromatography (FPLC) equipped with a MONO-Q column. The four major peaks were detected at 280 nm and each fraction was designated as N1, N2, N3, and N4 for normal pancreas and S1, S2, S3, and S4 for fractions of regenerating pancreatic extract, respectively. (From Ref. 20.) For color detail, please refer to the color plate section.

significantly up- or downregulated protein spots were identified by MALDI-TOF/MS. Moreover, the whole RPE was fractionated into four groups using anion-exchange chromatography (Fig. 16.3), and each fraction's cell proliferating activity was measured using an MTT assay. Compared to the normal pancreatic extract, fraction 3 and 4 of the RPE had the highest cell proliferating activity. A total of ten spots, which are differentially expressed after the pancreatectomy (Fig. 16.4), in the 2-DE of fractions 3 and 4 were identified by MS/MS. Of these identified proteins, Reg III, which may be functionally associated with the well-known regenerating factor (Reg I), was found. Taken together, these results demonstrated that the differential protein expression associated with pancreas regeneration could be determined by 2-DE and MS. In addition, these results suggested that the prefractionation method combined with an *in vitro* cell proliferation assay can effectively be used to pinpoint the active components for pancreas regeneration. In the study by Yang *et al.* [21], 2-DE revealed ninety-one spots that had at least a 1.5-fold increase in expression the third day after the pancreatectomy and fifty-three differentially expressed proteins were identified by PMF. These included cell growth-related, lipid and energy metabolism-related, protein and amino acid metabolism-related proteins, and signal transduction proteins. Vimentin, CK8, L-plastin, hnRNP A2/B1, and AGAT are associated with embryogenesis and cell differentiation, and may be new potential pancreatic stem cell markers.

Unfortunately, there was little correlation between the detected proteins in these two studies. Nevertheless, both studies clarified the global proteome during the pancreatic proliferation and differentiation processes, which is important for achieving a better understanding of pancreatic regeneration.

The islets of Langerhans are scattered throughout the pancreas and constitute only 1% of the total tissue [22]. β-cells, the principal cellular constituent of the islets, control whole-body metabolic fuel homeostasis by secreting insulin in response to elevations in plasma glucose concentration [23].

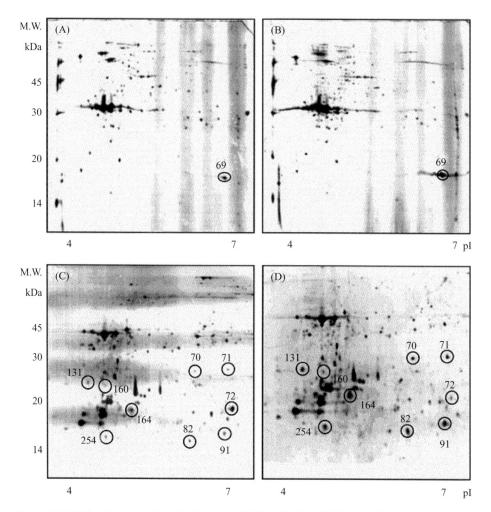

Figure 16.4 2-D gel maps of fraction 3 and 4. (A) N3, (B) S3, (C) N4, and (D) S4. One hundred micrograms of proteins were loaded and separated on a pH 4–7 linear IPG strip (24 cm). Fifteen percent acrylamide gels were used. (From Ref. 20.)

Proteomic studies have been performed on rat islet cells that were exposed to the cytokine interleukin (IL)-1β [24–26]. Cytokines released from macrophages are known to be toxic to islet β-cells, and this is thought to be one of the mechanisms involved in the development of type 1 diabetes in genetically predisposed individuals [27]. In this study, more than 2000 proteins were identified, of which 105 had an altered expression following exposure to IL-1β [25]. However, the functional relevance of these changes in relation to human disease is currently unclear. Proteomic analysis of murine pancreatic islet cells has identified proteins that are potentially involved in insulin resistance in mice because Lep/Lep mice were found to have altered actin-binding proteins, which may contribute to islet cell dysfunction and reduced insulin secretion [28].

Sanchez *et al.* [29] published the protein database, "The Mouse SWISS-2D PAGE Database," which is a descriptive analysis of protein expression in normal mouse pancreatic islets. In this study, they built the first detailed 2-DE reference map of normal mouse islet. These protein databases may serve as reference 2-DE maps. However, proteins that change expression cannot be identified on 2-DE databases alone, but rather have to be further identified using MS because colocalization of proteins on 2-DE is quite common. Although this may not be directly related to other type 1 diabetes studies, this map is a very valuable start for further investigations. Following this, Nicolls *et al.* identified eighty-eight normal mouse islet proteins [30]. An altered expression of mouse islet proteins in the hyperglycemic state has also been reported [31]. Recently, 130 spots, which correspond to 66 different protein entries, and a reference map of human islets were reported by Ahmed *et al.* [32]. In addition, Metz *et al.* used a two-dimensional LC-MS/MS-based analysis to characterize the human islet proteome, resulting in the identification of 29,021 different tryptic peptides covering 3365 proteins [33]. However, only a few studies have applied functional proteomics technologies to determine which islets proteins are relevant during the pathogenesis of type 1 diabetes [34]. Thus, to get a better understanding of the mechanisms underlying type 1 diabetes, Xie *et al.* [35] generated a mouse model by injecting multiple low doses of streptozotocin (MLD-STZ). The islets in the mouse pancreas were then analyzed by 2-DE. In this study, the expression levels of seven proteins were found to be significantly altered. These proteins were successfully identified by MALDI-TOF/MS and the changes in protein expression were further validated by quantitative RT-PCR and Western blotting. In this study, voltage-dependent anion-selective channel protein 1 and peroxiredoxin-4 were found, for the first time, to be associated with type 1 diabetes. These results suggest that glucose transport, beta cell proliferation/death, and oxidative stress play important roles in maintaining and balancing glucose levels. This study, which used the MLD-STZ-induced type 1 diabetes model, also provided novel insight into the mechanism of type 1 diabetes.

To further understand the link between insulin resistance and β-cell dysfunction in type 2 diabetes, Lu *et al.* [36] studied an insulin-resistant mouse model (transgenic MKR mice) that progressively developed diabetes [37]. In this study, global islet protein and gene expression changes were characterized in diabetic MKR versus nondiabetic control mice that were 10 weeks old. Using a quantitative proteomic approach (iTRAQ), a total of 159 proteins were found to be differentially expressed in MKR compared to control islets. Functional cluster analysis of these proteins revealed a marked upregulation of protein biosynthesis and endoplasmic reticulum (ER) stress pathways and a concomitant downregulation in insulin processing and secretion, as well as mitochondrial energy metabolism pathways in MKR islets. The mRNA expression level corresponding to these proteins was examined by microarray and then a small subset was validated using qPCR and Western blot analyses. From this analysis, $\sim$54% of the differentially expressed proteins in MKR islets (including proteins involved in proinsulin processing, protein biosynthesis, and mitochondrial oxidation) showed changes in the proteome but not transcriptome, suggesting posttranscriptional regulation. These results underscore the importance of integrating mRNA and protein expression analysis.

The earliest studies using 2-DE for protein separation in relation to type 1 diabetes were done in the 1980s. In this work, changes in the cellular protein expression pattern in β-cells of islets from mice with virus-induced hyperglycemia were described [38]. Other studies have investigated the expression of specific enzymes, such as hexokinase, glucokinase, and other proteins in β-cells, by combining 2-DE and specific antibodies for identification [39–41]. In addition, others have used 2-DE to describe insulin secretory granule biogenesis [42], exocytosis [43], and the effect of glucose on β-cells [44, 45].

Most recently, Fernandez *et al.* [46] conducted a metabolomic and proteomic analysis of a clonal insulin-producing β-cell line (INS-1 832/13). In this study, a 2-DE analysis of the 832/13 β-cells was performed to examine whether changes in metabolite levels could be attributed to changes in the level of proteins involved in cellular metabolism. The identities of eighty-six spots corresponding to seventy-five unique proteins that were significantly different in 832/13 β-cells cultured at 16.7 mM glucose were determined. Only five of these were found to be metabolic enzymes that could be involved in the metabolomic alterations that were observed. Anticipated changes in metabolite levels in cells exposed to increased glucose were observed, whereas changes in enzyme levels were much less profound. This suggests that substrate availability, allosteric regulation, and/or posttranslational modifications are more important determinants of metabolite levels than enzyme expression at the protein level.

PROTEOMICS IN TARGET ORGANS OR CELLS OF INSULIN

LIVER

The liver plays important roles in glucose absorption, metabolism, and excretion. Changes in this organ may reflect pathologic alterations during diabetes, because the serum has a direct connection with most organs and pathological changes. Analysis of liver protein expression profiles in connection to hepatic insulin resistance was performed by Morand et al. [47] using an insulin-resistant hamster model. In their search for mechanisms that linked development of hepatic insulin resistance and overproduction of atherogenic lipoproteins, the authors analyzed the proteome of the liver endoplasmic reticulum (ER). Using 2-DE, thirty-four differentially expressed proteins were identified. Expression of hepatic ER proteins ER60, ERp46, ERp29, glutamate dehydrogenase, and TAP1 were all decreased, whereas expression of alphaglucosidase, P-glycoprotein, fibrinogen, protein disulfide isomerase, GRP94, and apolipoprotein E were all increased in the insulin-resistant animal. As the authors point out, global proteomic analysis of the liver would also be useful in the context of insulin resistance.

The peroxisome proliferator-activated receptors (PPARs) are ligand-activated transcription factors that modulate lipid and glucose homeostasis. PPARα and PPARγ agonists are used to clinically treat hypertriglyceridemia and insulin resistance of diabetes, respectively. As a result of the clinical use of PPARα and PPARγ, several studies have been performed to investigate the proteomic effect of PPAR treatments on liver cells. First, the effects of the PPAR-α agonist were studied on ob/ob mouse liver cell proteins by Edvardsson et al. [48]. From this study, sixteen liver proteins were found to be upregulated, of which fourteen belonged to the peroxisomal fatty acid metabolism pathway. Based on these results, it was concluded that the therapeutic effect of PPAR-α agonist occurs through an upregulation of the peroxisomal fatty acid β-oxidation. Whereas PPAR-α agonists are used to treat hypertriglyceridemia, the PPAR-γ agonists are used to treat insulin resistance. PPAR-γ agonists are known to activate a nuclear receptor, which leads to increased insulin sensitivity. To gain further insight into the molecular mechanisms underlying the therapeutic actions of these drugs, Edvardsson et al. [49] performed a comparative analysis of the hepatic protein expression profiles of lean and obese (ob/ob) mice, and obese mice treated with WY14,643 (PPARα agonist) or rosiglitazone (PPARγ agonist) using 2-DE and MS. They found that livers from obese mice displayed higher levels of enzymes involved in fatty acid oxidation and lipogenesis compared to lean mice and these differences were further amplified by treatment with both PPAR activators. WY14,643 normalized the expression levels of several enzymes involved in glycolysis, gluconeogenesis, and amino acid metabolism in the obese mice, whereas rosiglitazone partially normalized the levels of enzymes involved in amino acid metabolism. In this study, a classical proteomics approach was successfully used to characterize differences between lean and obese diabetic mice at the hepatic proteome level, to map metabolic pathways affected by treatment, and to discriminate between effects caused by treatment with agonists of the closely related PPARα and PPARγ receptors.

Skeletal muscle

Skeletal muscle is principally responsible (>80%) for insulin-stimulated whole-body glucose disposal, and hence plays an important role in the pathogenesis of insulin resistance. A recent analysis of the proteome of human skeletal muscle in normal controls and in patients with type 2 diabetes illustrated the power of proteomics to be used as an analytical tool [50]. In this study, eight potential markers of type 2 diabetes were identified in the skeletal muscle.

From this analysis, the level of two proteins that have a crucial role in ATP synthesis was found to be lower in the muscles derived from patients with diabetes. However, as the authors indicate, it remains to be determined whether this downregulation of creatine kinase B and ATP synthase β-subunit represents a primary defect in type 2 diabetes or is the consequence of diabetes-induced alterations. Hittel *et al.* [51] characterized cytosolic skeletal muscle proteins in lean and obese individuals. The study identified increased levels of adenylate kinase, glyceraldehyd-3-phosphate dehydrogenase, and aldolase A in the skeletal muscle of obese women when compared to lean women. Proteomics analysis of human muscle is an important approach for gaining insight into the biochemical basis of normal and pathophysiological conditions [52]. Stentz *et al.* [53] analyzed the genes expressed (transcriptomes) and the proteins translated (proteomes) in muscle tissues and activated CD4(+) and CD8(+) T-lymphocytes (T-cells) of five type 2 diabetes subjects using Affymetrix microarrays and mass spectrometry, and compared them with matched nondiabetic controls. Gene expression of the insulin receptor (INSR), vitamin D receptor, insulin-degrading enzyme, Akt, insulin receptor substrate-1 (IRS-1), IRS-2, GLUT4, and enzymes of the glycolytic pathway were at least 50% lower in type 2 diabetes than in the controls. However, plasma cell glycoprotein-1, TNFα, and gluconeogenic enzymes in type 2 diabetes were upregulated by greater than twofold relative to the controls. Gene silencing of INSR or TNFα resulted in the inhibition or stimulation of GLUT4, respectively. Proteome profiles of the previously translated transcriptomes in type 2 diabetes and the controls had different patterns of change. Meanwhile, changes in the transcriptomes and proteomes between muscle and activated T cells of type 2 diabetes were comparable. Activated T cells expressed insulin signaling and glucose metabolism genes and gene products, which was analogous to muscle cells. In conclusion, the expression of certain genes and gene products in T cells and muscle in type 2 diabetes were different relative to the nondiabetic controls. These alterations in transcriptomes and proteomes in type 2 diabetes may be involved in insulin resistance.

Adipocytes

Because obesity is associated with insulin resistance and type 2 diabetes, several studies have focused on the specific functions of adipocytes, which play an important role in the development of diabetes. Separation of adipocyte proteins from 7-week-old lean and obese Zucker rats by 2-DE followed by laser densitometry revealed an obesity-related decrease in the concentration of a 28-kDa cytosolic protein in two lipogenic tissues, liver and white fat, but not in soleus muscle [54]. This protein was identified by microsequencing to be carbonic anhydrase III (CA III) and its activity was also found to be lower in obese Zucker rats [55]. Because a decrease in CA III activity was reversed following the suppression of the hyperinsulinemia by the beta cell toxin streptozotocin, the authors suggested that CA III levels may be decreased by hyperinsulinemia. The same group also used protein microsequencing to detect increased adipocyte (but not hepatic) expression of pyruvate carboxylase during the progression to obesity [55]. They suggested that this was related to the lipogenic capacity of obese Zucker rat adipocytes and demonstrated that the increased levels of pyruvate carboxylase were attenuated in streptozotocin-treated rats, thus implicating high insulin levels in the induction of the high levels of this enzyme [56].

The mouse 3T3-L1 fibroblastic cell line has been widely investigated in this regard, because it has been shown to differentiate rapidly into an adipocyte phenotype when treated with insulin. For example, 2-DE of ^{32}P-labeled proteins from 3T3-L1 cells, before and after immunoprecipitation with an anticalmodulin antibody, revealed that insulin does not stimulate calmodulin phosphorylation under conditions in which it stimulates the phosphorylation of other proteins [57]. Moreover, 2-DE delineated cellular proteins that were involved in the differentiation process of these cells [58]. However, none of these proteins were specifically identified. The same methodology was used to demonstrate that insulin stimulated the phosphorylation of multiple proteins in NIH 3T3 cells that expressed a high number of human insulin receptors, suggesting that these proteins act as substrates for the kinase activity of the receptor [59]. However, as in the previous studies none of these proteins were identified.

Under conditions of insulin resistance and type 2 diabetes, fat cells are subjected to increased levels of insulin, which may have a major impact on the secretion of adipokines. Using transcriptomics and proteomics, Wang et al. [60] investigated the effects of insulin on the transcription and protein secretion profile of mature 3T3-L1 adipocytes. In these studies, insulin was shown to have a large effect on the secretion of 3T3-L1 adipocytes but not on the transcription of secreted protein genes. This study was one of the first to use both transcriptomics and proteomics, and clearly demonstrated the added value of using a combined genomics approach in the study of biological mechanisms.

PROTEOMICS IN SERUM AND PLASMA FROM DIABETES

In type 1 diabetes

The SELDI-TOF technique was previously used to profile serum proteins from type 1 diabetic patients and healthy autoantibody-negative (AbN) controls [61]. Univariate and multivariate analyses were performed to identify putative biomarkers for type 1 diabetes and to assess the reproducibility of the SELDI technique. In this study, 146 protein/peptide peaks (581 total peaks discovered) were found in human serum that displayed statistical differences in expression levels between type 1 diabetic patients and the controls. This first attempt at high-throughput analyses of the human serum proteome in type 1 diabetic patients suggests that model averaging may be a viable method for developing biomarkers. However, the authors pointed out that the reproducibility of SELDI-TOF is currently not sufficient to be used to classify complex diseases like type 1 diabetes. More recently, Metz et al. [62] performed a comparative proteomics analysis of blood (plasma and serum) from recently diagnosed type 1 diabetic patients and controls with the goal of discovering protein biomarkers specific to type 1 diabetes. In this work, they applied the accurate mass and time (AMT) tag strategy and label-free quantitation in an initial proteomic study of a diabetes autoantibody standardization program (DASP) sample subset. Initial high-resolution capillary LC-MS/MS experiments were performed to augment an existing plasma peptide database, whereas subsequent LC-Fourier transform ion cyclotron resonance (LC-FT-ICR) studies identified quantitative differences in the abundance of plasma proteins. Five potential candidate protein biomarkers of type 1 diabetes were identified from the analysis of the LC-FT-ICR proteomic data. Although the discovery of these potential type 1 diabetes protein biomarkers is encouraging, follow-up studies are required for validation in a larger population of individuals and to determine the laboratory-defined sensitivity and specificity values using blinded samples.

In type 2 diabetes

The serum proteome has been the subject of many studies, frequently with the goal of identifying biomarkers. Matsumura et al. [63] reported on a differential proteomic analysis using male OLETF (Otsuka Long-Evans Tokushima Fatty) rats as an animal model of type 2 diabetes and associated metabolic disorders. Serum was analyzed by a new 2-D/LC system, which separated proteins by chromatofocusing and subsequent reversed-phase chromatography. Differentially expressed proteins, which were identified with MALDI-TOF/MS, included apolipoproteins and α2-HS-glycoprotein. This study demonstrated that the new LC system was a promising device in not only the discovery of serum biomarkers for various diseases but also in elucidating disease mechanisms.

In a study of serum samples from type 2 diabetic patients, Zhang et al. [64] used a combination of multidimensional liquid chromatography and gel electrophoresis coupled with MS to identify protein markers in the serum of insulin resistance/type 2 diabetes. Serum samples were fractionated using one or two columns and further separated on 1D or 2D gels and bands or spots were identified by MS analysis and database searching. In this study, the authors found that the levels of several other proteins, including fibrinogen and its fragments, α 2-macroglobulin, transthyretin, proplatelet basic protein, protease inhibitors clade A and C, as well as proteins involved in the classical complement pathway such as complement C3, C4, and C1 inhibitor, differed between the IR/D2 and the control serum. The biological meaning of using these changes as diagnostic tools or for therapeutic intervention will be discussed.

Sundsten *et al.* [4] attempted to optimize protocols for finding and identifying serum proteins that are differentially expressed in individuals with normal glucose tolerance (NGT) compared to individuals with type 2 diabetes. Serum from individuals with NGT and individuals with type 2 diabetes were profiled using ProteinChip arrays and SELDI-TOF MS. With this proteomic approach, differentially expressed proteins or biomarkers in individuals with NGT compared to type 2 diabetes were discovered. In addition, they developed a protocol for identifying these biomarkers. Because crude serum contains thousands of proteins [65], it was necessary to first reduce the sample complexity. This was achieved by fractionating the serum with anionic columns followed by either size fractionation or using reverse phase beads. Finally, the proteins were separated by one-dimensional SDS-PAGE. The identities of the differentially expressed serum proteins were obtained by PMF. Protocols for protein profiling by SELDI-TOF MS and protein identification by fractionation, SDS-PAGE, and PMF were optimized for serum from humans with type 2 diabetes. Using these protocols, differentially expressed proteins were discovered and identified when the serum from NGT and type 2 diabetes individuals was analyzed.

More recently, Sundsten *et al.* [66] compared the serum protein profiles from NGT and type 2 diabetic patients and determined the influence of the genetic background versus diabetic environment via SELDI-technique. Patients were selected based on high or low beta-cell function (HOMA-beta) and a family history of type 2 diabetes (FHD). Eight proteins were found to be elevated and five lowered ($p < 0.05$) in the serum of type 2 diabetic patients. In a second comparison, the NGT and type 2 diabetic groups were divided into subgroups of individuals with FHD and low HOMA-beta and individuals without FHD and high HOMA-beta. From this analysis, three proteins were rediscovered and found to be different due to genetic background. Two of these were identified as apolipoprotein C3 (apoC3) and albumin. Ten proteins were determined not to be related to FHD, one of which was identified as transthyretin.

In a more recent study by Sundsten *et al.*, [67] SELDI-TOF MS was used to investigate variations in plasma proteins in subjects with type 2 diabetes. The subjects were subsequently characterized by different EIRs (early insulin responses) after an oral glucose challenge as an indicator of β-cell function. Using this approach, these authors identified differentially expressed proteins when the plasma from subjects with NGT+high EIR were compared to subjects with type 2 diabetes+high or +low EIR. The results of this study showed that several plasma proteins between subjects with NGT and type 2 diabetes were different when characterized by differences in EIR. These differences are most likely due to manifestations of the disease state rather than being causative.

PROTEOMICS OF DIABETIC COMPLICATIONS

Diabetes can cause long-term complications such as kidney, eye, nerve, and heart disease. However, recently developed techniques have led to the discovery of biomarkers for diabetic complications [4]. The application of these biomarkers may lead to the detection of diabetic complications at an earlier stage and may also reduce the need to achieve hard clinical endpoints in clinical trials. This section will focus on the use of proteomics in the quest to better understand the pathogenesis of diabetic complication.

PROTEOMICS IN DIABETIC NEPHROPATHY

Diabetic nephropathy (DN) is also one of the main causes of morbidity and mortality in diabetic patients [68]. Approximately 40% of type 2 diabetes patients will develop diabetic kidney disease [69]. Lafitte *et al.* [70] previously compared the urine protein profiles from patients with DN and only a minimal change in the profiles was observed relative to normal urine; however, only one sample in each disease group was assessed. In this study, differences in protein spot patterns among groups were identified using 2-DE, but no disease-specific markers were reported. Mischak *et al.* [71] utilized an online combination of LC-MS/MS, which allowed for a rapid and accurate evaluation of up to 2000 polypeptides in the urine, to examine the urine samples of 39 healthy individuals and 119 patients with type 2 diabetes. In this study, the relationship between the polypeptide pattern in the urine and the different degrees to which albumin was excreted was closely examined. Using this method, these researchers established a normal polypeptide pattern in the urine samples of healthy subjects. However, in type 2 diabetic patients with normal albumin excretion rates, the polypeptide pattern in urine was significantly different than that of normal subjects; thus indicating that there is a specific pattern of diabetic polypeptide excretion. As a result of this work, these authors demonstrated that it was possible to detect a polypeptide pattern indicative of DN in patients with higher-grade albuminuria. This pattern was also detected in 35% of patients with low-grade albuminuria and in 4% of patients with normal albumin excretion. Similar studies were conducted in patients with type 1 diabetes [72]. Future studies involving the application of LC-MS/MS will allow for the earlier and more accurate detection of individuals at risk of developing DN.

Recently, Thongboonkerd *et al.* [73] evaluated the applicability of using microfluidic technology on a chip to profile the urinary proteome of human urine from thirty-one normal healthy individuals, six diabetics with nephropathy, and four patients with IgA nephropathy (IgAN). The results of this study showed that microfluidic technology could differentiate the urinary proteome profiles of the three different groups, and potential disease-specific urinary biomarkers could be determined using this technology. These findings suggest that microfluidic chip technology is applicable for urinary proteome profiling, with potential use in both clinical diagnostics and biomarker discovery. The principles of this LabChip microfluidics-based system are to some degree different from another chip-based technology, the "ProteinChip" or SELDI-TOF/MS system, which has been applied previously to urinary proteome profiling [74–77]. Whereas the LabChip microfluidic system is predicated based on electrophoretic separation, fluorescence detection, and the quantification of all proteins in the protein mixture or biofluid, SELDI-TOF/MS combines MALDI-TOF/MS with surface retentate chromatography. Unlike SELDI-TOF/MS, which analyzes only a subset of proteins retained on the chip, the LabChip microfluidic system can be utilized to analyze all proteins present in the sample [73]. However, the authors pointed out several technical concerns and limitations of the LabChip microfluidics-based system.

Otu *et al.* [78] attempted to determine whether proteomic technologies could be utilized to identify novel urine proteins associated with the development of DN in subjects with type 2 diabetes prior to the appearance of microalbuminuria, via SELDI-TOF/MS. A total of 714 unique urine protein peaks were detected via SELDI-TOF/MS. Among these, a 12-peak proteomic signature correctly predicted 89% of DN cases (93% sensitivity, 86% specificity) in the training set. The application of this signature to the independent validation set yielded a 74% accuracy rate (71% sensitivity, 76% specificity). In the multivariate analyses, the 12-peak signature was independently associated with subsequent DN when applied to the validation set, as well as the entire dataset. This study was the first to characterize the specific proteins involved in this early detection.

In a more recent study by Rossing *et al.* [79], differences in the urinary proteome between patients with normo-, micro-, and macroalbuminuric and type 1 diabetes were detected via CE-MS. Furthermore, they sought to evaluate whether CE-MS–defined patterns derived from the urinary polypeptides of patients with DN differed from those of patients with other chronic renal diseases. A panel of forty biomarkers distinguished patients with diabetes from healthy individuals with 89% sensitivity and 91% specificity. Among diabetic patients, 102 urinary biomarkers differed significantly between patients with normoalbuminuria and nephropathy, and the model used, which included 65 of these correctly identified DN, and had a 97% sensitivity and specificity. In addition, this panel of biomarkers identified patients who had microalbuminuria and diabetes that progressed toward overt DN over 3 years. Differentiation between DN and other chronic renal diseases reached 81% sensitivity and 91% specificity. To identify urine biomarkers for the early renal alterations in type 2 diabetes, Bellei *et al.* [80] performed urinary proteomic analysis in ten normoalbuminuric patients with type 2 diabetes, twelve patients with type 2 DN (T2DN), and twelve healthy subjects. Proteins were separated by 2-DE and identified using ESI-Q-TOF MS/MS. These studies showed that analysis of the urinary proteome may allow early detection of DN and may provide prognostic information.

To date, only two proteomic analyses of human sera in association with diabetic nephropathy have been reported [81,82]. Cho *et al.* [81] and Kim *et al.* [82] utilized proteomics techniques to identify human serum protein markers for a more specific and accurate prediction of progressive nephropathy in diabetes patients, which was monitored by determining albuminuria and creatinine concentrations in the blood. In addition, differentially expressed proteins were identified via ESI-MS/MS analysis, and the proteins, which could be utilized as diagnostic biomarkers for T2DN, were verified via Western blotting and ELISA. Cho *et al.* [81] analyzed the proteomes of sera collected from type 2 diabetes patients with normoalbuminuria ($n = 30$) and microalbuminuria ($n = 30$) by proteomic analysis (using 2-DE and ESI-Q-TOF MS) in order to identify a more accessible diagnostic marker and to delineate the pathogenic proteins of DN. A total of eighteen spots were expressed differentially in microalbuminuric patients, where twelve spots had ~50% lower protein level and six spots had ~100%–300% higher levels, compared to normoalbuminuric patients. Among these proteins, the presence of the vitamin D-binding protein (DBP) and pigment epithelium-derived factor (PEDF), which are found at various stages of diabetes, were confirmed via Western blotting. In particular, DBP expression levels were significantly higher in the microalbuminuric group than in the normoalbuminuric group. In addition, Kim *et al.* [82] conducted studies to identify other serum proteins that might enhance the accuracy of predicting the developmental and progressive stages of DN, by including chronic renal failure (CRF) patients. They assumed that the quantity of potential biomarker(s) would gradually increase or decrease in the serum proteome from the normoalbuminuric group ($n = 30$) to the microalbuminuric group ($n = 29$) and the CRF group ($n = 31$), based on the nature of the progressive development and the severity of nephropathy. In this study, proteins, which were differentially expressed at statistical significant level ($p < 0.05$) in the microalbuminuric and CRF patient groups compared to those in the normoalbuminuric group (Fig. 16.5), were selected and identified via ESI-Q-TOF MS/MS (Tables 16.1 and 16.2). Among these proteins, extracellular glutathione peroxidase (eGPx) and apolipoprotein (ApoE), which downregulated proteins in the early and progressive stages of DN, were confirmed by Western blotting. Notably, eGPx was further confirmed via ELISA using normoalbuminuric ($n = 100$) and microalbuminuric ($n = 96$) patient samples [82]. These results are completely consistent with the data from the 2-DE and Western blotting tests, and further suggest that eGPx may be utilized as a potential biomarker for the early diagnosis of DN.

In a follow-up study, Yang *et al.* [83] used SELDI-TOF-MS to define and validate a DN-specific protein pattern in the serum from sixty-five patients with DN and sixty-five non-DN subjects. From signatures based on protein/polypeptide mass, a decision tree model was established for diagnosing the presence of DN. The authors estimated the proportion of correct classifications from the model by applying it to a masked group of twenty-two patients with DN and twenty-eight non-DN subjects. The weak cationic exchange (CM10) ProteinChip arrays were performed on a ProteinChip PBS IIC reader. Based on spot intensities, twenty-two detected peaks appeared upregulated, whereas twenty-four peaks were downregulated by more than twofold ($P < 0.01$) in the DN group compared to the non-DN groups. The algorithm identified a diagnostic DN pattern

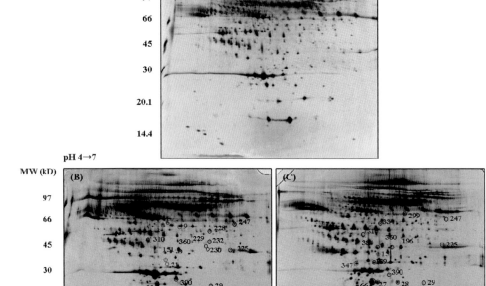

Figure 16.5 2-DE protein profiles of serum samples from the normoalbuminuric patient group (A), microalbuminuric patient group (B), and CRF patient group (C). Proteins (60 μg) were separated via IEF using 24 cm, pH 4–7 IPG strips, and 11%–16% gradient SDS-PAGE. The gels were visualized using silver staining, and their maps were analyzed with Image Master 2D Elite Software (GE Healthcare, Sweden). (From Ref. 82.) For color detail, please refer to the color plate section.

of six protein/polypeptide masses. In the masked assessment, prediction models, which were based on these protein/polypeptides, achieved a sensitivity of 90.9% and specificity of 89.3%. These observations suggest that DN patients have a unique cluster of molecular components in the serum, which are present in their SELDI profile. Identification and characterization of these molecular components will aid in achieving a better understanding of the pathogenesis of DN. The serum protein signature, combined with a tree analysis pattern, may provide a novel clinical diagnostic approach for detecting DN.

PROTEOMICS IN DIABETIC RETINOPATHY

Diabetic retinopathy (DR) is a major side effect of diabetes and is a leading cause of acquired blindness. Blood-retinal barrier breakdown is a hallmark of this microvascular disease and around 85% of all diabetics eventually develop retinopathy [84]. Few proteome analyses in human vitreous fluid have been performed in relation to diabetic eye disease [85–88], and no previous quantitative proteomic comparison has been reported. Yamane et al. [87] created a map of vitreous proteins and conducted a proteome analysis of human vitreous samples

Table 16.1 Identification of the downregulated proteins in the microalbuminuric and CRF patient groups compared with those in the normoalbuminuric patient group using ESI-Q-TOF MS/MS

Spot No.	Identification	MW (kDa)	pI	Accession No.	Score[a]
27	c-type lectin domain family 3, member B	22.9	5.5	NP_003269	68
28	c-type lectin domain family 3, member B	22.9	5.5	NP_003269	68
29	c-type lectin domain family 3, member B	22.9	5.5	NP_003269	68
38	Apolipoprotein CIII	8.8	4.7	1204193A	115
66	Retinol Binding Protein 4	23.2	5.4	CAA24959	344
69	Proapo-A-I protein	30.7	5.4	AAA35545	47
72	Proapo-A-I protein	30.7	5.5	AAA35545	172
139	Ficolin 3 precursor	33.4	6.2	O75636	76
196	Histidine-rich glycoprotein	40.2	3.9	CAA78925	61
225	Ficolin 3 precursor	33.4	6.2	O75636	121
228	Complement factor H-related 1	38.8	7.8	NP_002104	96
229	Ficolin 3 precursor	33.4	6.2	O75636	121
230	Ficolin 3 precursor	33.4	6.2	O75636	121
232	Haptoglobin-related protein precursor	39.3	6.7	AAC27433	111
247	Hemopexin precursor	52.3	6.6	AAA52704	55
299	Ficolin 3 precursor	33.4	6.2	O75636	76
310	Complement factor I light chain	22.9	5.2	1202205A	117
334	Sex hormone-binding globulin	43.9	6.2	AAC18778	122
360	Apolipoprotein E	36.3	5.7	AAB59518	194
390	Plasma glutathione peroxidase precursor	25.8	8.2	P22352	100

[a] Score is $-10 \times \log(P)$, where P is the absolute probability that the observed match between the experimental data and the database sequence is a random event. The NCBInr database is used through the MASCOT searching program (http://www.matrix.com) with ESI-Q-TOF MS/MS data as an input. (From Ref. 82.)

with the aim of understanding the underlying molecular mechanism of proliferative diabetic retinopathy (PDR) and to provide clues for the development of new therapeutic approaches to treat PDR. The expressed proteins in the vitreous samples were separated by 2-DE, and proteins that were expressed at different levels in the PDR groups were identified by MS. In PDR eyes, the observed increase in protein expression was due to a barrier dysfunction and/or

Table 16.2 Identification of the upregulated proteins in microalbuminuric and CRF patients using ESI Q-TOF MS/MS

Spot No.	Identification	MW (kDa)	pI	Accession No.	Score
6	Pigment epithelium-derived factor	46.5	5.8	AAA60058	54
9	Pigment epithelium-derived factor	46.5	5.8	AAA60058	135
21	Complement component C4B3	47.9	5.8	AAR89101	183
151	Adiponectin precursor	26.5	5.4	NP_004788	19
303	Beta 2-microglobulin	12.9	5.8	CAA23830	81
347	Complement component C4A	19.4	6.7	AAA51855	125

(From Ref. 82.)

production in the eye. Proteome analysis was found to be useful in systematically screening the expression of various proteins in human vitreous samples. In a similar study, García-Ramírez *et al.* [89] used DIGE to compare the protein profile of vitreous fluid from diabetic patients with PDR to that from nondiabetic patients with idiopathic macular holes (MH). This technique provided an accurate quantitative comparison of the two groups of samples because proteins whose levels of expression were significantly different between the two conditions were readily identified. In addition, Kim *et al.* [90] compared the proteome profiles of the vitreous humors of PDR and nondiabetic controls. In this study, the authors used proteomic methods, including immunoaffinity subtraction (IS)/2-DE/MALDI-MS, nano-LC-MALDI-MS/MS, and nano-LC-ESI-MS/MS, to comprehensively examine vitreous humor proteomes of PDR patients and nondiabetic controls. The identified proteins were subjected to the trans-proteomic pipeline validation process. In conclusion, these authors presented the possibility that differentially expressed proteins might be useful diagnostic or prognostic markers and aid in the management of DR.

Ahn *et al.* [91] screened diabetic sera for the presence of antiretinal autoantibodies with the aim of developing diagnostic markers for DR. Immunoblot analysis of the sera from DR patients with human retinal cytosolic proteins revealed a higher incidence of antiretinal autoantibodies when compared to normal blood donors or diabetic patients without DR. The antiretinal protein autoantibody profiles in the sera of DR patients were obtained by 2-DE immunoblot analysis. Specifically, twenty protein spots that were reactive with the sera from DR patients were identified by ESI-MS/MS. Of these spots, fourteen were specific for DR patients, and four reacted with both nonproliferative DR (non-PDR) and PDR sera. The antialdolase autoantibody was selected as a potential DR marker candidate, and the specific reactivity of the sera from DR patients was confirmed by immunoblot analysis with rabbit aldolase. The serum antialdolase autoantibody level was measured by ELISA. DR patients showed significantly higher autoantibody levels than normal donors or diabetic patients without retinopathy. However, no significant differences were observed between non-PDR and PDR patients, suggesting that the level of antialdolase autoantibody was not determined by the severity of retinopathy in diabetic patients. These findings may result in the development of an antialdolase autoantibody that can be employed as a useful marker for DR diagnosis.

More recently, Decanini *et al.* [92] conducted a comparative proteomics study of human donor RPE to begin defining molecular changes associated with preretinopathic diabetes. The RPE was dissected from diabetic human donor eyes that had no clinically apparent DR ($n = 6$) and from eyes of age-matched control donors ($n = 17$). Soluble proteins were separated based upon their mass and charge using 2-DE. Protein spots were visualized with a fluorescent dye and spot densities were compared between diabetic and control gels. Eighteen proteins with significant disease-related changes in density were identified using MS. The changes identified in the RPE proteome of preretinopathic diabetic donor eyes when compared with age-matched controls suggest specific cellular alterations that may contribute to DR. Defining the preretinopathic changes that affect RPE could provide important insight into the molecular events that lead to this disease.

Using retina from diabetic rats, Liu *et al.* [93] attempted to establish and optimize a 2-DE technique specific for retina proteomics. The goal of this work was to improve the resolution and reproducibility of the 2-DE technique, in addition to observing the proteomic changes of retinal tissues in diabetic and normal rats. Using this method, a difference in protein expression was observed in the retinas from diabetic and normal rats. This novel 2-DE technique could also be effectively applied for the proteomic analysis of all retina diseases. However, these authors did not identify differently expressed proteins. Quin *et al.* [94] employed proteomics to establish a proteome map of the normal rat retina. This baseline map was then compared to the proteome of an early diabetic rat retinal. The diabetic rat retina was obtained from Dark Agouti rats that had been subjected to STZ-induced hyperglycemia. Extracted proteins from normal and diabetic rat retina were separated and compared via 2-DE. A total of 145 protein spots were identified in the normal rat retina using MALDI-MS and database matching. LC-coupled ESI-MS increased the repertoire of identified proteins by 23 from 145 to 168. By comparing with early diabetic rat retina, 24 proteins unique to the diabetic gel and 37 proteins absent from the diabetic gels were found. Uniquely expressed proteins that were identified included the HSPs 70.1A and 8 and the platelet-activating factor. In the diabetic gels, there were eight spots that displayed an increased expression and twenty-seven displaying a decreased expression. Identifying such changes in protein expression has provided new insights and a more comprehensive understanding of the pathogenesis of diabetic retinopathy; thus, widening the scope of potential avenues for the development of new therapies to treat this common cause of blindness. Wang *et al.* [95] compared the protein expression profiles of the neural retina in an SD rat model of type 2 diabetes with nondiabetic rats via 2-DE. Some protein spots that exhibited statistically significant variations ($p < 0.05$) were randomly selected and identified by MS/MS. Some of the proteins that were identified in this study may lead to novel therapeutic targets for the treatment or prevention of DR.

PROTEOMICS IN DIABETIC CARDIOMYOPATHY

Diabetic cardiomyopathy (DC) is a common cardiac complication that leads to markedly in-creased heart failure and death among diabetics. The direct damaging effect of diabetes on the heart was first proposed in 1972 [96], and these conclusions were based on the postmortem findings that heart failure occurred in diabetic patients free of coronary artery disease or other known cardiac risk factors. That and numerous following studies, which reported similar find-ings [97–99], strongly support the existence of a specific cardiomyopathy with origins in diabetic cardiac muscle. Despite the negative implications of DC, the pathogenesis of this disease is still not well characterized. Shen *et al.* [100] performed proteomic analysis on total cardiac proteins from the OVE26 mouse model of type 1 diabetes to identify changes in protein expression that may contribute to DC using 2-DE. Protein spots that were significantly altered were identified by MALDI-TOF/MS using a Micromass Tof-Spec 2E instrument with a 337 mM N_2 laser. From this proteomic analysis, they found that more than eleven mitochondrial proteins were signif-icantly more abundant in the OVE26 heart. These results suggest that diabetes might induce mitochondrial biogenesis. In addition, to identify cardiac protein changes that may accompany DC, Hamblin *et al.* [101] analyzed the left ventricular protein expression profiles from STZ-diabetic rats using a proteomics approach (DIGE/MS). A comparison of the protein expression in the left ventricles from STZ-diabetic rats ($n = 5$ individual animals) with controls ($n = 5$ individual animals) by MS revealed differential expression of forty-two protein spots between the two groups. Interestingly, twelve of these proteins were localized to the mitochondria, and all but four were upregulated in the myocardium of STZ-diabetic rats. These results indicate that cardiac mitochondrial function may be modified during type 1 DC.

However, most proteomics studies on diabetic complication examined only DR and DN. Therefore, the use of a proteomic approach in studying other complications associated with diabetes is necessary to better understand the pathogenesis of diabetic complications.

CONCLUSIONS

Diabetes mellitus is a complex disorder that arises from various causes, including dysregulated glucose sensing or insulin secretion, autoimmune-mediated β-cell destruction, or insufficient compensation for peripheral insulin resistance. The signaling pathways linking obesity, periph-eral insulin action, and β-cell function are important to understand. Using proteomic approaches, it is possible to identify different biological pathways involved in the pathogenesis of diabetes in different key organs. Several new proteins, which have never been associated with diabetes before, were also identified using this technique, and these proteins may become new targets in the prevention of diabetes. In addition, proteomic analysis has provided new insights into the pathophysiology of diabetes. Finally, proteomic techniques will certainly help reach the ultimate goal of identifying all the proteomes of the pancreatic islet, β-cells, target tissues or cells and serum, involved in pathogenesis of diabetes, as well as helping to unravel the protein pathways involved in the development of diabetes and diabetic complications, which is an essential prerequisite for the development of new therapeutic agents.

REFERENCES

1. Mitrakou, A., Kelley, D., Mokan, M. *et al.* (1992) Role of reduced suppression of glucose production and diminished early insulin release in impaired glucose tolerance. *N. Engl. J. Med. 326*, 22–29.
2. Del, Prato, S., Tiengo, A. (2001) The importance of first-phase insulin secretion: implications for the therapy of type 2 diabetes mellitus. *Diabetes Metab. Res. Rev. 17*, 164–174.
3. Cerasi, E., Luft, R. (1967) The plasma insulin response to glucose infusion in healthy subjects and in diabetes mellitus. *Acta. Endocrinol. (Copenh) 55*, 278–304.
4. Sundsten, T., Eberhardson, M., Göransson, M., Bergsten, P. (2006) The use of proteomics in identifying differentially expressed serum proteins in humans with type 2 diabetes. *Proteome Sci. 4*, 22.
5. Trayhurn, P. (2000) Proteomics and nutrition – a science for the first decade of the new Millennium. *Br J Nutr 83*, 1–2.
6. O'Farrell, P.H. (1975) High resolution two-dimensional electrophoresis of proteins. *J. Biol. Chem. 250*, 4007–4021.
7. Cho, C.W., Kim, C.W. (2006) Toxicoproteomics in the study of aromatic hydrocarbon toxicity. *Biotechnol. Bioprocess Eng. 11*, 187–198.
8. Karas, M., Hillenkamp, F. (1998) Laser desorption ionization of proteins with molecular masses exceeding 10,000 daltons. *Anal. Chem. 60*, 2299–2301.
9. Fenn, J.B., Mann, M., Meng, C.K. *et al.* (1989) Electrospray ionization for mass spectrometry of large biomolecules. *Science 246*, 65–71.
10. Loboda, A.V., Krutchinsky, A.N., Bromirski, M. *et al.* (2000) A tandem quadrupole/time-of-flight mass spectrometer with a matrix-assisted laser desorption/ionization source: Design and performance. *Rapid. Commun. Mass Spectrom. 14*, 1047–1057.
11. Joo, W.A., Lee, J.B., Park, M. *et al.* (2007) Comparison of search engine contributions in protein mass fingerprinting for protein identification. *Biotechnol. Bioprocess Eng. 12*, 125–130.
12. Baldwin, M.A., Medzihradszky, K.F., Lock, C.M. *et al.* (2001) Matrix-assisted laser desorption/ionization coupled with quadrupole/orthogonal acceleration time-of-flight mass spectrometry for protein discovery, identification, and structural analysis. *Anal. Chem. 73*, 1707–1720.
13. No authors listed (1997) Report of the Expert Committee on the Diagnosis and Classification of Diabetes Mellitus. *Diabetes Care. 20*, 1183–1197.
14. National Diabetes Data Group (1995) *Diabetes in America.* (2d ed.) Bethesda, MD. National Institutes of Health, National Institute of Diabetes and Digestive and Kidney Diseases. NIH publication no. 95-1468.
15. Eberhart, M.S., Ogden, C., Engelgau, M. *et al.* (2004) Prevalence of Overweight and Obesity Among Adults with Diagnosed Diabetes — United States, 1988–1994 and 1999–2002. *MMWR 53*, 1066–1068.
16. Peraldi, P., Spiegelman, B. (1998) TNF-alpha and insulin resistance: summary and future prospects. *Mol. Cell. Biochem. 182*, 169–175.
17. Le Roith, D., Zick, Y. (2001) Recent advances in our understanding of insulin action and insulin resistance. *Diabetes Care 24*, 588–597.
18. Weyer, C., Funahashi, T., Tanaka, S. *et al.* (2001) Hypoadiponectinemia in obesity and type 2 diabetes: Close association with insulin resistance and hyperinsulinemia. *J. Clin. Endocrinol. Metab. 86*, 1930–1935.

19. Kim, S.W., Hwang, H.J., Baek, Y.M. *et al.* (2008) Proteomic and transcriptomic analysis for streptozotocin-induced diabetic rat pancreas in response to fungal polysaccharide treatments. *Proteomics 8*, 2344–2361.

20. Shin, J.S., Lee, J.J., Lee, E.J. *et al.* (2005) Proteome analysis of rat pancreas induced by pancreatectomy. *Biochim. Biophys. Acta.* 1749, 23–32.

21. Yang, M., Liu, W., Wang, C.Y. *et al.* (2006) Proteomic analysis of differential protein expression in early process of pancreatic regeneration in pancreatectomized rats. *Acta. Pharmacol. Sin. 27*, 568–578.

22. Hohmeier, H.E., Newgard, C.B. (2004) Cell lines derived from pancreatic islets. *Mol. Cell. Endocrinol. 228*, 121–128.

23. Tomita, T., Kimmel, J.R., Friesen S.R., Mantz Jr., F.A. (1980) Pancreatic polypeptide cell hyperplasia with and without watery diarrhea syndrome. *J. Surg. Oncol. 14*, 11–20.

24. Mose Larsen, P., Fey, S.J., Larsen, M.R. *et al.* (2001) Proteome analysis of IL-1 ß induced changes in protein expression in rat islets of Langerhans. *Diabetes 50*, 1056–1063.

25. Sparre, T., Christensen, U.B., Larsen, P.M. *et al.* (2002) IL-1 ß induced protein changes in diabetes prone BB rat islets of Langerhans identified by proteome analysis. *Diabetologia 45*, 1550–1561.

26. Andersen, H.U., Fey, S.J., Mose Larsen, P. *et al.* (1997) Interleukin-1 ß induced changes in the protein expression of rat islets: A computerized database. *Electrophoresis 18*, 2091–2103.

27. Nerup, J., Mandrup-Poulsen, T., Helqvist, S. *et al.* (1994) On the pathogenesis of IDDM. *Diabetologia 37*(Suppl 2), S82–S89.

28. Sanchez, J.C., Converset, V., Nolan, A. *et al.* (2002) Effect of rosiglitazone on the differential expression of diabetes-associated proteins in pancreatic islets of C57Bl/6 lep/lep mice. *Mol. Cell. Proteomics 1*, 509–516.

29. Sanchez, J.C., Chiappe, D., Converset, V. *et al.* (2001) The mouse SWISS-2D PAGE database: A tool for proteomics study of diabetes and obesity. *Proteomics 1*, 136–163.

30. Nicolls, M.R., D'Antonio, J.M., Hutton, J.C. *et al.* (2003) Proteomics as a tool for discovery: proteins implicated in Alzheimer's disease are highly expressed in normal pancreatic islets. *J. Proteome Res. 2*, 199–205.

31. Ahmed, M., Bergsten, P. (2005) Glucose-induced changes of multiple mouse islet proteins analysed by two-dimensional gel electrophoresis and mass spectrometry. *Diabetologia 48*, 477–485.

32. Ahmed, M., Forsberg, J., Bergsten, P. (2005) Protein profiling of human pancreatic islets by two-dimensional gel electrophoresis and mass spectrometry. *J. Proteome Res. 4*, 931–940.

33. Metz, T.O., Jacobs, J.M., Gritsenko, M.A. *et al.* (2006) Characterization of the human pancreatic islet proteome by two-dimensional LC/MS/MS. *J. Proteome Res. 5*, 3345–3354.

34. Sparre, T., Bergholdt, R., Nerup, J., Pociot, F. (2003) Application of genomics and proteomics in type 1 diabetes pathogenesis research. *Expert Rev. Mol. Diagn. 3*, 743–757.

35. Xie, X., Li, S., Liu, S. *et al.* (2008) Proteomic analysis of mouse islets after multiple low-dose streptozotocin injection. *Biochim. Biophys. Acta.* 1784, 276–284.

36. Lu, H., Yang, Y., Allister, E.M. *et al.* (2008) The identification of potential factors associated with the development of type 2 diabetes: A quantitative proteomic approach. *Mol. Cell. Proteomics 7*, 1434–1451.

37. Fernandez, A.M., Kim, J.K., Yakar, S. *et al.* (2001) Functional inactivation of the IGF-I and insulin receptors in skeletal muscle causes type 2 diabetes. *Genes Dev. 15*, 1926–1934.

38. Chatterjee, N.K., Haley, T.M., Nejman, C. (1985) Functional alterations in pancreatic ß cells as a factor in virus-induced hyperglycemia in mice. *J. Biol. Chem. 260*, 12786–12791.

39. Iynedjian, P.B., M.öbius, G., Seitz, H.J. *et al.* (1986) Tissue-specific expression of glucokinase: Identification of the gene product in liver and pancreatic islets. *Proc. Natl. Acad. Sci. U. S. A. 83*, 1998–2001.

40. Vischer, U., Blondel, B., Wollheim, C.B. *et al.* (1987) Hexokinase isoenzymes of RIN-m5F insulinoma cells. Expression of glucokinase gene in insulin-producing cells. *Biochem. J. 241*, 249–255.

41. Escurat, M., Djabali, K., Huc, C. *et al.* (1991) Origin of the ß cells of the islets of Langerhans is further questioned by the expression of neuronal intermediate filament proteins, peripherin and NF-L, in the rat insulinoma RIN5F cell line. *Dev. Neurosci. 13*, 424–432.

42. Guest, P.C., Bailyes, E.M., Rutherford, N.G., Hutton, J.C. Insulin secretory granule biogenesis. Co-ordinate regulation of the biosynthesis of the majority of constituent proteins. *Biochem. J. 274* (Pt. 1), 73–78.

43. Regazzi, R., Vallar, L., Ullrich, S., Ravazzola, M. *et al.* (1992) Characterization of small-molecular-mass guanine-nucleotide-binding regulatory proteins in insulin-secreting cells and PC12 cells. *Eur. J. Biochem. 208*, 729–737.

44. Collins, H.W., Buettger, C., Matschinsky, F. (1990) High-resolution two-dimensional polyacrylamide gel electrophoresis reveals a glucose-response protein of 65 kDa in pancreatic islet cells. *Proc. Natl. Acad. Sci. U. S. A. 87*, 5494–5498.

45. Collins, H., Najafi, H., Buettger, C. *et al.* (1992) Identification of glucose response proteins in two biological models of ß-cell adaptation to chronic high glucose exposure. *J. Biol. Chem. 267*, 1357–1366.

46. Fernandez, C., Fransson, U., Hallgard, E. *et al.* (2008) Metabolomic and proteomic analysis of a clonal insulin-producing beta-cell line (INS-1 832/13). *J. Proteome Res. 7*, 400–411.

47. Morand, J.P., Macri, J., Adeli, K. (2005) Proteomic profiling of hepatic endoplasmic reticulumassociated proteins in an animal model of insulin resistance and metabolic dyslipidemia. *J. Biol. Chem. 280*, 17626–17633.

48. Edvardsson, U., Alexandersson, M., Brockenhuus, von L. H. *et al.* (1999) A proteome analysis of livers from obese (ob/ob) mice treated with the peroxisome proliferator WY14,643. *Electrophoresis 20*, 935–942.

49. Edvardsson, U., von Löwenhielm, H.B., Panfilov, O. *et al.* (2003) Hepatic protein expression of lean mice and obese diabetic mice treated with peroxisome proliferator-activated receptor activators. *Proteomics 3*, 468–478.

50. Fey, S.J., Larsen, P.M. (2001) 2D or not 2D. Two-dimensional gel electrophoresis. *Curr. Opin. Chem. Biol. 5*, 26–33.

51. Hittel, D.S., Hathout, Y., Hoffman, E.P., Houmard, J.A. (2005) Proteome analysis of skeletal muscle from obese and morbidly obese women. *Diabetes 54*, 1283–1288.

52. Sundsten, T., Ortsäter, H. (2008) Proteomics in diabetes research. *Mol. Cell. Endocrinol.* In press.

53. Stentz, F.B., Kitabchi, A.E. (2007) Transcriptome and proteome expressions involved in insulin resistance in muscle and activated T-lymphocytes of patients with type 2 diabetes. *Genomics Proteomics Bioinformatics. 5*, 216–235.

54. Lynch, C.J., Brennan, W.A. Jr., Vary, T.C. *et al.* (1993) Carbonic anhydrase III in obese Zucker rats. *Am. J. Physiol. 264*, E621–E630.

55. Edvardsson, U., Alexandersson, M., Brockenhuus von Lowenhielm, H. *et al.* (1999) A proteome analysis of livers from obese (ob/ob) mice treated with the peroxisome proliferator WY14, 643. *Electrophoresis 20*, 935–942.

56. Korc, M. (2003) Diabetes mellitus in the era of proteomics. *Mol. Cell. Proteomics 2*, 399–404.

57. Blackshear, P.J., Haupt, D.M. (1989) Evidence against insulin-stimulated phosphorylation of calmodulin in 3T3-L1 adipocytes. *J. Biol. Chem. 264*, 3854–3858.

58. Sadowski, H.B., Wheeler, T.T., Young, D.A. (1992) Characterization of initial responses to the inducing agents and changes during commitment to differentiation. *J. Biol. Chem. 267*, 4722–4731.

59. Levenson, R.M., Blackshear, P.J. (1989) Insulin-stimulated protein tyrosine phosphorylation in intact cells evaluated by giant two-dimensional gel electrophoresis. *J. Biol. Chem. 264*, 19984–19993.

60. Wang, P., Keijer, J., Bunschoten, A. *et al.* (2006) Insulin modulates the secretion of proteins from mature 3T3-L1 adipocytes: a role for transcriptional regulation of processing. *Diabetologia 49*, 2453–2462.

61. Purohit, S., Podolsky, R., Schatz, D. *et al.* (2006) Assessing the utility of SELDI-TOF and model averaging for serum proteomic biomarker discovery. *Proteomics 6*, 6405–6415.

62. Metz, T.O., Qian, W.J., Jacobs, J.M. *et al.* (2008) Application of proteomics in the discovery of candidate protein biomarkers in a diabetes autoantibody standardization program sample subset. *J. Proteome Res. 7*, 698–707.

63. Matsumura, T., Suzuki, T., Kada, N. *et al.* (2006) Differential serum proteomic analysis in a model of metabolic disease. *Biochem. Biophys. Res. Commun. 351*, 965–971.

64. Zhang, R., Barker, L., Pinchev, D. et al. (2004) Mining biomarkers in human sera using proteomic tools. *Proteomics 4*, 244–256.

65. Pieper, R., Gatlin, C.L., Makusky, A.J. *et al.* (2003) The human serum proteome: display of nearly 3700 chromatographically separated protein spots on two-dimensional electrophoresis gels and identification of 325 distinct proteins. *Proteomics 3*, 1345–1364.

66. Sundsten, T., Ostenson, C.G., Bergsten, P. (2008) Serum protein patterns in newly diagnosed type 2 diabetes mellitus—influence of diabetic environment and family history of diabetes. *Diabetes Metab. Res. Rev. 24*, 148–154.

67. Sundsten, T., Zethelius, B., Berne, C., Bergsten, P. (2008) Plasma proteome changes in subjects with Type 2 diabetes mellitus with a low or high early insulin response. *Clin. Sci. (Lond) 114*, 499–507.

68. Phillips, C.A., Molitch, M.E. (2002) The relationship between glucose control and the development and progression of diabetic nephropathy. *Curr. Diabetes Rep. 2*, 523–529.

69. Lewis, E.J., Lewis, J.B. (2003) Treatment of diabetic nephropathy with angiotensin II receptor antagonist. *Clin. Exp. Nephrol. 7*, 1–8.

70. Lafitte, D., Dussol, B., Andersen, S. *et al.* (2002) Optimized preparation of urine samples for two-dimensional electrophoresis and initial application to patient samples. *Clin. Biochem. 35*, 581–589.

71. Mischak, H., Kaiser, T., Walden, M. *et al.* (2004) Proteomic analysis for the assessment of diabetic renal damage in humans. *Clin. Sci. 107*, 485–495

72. Meier, M., Kaiser, T., Herrmann, A. *et al.* (2005) Identification of urinary protein pattern in type 1 diabetic adolescents with early diabetic nephropathy by a novel combined proteome analysis. *J. Diabetes Complications 19*, 223–232.

73. Thongboonkerd, V., Songtawee, N., Sritippayawan, S. (2007) Urinary proteome profiling using microfluidic technology on a chip. *J. Proteome Res. 6*, 2011–2018.

74. Wright, G.L. Jr. (2002) SELDI proteinchip MS: a platform for biomarker discovery and cancer diagnosis. *Expert. Rev. Mol. Diagn. 2*, 549–563.

75. Fung, E., Diamond, D., Simonsesn, A.H., Weinberger, S.R. (2003) The use of SELDI ProteinChip array technology in renal disease research, Methods Mol. *Med. 86*, 295–312.

76. Tang, N., Tornatore, P., Weinberger, S.R. (2004) Current developments in SELDI affinity technology. *Mass Spectrom. Rev. 23*, 34–44.

77. Seibert, V., Wiesner, A., Buschmann, T., Meuer, J. (2004) Surface-enhanced laser desorption ionization time-of-flight mass spectrometry (SELDI TOF-MS) and ProteinChip technology in proteomics research. *Pathol. Res. Pract. 200*, 83–94.

78. Out, H.H., Can, H., Spentzos, D. *et al.* (2007) Prediction of diabetic nephropathy using urine proteomic profiling 10 years prior to development of nephropathy. *Diabetes Care 30*, 638–643.

79. Rossing, K., Mischak, H., Dakna, M. *et al.* (2008) Urinary proteomics in diabetes and CKD. *J. Am. Soc. Nephrol. 19*, 1283–1290.

80. Bellei, E., Rossi, E., Lucchi, L., Uggeri, S. et al. (2008) Proteomic analysis of early urinary biomarkers of renal changes in type 2 diabetic patients. *Proteomics Clin. Appl. 2*, 478–491.

81. Cho, E.H., Kim, M.R., Kim, H.J. *et al.* (2007) The discovery of biomarkers for Type 2 diabetic nephropathy by serum proteome analysis. *Proteomics Clin. Appl. 1*, 362–372.

82. Kim, H.J., Cho, E.H., Yoo, J.H. *et al.* (2007) Proteome analysis of serum from type 2 diabetics with nephropathy. *J. Proteome Res. 6*, 735–743.

83. Yang, Y.H., Zhang, S., Cui, J.F. *et al.* (2007) Diagnostic potential of serum protein pattern in Type 2 diabetic nephropathy. *Diabet Med. 24*, 1386–1392.
84. Tewari, H.K., Venkatesh, P. (2004) Diabetic retinopathy for general practitioners. *J. Indian. Med. Assoc. 102*, 722–723.
85. Nakanishi, T., Koyama, R., Ikeda, T., Shimizu, A. (2003) Catalogue of soluble proteins in the human vitreous humor: comparison between diabetic retinopathy and macular hole. *J Chromatogr. B Analyt. Technol. Biomed. Life Sci. 776*, 89–100.
86. Koyama, R., Nakanishi, T., Ikeda, T., Shizimu, A. (2003) Catalogue of soluble proteins in human vitreous humor by one-dimensional sodium dodecyl sulfate-polyacrylamide gel electrophoresis and electrospray ionization mass spectrometry including seven angiogenesis-regulating factors. *Chromatogr. B Analyt. Technol. Biomed. Life Sci. 792*, 5–21.
87. Yamane, K., Minamoto, A., Yamashita, H. *et al.* (2003) Proteome analysis of human vitreous proteins. *Mol. Cell. Proteomics 2*, 1177–1187.
88. Wu, C.H., Sauter, J.L., Johnson, P.K. *et al.* (2004) Identification and localization of major soluble vitreous proteins in human ocular tissue. *Am. J. Ophthalmol. 137*, 655–661.
89. García-Ramírez, M., Canals, F., Hernández, C. *et al.* (2007) Proteomic analysis of human vitreous fluid by fluorescence-based difference gel electrophoresis (DIGE): a new strategy for identifying potential candidates in the pathogenesis of proliferative diabetic retinopathy. *Diabetologia 50*, 1294–1303.
90. Kim, T., Kim, S.J., Kim, K. *et al.* (2007) Profiling of vitreous proteomes from proliferative diabetic retinopathy and nondiabetic patients. *Proteomics 7*, 4203–4215.
91. Ahn, B.Y., Song, E.S., Cho, Y.J. *et al.* (2006) Identification of an anti-aldolase autoantibody as a diagnostic marker for diabetic retinopathy by immunoproteomic analysis. *Proteomics 6*, 1200–1209.
92. Decanini, A., Karunadharma, P.R., Nordgaard, C.L. *et al.* (2008) Human retinal pigment epithelium proteome changes in early diabetes. *Diabetologia 51*, 1051–1061.
93. Liu, S., Zhang, Y., Xie, X. (2007) Application of two-dimensional electrophoresis in the research of retinal proteins of diabetic rat. *Cell Mol. Immunol. 4*, 65–70.
94. Quin, G.G., Len, A.C., Billson, F.A., Gillies, M.C. (2007) Proteome map of normal rat retina and comparison with the proteome of diabetic rat retina: new insight in the pathogenesis of diabetic retinopathy. *Proteomics 7*, 2636–2650.
95. Wang, Y.D., Wu, J.D., Jiang, Z.L. *et al.* (2007) Comparative proteome analysis of neural retinas from type 2 diabetic rats by two-dimensional electrophoresis. *Curr. Eye Res. 32*, 891–901.
96. Rubler, S., Dlugash, J., Yuceoglu, Y.Z. *et al.* (1972) New type of cardiomyopathy associated with diabetic glomerulosclerosis. *Am. J. Cardiol. 30*, 595–602.
97. Cohen, A. (1995) Diabetic cardiomyopathy. *Arch. des Maladies du Coeur. et des Vaisseaux 88*, 479–486.
98. Gargiulo, P., Jacobellis, G., Vaccari, V., Andreani, D. (1998) Diabetic cardiomyopathy: pathophysiological and clinical aspects. *Diabetes Nutr. Metab. 11*, 336–346.
99. Spector, K. S. (1998) Diabetic cardiomyopathy. *Clin. Cardiol. 21*, 885–887.
100. Shen, X., Zheng, S., Thongboonkerd, V., Xu, M. *et al.* (2004) Cardiac mitochondrial damage and biogenesis in a chronic model of type 1 diabetes. *Am. J. Physiol. Endocrinol. Metab. 287*, E896–E905.
101. Hamblin, M., Friedman, D.B., Hill, S. *et al.* (2007) Alterations in the diabetic myocardial proteome coupled with increased myocardial oxidative stress underlies diabetic cardiomyopathy. *J. Mol. Cell Cardiol. 42*, 884–895.

Section 4

Metabolomics

17

NMR-Based Metabolomics Strategy for the Classification and Quality Control of Nutraceuticals and Functional Foods

Yulan Wang and Huiru Tang

INTRODUCTION

It is widely accepted that plant-based foods, such as fruits, vegetables, soybeans, and spices, have beneficial effects on human health, especially on age-related degenerative conditions [1–6]. As a result, there has been a strong growth in consumer demand for plant-based products that are identified as beneficial to health. The current global market for plant-based health products is estimated to be between US$70 billion and US$250 billion per annum [7]. Products derived from the plants can be divided into two categories, nutraceuticals and functional foods. Nutraceuticals are diet supplements that deliver a concentrated form of presumed bioactive agents from a food in tablet form, and are used to enhance health in dosages that exceed that which is obtained from normal foods [8]. Functional foods are those that when consumed regularly exert a specific health-beneficial effect beyond their nutritional properties, and this effect must be scientifically proven (International Life Science Institute; http://www.ilsi.org). It is clear that all nutraceuticals and functional foods are mixtures of many compounds, of which many are often unknown, and these products are often marketed based on declared bioactive ingredients. Hence, those with nondeclared components may have biological effects or synergistic effects that are not known. The quality of nutraceuticals and functional foods on the market is not controlled or regulated in a holistic manner. It is therefore necessary to develop some analytical methods capable of detecting all compounds simultaneously, thus providing a means for standardizing and controlling the quality of nutraceuticals based on the entire biochemical composition of the preparation, not just bioactive ingredients only. Newly emerged metabonomics/metabolomics technologies have the potential to provide such an analytical tool for quality control of nutraceuticals.

Metabonomics involves the study of multivariate metabolic responses of complex organisms to physiological and/or pathological stressors, including the consequent disruption of systems regulation [9–12]. Metabonomics analyzes data from nuclear magnetic resonance (NMR) or mass spectroscopy (MS) using appropriate multivariate data analysis. With ^{1}H NMR spectroscopy, a wide range of plant metabolites can be detected, including sugars, amino acids, organic acids, and polyphenols. In such cases, all the chemical components present in a plant extract are viewed simultaneously as a "metabolic fingerprint." The complex "metabolic fingerprint" generated can subsequently be visualized by application of multivariate statistical data analysis to reduce the complexity of the data and to detect the pattern of changes related to environmentally or genetically induced variations in metabolite composition. In this chapter, the basic NMR techniques involved in metabonomics/metablomics, such as profiling methods and principle of data analysis, and the roles of metabonomics in quality control of nutraceuticals and functional foods, will be discussed.

NMR-BASED METABOLIC PROFILING TECHNIQUES

For nuclei having an odd number of protons and/or neutrons, an overall nuclear spin value I (e.g., $I = 1/2$, 1, 3/2, 2, 5/2, and so on) will be present. Such nuclei behave like bar magnets, hence they produce a nuclear magnetic moment. When placed in an external magnetic field, the nuclear magnetic moment will take up $2I + 1$ orientations with respect to the external field. For example, for spin $I = 1/2$ nuclei, such as proton, there will be two energy levels with a higher energy level as opposed to an external magnetic field and a lower energy level aligned with an external magnetic field. At equilibrium, for a spin $I = 1/2$ nuclei, a population of spin states will obey the Boltzmann distribution and the energy difference between the higher and lower energy states facilitates the observation of an NMR signal when such equilibrium is pertubated by a radio frequency pulse. The signal intensity depends on the external magnetic field strength and number of protons involved; hence, the area of an NMR signal is quantitatively proportional to the number of nuclei contributing to it and thus the concentrations of analytes.

In NMR experiments, the electronic environments of different proton nuclei are slightly different due to the different functional groups protons attach to. Therefore, the local magnetic field experienced by any proton will not correspond absolutely with the external field, but will diverge slightly and hence resonate at slightly different frequency values termed as chemical shift. In addition, neighboring nuclei interact via bonding electrons and carry information regarding the neighboring nuclei, hence providing molecular structure information or atomic connectivity.

A catalog of two-dimensional NMR techniques is also available to obtain more direct atomic connectivity. Among them, the most frequently employed ones include ^{1}H-^{1}H COSY, TOCSY which provides two- to three-band coupling and more extensive coupling respectively (four carbon chains for organic compounds), ^{1}H-^{13}C HSQC, which provides direct H-C bonding, and ^{1}H-^{13}C HMBC provides long-range H-C coupling information. When interspace information is required for a three-dimensiaonal structure, NOESY and ROSEY 2D NMR methods are extremely useful. Furthermore, information on interactions between analytes can also be readily extracted by measuring chemical shift changes, relaxation times, and diffusion coefficients. For complex mixtures, numerous spectral editing techniques have been developed.

DATA ANALYSIS

Because NMR metabolic profiling data are of high density and complexity, it is often desirable to extract information by application of multivariate statistical data analysis to reduce the complexity of these data and to facilitate visualization of inherent patterns in the data. Various multivariate statistical data analysis tools, including both the unsupervised and supervised methods, have been extensively applied to metabolic profiling data. Principal component analysis (PCA) [13,14], nonlinear mapping procedures (NLM) [15,16], and hierarchical cluster analysis (HCA) [13] are examples of unsupervised multivariate techniques; they are employed to establish a model without a prior knowledge of class membership and can show the intrinsic similarities or differences within a dataset. Supervised multivariate techniques include partial least squares (PLS), partial least squares-discriminant analysis (PLS-DA) [17,18], and K-nearest neighbor (KNN), which require prior knowledge of class membership. The supervised techniques use the mathematical algorithms to maximize the separation between different classes, and information on the analytes contributed to such classifications.

UNSUPERVISED ANALYSIS

Principal component analysis is one of the widely used unsupervised methods for analyzing a large dataset that shows the intrinsic similarities and differences within the dataset. A dataset normally organized as a matrix (X) contains N rows and K columns, where each row represents an NMR spectrum obtained from one sample called observation and each column denotes an integral of chemical shift region called variable. PCA is mathematically defined as an orthogonal linear transformation that transforms the data to a new coordinate system. In practice, mean centering of data matrix is required prior to PCA calculation and all the principal components must go through the center of the dataset. The first principal component (PC1) normally finds an axis that minimizes the square of the distance of each point to that axis. In other words, PC1 explains the maximum amount of variance possible in the dataset. The second PC is orthogonal to the first principal component (PC) and also describes the dataset the best, whereas the third PC must meet two more conditions: it must be orthogonal to the first two PCs and explain/show the maximum amount of variance in the dataset. Hence, each PC is a linear combination of the original variables and each successive PC is orthogonal to every other PC [19]. Data can be visualized by plotting the PC scores where each point in the scores plot represents an individual sample. Similarities and differences between samples can be detected in the score plots, for example, clustered points suggest a high similarity between the observations, whereas points far apart suggest very different profiles between the observations. The PC loadings, where each point represents one spectral region, gives an indication of the spectral variables of metabolites that most strongly influence the patterns in the scores plot. Hence, the corresponding loading plot displays the variables that dominate the score plot and reveals information about the relationship between the separation and the variables.

SUPERVISED METHOD

One widely used supervised method is partial least squares (PLS) regression, also known as "projection to latent structure." PLS is a regression extension of PCA and requires two matrices; one is the X matrix, which is identical to the matrix in PCA, and the other one is a Y matrix, which corresponds either class membership information or variables associated with the X matrix. PLS regression models the covariance structures in these two data matrixes and finds the linkage between dataset X and variable response Y. This results in a PLS model expressed in terms of weights, referring to the residuals of previous dimensions. These weights are useful for interpreting the observable variables that are influential for the modeled Y dataset.

Because PLS algorithm is a supervised method, to avoid overfitting the dataset, cross-validation must be carried out. This can be performed by a sevenfold cross validation strategy, that is, iterative construction of models by repeatedly leaving out one-seventh of the samples and predicting them back into the model.

Another commonly used supervised method of multivariable data analysis is orthogonal partial least squares (O-PLS), which is an extension of the partial least squares regression method [20] featuring an integrated orthogonal signal correction filter (OSC) [21]. OSC can remove or suppress systematic variations from the X or Y matrices that are nonrelevant to X or Y matrices [22]. The advantage of applying OSC to ^{1}H NMR data is that the resulting multivariate models are solely focused on class discrimination.

APPLICATION OF QUALITY CONTROL IN NUTRACEUTICALS AND FUNCTIONAL FOODS

Nutraceuticals and functional foods are often mixtures from a number of plant sources with incomplete defined composition. It is well known that environmental factors such as soil, climate, and cultivation methods, including origin of plant and harvest time, have a great impact on the composition of a plant as well as the nutraceutical raw materials and functional foods. In addition, extraction methods also critically affect the composition of these products. Currently, the quality control of these products is carried out based on the active ingredients presented. A drawback of this method is that some of the unknown ingredients may have potential synergic interactions with each other and may have certain biological effects. As a consequence, information on the mechanism of action of these products at the molecular levels is unknown. Thus, the quality control of both raw and final products in a holistic manner is necessary to ensure the consistence of these products and to provide a means to further understand the molecular mechanism of these products.

Metabolomics that employs ^{1}H NMR spectroscopy facilitates the simultaneous detection of chemical components present in a plant extract as a "metabolic fingerprint," and meets the requirement(s) for quality control in a holistic manner. An example of metabonomics utilization as a quality control tool for plant-based mixtures includes common herbs that have been subjected to metabonomics studies. Multiple component analysis based on the combination of high-resolution NMR spectroscopy with pattern recognition has been employed to investigate batches of fourteen commercially available feverfew samples. The results showed [23] that two batches are significantly different in chemical compositions from each other and the remaining twelve batches can also be classified into discrete groups by PCI on the basis of differences in overall chemical compositions. This investigation highlighted the powerfulness of such methods in quality control of plant extracts.

In another study, chamomile (*Matricaria recutita*) flowers were chosen because the herb is often marketed in Europe as a beverage. Previously, the quality control of chamomile usually relied on a practitioner's experience and monitoring of limited "markers" using thin layer chromatography (TLC) and high-performance liquid chromatography (HPLC) methods. In this study, metabonomics methods have been employed to profile its global compositions and show [24] clear differences between samples from Northern Africa (Egypt) and Eastern Europe (Hungary and Slovak Republic). Chamomile also has distinguished profiles from Hungary and Slovak Republic based on its metabolomic compositions despite being close in terms of geographic location [24]. Furthermore, this method is effective to monitor the "purity" of chamomile samples such as the percentage of stalks mixed with flowers [24], suggesting that this is an excellent method for authenticity and quality control. From a processing point of view, NMR-based metabonomics methods have also been extremely powerful in distinguishing samples extracted with different methods, providing a potential tool for quality control in processing. In addition, the metabonomic approach has also been employed to investigate the metabolic effects of chamomile ingestion on humans [25]. The results suggsted that chamomile ingestion is a mild intervention to the human body in general, as it causes a reduction in oxidative stress and alters the state of gut microflora. The effects of chamomile ingestion on human metabolism were also found to not be completely recoverable within a successive week after ingestion [25]. It is unknown what significance this slow recovery has at present.

Metabonomics studies have also been carried out on the extracts of *Artemisia annua* to discriminate samples from different sources and classify them according to their antiplasmodial activity, without preknowledge of this activity [26]. The use of PLS analysis also allows the predictions of actual values of such activities for some independent samples not used in the model construction.

Another study was conducted on the complex pharmaceutical preparations, St. John's Wort, using multivariate analysis of full-resolution ^{1}H-NMR spectral data [27]. The results showed that ten preparations from markets were compositionally diverse and such diversity resulted from plant extract preparation rather than postextraction processes, revealing the seriousness and urgency for quality control and powerfulness of NMR-based methods in such cases.

More recently, the NMR-based metabonomics approach has been successfully employed to reveal the variations of rosemary chemical compositions resulting from the extraction solvents, seasons, and drying processes [28]. It was found that the rosemary metabonome was dominated by thirty-three metabolites including sugars, amino acids, organic acids, polyphenolic acids, and diterpenes. Three metabolites were found in rosemary for the first time including quinate, 3,4,5-trimethoxyphenylmethanol, and *cis*-4-glucosyloxycinnamic acid. Compared with water extracts, the 50% aqueous methanol extracts of rosemary contained higher levels of sucrose, succinate, fumarate, malonate, shikimate, and phenolic acids, but lower levels of fructose, glucose, citrate, and quinate. Chloroform-methanol was an excellent solvent for selective extraction of diterpenes. From February to August, the levels of rosmarinate and quinate increased, whereas the sucrose level decreased. The sun-dried samples contained higher concentrations of rosmarinate, sucrose, and some amino acids, but lower concentrations of glucose, fructose, malate, succinate, lactate, and quinate compared to freeze-dried samples. These findings indicate that NMR-based metabonomics offers an excellent holistic method to monitor variations resulting from various environmental and postharvest processing factors. These examples are only a reflection on the developments in this area and are by no means exhaustive. In fact, many studies have also been carried out in terms of phytomedicines and authenticity and it is conceivable that such applications of metabonomics technology will be extended much further in the near future.

REFERENCES

1. Yeung, M. T.; Hobbs, J. E.; Kerr, W. A. *Journal of International Food & Agribusiness Marketing* 2007, *19*, 53–79.
2. Espin, J. C.; Garcia-Conesa, M. T.; Tomas-Barberan, F. A. *Phytochemistry* 2007, *68*, 2986–3008.
3. Ramaa, C. S.; Shirode, A. R.; Mundada, A. S.; Kadam, V. J. *Current Pharmaceutical Biotechnology* 2006, *7*, 15–23.
4. Tripathi, Y. B.; Tripathi, P.; Arjmandi, B. H. *Frontiers in Bioscience* 2005, *10*, 1607–18.
5. Huang, M. T.; Ghai, G.; Ho, C. T. *Comprehensive Reviews in Food Science and Food Safety* 2004, *3*, 127–39.
6. Hardy, G.; Hardy, I.; Ball, P. A. *Curr Opin Clin Nutr Metab Care* 2003, *6*, 661–71.
7. Losso, J. N. *Trends in Food Science & Technology* 2003, *14*, 455–68.
8. Zeisel, S. H. *Science* 1999, *285*, 1853–55.
9. Nicholson, J. K.; Lindon, J. C.; Holmes, E. *Xenobiotica* 1999, *29*, 1181–89.
10. Nicholson, J. K.; Connelly, J.; Lindon, J. C.; Holmes, E. *Nat Rev Drug Discov* 2002, *1*, 153–61.
11. Nicholson, J. K.; Wilson, I. D. *Nat Rev Drug Discov* 2003, *2*, 668–76.
12. Nicholson, J. K.; Holmes, E.; Lindon, J. C.; Wilson, I. D. *Nat Biotechnol* 2004, *22*, 1268–74.
13. Li, X. H.; Zhang, X. Z.; Cheng, X. L.; Yang, X. D.; Zh, Z. L. *Chinese Journal of Chemical Physics* 2006, *19*, 143–48.
14. Ciosek, P.; Brzozka, Z.; Wroblewski, W.; Martinelli, E.; Di Natale, C.; D'Amico, A. *Talanta* 2005, *67*, 590–96.
15. Jiang, J. H.; Wang, J. H.; Liang, Y. Z.; Yu, R. Q. *Journal of Chemometrics* 1996, *10*, 241–52.
16. Domine, D.; Devillers, J.; Chastrette, M. *Journal of Medicinal Chemistry* 1994, *37*, 973–80.
17. Holmes, E.; Nicholls, A. W.; Lindon, J. C.; Ramos, S.; Spraul, M.; Neidig, P.; Connor, S. C.; Connelly, J.; Damment, S. J. P.; Haselden, J.; Nicholson, J. K. *NMR Biomed* 1998, *11*, 235–44.
18. Tate, A. R.; Foxall, P. J. D.; Holmes, E.; Moka, D.; Spraul, M.; Nicholson, J. K.; Lindon, J. C. *NMR Biomed* 2000, *13*, 64–71.
19. Wold, S.; Esbensen, K.; Geladi, P. *Chemometr Intell Lab* 1987, *2*, 37–52.
20. Wold, S.; Ruhe, A.; Wold, H.; Dunn, W. J. *SIAM J Sci Stat Comput* 1984, *5*, 735–43.
21. Wold, S.; Antti, H.; Lindgren, F.; Ohman, J. *Chemometr Intell Lab* 1998, *44*, 175–85.
22. Trygg, J.; Wold, S. *Journal of Chemometrics* 2002, *16*, 119–28.
23. Bailey, N. J. C.; Sampson, J.; Hylands, P. J.; Nicholson, J. K.; Holmes, E. *Plant Med* 2002, *68*, 734–38.
24. Wang, Y. L.; Tang, H. R.; Nicholson, J. K.; Hylands, P. J.; Sampson, J.; Whitcombe, I.; Stewart, C. G.; Caiger, S.; Oru, I.; Holmes, E. *Plant Med* 2004, *70*, 250–55.

25. Wang, Y. L.; Tang, H. R.; Nicholson, J. K.; Hylands, P. J.; Sampson, J.; Holmes, E. *J Agri Food Chem* 2005, *53*, 191–96.
26. Bailey, N. J. C.; Wang, Y. L.; Sampson, J.; Davis, W.; Whitcombe, I.; Hylands, P. J.; Croft, S. L.; Holmes, E. *J Pharm Biomed Anal* 2004, *35*, 117–26.
27. Rasmussen, B.; Cloarec, O.; Tang, L.; Staerk D; Jaroszewski, J. W. *Planta Med* 2006, *72*, 556–63.
28. Xiao, C. N.; Dai, H.; Liu, H. B.; Wang, Y. L.; Tang, H. R. *J Agri Food Chem* 2008, *56*, 10142–53.

18
Metabolomics: An Emerging Postgenomic Tool For Nutrition

Phillip Whitfield and Jennifer Kirwan

INTRODUCTION

The interaction between nutrition and metabolism involves complex molecular processes, and understanding how food components can modulate health is an important goal. The advent of the postgenomic era has led to strategies aimed at relating gene expression to phenotypic outcome at different levels of biological organization. Transcriptomics monitors the expression levels of thousands of genes; proteomics defines changes in protein expression, posttranslational modifications, and protein dynamics, whereas metabolomics is concerned with the analysis of the metabolite complement of a biological system. The development of these technologies has created a unique opportunity to obtain an integrative view of the molecular regulation of cells and whole organisms with the aim of understanding the fundamental mechanisms of physiology and disease.

Because low molecular weight metabolites represent the end products of gene expression they can be regarded as important indicators, and indeed, integrators, of phenotype (Nicholson and Wilson 2003; Oresic *et al.* 2006). Metabolomics monitors alterations in cell function that are perhaps most evident at the level of small molecule metabolism and can provide a coherent view of the response of biological systems to a variety of genetic and environmental influences (Schnackenberg and Beger 2006). Metabolomic strategies involve the use of modern analytical techniques to measure global populations of metabolites in biological samples including amino acids, lipids, organic acids, carbohydrates, and nucleotides. Advanced bioinformatic and statistical tools are then employed to maximize the recovery of information and to aid interpretation of the large datasets that are generated.

Metabolomics is beginning to make a significant impact on the landscape of biological and biomedical research. Metabolomic approaches have been employed to study the biology of plants (Sumner *et al.* 2003; Schauer and Fernie 2006; Seger and Sturm 2007) and microbial systems (Mashego *et al.* 2007). Investigators have also used metabolomics to investigate the mechanisms of xenobiotic toxicity (Lindon *et al.* 2005; Kaddurah-Daouk *et al.* 2008), explore pathophysiological processes (Griffin and Shockor 2004; Denkert *et al.* 2006), and improve the diagnosis of disease states (Whitfield *et al.* 2005; Marchesi *et al.* 2007; Kenny *et al.* 2008). Metabolomic strategies are now being applied to the fields of nutrition and food science (Whitfield *et al.* 2004; Gibney *et al.* 2005; Rezzi *et al.* 2007; Wishart 2008).

This chapter is focussed on the developing role of metabolomics within the nutritional sciences. The experimental strategies used in the profiling of metabolite populations and the applications of metabolomics to investigate the complex relationships between nutrition and metabolism will be outlined. The ability of metabolomics to characterize functional foods and

Table 18.1 Terminology of metabolomics. Reproduce from (Ellis *et al.* 2007)

Term	Definition
Metabolomics	The nonbiased identification and quantification of all metabolites in a biological system
Metabonomics	The quantitative measurement of time-related multiparametric metabolic responses of multicellular systems to pathophysiological stimuli or genetic modification
Metabolome	The complete set of low molecular weight metabolites within, or that can be secreted by, a given cell type or tissue
Metabolic Profiling	Identification and quantification of a selective number of predefined metabolites, which are generally related to a specific metabolic pathway
Metabolic Fingerprinting	Global, high-throughput analysis to provide sample classification

to discover the roles that dietary components play in many aspects of health and disease will also be discussed.

TERMINOLOGY OF METABOLOMICS

The field of metabolomics has already produced an array of specialist terms and as with any emerging science their precise definitions are still evolving. Perhaps the best example of this relates to the terms metabolomics and metabonomics. Metabonomics was originally defined as the quantitative measurement of time-related multiparametric metabolic responses of multicellular systems to pathophysiological stimuli or genetic modification (Nicholson *et al.* 1999). Metabolomics describes the nonbiased identification and quantification of all metabolites in a biological system under a given set of conditions (Fiehn 2002). The difference might be valued by specialists within the field; however, the methods and approaches used in the two disciplines are highly convergent and the terms are often used interchangeably. Full definitions of terms used in metabolomics and related disciplines are listed in Table 18.1 (Ellis *et al.* 2007).

EXPERIMENTAL DESIGN AND METABOLOMIC TECHNOLOGIES

The experimental approaches employed in metabolomic studies in general involve several stages including sample collection, sample processing, metabolite detection, data analysis and interpretation (Fig. 18.1).

ANALYSIS OF BIOFLUIDS AND TISSUES

In mammalian systems, cerebrospinal fluid, amniotic fluid, synovial fluid, seminal fluids, digestive fluids, bile, saliva and feces, tumor biopsies, and body tissues have all been analyzed. However, as might be expected, plasma and urine are the most commonly used materials for studies involving humans and animals, as they are readily accessible and provide a cumulative picture of metabolic events.

Body fluids and tissues are estimated to contain a large number (hundreds to thousands) of endogenous metabolites that belong to multiple compound classes, such as amino acids, organic

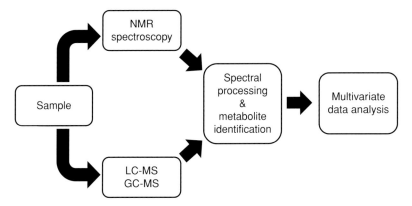

Figure 18.1 Overview of metabolomic experimental strategies. Analytical technologies such as NMR spectroscopy, LC-MS, and GC-MS are employed to characterize populations of low molecular weight metabolites in biofluids or tissues. Advanced statistical methods are used to maximize the recovery of information and interpret the complex datasets that are generated.

acids, lipids, and carbohydrates. An additional level of complexity comes from gut microflora and the consumption of foods and beverages that contribute to and alter the metabolome of mammalian biofluids and tissues.

Biological and environmental influences can also lead to marked differences in the baseline composition of body fluids and tissues (Bollard *et al.* 2005), and it is important to define metabolite variance within a normal population (Lenz *et al.* 2003). A wide range of endogenous (intrinsic) factors including diurnal cycles, gender, age, and health status (Kochhar *et al.* 2006; Slupsky *et al.* 2007) and exogenous (extrinsic) properties such as diet, lifestyle choices (e.g., alcohol consumption and smoking), and the effect of gut microflora need to be considered (Wang *et al.* 2006; Gu *et al.* 2007). Therefore, careful experimental design is critical and a standardized protocol for sample collection, storage, processing, and analysis should be established to produce valid and reproducible data for comparison in basic research or clinical investigations (Teahan *et al.* 2006; Saude and Sykes 2007; Maher *et al.* 2007).

The introduction of bias is a significant concern in the design, conduct, interpretation, and reporting of metabolomic studies. In order to provide guidance on these issues the Metabolomics Standards Initiative (http://msi-workgroups.sourceforge.net) has been established (Sansone *et al.* 2007). This international program aims to recommend standards covering all aspects of metabolomic studies from sample processing to data acquisition and reporting arrangements and has recently published a series of articles (Metabolomics Standard Initiative 2007). These guidelines are equivalent to the reporting standards currently applied in transcriptomics and proteomics. They do not detail how to perform metabolomic investigations but rather offer a framework with which to describe and report experimental work and facilitate the exchange of information between laboratories. The consultations are continuing within the metabolomics community and will develop further as the field matures.

Conventional methods for the analysis of low molecular weight metabolites target specific compound classes with isolation and detection protocols being tailored to the metabolites of interest. Metabolomics represents a radical shift from individual metabolite monitoring to complex metabolite profiling and pattern recognition and is concerned with characterizing global populations of low molecular metabolites. This is a considerable analytical challenge and involves detecting a broad spectrum of molecules with diverse chemical and physical properties and a wide dynamic range of concentrations (from subpicomolar to millimolar).

However, it is not possible to measure all low molecular weight metabolites in a biological sample with a single analytical platform. As a result, metabolomic strategies require a variety of complementary approaches for the detection, identification, and quantification of low molecular weight metabolites. Suitable techniques must be sensitive, robust, and have the capacity to acquire data on large populations of metabolites. The techniques most frequently employed for metabolomic studies are nuclear magnetic resonance (NMR) spectroscopy and mass spectrometry. The analytical technologies used in metabolomic studies have been extensively reviewed (e.g., Dunn and Ellis 2005; Moco et al. 2007; Lindon and Nicholson 2008), and as such only a brief outline of these approaches will be described in this section.

NMR SPECTROSCOPY

NMR spectroscopy has been widely used to obtain comprehensive profiles of low molecular weight metabolites in biological samples. The technique is not biased toward specific metabolites and is highly reproducible and potentially fully quantitative. As most low molecular weight metabolites contain hydrogen, ^{1}H NMR spectroscopy is commonly used in metabolomics applications. NMR can also be used to detect ^{13}C and ^{15}N nuclei, although these approaches tend to be much less sensitive due to the lower natural abundance of these isotopes in biologically important molecules. Similarly, NMR can detect ^{31}P, but this is only present in a relatively small number of metabolites in mammalian systems such as nucleotides.

A major advantage of NMR spectroscopy is that it requires little or no sample preparation and any biofluid is amenable to analysis. It is even possible to generate metabolic profiles from intact tissues using a technique known as high-resolution magic angle spinning (HRMAS) (Tsang et al. 2005). NMR spectroscopy is insensitive compared to other analytical approaches used in metabolomic studies and requires relatively large sample volumes. The issue of sensitivity has been addressed by the production of larger field magnets and cryogenic probes (Griffin 2003); however, the analysis of low abundance metabolites by NMR spectroscopy can still prove problematic. In addition, ^{1}H NMR spectra of plasma or serum samples can contain broad peaks and overlapping signals due to the presence of proteins and lipids (Nicholson et al. 1995). In such cases, two-dimensional NMR spectroscopic approaches can be useful to resolve spectral peaks and improve metabolite analysis (Viant 2003). The online coupling of NMR with high-pressure liquid chromatography (HPLC) has also been developed and applied to biological samples for enhanced metabolite identification (Lindon et al. 2000).

MASS SPECTROMETRY

The high sensitivity of mass spectrometry combined with its wide dynamic range makes the technique an extremely powerful tool for the analysis of large populations of metabolites. Low molecular weight metabolites can be directly analyzed by mass spectrometry; however, ion suppression and matrix effects can lead to difficulties in detecting certain compounds (Villas-Boas et al. 2005; Dettmer 2007). As a result, mass spectrometer systems are typically interfaced with chromatographic platforms to separate complex mixtures of metabolites in biological samples.

Liquid chromatography–mass spectrometry (LC-MS) is used for the profiling of thermally labile or involatile molecules (Lu et al. 2008), whereas gas chromatography–mass spectrometry (GC-MS) is employed for the analysis of volatile molecules and appropriately derivatized metabolites (Pasikanti et al. 2008). The recent introduction of ultraperformance liquid chromatography (UPLC) (Plumb et al. 2004), the Orbitrap mass spectrometer (Hu et al. 2005), and multidimensional GC-MS (Fiehn 2008) has aroused considerable interest due to the capability of these techniques to generate high-resolution data.

There are a number of weaknesses associated with mass spectrometry. Metabolites from biofluids and tissues typically need to be extracted in organic solvents, which can result in variable losses of certain compounds. In addition, with GC-MS methods many biologically important metabolites, for example, amino acids, must undergo chemical derivatization procedures prior to their analysis. Furthermore, mass spectrometry is unable to provide absolute quantification, although the concentrations of low molecular weight metabolites can be accurately determined through the inclusion of appropriate stable isotope internal standards in the experimental system. Because it is not feasible to obtain internal standards for all low molecular weight metabolites in biological samples, label-free relative quantification strategies have been employed in mass spectrometry-based metabolomic studies.

OTHER TECHNIQUES

Alternative analytical techniques have also been utilized in metabolomic studies. Fourier transform infrared (FT-IR) spectroscopy is able to rapidly provide a metabolic fingerprint of biological samples and detect small phenotypic differences within sample sets (Ellis *et al.* 2002). However, the technique only affords a low level of chemical selectivity and is limited in its ability to identify individual metabolites. Similarly, capillary electrophoresis (CE) has shown promise for the profiling of low molecular weight metabolites in body fluids and tissues (Monton and Soga 2007; Barbas *et al.* 2008), although it has not been used extensively for metabolomic applications.

DATA PROCESSING

The raw data generated in metabolomic studies need to be processed prior to statistical analysis. One approach used for data acquired by NMR spectroscopy is to divide each spectrum into smaller bins of defined spectral ranges and to sum all the signal intensities within each bin (Holmes *et al.* 1994). However, following recent improvements in computational approaches, spectral alignment is now starting to replace the need for spectral binning (Stoyanova *et al.* 2004; Forshed *et al.* 2005). The transformation of LC-MS and GC-MS raw data typically entails the automatic detection, alignment, and normalization of peaks in a chromatogram (Dunn 2008). A variety of open source packages are available for the data processing of NMR spectroscopy and mass spectral data, and many instrument manufacturers have also produced their own software.

STATISTICAL ANALYSIS

One of the key issues with metabolomics studies is data interpretation, and researchers require strategies to efficiently extract information. In metabolomic studies, multivariate data analysis is often used to characterize the metabolic responses in biological systems (Holmes and Antti 2002; Kemsley *et al.* 2007).

These methods fall into two general classes – supervised and unsupervised. Both techniques reduce data complexity and search for pattern sets in the metabolite profiles according to their properties. However, unsupervised techniques do so without knowledge of the sample classes to which the individual samples belong. Unsupervised methods include hierarchical cluster analysis (HCA) and principal component analysis (PCA). PCA is often used as a first explorative step in metabolomic analysis and can reveal relationships between sample groups, which may reflect differences in metabolic pathways.

Supervised methods are more powerful statistical tools that use additional information on the datasets, such as clinical data, to determine similarities and differences between predefined groups.

Examples of supervised methods include partial least squares discriminant analysis (PLS-DA), orthogonal projection to latent structures (OPLS), genetic algorithms, and neural networks. All have been employed in metabolomics-based applications. It is important to note that appropriate validation steps are required when using supervised approaches (Rubingh *et al.* 2006).

Furthermore, investigators routinely use univariate statistical analysis such as Student's t-test following, or in conjunction with, multivariate statistical techniques to confirm the significance of results.

IDENTIFICATION OF LOW MOLECULAR WEIGHT METABOLITES

The identification of metabolites in biological systems can lead to the targeting of specific metabolic pathways, which may provide greater insight into the mechanisms of metabolic disturbances. Although investigators in the fields of genomics and proteomics can access well-annotated databases to rapidly determine the identity of a gene or protein, the bioinformatic tools required to identify large numbers of low molecular weight metabolites are still being established.

Metabolites can be identified through the production of in-house spectral libraries or through comparison with authentic standards. However, this is neither practical nor affordable for every laboratory, with many metabolites not available from commercial sources, and as a result public metabolite databases have started to be constructed. One of the largest initiatives is the Human Metabolome Database (http://www.hmdb.ca/) (Wishart 2007), which includes NMR and mass spectra of biologically important metabolites. Similar databases have also been developed such as the Madison Metabolomics Consortium Database (http://mmcd.nmrfam.wisc.edu) (Cui et al. 2008) and METLIN Metabolite Database (http://metlin.scripps.edu) (Smith et al. 2005). A variety of commercially available libraries also exist, for example, the National Institute of Standards and Technology (NIST) library, which includes mass spectral data for a number of endogenous metabolites (Halket et al. 2005).

Although these libraries and databases are a good starting point for metabolite characterization, their usefulness is often limited by their incompleteness. For nutritional metabolomics, it will be necessary to develop resources that not only permit the identification of endogenous metabolites but also intermediates derived from the metabolism of dietary components.

METABOLOMICS AND NUTRITIONAL SCIENCES

Metabolomics has wide applications in the field of nutrition. Metabolomic strategies can be employed to explore homeostatic control and how this metabolic balance may be disturbed by deficiencies or excesses of dietary components (German et al. 2003). Metabolomic approaches can also provide insights of the interaction between intestinal microflora and mammalian metabolism and be used to determine the composition of foods with proposed health benefits.

Nutritional metabolomics may prove useful in the identification of molecular biomarkers of health and disease and in assessing the impact induced by whole diets or individual dietary constituents on human health (Zeisel et al. 2005; Kussmann et al. 2008; Zhang et al. 2008). This has evolved into the concept of personalized nutrition, which offers the potential to guide dietary recommendations and lead to more effective treatment and management of diseases (Watkins and German 2002; Joost et al. 2007). As a result, metabolomic strategies are playing a central role in programs focused on development and promotion of mechanistic nutrition and health such as the European Nutrigenomics Organization (NuGO) (http://www.nugo.org).

DIETARY INFLUENCES ON METABOLITE PROFILES

Metabolomic approaches have been used to monitor the influence of diet to understand biological variation in clinical studies. In one investigation, metabolomic analysis was performed on urine and plasma samples collected from a group of thirty healthy subjects (Walsh et al. 2006). NMR spectroscopy in conjunction with multivariate data analysis was used to characterize the metabolite profiles and the results showed considerable intra- and interindividual variation for each biofluid. However, the authors reported that consumption of a standard diet 24 hours prior to sample collection was able to reduce the variation in the urinary profiles, although this failed to eliminate the metabolite variation observed in plasma and saliva samples.

Other studies have investigated changes in human metabolite profiles resulting from different dietary regimens or specific dietary components. Stella *et al.* (2006) examined the metabolite changes in urine associated with vegetarian and meat diets. The study revealed that a high meat diet was associated with increased urinary levels of creatine, creatinine, carnitine, acetyl-carnitine, taurine, trimethylamine-N-oxide (TMAO), and glutamine. The consumption of the vegetarian diet was characterized by higher levels of p-hydroxyphenylacetate in urine but a decrease in the urinary excretion of N,N,N-trimethyllysine.

Metabolomic strategies have also been employed to define the biochemical changes in serum and urine of children whose diet had been supplemented with animal protein from milk or meat (Bertram *et al.* 2007) and to investigate the role of dietary phytochemicals in shaping urinary metabolic profiles (Walsh *et al.* 2007). Similarly, these experimental approaches have been used to examine metabolic responses in humans to the intake of soy (Solanky *et al.* 2005), ingestion of a variety of teas (Daykin *et al.* 2005; Wang *et al.* 2005; van Dorsten *et al.* 2006; Law *et al.* 2008), and dietary exposure to rice wine (Teague *et al.* 2004). Metabolomic strategies have revealed changes in metabolite profiles resulting from the consumption of whole grain cereals by animals (Bertram *et al.* 2006; Fardet *et al.* 2007) and have successfully discriminated between sheep maintained on different plant-based diets (Parveen *et al.* 2007, 2008).

Metabolomics has also been able to characterize the effect of dietary influences between populations. Lenz *et al.* (2004) showed that the high dietary intake of fish in the Swedish population resulted in increased urinary excretion of TMAO. In comparison, British volunteers were found to have unusually high levels of urinary taurine due to the prevalence of the Atkins diet within the study cohort. This highlights how changes in the dietary habits of volunteers during trials can have dramatic effects and that specific dietary restrictions may improve data interpretation. As such, metabolomic analysis could prove useful in accurately monitoring dietary intakes (Mathers 2006).

DIET AND DISEASE

There are well-established associations between nutrition and health. Diet is known to play a key role in the development of important diseases such as cancer and cardiovascular disease. Metabolomics can report directly on the metabolic and physiological status of humans and animals. As a result, investigators are increasingly employing metabolomic strategies to provide additional insights into the molecular basis of disease processes and to identify biomarkers of disease states.

A number of studies have used metabolomics to explore the metabolite disturbances in the body fluids of obese Zucker rats (Serkova *et al.* 2006; Williams *et al.* 2006; Salek *et al.* 2007). These rats lack the leptin receptor and are widely used to model type 2 diabetes mellitus and study insulin resistance. In one study, Serkova *et al.* (2006) determined changes in a variety of lipid metabolites in the plasma of Zucker rats. Furthermore, analysis of liver extracts from the same animals showed signs of disturbed energy balance, decreased hepatic glutathione concentrations, and altered levels of unsaturated fatty acids. These findings suggested that obesity may lead to increased oxidative stress and lipid peroxidation.

In an additional study, metabolite changes in human patients with type 2 diabetes mellitus were compared with obese Zucker rats and another animal model with a defective leptin receptor, the *db/db* mouse (Salek *et al.* 2007). Urinary metabolite profiles were analysed by NMR spectroscopy and the data were then subjected to statistical analyses. The study highlighted disease-related metabolite disturbances that were common to affected individuals and the rodent models. These included changes in methylamine metabolism, nucleotide pathways, and the tricarboxylic acid (TCA) cycle.

Metabolomics strategies are now also being used as part of large international epidemiological studies to help understand the link between diet and diseases. Holmes and colleagues (2008) recently reported the use of metabolomics to investigate metabolite disturbances associated with increased blood pressure, a major risk factor for coronary heart disease and stroke. Profiling of urine samples from more than 4000 individuals from China, Japan, The United Kingdom, and the United States demonstrated significant differences in metabolism both from country to country and within populations. Formate, alanine, and hippurate excretion emerged as markers related to blood pressure. Interestingly, these changes could be linked to the diet and lifestyles of the study participants.

GUT MICROFLORA

Many metabolomic studies have reported altered levels of metabolites, which are generated by gut microbial activity. For example, hippurate is mainly derived from the breakdown of phenolics and aromatic amino acids by gut microflora. Humans and animals have evolved a symbiotic relationship with a consortium of gut microbes (microbiome) that play a key role in maintaining health and can provide essential nutrients for the host (Nicholson *et al.* 2005).

Diet in particular can influence the complex internal community of gut microorganisms. Gut microbes interact closely with the metabolism of their host and can drastically modify nutrient bioavailability (Goodacre 2007). These microbes have also been implicated in pathogenesis of diet-induced disease states (Dumas *et al.* 2006; Backhed *et al.* 2007). It is therefore important to understand and characterize the interactive molecular processes between the host and its microbiome. This has particular relevance to functional foods or nutraceuticals, the action of which may be altered by microbial metabolism. Therefore, investigators have begun to use metabolomic strategies to assess the effects of dietary intake on gut microflora activity (Jacobs *et al.* 2008).

Another area of interest has focused on the mechanism of action of probiotics and prebiotics. Martin *et al.* (2007) have investigated the impact of a bacterial probiotic *Lactobacillus paracasei* on the intestinal physiology and metabolism of germ-free mice. The mice were raised without any resident microorganisms and their diet was supplemented with either live *Lactobacillus paracasei* or an irradiated form of the bacteria. HRMAS NMR spectroscopy was used to generate metabolite profiles of a variety of gut tissues including duodenum, jejunum, ileum, and colon. Multivariate data analysis revealed that colonization of the mice with the live *Lactobacillus paracasei* probiotic led to metabolite changes in the intestinal tissues that were associated with digestion, absorption of nutrients, energy metabolism, lipid synthesis, and the antioxidant capacity of the gut. In contrast, consumption of the irradiated probiotic failed to alter the profiles of the intestinal metabolites.

More recent studies have examined metabolic effects of exposing humanized microbiome mice to different probiotics and prebiotics (Martin *et al.* 2008a, 2008b). These investigations revealed that the treatments produce changes in gut microbial populations of the mice that were reflected in alterations to a variety of host pathways involved in lipid, carbohydrate, and amino acid metabolism.

METABOLOMICS AND FUNCTIONAL FOODS

Metabolomic approaches are also playing an important role at the interface between food production and human nutrition (Brown and van der Ouderaa 2007). This is reflected through research programs such as META-PhoR (http://www.meta-phor.eu). A range of foods and food products have been proposed to confer specific health-promoting effects. These benefits have been widely attributed to bioactive metabolites such as those found in fruit and vegetables including carotenoids and phytosterols (Hall *et al.* 2008). These components are increasingly becoming the focus of studies that suggest that their consumption may prevent or reduce the risk of a variety of common diseases (Davis and Hord 2005).

Metabolomics may be of use in monitoring the effects of genetic engineering and selective breeding on the nutritional value of foodstuffs (DellaPenna 1999; Trethewey 2004). Metabolomic technologies have been used to examine the effect of environmental influences on the level of bioactive compounds in breeding populations of raspberries (Stewart *et al.* 2007). The plants were grown in separate fields and were either well maintained or received little management. The ripe raspberries were collected and their metabolites were then extracted and analyzed by direct injection mass spectrometry. Multivariate data analysis showed that the different growth conditions failed to alter the metabolite profile of the raspberries. However, there were changes in the relative levels of a range of polyphenols, which have potential health benefits.

Similar approaches have also been employed to assess the effects of food processing on the metabolite composition of fruits and vegetables. Capanoglu *et al.* (2008) recently determined the metabolite changes associated with the industrial production of tomato paste. Material was analyzed from five individual steps in the production chain from the fresh fruit to final paste. The results revealed that each processing step resulted in alterations in the metabolic profile of the tomatoes. In particular, the breaking step and pulping step, in which the seed and skin are removed, led to significant changes in the concentrations of several alkaloids and antioxidants such as flavenoids.

Researchers are also using metabolomics to explore the biological activities of dietary components. The effect of catechin, a major flavenoid found in fruits, wine, and chocolate, was examined in rats fed a high-fat diet (Fardet *et al.* 2008). Although catechin supplementation had no effect on the oxidative stress induced in hepatic and cardiac tissues by the hyperlipidic diet, it was able to normalize metabolite disturbances observed in the urinary metabolome of the rats.

In another study, Pettersson *et al.* (2008) characterized the metabolite profiles of fecal water from vegetarians. Analysis by NMR spectroscopy showed the presence of significant quantities of amino acids, organic acids, and short-chain fatty acids. The development of colon cancer has been linked to an increased production of prostaglandins. However, a protective effect has been proposed to be exerted by a range of phytochemicals. The investigators therefore treated the human colon adenocarcinoma cell line HT-29 with the individual fecal water samples. PCA was used to correlate the effect of metabolite composition of the fecal water samples on the activity of the enzyme cyclooxygenase-2 (COX-2), which plays a key role in prostaglandin biosynthesis. The findings indicated that samples with high levels of phenylalanine, alanine, glutamic acid, glycine, lactate, and short-chain fatty acids inhibited COX-2 protein expression.

CONCLUSIONS AND FUTURE DIRECTIONS

Metabolomics is increasingly playing a major role in the nutritional sciences. Metabolomic strategies are capable of defining the biochemical responses and have created a unique opportunity to understand the molecular basis of metabolic regulation and dysfunction. However, key challenges remain and it is highly probable that the field will continue to be driven by technological developments for the foreseeable future.

Nonetheless, the ability to measure and interpret complex metabolite profiles will provide a more comprehensive view of cellular control mechanisms, which may have direct outcomes in improving our understanding of dietary influences on individuals. Similarly, metabolomics has the potential to assist in the development of functional foods and define how food components affect health status. This may ultimately lead to personalized nutrition where dietary recommendations prevent the onset of chronic diseases.

REFERENCES

Backhed, Fredrick, Manchester, Jill K., Semenkovich, Clay F., Gordon, Jeffrey I. 2007. Mechanisms underlying the resistance to diet-induced obesity in germ-free mice. *Proceedings to the National Academy of Sciences of the United States of America 104*(3):979–984.

Barbas, C., Vallejo, M., Garcia, A., Barlow, D., Hanna-Brown, M. 2008. Capillary electrophoresis as a metabolomic tool in antioxidant therapy studies. *Journal of Pharmaceutical and Biomedical Analysis* 47(2):388–398.

Bertram, Hanne C., Bach Knudsen, Knud E., Serena, Anja, Malmendal, Anders, Nielsen, Niels Chr., Frette, Xavier C., Andersen, Henrik J. 2006. NMR-based metabonomic studies reveal changes in the biochemical profile of plasma and urine from pigs fed high-fibre rye bread. *British Journal of Nutrition* 95(5):955–962.

Bertram, Hanne C., Hoppe, Camilla, Petersen, Bent O., Duus, Jens O., Molgaard, Christian, Michaelsen, Kim F. 2007. An NMR-based metabonomic investigation on effects of milk and meat protein diets given to 8-year old boys. *British Journal of Nutrition* 97(4):758–763.

Bollard, Mary E., Stanley, Elizabeth G., Lindon, John C., Nicholson, Jeremy K., Holmes, Elaine. 2005. NMR-based metabonomic approaches for evaluating physiological influences on biofluid composition. *NMR in Biomedicine* 18(3):143–162.

Brown, Louise, van der Ouderaa, Frans. 2007. Nutritional genomics: food industry applications from farm to fork. *British Journal of Nutrition* 97(6):1027–1035.

Capanoglu, Esra, Beekwilder, Jules, Boyacioglu, Dilek, Hall, Robert, De Vos, Ric. 2008. Changes in antioxidant and metabolite profiles during production of tomato paste. 2008. *Journal of Agricultural and Food Chemistry* 56(3):964–973.

Cui, Qiu, Lewis, Ian A., Hegeman, Adrian D., Anderson, Mark E., Li, Jing, Schulte, Christopher F., Westler, William M., Eghbalnia, Hamid R., Sussman, Michael R., Markley, John L. 2008. Metabolite identification via the Madison Metabolomics Consortium Database. *Nature Biotechnology* 26(2):162–164.

Davis, Cindy D., Hord, Norman G. 2005. Nutritional "omics" technologies for elucidating the role(s) of bioactive food components in colon cancer prevention. *Journal of Nutrition* 135(11):2694–2697.

Daykin, Claire A., van Duynhoven, John P.M., Groenewegen, Anneke, Dachtler, Markus, van Amelsvoort, Johan M.M., Mulder, Theo P.J. 2005. Nuclear magnetic resonance spectroscopic based studies of the metabolism of black tea polyphenols in humans. *Journal of Agricultural and Food Chemistry* 53(5):1428–1434.

DellaPenna, D. 1999. Nutritional genomics: manipulating plant micronutrients to improve human health. *Science* 285(5426):375–379.

Denkert, Carsten, Budczies, Jan, Kind, Tobias, Weichert, Wilko, Tablack, Peter, Sehouli, Jalid, Niesporek, Silvia, Konsgen, Dominique, Dietel, Manfred, Fiehn, Oliver. 2006. Mass spectrometry-based metabolic profiling reveals different metabolite patterns in invasive ovarian carcinomas and ovarian borderline tumors. *Cancer Research* 66(22):10795–10804.

Dettmer, Katja, Aronov, Pavel A., Hammock, Bruce D. 2007. Mass spectrometry-based metabolomics. *Mass Spectrometry Reviews* 26(1):51–78.

Dumas, Marc-Emmanuel, Barton, Richard H., Toye, Ayo, Cloarec, Olivier, Blancher, Christine, Rothwell, Alice, Fearnside, Jane, Tatoud, Roger, Blanc, Veronique, Lindon, John C., Mitchell, Steve C., Holmes, Elaine, McCarthy, Mark I., Scott, James, Gauguier, Dominique, Nicholson, Jeremy K. 2006. Metabolic profiling reveals a contribution of gut microbiota to fatty liver phenotype in insulin resistant mice. *Proceedings to the National Academy of Sciences of the United States of America* 103(33):12511–12516.

Dunn, Warwick B. 2008. Current trends and future requirements for the mass spectrometric investigation of microbial, mammalian and plant metabolomes. *Physical Biology* 5(1):11001.

Dunn, Warwick B., Ellis, David I. 2005. Metabolomics: current analytical platforms and methodologies. *TrAC Trends in Analytical Chemistry* 24(4):285–294.

Ellis, David I., Broadhurst, David, Kell, Douglas B., Rowland, Jem J., Goodacre, Royston. 2002. Rapid and quantitative detection of the microbial spoilage of meat by Fourier transform infrared spectroscopy and machine learning. *Applied and Environmental Microbiology* 68(6):2822–2828.

Ellis, David I., Dunn, Warwick B., Griffin, Julian L., Allwood, J. William, Goodacre, Royston. 2007. Metabolic fingerprinting as a diagnostic tool. *Pharmacogenomics* 8(9):1243–1266.

Fardet, Antony, Canlet, Cecile, Gottardi, Gaelle, Lyan, Bernard, Llorach, Rafael, Remesy, Christian, Mazur, Andre, Paris, Alain, Scalbert, Augustin. 2007. Whole-grain and refined wheat flours show distinct metabolic profiles in rats as assessed by a ^{1}H NMR-based metabonomic approach. *Journal of Nutrition 137*(4):923–929.

Fardet, Antony, Llorach, Rafael, Martin, Jean-Francois, Besson, Catherine, Lyan, Bernard, Pujos-Guillot, Estelle, Scalbert, Augustin. 2008. A liquid chromatography-quadrupole time-of-flight (LC-QTOF)-based metabolomic approach reveals new metabolic effects of catechin in rats fed high-fat diets. *Journal of Proteome Research 7*(6):2388–2398.

Fiehn, Oliver. 2002. Metabolomics - the link between genotypes and phenotypes. *Plant Molecular Biology 48*(1–2):155–171.

Fiehn, Oliver. 2008. Extending the breadth of metabolite profiling by gas chromatography coupled to mass spectrometry. *TrAC Trends in Analytical Chemistry 27*(3):261–269.

Forshed, Jenny, Torgripa, Ralf J.O., Aberg, K. Magnus, Karlberg, Bo, Lindberg Johan, Jacobssona, Sven P. 2005. A comparison of methods for alignment of NMR peaks in the context of cluster analysis. *Journal of Pharmaceutical and Biomedical Analysis 38*(5):824–832.

German, J. Bruce, Roberts, Matthew-Alan, Watkins, Steven M. 2003. Personal metabolomics as a next generation nutritional assessment. *Journal of Nutrition 133*(12):4260–4266.

Gibney, Michael J., Walsh, Marianne, Brennan, Lorraine, Roche, Helen M., German, Bruce, van Ommen, Ben. 2005. Metabolomics in human nutrition: opportunities and challenges. *American Journal of Clinical Nutrition 82*(3):497–503.

Goodacre, Royston. 2007. Metabolomics of a superorganism. *Journal of Nutrition 137*(1):259S-266S.

Griffin, Julian L. 2003. Metabonomics: NMR spectroscopy and pattern recognition analysis of body fluids and tissues for characterisation of xenobiotic toxicity and disease diagnosis. *Current Opinion in Chemical Biology 7*(5):648–654.

Griffin, Julian L., Shockor, John P. 2004. Metabolic profiles of cancer cells. *Nature Reviews Cancer 4*(7):551–561.

Gu, Haiwei, Chen, Huanwen, Pan, Zhengzheng, Jackson, Ayanna U, Talaty, Nari, Xi Bowei, Kissinger, Candice, Duda, Chester, Mann, Doug, Raftery, Daniel, Cooks, R. Graham. 2007. Monitoring diet effects via biofluids and their implications for metabolomics studies. *Analytical Chemistry 79*(1):89–97.

Halket, John M., Waterman, Daniel, Przyborowska, Anna M., Patel, Raj K.P., Fraser, Paul D., Bramley, Peter M. 2005. Chemical derivatization and mass spectral libraries in metabolic profiling by GC/MS and LC/MS/MS. *Journal Experimental Botany 56*(410):219–243.

Hall, Robert D., Brouwer, Inge D., Fitzgerald, Melissa A. 2008. Plant metabolomics and its potential application for human nutrition. *Physiologia Plantarum 132*(2):162–175.

Holmes, E., Antti, H. 2002. Chemometric contributions to the evolution of metabonomics: mathematical solutions to characterising and interpreting complex biological NMR spectra. *The Analyst 127*(12):1549–1557.

Holmes, E., Foxall, P.J.D., Nicholson, J.K., Neild, G.H., Brown, S.M., Beddell, C.R., Sweatman, B.C., Rahr, E., Lindon, J.C., Spraul, M., Neidig, P. 1994. Automatic data reduction and pattern recognition methods for analysis of ^{1}H nuclear magnetic resonance spectra of human urine from normal and pathological states. *Analytical Biochemistry 220*(2):284–296.

Holmes, Elaine, Loo, Ruey Leng, Stamler, Jeremiah, Bictash, Magda, Yap, Ivan K.S., Chan, Queenie, Ebbels, Tim, De Iorio, Maria, Brown, Ian J., Veselkov, Kirill A., Daviglus, Martha L., Kesteloot, Hugo, Ueshima, Hirotsugu, Zhao, Liancheng, Nicholson, Jeremy K., Elliott, Paul. 2008. Human metabolic phenotype diversity and its association with diet and blood pressure. *Nature 453*(7193):396–401.

Hu, Qizhi, Noll, Robert J., Li, Hongyan, Makarov, Alexander, Hardman, Mark, Cooks, R. Graham. 2005. The Orbitrap: a new mass spectrometer. *Journal of Mass Spectrometry 40*(4):430–443.

Jacobs, Doris M., Deltimple, Nancy, van Velzen, Ewoud, van Dorsten, Ferdi A., Bingham, Max, Vaughan, Elaine E., van Duynhoven, John. 2008. ^{1}H NMR metabolite profiling of feces as a tool to assess the impact of nutrition on the human microbiome. *NMR in Biomedicine 21*(6): 615–626.

Joost, Hans-Georg, Gibney, Michael J., Cashman, Kevin D., Gorman, Ulf, Hesketh, John E., Mueller, Michael, van Ommen, Ben, Williams, Christine M., Mathers, John C. 2007. Personalised nutrition: status and perspectives. *British Journal of Nutrition 98*(1): 26–31.

Kaddurah-Daouk, Rima, Kristal, Bruce S., Weinshilboum, Richard M. 2008. Metabolomics: a global biochemical approach to drug response and disease. *Annual Review of Pharmacology and Toxicology 48*:653–683.

Kemsley, E. Katherine, Le Gall, Gwenaelle, Dainty, Jack R., Watson, Andrew D., Harvey, Linda J., Tapp, Henri S., Colquhoun, Ian J. 2007. Multivariate techniques and their application in nutrition: a metabolomics case study. *British Journal of Nutrition 98*(1):1–14.

Kenny, Louise C., Broadhurst, David, Brown, Marie, Dunn, Warwick B., Redman, Christopher W.G., Kell, Douglas B., Baker, Philip N. 2008. Detection and identification of novel metabolomic biomarkers in preeclampsia. *Reproductive Sciences* 15(6):591–597.

Kochhar, Sunil, Jacobs, Doris M., Ramadan, Ziad, Berruex, France, Fuerholz, Andreas, Fay, Laurent B. 2006. Probing gender-specific metabolism differences in humans by nuclear magnetic resonance-based metabonomics. *Analytical Biochemistry 352*(2):274–281.

Kussmann, Martin, Rezzi, Serge, Daniel, Hannelore. 2008. Profiling techniques in nutrition and health research. *Current Opinion in Biotechnology 19*(2):83–99.

Law, Wai Siang, Huang, Pei Yun, Ong, Eng Shi, Ong, Choon Nam, Li, Sam Fong Yau, Pasikanti, Kishore Kumar, Chan, Eric Chung Yong. 2008. Metabonomics investigation of human urine after ingestion of green tea with gas chromatography/mass spectrometry, liquid chromatography/mass spectrometry and ^{1}H NMR spectroscopy. *Rapid Communications in Mass Spectrometry 22*(16):2436–2446.

Lenz E.M., Bright, J., Wilson, I.D., Hughes, A., Morrisson, J., Lindberg, H., Lockton, A. 2004. Metabonomics, dietary influences and cultural differences: a ^{1}H NMR-based study of urine samples obtained from healthy British and Swedish subjects. *Journal of Pharmaceutical and Biomedical Analysis 36*(4):841–849.

Lenz, Eva M., Bright, J., Wilson, Ian D., Morgan, S.R., Nash, A.F.P. 2003. A ^{1}H NMR-based metabonomic study of urine and plasma samples obtained from healthy human subjects. *Journal of Pharmaceutical and Biomedical Analysis 33*(5):1103–1115.

Lindon, John C., Keun, Hector C., Ebbels, Timothy M.D., Pearce, Jake M.T., Holmes, Elaine, Nicholson, Jeremy K. 2005. The Consortium for Metabonomic Toxicology (COMET): aims, activities and achievements. *Pharmacogenomics 6*(7):691–699.

Lindon, John C., Nicholson, Jeremy K. 2008. Analytical technologies for metabonomics and metabolomics, and multi-omic information recovery. *TrAC Trends in Analytical Chemistry 27*(3):194–204.

Lindon, John C., Nicholson, Jeremy K., Wilson, Ian D. 2000. Directly coupled HPLC-NMR and HPLC-NMR-MS in pharmaceutical research and development. *Journal of Chromatography B. Analytical Technologies in the Biomedical and Life Sciences 748*(1):233–258.

Lu, Xin, Zhao, Xinjie, Bai, Changmin, Zhao, Chunxia, Lu, Guo, Xu, Guowang. 2008. LC-MS based metabonomics analysis. *Journal of Chromatography B. Analytical Technologies in the Biomedical and Life Sciences 866*(1–2):64–76.

Maher, Anthony D., Zirah, Severine F.M., Holmes, Elaine, Nicholson, Jeremy K. 2007. Experimental and analytical variation in human urine in ^{1}H NMR spectroscopy-based metabolic phenotyping studies. *Analytical Chemistry 79*(14):5204–5211.

Marchesi, Julian R., Holmes, Elaine, Khan, Fatima, Kochhar, Sunil, Scanlan, Pauline, Shanahan, Fergus, Wilson, Ian D., Wang, Yulan. 2007. Rapid and noninvasive metabonomic characterization of inflammatory bowel disease. *Journal of Proteome Research 6*(2):546–551.

Martin, Francois-Pierre J., Wang, Yulan, Sprenger, Norbert, Holmes, Elaine, Lindon, John C., Kochhar, Sunil, Nicholson, Jeremy K. 2007. Effects of probiotic Lactobacillus paracasei treatment on the host gut tissue metabolic profiles probed via magic-angle-spinning NMR spectroscopy. *Journal of Proteome Research* 6(4):1471–1481.

Martin, Francois-Pierre J., Wang, Yulan, Sprenger, Norbert, Yap, Ivan K.S., Lundstedt, Torbjorn, Lek, Per, Rezzi, Serge, Ramadan, Ziad, van Bladeren, Peter, Fay, Laurent B., Kochhar, Sunil, Lindon, John C., Holmes, Elaine, Nicholson, Jeremy K. 2008a. Probiotic modulation of symbiotic gut microbial-host metabolic interactions in a humanized microbiome mouse model. *Molecular Systems Biology* 4:157.

Martin, Francois-Pierre J., Wang, Yulan, Sprenger, Norbert, Yap, Ivan K.S., Rezzi, Serge, Ramadan Ziad, Pere-Trepat, Emma, Rochat, Florence, Cherbut, Christine, van Bladeren, Peter, Fay, Laurent B., Kochhar, Sunil, Lindon, John C., Holmes, Elaine, Nicholson, Jeremy K. 2008b. Top-down systems biology integration of conditional prebiotic modulated transgenomic interactions in a humanized microbiome mouse model. *Molecular Systems Biology* 4: 205.

Mashego, Mlawule R., Rumbold, Karl, De Mey, Marjan, Vandamme, Erick, Soetaert, Wim, Heijnen, Joseph J. 2007. Microbial metabolomics: past, present and future methodologies. *Biotechnology Letters* 29(1):1–16.

Mathers, John C. 2006. Plant foods for human health: research challenges. *Proceedings of the Nutrition Society* 65(2):198–203.

Metabolomics Standards Initiative articles. 2007. *Metabolomics* 3(3):175–256.

Moco, Sofia, Bino, Raoul J., De Vos, Ric C.H., Vervoort, Jacques. 2007. Metabolomics technologies and metabolite identification. *TrAC Trends in Analytical Chemistry* 26(9):855–866.

Monton, Maria Rowena N., Soga, Tomoyoshi. 2007. Metabolome analysis by capillary electrophoresis-mass spectrometry. *Journal of Chromatography A* 1168(1–2):237–246.

Nicholson, Jeremy K., Foxall, Peta J.D., Spraul, Manfred, Farrant, R. Duncan, and John C. Lindon. 1995. 750 MHz ^{1}H and ^{1}H-^{13}C NMR spectroscopy of human blood plasma. *Analytical Chemistry* 67(5):793–811.

Nicholson, Jeremy K., Holmes, Elaine, Wilson, Ian D. 2005. Gut microorganisms, mammalian metabolism and personalized health care. *Nature Reviews Microbiology* 3(5):431–438.

Nicholson, Jeremy K., Lindon, John C., Holmes, Elaine. 1999. 'Metabonomics': understanding the metabolic responses of living systems to pathophysiological stimuli via multivariate statistical analysis of biological NMR spectroscopic data. *Xenobiotica* 29(11):1181–1189.

Nicholson, Jeremy K., Wilson, Ian D. 2003. Understanding 'global' systems biology: metabonomics and the continuum of metabolism. *Nature Reviews Drug Discovery* 2(8):668–676.

Oresic, Matej, Vidal-Puig, Antonio, Hanninen, Virve. 2006. Metabolomic approaches to phenotype characterization and applications to complex diseases. *Expert Review of Molecular Diagnosis* 6(4):575–585.

Parveen, Ifat, Moorby, Jon M., Fraser, Mariecia D., Allison, Gordon G., Kopka, Joachim. 2007. Application of gas chromatography-mass spectrometry metabolite profiling techniques to the analysis of heathland plant diets of sheep. *Journal of Agricultural Food Chemistry* 55(4):1129–1138.

Parveen, I., Moorby, J.M., Hirst, W.M., Morris, S.M., Fraser, M.D. 2008. Profiling of plasma and faeces by FT-IR to differentiate between heathland plant diets offered to zero-grazed sheep. *Animal Feed Science and Technology* 144(1–2):65–81.

Pasikanti, Kishore K., Ho, P.C., Chan, E.C.Y. 2008. Gas chromatography/mass spectrometry in metabolic profiling of biological fluids. *Journal of Chromatography B. Analytical Technologies in the Biomedical and Life Sciences* 871(2):202–211.

Pettersson, Jenny, Karlsson, Pernilla, Christina, Choi, Young, Hae, Verpoorte, Robert, Rafter, Joseph James, Bohlin, Lars. 2008. NMR Metabolomic analysis of fecal water from subjects on a vegetarian diet. *Biological and Pharmaceutical Bulletin* 31(6):1192–1198.

Plumb, Robert, Castro-Perez, Jose, Granger, Jennifer, Beattie, Iain, Joncour, Karine, Wright, Andrew. 2004. Ultra-performance liquid chromatography coupled to quadrupole-orthogonal time-of-flight mass spectrometry. *Rapid Communications in Mass Spectrometry* 18(19):2331–2337.

Rezzi, Serge, Ramadan, Ziad, Fay, Laurent B., Kochhar, Sunil. 2007. Nutritional metabonomics: Applications and perspectives. *Journal of Proteome Research* 6(2):513–525.

Rubingh, Carina M., Bijlsma, Sabina, Derks, Eduard P.P.A., Bobeldijk, Ivana, Verheij, Elwin R., Kochhar, Sunil, Smilde, Age K. 2006. Assessing the performance of statistical validation tools for megavariate metabolomics data. *Metabolomics* 2(2):53–61.

Salek, R.M., Maguire, M.L., Bentley, E., Rubtsov, D.V., Hough, T., Cheeseman, M., Nunez, D., Sweatman, B.C., Haselden, J.N., Cox, R.D., Connor, S.C., Griffin, J.L. 2007. A metabolomic comparison of urinary changes in type 2 diabetes in mouse, rat, and human. *Physiological Genomics* 29(2):99–108.

Sansone, Susanna-Assunta, Fan, Teresa, Goodacre, Royston, Griffin, Julian L., Hardy, Nigel W., Kaddurah-Daouk, Rima, Kristal, Bruce S., Lindon, John, Mendes, Pedro, Morrison, Norman, Nikolau, Basil, Robertson, Don, Sumner, Lloyd W., Taylor, Chris, van der Werf, Mariet, van Ommen, Ben, Fiehn, Oliver. 2007. The metabolomics standards initiative. *Nature Biotechnology* 25(8):846–848.

Saude, Erik J., Sykes, Brian D. 2007. Urine stability for metabolomic studies: effects of preparation and storage. *Metabolomics* 3(1):19–27.

Schauer, Nicolas, Fernie, Alisdair R. 2006. Plant metabolomics: towards biological function and mechanism. *Trends in Plant Science* 11(10):508–516.

Schnackenberg, Laura K., Beger, Richard D. 2006. Monitoring the health to disease continuum with global metabolic profiling and systems biology. *Pharmacogenomics* 7(7):1077–1086.

Seger, Christoph, Sturm, Sonja. 2007. Analytical aspects of plant metabolite profiling platforms: current standings and future aims. *Journal of Proteome Research* 6(2):480–497.

Serkova, Natalie J., Jackman, Matthew, Brown, Jaimi L., Liu, Tao, Hirose, Ryutaro, Roberts, John P., Maher, Jacquelyn J., Niemann, Clauss U. 2006. Metabolic profiling of livers and blood from obese Zucker rats. *Journal of Hepatology* 44(5):956–962.

Slupsky, Carolyn M., Rankin, Kathryn N., Wagner, James, Fu, Hao, Chang, David, Weljie, Aalim M., Saude, Erik J., Lix, Bruce, Adamko, Darryl J., Shah, Sirish, Greiner, Russ, Sykes, Brian D., Marrie, Thomas J. 2007. Investigations of the effects of gender, diurnal variation, and age in human urinary metabolomic profiles. *Analytical Chemistry* 79(18):6995–7004.

Smith, Colin A., O'Maille, Grace, Want, Elizabeth J., Qin, Chuan, Trauger, Sunia A., Brandon, Theodore R., Custodio, Darlene E., Abagyan, Ruben, Siuzdak, Gary. 2005. METLIN: a metabolite mass spectral database. *Therapeutic Drug Monitoring* 27(6):747–751.

Solanky, Kirty S., Bailey, Nigel J., Beckwith-Hall, Bridgette M., Bingham, Sheila, Davis, Adrienne, Holmes, Elaine, Nicholson, Jeremy K., Cassidy, Aedin. 2005. Biofluid ^{1}H NMR-based metabonomic techniques in nutrition research - metabolic effects of dietary isoflavones in humans. *Journal of Nutritional Biochemistry* 16(4):236–244.

Stella, Cinzia, Beckwith-Hall, Bridgette, Cloarec, Olivier, Holmes, Elaine, Lindon, John C., Powell, Jonathan, van der Ouderaa, Frans, Bingham, Sheila, Cross, Amanda J., Nicholson, Jeremy K. 2006. Susceptibility of human metabolic phenotypes to dietary modulation. *Journal of Proteome Research* 5(10):2780–2788.

Stewart, Derek, McDougall, Gordan J., Sungurtas, Julie, Verrall, Susan, Graham, Julie, Martinussen, Inger. 2007. Metabolomic approach to identifying bioactive compounds in berries: Advances toward fruit nutritional enhancement. *Molecular Nutrition and Food Research* 51(6):645–651.

Stoyanova, Radka, Nicholson, Jeremy K., Lindon, John C., Brown, Truman R. 2004. Sample classification based on Bayesian spectral decomposition of metabonomic NMR data sets. *Analytical Chemistry* 76(13):3666–3674.

Sumner, Lloyd W., Mendes, Pedro, Dixon, Richard A. 2003. Plant metabolomics: large-scale phytochemistry in the functional genomics era. *Phytochemistry* 62(6):817–836.

Teague, Claire, Holmes, Elaine, Maibaum, Elaine, Nicholson, Jeremy, Tang, Huiru, Queenie, Chan, Elliott, Paul, Wilson, Ian. 2004. Ethyl glucoside in human urine following dietary exposure: detection by ^{1}H NMR spectroscopy as a result of metabonomic screening of humans. *The Analyst* 129(3):259–264.

Teahan, Orla, Gamble, Simon, Holmes, Elaine, Waxman, Jonathan, Nicholson, Jeremy K., Bevan, Charlotte, Keun, Hector C. 2006. Impact of analytical bias in metabonomic studies of human serum and plasma. *Analytical Chemistry* 78(13):4307–4318.

Trethewey, Richard N. 2004. Metabolite profiling as an aid to metabolic engineering in plants. *Current Opinion in Plant Biology* 7(2):196–201.

Tsang, T.M., Griffin, J.L., Haselden, J., Fish, C., Holmes, E. 2005. Metabolic characterization of distinct neuroanatomical regions in rats by magic angle spinning ^{1}H nuclear magnetic resonance spectroscopy. *Magnetic Resonance in Medicine* 53(5):1018–1024.

van Dorsten, Ferdi A., Daykin, Claire A., Mulder, Theo P.J., van Duynhoven, John P.M. 2006. Metabonomics approach to determine metabolic differences between green tea and black tea consumption. *Journal of Agricultural and Food Chemistry* 54(18):6929–6938.

Viant, Mark R. 2003. Improved methods for the acquisition and interpretation of NMR metabolomic data. *Biochemical and Biophysical Research Communications* 310(3):943–948.

Villas-Boas, S.G., Mas, S., Akesson, M., Smedsgaard, J., Nielsen, J. 2005. Mass spectrometry in metabolome analysis. *Mass Spectrometry Reviews* 24(5):613–646.

Walsh, Marianne C., Brennan, Lorraine, Malthouse, J. Paul G., Roche, Helen M., Gibney, Michael J. 2006. Effect of acute dietary standardization on the urinary, plasma, and salivary metabolomic profiles of healthy humans. *American Journal of Clinical Nutrition* 84(3):531–539.

Walsh, Marianne C., Brennan, Lorraine, Pujos-Guillot, Estelle, Sebedio, Jean-Louis, Scalbert, Augustin, Fagan, Ailis, Higgins Desmond G., Gibney, Michael J. 2007. Influence of acute phytochemical intake on human urinary metabolomic profiles. *American Journal of Clinical Nutrition* 86(6):1687–1693.

Wang, Yulan, Holmes, Elaine, Tang, Huiru, Lindon, John C., Sprenger, Norbert, Turini, Marco E., Bergonzelli, Gabriela, Fay, Laurent B., Kochhar, Sunil, Nicholson, Jeremy K. 2006. Experimental metabonomic model of dietary variation and stress interactions. *Journal of Proteome Research* 5(7):1535–1542.

Wang, Yulan, Tang, Huiru, Nicholson, Jeremy K., Hylands, Peter J., Sampson J., Holmes, Elaine. 2005. A metabonomic strategy for the detection of the metabolic effects of chamomile (Matricaria recutita L.) ingestion. *Journal of Agricultural and Food Chemistry* 53(2):191–196.

Watkins, Steven M., German, J. Bruce. 2002. Toward the implementation of metabolomic assessments of human health and nutrition. *Current Opinion in Biotechnology* 13(5):512–516.

Whitfield, Phillip D., German, Alexander J., Noble, Peter-John M. 2004. Metabolomics: an emerging post-genomic tool for nutrition. *British Journal of Nutrition* 92(4):549–555.

Whitfield, Phillip D., Noble, Peter-John M., Major, Hilary, Beynon, Robert J., Burrow, Rachel, Freeman, Alistair I., German, Alexander J. 2005. Metabolomics as a diagnostic tool for hepatology: validation in a naturally occurring canine model. *Metabolomics* 1(3):215–225.

Williams, R, Lenz, E.M., Wilson, A.J., Granger, J., Wilson, I.D., Major, H., Stumpf, C., Plumb, R. 2006. A multi-analytical platform approach to the metabonomic analysis of plasma from normal and zucker (fa/fa) obese rats. *Molecular Biosystems* 2(3–4):174–183.

Wishart, David S. 2007. Human Metabolome Database: completing the 'human parts list.' *Pharmacogenomics* 8(7):683–686.

Wishart, David S. 2008. Metabolomics: applications to food science and nutrition research. *Trends in Food Science and Technology* 19(9):482–493.

Zeisel, S.H., Freake, H.C., Bauman, D.E., Bier, D.M., Burrin, D.G., German, J.B., Klein, S., Marquis, G.S., Milner, J.A., Pelto, G.H., Rasmussen, K.M. 2005. The nutritional phenotype in the age of metabolomics. *Journal of Nutrition* 135(7):1613–1616.

Zhang, Xuewu, Yap, Yeeleng, Wei, Dong, Chen, Gu, Chen, Feng. 2008. Novel omics technologies in nutrition research. *Biotechnology Advances* 26(2):169–176.

19

Evaluation of the Beneficial Effects of Phytonutrients by Metabolomics

Katia Nones and Silas G. Villas-Bôas

Fruits and vegetables are known to be rich in fiber, vitamins, phytosterols, sulfur compounds, carotenoids, and organic acids, which have health benefits. However, fruits, vegetables, teas, and herbal extracts contain other less well-characterized active components designated as phytochemicals or phytonutrients. This chapter will present recent studies indicating the potential health benefits of phytonutrients and how metabolomics can help us gain more insight into the diversity and potential protective effects of these compounds.

PHYTONUTRIENTS OR PHYTOCHEMICALS

Phytochemicals are mainly plant secondary metabolites usually presenting antioxidant properties. Some of these phytochemicals, such as polyphenols, carotenoids, glucosinolates, and sulfur-containing compounds (Table 19.1) are believed to promote heath and prevent degenerative diseases. These metabolites are generally involved in plant protection mechanisms in response to a variety of environmental stress factors such as ultraviolet radiation and pathogen invasion (Manach and Williamson 2005; Kluth *et al.* 2007). Polyphenols are the most studied group of phytochemicals and are grouped into four major classes: (i) phenolic acids (e.g., gallic acid, caffeic acid, ferulic acid, hydroxyl cinnamic acid, etc.); (ii) flavonoids (e.g., flavonols, flavones, isoflavones, flavonones, anthocyanidins, etc.); (iii) stilbenes; and (iv) lignans and polymeric lignins (Kluth *et al.* 2007). Table 19.1 lists the main groups of phytochemicals and their subdivisions.

Over the past few decades, there was an increased interest from the scientific community in phytochemicals or phytonutrients mainly because of their potential role in the maintenance of health and prevention of diseases. A search in the National Center for Biotechnology Information (NCBI) database (http://www.ncbi.nlm.nih.gov/pubmed) to access scientific publications by using the keywords "phytochemical" or "phytonutrient" or "polyphenol" presented 4748 articles involving these compounds.

Studies suggest that populations on a Western diet, which is characterized by high intakes of high-fat red meat, processed meat, butter, processed and fried foods, eggs and refined grains, but low in fruits and vegetables, are at a high risk for the development of degenerative diseases such as cancer and cardiovascular diseases (Riboli and Norat 2003; Manach and Williamson 2005; Kanner 2007). One of the challenges of modern nutrition research is the development of functional and health-promoting foods, dietary recommendations for health maintenance and well-being throughout life, as well as diets for special groups of the population (Holst and Williamson 2008). Phytochemicals are good candidates for the development of functional foods, but further studies are needed to confirm their health benefits to humans.

Table 19.1 Phytochemicals: main groups & subgroups

Groups	Subgroups	Examples
Polyphenols	Phenolic acids	
	Flavonoids	
	Stilbenes	
	Lignans & lignins	
Carotenoids	Xanthophylls	
	Carotenes	

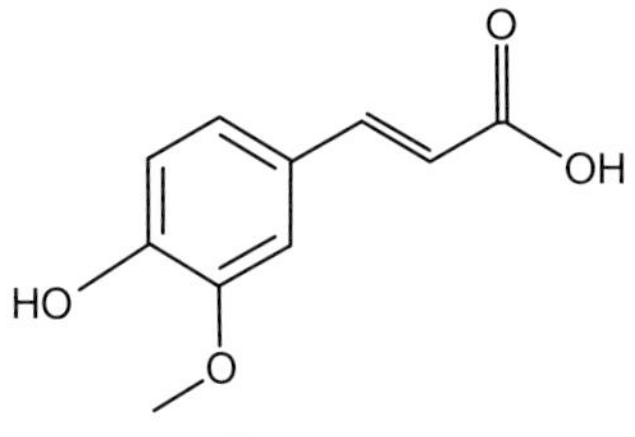

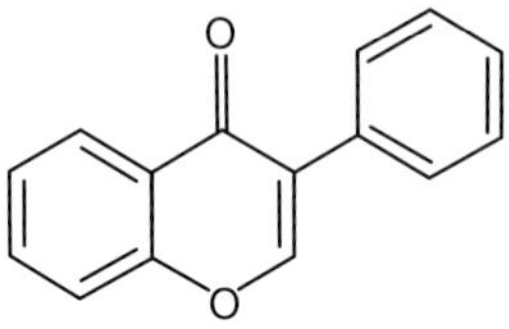

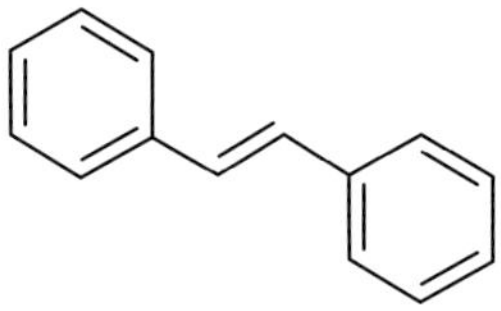

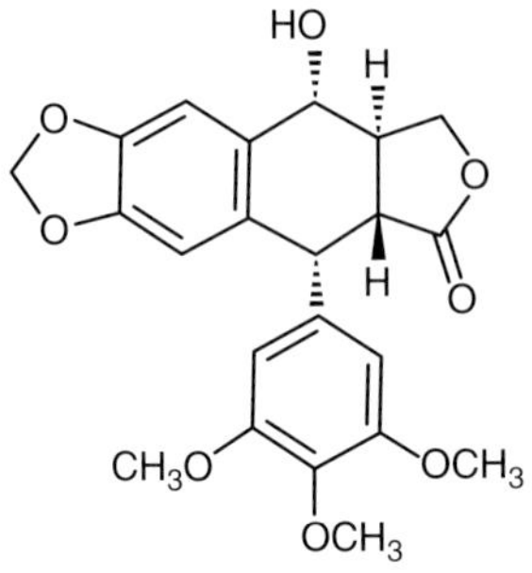

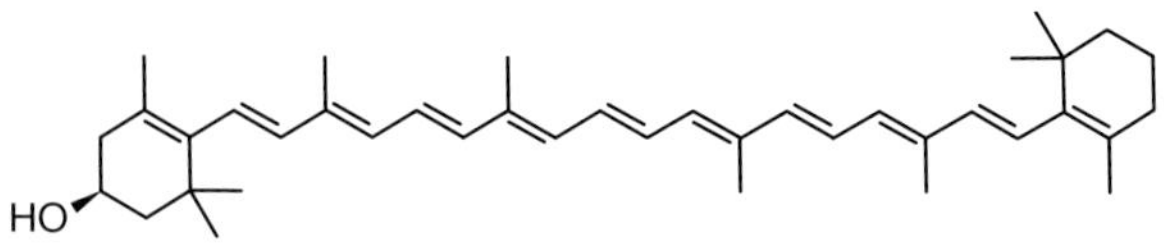

Ferulic acid

Isoflavone

(E)-Stilbene

Podophyllotoxin

Cryptoxanthin

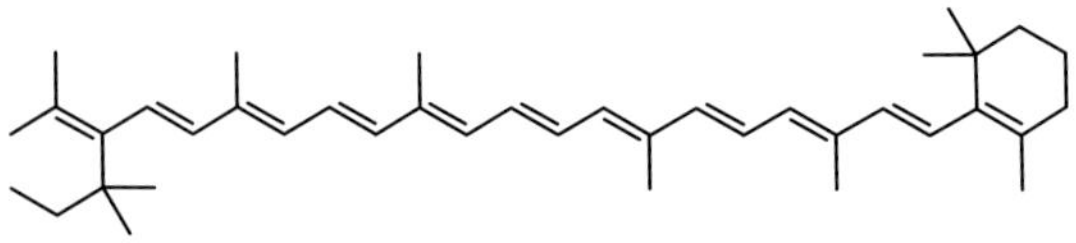

Beta-carotene

Table 19.1 (*Continued*)

Groups	Subgroups	Examples
Glucosinolates	Aliphatic	

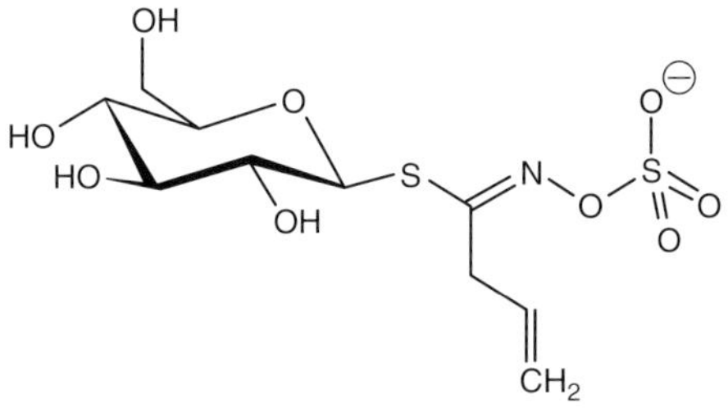

Sinigrin

| | Aromatic | |

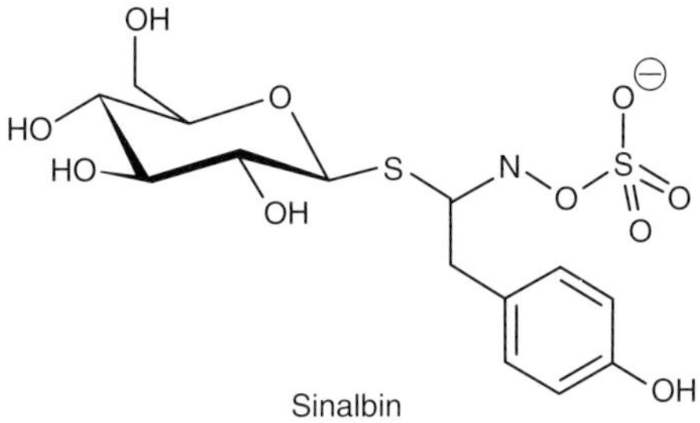

Sinalbin

| Sulphur-containing compounds | | |

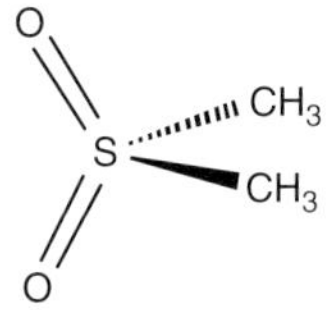

Methylsulfonylmethane

HEALTH BENEFITS OF PHYTONUTRIENTS OR PHYTOCHEMICALS

There are studies suggesting correlations between the frequency and/or amount of fruits and vegetables consumed and longevity and/or the delay in the onset of chronic degenerative diseases such as cancer, type 2 diabetes, and heart diseases (Temple and Gladwin 2003; Birt 2006; Kluth *et al.* 2007; Kale *et al.* 2008; Dembinska-Kiec *et al.* 2008). Some of these beneficial effects have been associated with the consumption of polyphenols–flavonoids (D'Ambrosio 2007; Geleijnse and Hollman 2008). Different flavonoids compounds have been reported to influence a series of biological processes such as inhibition of activation of pro-carcinogens, inhibition of proliferation of cancer cells, apoptosis, inhibition of metastasis and angiogenesis, protection against oxidative damage (antioxidant effect), activation of immune response, modulation of the inflammatory response, and modulation of drug resistance (Middleton 1998; Ofer *et al.* 2005; Yoon and Liu 2007; Walle 2007; Dembinska-Kiec *et al.* 2008; Hatcher *et al.* 2008; Kale *et al.* 2008). These biological processes are known to be involved in the development of chronic degenerative diseases such as cancer, type 2 diabetes, and cardiovascular diseases. Most of the studies have been carried out using cell culture or animal models to predict the pharmacological aspects of absorption, distribution, and metabolism, and beneficial effects of phytonutrients.

However, due to pharmacological and genetic differences, animal models are limited in their ability to predict phytonutrient responses in humans (D'Ambrosio 2007). For this reason further research is needed to determine the bioavailability and metabolism of phytonutrients and confirm their influence and activity in humans. A better understanding of the protective mechanisms exerted by phytonutrients is hindered by the lack of complete knowledge on their bioavailability *in vivo*. More information about absorption, distribution, metabolism, and excretion of individual phytonutrients is needed (Canali *et al.* 2007). Therefore, metabolomics strategies can help unravel the beneficial effects of phytonutrients to human health in a comprehensive manner.

LIMITATION OF ASSESSING HEALTH BENEFITS

Assigning health benefits of an individual food or food constituents is difficult because of the complexity of the diet, the co-occurrence and interactions of nutrients and phytochemicals in foods, interactions between diet, and human genetic background and environmental factors (Kluth *et al.* 2007; Holst and Williamson 2008; Kale *et al.* 2008).

Another challenge in assessing the health benefits of phytonutrients is related to the difficulty in controlling the amount and quality of phytochemical consumed. The levels of these compounds in plant-derived foods are highly variable as they are affected by seasonal and agronomic factors, plant variety, age and part of the plant used, food preparation conditions, etc. (Holst and Williamson 2008). One way to overcome this limitation is to monitor the metabolite profile of different individuals during the study. Metabolite profiles represent the real endpoints of metabolism, and any physiological change underlying regulatory processes will alter this profile. Therefore, metabolite profiling analysis (metabolomics) can provide relevant information on the adaptations of the body due to increase or decrease in the flux of certain nutrients through the metabolic pathways in the body (Kussmann *et al.* 2008).

Metabolomics enables a quantitative, noninvasive analysis of easily accessible human body fluids (urine, blood, saliva, tears) and has therefore gained popularity in current nutrition research (Holst and Williamson 2008). In humans and animals, metabolite profiles can be used as important indicators of normal phenotype and pathology and offer the possibility of identifying biomarkers of disease states.

EFFECT OF THE GUT MICROFLORA

During studies evaluating health benefits of phytochemicals, another important point to take in consideration is related to compounds that are also active in the lumen of the intestinal tract, rather than just systemically (Halliwell *et al.* 2000; Gorelik *et al.* 2007; Grün *et al.* 2008). There is a complex mixture of antioxidants such as flavonoids in the diets, which may have an important role in protecting the gastrointestinal tract itself from oxidative damage (Halliwell *et al.* 2000). The intestine is the major organ involved in immune response. The possible effects of phytochemicals on the gut microflora may affect systemic health parameters.

The intestinal microflora is estimated to present around 10^{10} to 10^{12} bacterial cells/g colonic content (Simon and Gorbach 1984; Yuan *et al.* 2007). The diet can alter the bacterial profile and determine which bacterial groups become dominant in the gastrointestinal tract. Intestinal microflora can also break down phytonutrients and form a wide range of bioactive products, which can be responsible for some of the health benefits ascribed to the original phytochemical (Grün *et al.* 2008). For example, Yuan *et al.* (2007) showed that the clinical effectiveness of soy isoflavones depends on the microbial biotransformation of isoflavones into the stronger oestrogen, equol (Fig. 19.1). Absorption of polyphenols, for instance, normally occurs after the compounds are degraded into smaller phenolic acids by the colonic microbiota (Grün *et al.* 2008). These authors have studied the metabolic impact of flavonoid intake and presented

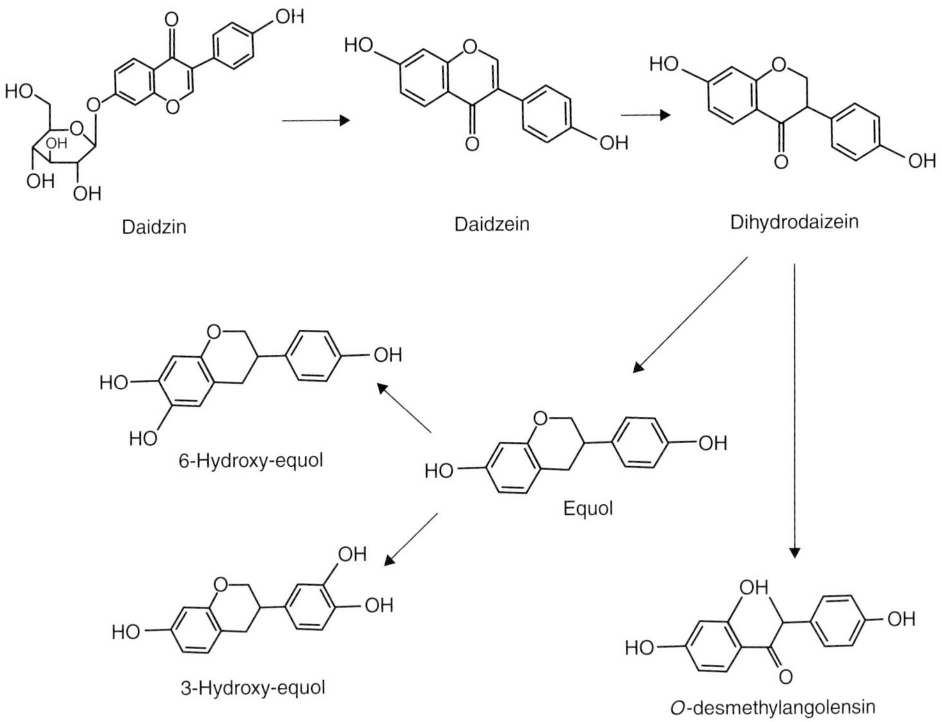

Figure 19.1 Biotransformation of soy isoflavones by gut microorganisms, according to Yuan *et al.* (2007). Gut microbes are able to convert isoflavones into molecules with higher biological activity such as equol and derivatives.

methods applicable to different biological fluids to investigate at a large scale the gut microbiome-mediated bioavailability of flavonoids (Grün *et al.* 2008).

Phytochemicals are largely excreted in urine as catabolites that are formed in host tissues or by the intestinal microbiota and can be characterized by metabolomics (Fardet *et al.* 2008).

METABOLOMICS TOOLS FOR PHYTONUTRIENTS' RESEARCH

Analytical methods optimized for metabolomics research are capable of detecting changes in the distribution and concentration of a wide range of low molecular weight metabolites in biological samples (Villas-Bôas *et al.* 2007). Metabolites can be profiled in different levels of biological organization from cell culture to whole organisms (Whitfield *et al.* 2004). Within this context, extracellular metabolites present in body fluids such as milk, urine, and plasma, are an important source of metabolic information and can be handled easier than samples from solid tissues. For instance, metabolites present in the blood provide metabolic information on all tissues that deliver metabolites to the blood and uptake metabolites from it (Villas-Bôas 2007). Therefore, body fluids are preferable biological samples to study the effects of phytonutrients on animal metabolism using metabolomics.

The potential value of the metabolomics approach to phytonutrients research has been demonstrated in several studies such as the metabolism of isoflavones (Solanky *et al.* 2003); ethyl glucoside (Teague *et al.* 2004); lignins, ferulic acid, and sinapic acid (Fardet *et al.* 2008), and others. Both nuclear magnetic resonance (NMR) and mass spectrometry-based techniques have been used to profile metabolites in body fluids and animal tissues (Walsh *et al.* 2007; Guy *et al.* 2008). However, most phytochemical compounds such as phenolic acids are found in animal body fluids at very low levels (ng/mL to μg/mL levels). Thus, mass spectrometry-based analytical methods are more sensitive, and therefore preferable, to profile these compounds. On the other hand, ^{1}H-NMR-based techniques can easily assess the effect of different dietetic phytonutrients on global metabolite profile of individuals or animal models (Walsh *et al.* 2007; Jacobs *et al.* 2008).

There are two main metabolomics approaches suitable to evaluate the effects of photochemicals using humans or animal models: (i) metabolite profiling without identification of detected compounds (sometimes referred to as metabolic fingerprinting), and (ii) global metabolite profiling. Both strategies have their advantages and disadvantages and are discussed in the following:

1. Metabolite profiling without identifications

    Metabolite profiling without identifying the compounds analyzed is a high-throughput approach based on a rapid screening of different metabolites in a sample using powerful detectors such as mass spectrometers or NMR spectroscopy. In these analyses only the raw analytical data (e.g., MS ions or fragments, NMR peaks, chromatographic peaks) are used to characterize a set of biological samples (Smedsgaard 2007). The raw data are statistically treated and unidentified components that are significantly associated with different data classes can be short listed using a wide range of multivariate data analysis techniques (Hansen 2007). This approach has been widely used in metabolomics research (Villas-Bôas *et al.* 2005; Mas *et al.* 2007; Cao *et al.* 2008, and many others) and several examples can be found applied to phytonutrients research (Fardet *et al.* 2008; Guy *et al.* 2008; Jacobs *et al.* 2008; McDougall *et al.* 2008).

    The main advantage of metabolite profiling without identifications is the rapid analysis of metabolites within a few minutes and with minor sample preparation requirements (high-throughput metabolome analysis). A drawback of this approach is, however, the sample matrix-effect due to interference of salts and strong ions that decrease considerably the sensitivity as well as the formation of artifacts and the limitation in identifying the metabolites detected (Mas *et al.* 2007). Without identification of a large number of metabolites, the biological information obtained is poor and can only be used to determine how similar or different the data classes are. Identification of compounds analyzed by metabolite fingerprinting approach is very time-consuming and can be inaccurate depending of the analytical technique used (e.g., metabolite isomers when analyzed by direct infusion mass spectrometry).

2. Global metabolite profiling

    Global metabolite profiling involves the analysis and identification of as many metabolites as possible present in a biological sample using one or more analytical techniques. Gas chromatography coupled with mass spectrometry (GC-MS) is the preferable analytical method for this purpose. GC provides the best resolution to separate hundreds of metabolites in a single analysis, and coeluted peaks can be further deconvoluted automatically by using deconvolution software (Villas-Bôas *et al.* 2005). In addition, the reproducibility of electron impact ionization used by common GC-MS quadrupole setups makes the use and exchange of MS library possible, which enhances tremendously the compound identification potential of this technique (Villas-Bôas *et al.* 2005). However, there are very few reported works using GC-MS for phytonutrients' research (e.g., Grün *et al.* 2008). Nevertheless, global metabolite profiling can also be achieved by combining data obtained from different analytical methods. In this case, liquid chromatography coupled with tandem mass spectrometry (LC-MS/MS^n) or high-mass resolution mass spectrometry (e.g., time-of-flight, TOF; Fourier transform ion cyclotron resonance, FTICR; orbitrap) can be an alternative despite presenting poor chromatographic separation power (Fardet *et al.* 2008).

PHYTOCHEMICAL RESEARCH REQUIRES A "METAMETABOLOMICS" APPROACH

Most metabolomics studies have been focused on the analysis of metabolites produced or transformed by a living organism. However, to evaluate the effect of phytochemicals on the health of individuals the original concept of metabolomics as the set of metabolites produced a single organism in a given environmental condition is challenged because phytochemical research involves metabolites from at least two (but often more than two) different organisms as illustrated in Fig. 19.2: (1) the animal consuming the phytochemicals in its diet; (2) the plant tissues rich in phytochemicals; and (3) the microorganisms in the bowel. Therefore, this kind of metabolomics study involves multiple organisms and, thus, multiple metabolomes. We suggest referring to this type of approach as "metametabolomics," which is focused on the interactions of different metabolomes.

The main challenge for a metametabolome approach, however, is our limited ability in distinguishing metabolites originating from different metabolomes. Just a few metabolites are genuinely produced by a single group of organisms (e.g., some plant secondary metabolites). Most metabolites, however, are ubiquitous to all living cells. Therefore, metametabolomics rely on the comparison of data obtained from single metabolomes (e.g., germ-free mice) to data

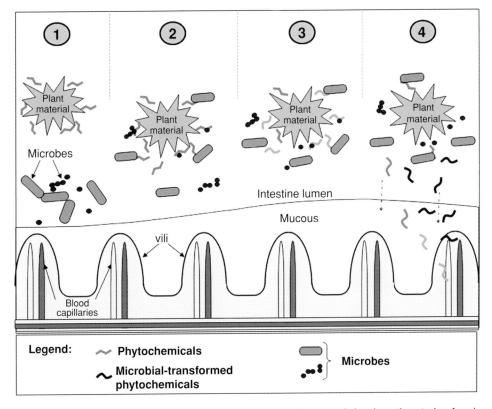

Figure 19.2 Metametabolomics approach – Phytochemical research involves the study of multiple metabolomes and their interaction. Metabolites from plant material containing phytochemicals (as part of the diet) interact with the gut microflora and with the animal metabolic processes. Microorganisms in the bowel can biotransform or degrade some phytochemical compounds enhancing their absorption or changing their biological activity. (Kindly prepared by Xavier Duportet.) For color detail, please refer to the color plate section.

obtained from combined metabolomes. Alternatively, phytochemicals or plant metabolites can be labeled with isotope atoms, and their fate within the animal and/or animal + gut microbial metabolism can be potentially traced using mass spectrometry or NMR techniques.

In conclusion, metabolomics present a great potential to evaluate the effect and fate of phytochemicals on animal health and metabolism as well as to understand the role of the gut microflora on the absorption and transformation of these phytonutrients. Better knowledge in this field will assist in developing novel functional food ingredients and novel nutrition concepts targeted to enhanced health and well-being.

REFERENCES

Birt, Diane F. 2006. Phytochemicals and cancer prevention: from epidemiology to mechanism of action. *Journal of the American Dietetic Association 106*(1):20–21.

Canali, Raffaella; Ambra, Roberto; Stelitano, Cecilia; Mattivi, Fulvio; Scaccini, Cristina; Virgili, Fabio. 2007. A novel model to study the biological effects of red wine at the molecular level. *British Journal of Nutrition 97*(6):1053–1058.

D'Ambrosio, Steven M. 2007. Phytonutrients: A more natural approach toward cancer prevention. *Seminars in Cancer Biology 17*(5):345–346.

Dembinska-Kiec, Aldona; Mykkänen, Otto; Kiec-Wilk, Beata; Mykkänen, Hannu. 2008. Antioxidant phytochemicals against type 2 diabetes. *British Journal of Nutrition 99*(E-S1):ES109–ES117.

Fardet, Anthony; Llorach, Rafael; Orsoni, Alexina; Martin, Jean-François; Pujos-Guillot, Estelle; Lapierre, Catherine; Scalbert, Augustin. 2008. Metabolomics provide new insight on the metabolism of dietary phytochemicals in rats. *The Journal of Nutrition 138*:1282–1287.

Geleijnse, Johanna M.; Hollman, Peter C.H. 2008. Flavonoids and cardiovascular health: which compounds, what mechanisms? *American Journal of Clinical Nutrition 88*(1):12–13.

Gorelik, S., Kohen, R., Ligumsky, M., Kanner, J. 2007. Saliva plays a dual role in oxidation process in stomach medium. *Archives of Biochemistry and Biophysics 458*(2):236–243.

Grün, Christian H.; van Dorsten, Ferdi A.; Jacobs, Doris M.; Le Belleguic, Marie; van Velzen, Ewoud J.; Bingham, Max O.; Janssen, Hans-Gerd; van Duynhoven, John P.M. 2008. GC-MS methods for metabolic profiling of microbial fermentation products of dietary polyphenols in human and in vitro intervention studies. *Journal of Chromatography B 871*(2):212–219.

Guy, Philippe A.; Tavazzi, Isabelle; Bruce, Stephen J.; Ramadan, Ziad; Kochhar, Sunil. 2008. Global metabolite profiling analysis on human urine by UPLC-TOFMS: Issues and method validation in nutritional metabolomics. *Journal of Chromatography B 871*(2):253–260.

Halliwell, Barry; Zhao, Kaicun; Whiteman, Matthew. 2000. The gastrointestinal tract: a major site of antioxidant action? *Free Radical Research 33*(6):819–830.

Hansen, Michael A.E. 2007. "Data analysis." In *Metabolome Analysis: an Introduction*, edited by Silas G. Villas-Bôas, Ute Roessner, Michael A.E. Hansen, Jørn Smedsgaard & Jens Nielsen; pp. 146–190. John Wiley & Sons.

Hatcher, H.; Planalp, R.; Cho, J.; Torti, F.M.; Torti, S.V. 2008. Curcumin: from ancient medicine to current clinical trials. *Cellular and Molecular Life Sciences 65*(11):1631–1652.

Holst, Birgit; Williamson, Gary. 2008. Nutrients and phytochemicals: from bioavailability to bioefficacy beyond antioxidants. *Current Opinion in Biotechnology 19*(2):73–82.

Jacobs, Doris M.; Deltimple, Nancy; van Velze, Ewoud; van Dorsten, Ferdi A.; Bingham, Max; Vaughan, Elaine E.; van Duynhoven, John. 2008. ^{1}H NMR metabolite profiling of feces as a tool to assess the impact of nutrition on the human microbiome. *NMR in Biomedicine 21*:615–626.

Kale, Anup; Gawande, Sonia; Kotwal, Swati. 2008. Cancer phytotherapeutics: role for flavonoids at the cellular level. *Phytotherapy Research 22*(5):567–577.

Kanner, Joseph. 2007. Dietary advanced lipid oxidation endproducts are risk factors to human health. *Molecular Nutrition and Food Research 51*(9):1094–101.

Kluth, Dirk; Banning, Antje; Paur, Ingvild; Blomhoff, Rune; Brigelius-Flohé, Regina. 2007. Modulation of pregnane X receptor-and electrophile responsive element-mediated gene expression by dietary polyphenolic compounds. *Free Radical Biology & Medicine 42*:315–325.

Kussmann, Martin; Rezzi, Serge; Daniel, Hannelore. 2008. Profiling techniques in nutrition and health research. *Current Opinion in Biotechnology 19*(2):83–99.

Manach, Claudine; Williamson, Gary. 2005. Bioavailability and bioefficacy of polyphenols in humans. II. Review of 93 bioavailability studies. *American Journal of Clinical Nutrition 81*(Suppl):243S–255S.

Mas, Sandrine; Villas-Bôas, Silas G.; Hansen, Michael E.; Åkesson, Mats; Nielsen, Jens. 2007. A comparison of direct infusion MS with GC-MS for metabolic footprinting of yeast mutants. *Biotechnology & Bioengineering 96*(5):1014–1022.

McDougall, Gordon; Martinussen, Inger; Stewart, Derek. 2008. Towards fruitful metabolomics: High throughput analyses of polyphenol composition in berries using direct infusion mass spectrometry. *Journal of Chromatography B 871*:362–369.

Middleton, E. Jr. 1998. Effect of plant flavonoids on immune and inflammatory cell function. *Advances in Experimental Medical Biology 439*:175–182.

Ofer, Monikap; Wolffram, Siegfried; Koggel, Annette; Spahn-Langguth, Hilde; Langguth, Peter. 2005. Modulation of drug transport by selected flavonoids: Involvement of P-gp and OCT? *European Journal of Pharmaceutical Science 25*; 263–271.

Riboli, Elio; Norat, Teresa. 2003. Epidemiologic evidence of the protective effect of fruit and vegetables on cancer risk. *American Journal of Clinical Nutrition 78*(3):559S–569S.

Simon, G.L.; Gorbach, S.L. 1984. Intestinal flora in health and disease. *Gastroenterology 86*:174–193.

Smedsgaard, Jørn. 2007. "Analytical tools." In *Metabolome Analysis: an Introduction*, edited by Silas G. Villas-Bôas, Ute Roessner, Michael A.E. Hansen, Jørn Smedsgaard & Jens Nielsen; pp. 146–190. John Wiley & Sons.

Solanky, Kirty S.; Bailey, Nigel J.C.; Beckwith-Hall, Bridgette M.; Davis, Adrienne; Bingham, Sheila; Holmes, Elaine; Nicholson, Jeremy K.; Cassidy, Aedin. 2003. Application of biofluid ^{1}H nuclear magnetic resonance-based metabonomic techniques for the analysis of the biochemical effects of dietary isoflavones on human plasma profile. *Analytical Biochemistry 323*(2):197–204.

Teague, Claire; Holmes, Elaine; Maibaum, Elaine; Nicholson, Jeremy; Tang, Huiru; Chan, Queenie; Elliott, Paul; Wilson, Ian. 2004. Ethyl glucoside in human urine following dietary exposure: detection by ^{1}H NMR spectroscopy as a result of metabonomic screening of humans. *Analyst 129*(3):259–264.

Temple, Norman J.; Gladwin, Kerri K. 2003. Fruit, vegetables, and the prevention of cancer: research challenges. *Nutrition 19*(5):467–470.

Villas-Bôas, Silas G. 2007. "Sampling and sample preparation." In *Metabolome Analysis: an Introduction*, edited by Silas G. Villas-Bôas, Ute Roessner, Michael A.E. Hansen, Jørn Smedsgaard & Jens Nielsen; pp. 146–190. John Wiley & Sons.

Villas-Bôas, Silas G.; Mas, Sandrine; Åkesson, Mats; Smedsgaard, Jørn; Nielsen, Jens. Mass spectrometry in metabolome analysis. *Mass Spectrometry Reviews 24*:613–646.

Villas-Bôas, Silas G.; Roessner, Ute; Hansen, Michael A.E.; Smedsgaard, Jørn; Nielsen, Jens. 2007. *Metabolome analysis, an introduction*. John Wiley & Sons.

Walle, Thomas. 2007. Methoxylated flavones, a superior cancer chemopreventive flavonoid subclass? *Seminars in Cancer Biology 17*(5):354–362.

Walsh, Marianne C.; Brennan, Lorraine; Pujos-Guillot, Estelle; Sébédio, Jean-Louis; Scalbert, Augustin; Fagan, Ailís; Higgins, Desmond G.; Gibney, Michael J. 2007. Influence of acute phytochemical intake on human urinary metabolomic profiles. *American Journal of Clinical Nutrition 86*:1687–1693.

Whitfield, Phillip D.; German, Alexander J.; Noble, Peter J. 2004. Metabolomics: an emerging post-genomic tool for nutrition. *British Journal of Nutrition 92*(4):549–555.

Yoon, Hyungeun; Liu, Rui H. 2007. Effect of selected phytochemicals and apple extracts on NF-KB activation in human breast cancer MCF-7 cells. *Journal of Agricultural and Food Chemistry* 55(8):3167–3173.

Yuan, Jian-Ping; Wang, Jiang-Hai; Liu, Xin. 2007. Metabolism of dietary soy isoflavones to equol by human intestinal microflora-implications for health. *Molecular Nutrition & Food Research* 51(7):765–781.

Section 5

Nutrigenomics in Human Health

20

Omics for the Development of Novel Phytomedicines

Kandan Aravindaram, Harry Wilson, and Ning-Sun Yang

INTRODUCTION

It is well appreciated now that food provides not only various essential nutrients but also a spectrum of bioactive compounds that promote health and prevent disease. The routine and adequate consumption of fruits, vegetables, and spices, as well as grains, is strongly associated with a reduced risk of cancer, diabetes, cardiovascular disease, and aging-related functional decline (Willett 1994; Temple 2000; Chang *et al*. 2007; Hou *et al*. 2007). The use of fruits, vegetables, whole grains, and "medicinal" herbs as dietary supplements or as "functional foods" is a new dietary trend for providing and/or maintaining good health in many developed countries (e.g., Western Europe and Northern America). Prominent in this new "health" food culture are medicinal or health-promoting herbal materials, especially phytoremedies from traditional Chinese medicine (TCM), Ayurveda, and European herbal traditions, which contain significant amounts of bioactive phytocompounds. As a consequence, the potential health benefits of a broad spectrum of botanical substances, herbal remedies, foods, and nutritional supplements have been actively investigated in recent years.

The word "phytomedicine" can be defined in a number of different ways. These may include the single medicinal compounds derived from plant materials (e.g., taxol, salicylic acid), the plant extracts, either crude (e.g., *Echinacea purpurea*), solvent-fractionated, or partially purified phytochemical mixtures (e.g., curcumins, shikonins), and plant extracts prepared from a combination of several medicinal plants as found in most traditional Chinese medicines. In this chapter, though we do not intend to be exclusive, we shall mainly use a definition of phytomedicinal plant extracts as partitioned fractions, or the derived phytocompounds from a single plant species.

Although there has been a long history of the use of medicinal herbs or phytomedicines, these traditional therapies are now recognized to have a number of drawbacks compared to modern "Western medicines." However, traditional phytomedicines also have some clear strengths, especially (1) their apparent low acute cytotoxicity, (2) the fact that long experience of their use in humans can identify candidate botanical drugs; and (3) the combinational or combinatorial use of multiple plant extracts, as in TCM, may indeed be a useful alternative or adjunct to experimentation as an approach for developing novel phytomedicines. There is much social interest and potential public health benefit in the pursuit of systematic research into the development of phytomedicines, either as new drugs or as alternative and supplementary medicines. Among the many research tools available in modern medical biology, the "omics" approach will be a key experimental avenue for such an important research task. The application of the various "omics" approaches to phytomedicine development is the theme of this review.

Unlike the "standard" synthesized drugs, phytomedicines, as broadly defined previously, have in recent years been used as food supplements, health foods, and nutraceuticals. In some countries (e.g., in Germany but not in the United States), medicinal herbal products are commercially available as botanical drugs/pharmaceuticals or as functional food or health food supplements. Though there has been extensive worldwide use of St. John's wort, saw palmetto, *Ginkgo biloba*, milk thistle, *Echinacea*, and others, the lack of useful information on their medicinal or therapeutic efficacy is a potential health care problem. Another category of phytomedicines is plant extracts from turmeric (curcumin), ginger, garlic, scallion, and other common spices, or even from specialty fruits and vegetables, including tomato (lycopene), grape (resveratrol), olive oil, and others that have antioxidant, antipathogen, and anti-inflammatory activities, and are thus used for prevention of various cancers, cardiovascular disease, and aging-related diseases (Anand *et al.* 2008). The need to improve the appropriate use of these two categories of plant materials, that is, medicinal herb extracts (e.g., *Ginkgo*, *Echinacea*, etc.) as botanical drugs, and the specialty spices, fruits, and vegetables as fresh produce, also highlights the need for evidence-based research. One line of evidence comes from clinical trials of plant extracts as botanical drugs, and another is to characterize their specific physiological effects, and if possible, isolate the active compound(s) using bioactivity-guided assays. It is in this latter task that the "omics" approach may be most useful. By expediting the identification of "functional activity" at the molecular, cellular, and tissue/organ levels, it will advance our understanding of various important traditional medicinal plants and help future development of high-quality phytomedicines.

Many herbal extract preparations or derived phytocompounds have been claimed to stimulate or enhance immune function and to display an array of immuomodulatory effects. A number of *in vitro* and *in vivo* studies of these candidate phytomedicines have shown that they can modulate cytokine/chemokine secretion, cellular factor or receptor expression, phagocytotic activity, immune cell modification, and innate immunity. A well-known example is the still controversial use of *Echinacea* plant extracts in the treatment of the common cold and flu (Plaeger 2003; Wang *et al.* 2006). A diverse spectrum of natural products, chiefly plant secondary metabolites, possess antimicrobial and immunomodulatory potential, including indoles, isoflavonoids, galactolipids, polysaccharides, phytosterols, alkaloids, sesquiterpenes, glucans, terpenoids, and tannins. Several plant extracts or fractionated metabolite preparations are now undergoing serious evaluation as alternative or complementary agents for costly immunotherapeutic agents, often recombinant cytokines or chemokines, for various diseases.

Cells of the human body are constantly exposed to a variety of oxidizing agents, which cause an imbalance in homeostasis and normal physiology, leading to oxidative stress and increased susceptibility to various infections. Various plant-derived antioxidants can reduce the level of oxidative stress induced by free radicals and help lower the risk of various diseases caused or associated with such stress. Because microbial infection and various physiological and environmental stresses also often induce an inflammatory response in the host, dietary anti-inflammatory phytocompounds such as epigallocatechin gallate (EGCG), curcumin, and emodin may also contribute to the prevention of chronic diseases or cancer (Table 20.1). There is an urgent need for scientists to identify various active phytocompounds that can be developed as medicines or nutraceuticals to provide cost-effective and acceptable dietary avenues for disease control and prevention.

Recent advances in the development of biotechnology platforms in genomics, proteomics, and metabolomics will play a major role in phytomedicine research. The so-called -omics view is part of a systems biology concept that looks at effects on both the cellular and organismic levels. Functional genomics has evolved into a science that deals with temporal and spatial differences in gene expression at the RNA level, of a specific biological activity, drug effect, or physiological response. The proteomics approach explores the different expression of all proteins, their organellar distribution, and signaling pathway involvement and interactions of groups of proteins. Metabolomics is a systems study of the overall metabolite profile or "chemical fingerprint" of

Table 20.1 Some examples of herbal extracts or/and the derived phytochemicals that exhibit specific bioactivities and may serve as future candidate phytomedicines

Common Name	Formal Name	Activity	References
Ginseng	*Panax ginseng*	Anticancer and improves physical performance	Song *et al.* (2002); Tyler & Foster (1999)
Black cohosh	*Cimicifuga racemosa*	Improves premenstrual syndrome, menopausal symptoms	Tyler & Robbers (1999)
Cone flower	*Echinacea angustifolia*	Immune modulation, promotes wound healing	Wang *et al.* (2006); Blumenthal *et al.* (2000)
Lei gong teng	*Tripterygium wilfordii*	Antihepatitis, antirheumatoid arthritis, and leprosy	Chen (2001)
Comfrey	*Symphytum officinale*	Wound healing	Tyler & Robbers (1999)
Ma huang	*Ephedra sinica*	Relieves asthma and nasal congestion	US-FDA (1997)
European mistletoe	*Viscum album*	Improves immune system, cytostatic and apoptotic	Fritz (2004)
Garlic	*Allium sativum*	Anti-inflammatory and reduces cardiovascular risk	Blumenthal *et al.* (2000)
Long pepper	*Piper longum*	Powerful digestive stimulant and anticancer	Sunila & Kuttan (2004)
Ginkgo	*Ginkgo biloba*	Improves memory, cognition, vertigo, and intermittent claudication	Tyler & Foster (1999)
Fox tail	*Sophora alopecuoides*	Antiviral	Liu *et al.* (2003)
Kava-kava	*Piper methysticum*	Reduces anxiety and depression	US-FDA (2002)
Snail seed	*Sinomenium acutum*	Antiarthritis and rheumatoid arthritis	Liu *et al.* (2003)
Saw palmetto	*Serenoa repens*	Builds sexual vigor, reduces hair loss, reduces prostatic hypertrophy and bladder disorders	Tyler & Robbers (1999)
Sweet wormwood	*Artemisia annua*	Psoriasis and autoimmune disorders	Zhang *et al.* (1995)
St. John's wort	*Hypericum performatum*	Anti-inflammatory, antiseptic, antiviral, and antidepressive	Tyler & Robbers (1999)
Devil's bush	*Acanthopanax senticosus*	Anticancer	Han *et al.* (2003)

(Continued)

Table 20.1 *(Continued)*

Common Name	Formal Name	Activity	References
Yohimbe	*Pausinystalia yohimbe*	Treats male impotence, blood pressure boosting agent, body building	Blumenthal *et al.* (2000)
Kutki	*Picrorhiza kurroa*	Hepatoprotective, antioxidant, anti-inflammatory, and antiallergic	Vanden *et al.* (2001)
Beggar's ticks	*Bidens pilosa*	Antidiabetes	Chang *et al.* (2007)
Satavar	*Asparagus racemosus*	Used as anodyne, aphrodisiac, galactogogue	Gautam *et al.* (2004)
Nodding burnweed	*Crassocephalum rabens*	Anticancer	Hou *et al.* (2007)
Marine sponge	*Agelas mauritianus*	Anticancer	Tan & Vanitha (2004)
Yellow cypress	*Chamaecyparis obtuse var. formosana*	Antiviral	Wen *et al.* (2007)

a given test system, which in the future will be capable of investigating a broad spectrum of relatively low molecular weight metabolites. The omics approach is expected to be developed into a powerful tool for the identification of novel phytomedicine from medicinal herbs, foods, and spices. We will here present an overview of the various approaches and technologies of the omics experimental systems, based on experience gained in our own laboratory. Major aspects of the "-omics" technologies used in phytomedicine research are outlined in Fig. 20.1.

GENOMICS

The term "genomics" was coined by Tom Roderick in the year 1986 to describe the discipline of "mapping/sequencing (including analysis of the information)" of the entire genome (McKusick & Ruddle 1997), and came to include the analysis of the genomic effects of specific biological activities of defined cells, tissues, organs, or even an organism. These biological activities can be a response to specific stimulation, stress, or treatment of the test system. For decades, understanding the functions of specific genes in the context of a defined biological system or activity, for example, hormonal effect, drug response, or nutritional deficiency, has been a research challenge, mainly because in most, if not all cases, experimenters were only able to follow the gene expression of one, two, or several genes at a time. This "single-minded" and "simple-minded" approach was nonetheless extremely powerful during the 1980s and 1990s. However, it soon became apparent that most biochemical, cellular, and tissue functions are not controlled by a small number of genes (≤ 5), but rather by a large number, often in the hundreds. These genes are also often involved in one or several cellular physiological pathways such as the insulin, the G protein, or the toll-like receptor signaling pathways. Therefore, instead of sporadically pinpointing differential gene expressions with limited, known, or obvious target genes, an integrated, global view of changes in gene expression (at both RNA transcription and protein translation levels) is necessary. DNA microarray technology was developed in

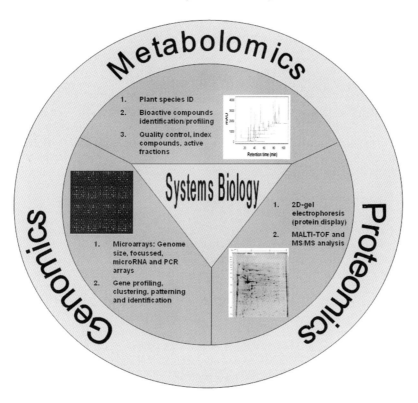

Figure 20.1 Major key aspects of the omics technologies used in phytomedicine research. For color detail, please refer to the color plate section.

response to the need for a high-throughput, comprehensive, and efficient means of analyzing gene expression profiles: measuring the expression of all genes, or of a defined subset thereof, in a genome (Schena et al. 1995; Schena et al. 1996).

Affymetrix and Agilent are the two manufacturers that dominate this field today. DNA microarrays consist of an arrayed series of thousands to tens of thousands of oligonucleotides, or short cDNA sequences, printed onto an impermeable solid support, usually silicon chips or glass. This short section of a DNA fragment with unique gene sequence is used as a probe to hybridize with test DNA, under conditions of high stringency. DNA microarrays or "gene chips" come in different sizes: in addition to the full genome size (30,000 genes/chip) (Wang et al. 2008), there are also subgenomes, and focused or functionally grouped DNA microarrays that test only hundreds to a couple of thousand genes (Wang et al. 2006) at a time for functionally defined gene subsets.

DNA microarray technology was developed with a combination of several different technologies, and it is now the most often used tool for studies of functional genomics at the RNA level, also called transcriptomics (i.e., the "-omics" of transcripts). In addition to DNA microarray system, functional genomics, the study of global gene expression profiles, and key differential gene expression networks or signaling pathways, is also supported by automated DNA sequencing, polymerase chain reaction (PCR) amplification, synthesis and labeling of oligonucleotides and bioinformatics tools. Based on the target DNA sequences embedded onto the arrays, hundreds, thousands, or tens of thousands of genes can be targeted and significant changes in their mRNA, microRNA, or other transcripts can be determined simultaneously. Biostatistical and bioinformatic analyses and computation of the detection signal data, and functional characterization of gene patterns, clusters, and groupings with significant changes in expression, allow the prediction of major physiological changes in response to drug or herbal medicine treatment.

As an example of this kind of functional genomic study, we used a subgenomic size (~4000 genes) human cDNA microarray assay system to demonstrate that specific genes related to cytoskeleton rearrangement, apoptotic signal transduction, and various transcription factors were differentially regulated by an ethyl acetate (EA)-partitioned fraction of *Anoectochilus formosanus* in breast cancer MCF-7 cells (Yang *et al.* 2004). This study shows that, simply by using the DNA microarray approach alone, we were able to rapidly analyze the complex interaction and multifunctional effects of a mixture of phytocompounds, a technique that can easily be applied to either a single purified compound or to complicated mixtures of hundreds of phytocompounds as present in most plant-derived phytomedicines. The single experiment/global result is the real power of transcriptome analysis.

We developed a function-targeted DNA microarray system to characterize the effects of phytocompounds on human dendritic cells (DCs). *Echinacea* extracts have been used for decades in European and American countries as a botanical drug or nutraceutical, and are anecdotally claimed to boost immunity. We investigated the effect of crude extracts of *E. purpurea* on human DCs in primary culture using a focused cDNA microarray displaying ~250 immunoregulatory genes, either made in-house or commercially obtained (Wang *et al.* 2006). Downregulation of mRNA expression of specific chemokines (e.g., CCL3 and CCL8) and their receptors (e.g., CCR1 and CCR9) was observed in DCs treated with stem and leaf extract, whereas other chemokines and regulatory molecules (e.g., CCL4 and CCL2) involved in the c-Jun pathway were found to be upregulated in root extract-treated DCs (Wang *et al.* 2006).

More recently, we observed that the butanol-partition of the water-soluble fraction of the stem and leaf tissues of *E. purpurea* plants, a phytocompound mixture we have labeled [BF/S+L/Ep], has a number of specific effects on human DCs, as detected with an Affymetrix human DNA microarray gene chip, testing approximately 28,300 genes in one assay (Wang *et al.* 2008). The previously noted findings, exploring fractionated medicinal plant extracts as candidate therapeutic agent for immune-enhancing effect, may serve as an example for future development of phytomedicines, in terms of defining specific molecular, genomic, and cellular functions in targeted cells or molecular targets.

The vast amount of information generated by such large-scale studies makes the modern techniques of bioinformatics and data management crucial. For streamlined analyses of microarray raw data, gene expression pattern, profiling, and clustering, gene crosstalk networks and signaling pathway identification or characterization, various software systems, including Spotfire, TRANSPATH, and Ingenuity have often been used. An example of the gene expression/interaction network in mouse bone marrow-derived dendritic cells in response to phytomedicine treatment, analyzed by Ingenuity software, is shown in Fig. 20.2.

PROTEOMICS

Proteomics involves the large-scale study of the entire cellular array of proteins, particularly their biochemical identity, structure, and functions. The term proteome was proposed by Wilkins *et al.* (1995). Proteomics can be defined as the qualitative and quantitative comparison of protein profiles and their constituents under test conditions to unravel biological processes. The main technologies used in proteomics are two-dimensional sodium dodecyl sulfate gel electrophoresis to separate different proteins according to variations in isoelectric point and relative mass (Williams *et al.* 2003). Currently, when using silver stain or other dyes, approximately 2500 protein spots can be visualized for human proteins in most cell systems. The resolution capacity of such 2D gels is presently limited due to the low abundance of many regulatory proteins, (e.g., cytokines, transcription factors) or difficulties in extraction and separation of many membrane-bound proteins. The visible protein spots on a 2D gel can be digested *in gel* with trypsin

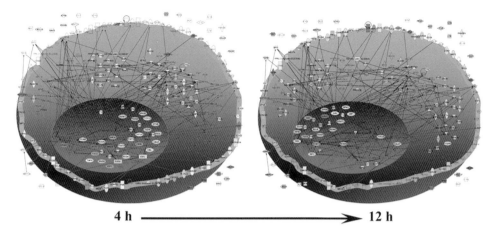

4 h ————————————————————————→ **12 h**

Figure 20.2 A typical display of signaling networks, connecting cell surface, cytosolic, and nuclear compartments, for functional and differential immunity gene expression in mouse bone marrow-derived dendritic cells in response to treatment with specific medicinal phytocompounds. A prototypical cell was constructed from 293 representative genes involved in immunity of immature mouse bone marrow dendritic cells. Genes for which the expression statistically increased are colored red, those for which expression decreased are shown in green. Temporal changes in apparent expression. The response to phytocompounds extract [BF/S+L/Ep] administration in mouse immature dendritic cells can be viewed as an integrated cell-wide response, propagating and resolving over time. Result of genes networks are analysis by the Ingenuity Pathways program. (Yin *et al.*, unpublished data.) For color detail, please refer to the color plate section.

or some other proteolytic enzyme, collected, sequenced, and analyzed by MALDI-TOF mass spectrometry (MS) to identify proteins that are already known and registered in various protein/peptide databases. In addition to the 2D gel protein profiling system, other methods of systematic and/or comparative analysis of protein expression patterns include liquid chromatography followed by MS (LC/MS) or sequential MS analyses (MS/MS). Single fluorescence and FRET-based sensors have been successful in reporting small molecule and ion concentration, protein-ligand binding, and protein–protein interactions. Particularly exciting are protein biosensors that act as surrogate substrates for enzymes, reporting localized activities of kinases, protease, and small GTPases (van Engelenburg and Palmer 2008).

As most proteins function in collaboration with other proteins in the cell, one goal of proteomics is to identify the interactions between various large key protein complexes such as organelles or suborganelles (proteosome, nucleosome, endosome, etc.). This protein complex study is especially useful in determining potential regulatory partners in cell signaling cascades. Several methods are available to probe protein–protein interactions. One common method is yeast two-hybrid analysis. Other newer methods, including protein microarrays, immunoaffinity chromatography followed by MS, and combinations of diverse experimental methods such as phage display and computational methods have also been developed for proteomics studies.

Using the human DC as model immune cells, we characterized the proteomics of their response to treatment with the [BF/S+L/Ep] fraction of *Echinacea purpurea* (Wang *et al.* 2008). We observed that the proteins Mn-SOD and cofilin, which are involved in human DC cell maturation and tissue migration, were sensitive to the phytocompound treatment. On the other hand, however, various cytokines, chemokines that were significantly affected in our transcriptome study of the same system, were not found to have major changes in protein levels on 2D gel profiles, probably due to the normal low abundance of cytokine and chemokine gene products. Hence, cytokine/chemokine protein arrays or panel ELISAs are likely needed for further detailed studies. Proteomics does have some technical limitations, particularly with proteins expressed at low levels, as their detection resolution is still low.

METABOLOMICS

Metabolomics is an emerging field that originates from metabolite profiling, either as a targeted subset of related compounds or as a mapping of all extractable metabolites. The approach is expected to have diverse applications in toxicology, disease diagnosis, drug discovery and development, and in phytomedicine research (Ellis *et al.* 2007; Lindon *et al.* 2007; Chen *et al.* 2007a; Claudino *et al.* 2007; Ulrich-Merzenich *et al.* 2007; Kell 2006a; Griffin & Kauppinen 2007). Early pioneering research in metabolomics involved the quantitative metabolite profiling of urine or urinary drug metabolites of humans (Pauling *et al.* 1971; Horning & Horning 1971). Recently, the idea was extended to all other similar metabolite systems as an integral part of the omics experimental approach. In the plant metabolite research area, secondary metabolite profiling has been used in a pioneer study as a diagnostic technique to determine the mode of action of herbicides (Sauter *et al.* 1991). The use of the word "metabolome" was first reported in 1998 as a way to quantitatively and qualitatively measure the defined phenotypes to assess gene functions in yeast (Oliver *et al.* 1998) and to discuss the interplay between the global metabolite pool and specific environmental conditions in *Escherichia coli* (Tweeddale *et al.* 1998). Generally, the metabolome refers to the complete set of small-molecule metabolites (such as metabolic hormones and other signaling molecules and secondary metabolites) to be found within a biological sample, such as a specific tissue, organ, or an organism (Oliver *et al.* 1998). The word was coined in analogy with transcriptomics and proteomics; as is the case in the transcriptome and the proteome, the metabolome may also be quite dynamic, changing significantly in a time- and tissue-dependent way. Although the metabolome can be defined conceptually, it is not technically possible to observe, reveal, and analyze by a single analytical method (Shyur & Yang 2008) (e.g., via a 2D exhibition mode) the entire spectrum of a metabolome from the crude extract of a test tissue or organ. In early 2007, scientists at the University of Alberta and the University of Calgary completed the first draft of a human metabolome. They catalogued approximately 2500 metabolites, 1200 drugs, and 3500 food components that can be detected in a human body, as recently reported (Kell 2004; Ellis & Goodacre 2006; Wishart *et al.* 2007). This information is available from the Human Metabolome Database (www.hmdb.ca). Metabonomics is another word commonly substituted by the term metabolomics. Metabonomics, however, is the term used for the measurement of metabolite profiles, activities, and reactions toward the environment, medication, or disease of a given tissue or biological fluid (Dunn & Ellis 2005). Metabolic fingerprinting, like metabonomics, usually refers to the high-throughput global analysis of metabolites with minimal sample preparation, that is, using crude extract to contain the "metabolome." The two major approaches in metabolomics are the targeted and the global (or unbiased) metabolite analyses. Targeted metabolite analysis or metabolite profiling, as the name implies, targets a subset (e.g., with similar solvent affinity or chemical grouping) of metabolites in a sample, instead of a complete metabolome analysis. This is usually carried out by using a particular set of analytical techniques such as gas chromatography-MS (GC-MS) or LC-MS, together with an estimate of quantity. The hyphenated or inline technology approach established by K. Hostettmann and coworkers (Ferrari *et al.* 2000) provided a good strategy for such an effort. Alternatively, Kim and his coworkers (Kim *et al.* 2008) showed that high-performance liquid chromatography (HPLC)-based profiling of candidate phytomedicines, assayed for GABA$_A$ receptor modulators and PI3 kinase inhibitors, provides another good example for this approach. Various other techniques including thin layer chromatography (TLC), Fourier transform infrared spectroscopy (FT-IR), Raman spectroscopy, and nuclear magnetic resonance (NMR) are also established as good tools of the metabolite analysis arsenal (Shyur & Yang 2008).

SYSTEMS BIOLOGY

Systems biology is an emerging science field that aims to understand a biological system as an integrated whole, particularly with respect to the dynamic interactions of the various components of that system. The combination of information from genomics, proteomics, and metabolomics studies offers a direct route to such system-wide understanding, from the cellular to the whole animal level. With the rapid progress of various genome sequencing and molecular biology projects that are generating in-depth knowledge of the molecular and developmental biology of many systems, we are now ready to seriously consider the possibilities offered by systems biology. A complete system-level analysis of a specific biological activity and its regulation is, however, still beyond the scope of our current experimental practices and new technologies and methodologies must be developed. Some of the newer high-throughput analytical platforms have indeed substantially enhanced the detection of a dynamic range of metabolites, proteins, and their encoding genes (Dunn & Ellis 2005; Kell 2006b). Technical innovations in experimental devices, such as single-molecule measurements, femto-lasers that allow visualization of molecular interactions, confocal and 3D imaging of crosstalk in cells, and nanotechnology are also providing powerful new tools for studies of systems biology (Kitano 2002).

The coordinated broad-scale view of systems biology depends on the combined findings from different fields of study to develop an integrated, kinetic, and real-time understanding of a defined biology system. Even at the limited scale of the cellular level, this will require the development of sophisticated mathematical and computational modeling to analyze the many networks and signaling pathways, and their crosstalk and interplay that comprise the complex dynamic interactions between the chromosomes, organelles, genes, transcripts, proteins, and various metabolites. Although systems biology is still very much in its infancy, it will be invaluable in the modern attempts to unravel disease mechanisms, to reveal and validate drug targets, and to lead the discovery of new drugs (Keusch 2006), and we believe this systems biology approach is also applicable to the development of future phytomedicines.

OMICS IN PHYTOMEDICINE

Traditional medicinal plants are now considered by many scientists to be a more promising source of novel medicines than the screening of libraries of synthetic compounds for drug development. This is partly due to their long-term use with limited toxicity but also to their immense structural diversity. The limited success of combinatorial chemistry in novel drug development has been attributed in part to the fact that the many newly synthesized chemicals did not "co-evolve" with various protein folds of our biological systems, and are therefore often either quite toxic or quite inert. Natural product research for drug development has, to date, been mainly focused on the isolation of specific compounds from crude or prefractionated extracts using bioactivity-guided assays. Finally, pure compounds are isolated, which are then screened for bioactivity toward drug target molecules or activity against various cell lines or primary cultures, especially leukocytes. On the other hand, herbal or botanical products are usually produced in the form of crude extracts or formulations consisting of several or many bioactive compounds. Such complexity is, in one sense, troublesome to analzye but may reflect the reality of the use of any drug or extract and should not deter us from the use of phytomedicine mixtures rather than pure compounds (Yang *et al.* 2004). The various techniques of genomics, proteomics, and metabolomics may well provide a suitable platform for future herbal medicine research and the more conventional single drug investigations.

The signaling pathways and molecular networks underlying the diverse biological activities of several selected, purified single phytocompounds including shikonin, cytopiloyne, and emodin, and the complex [BF/S+L/Ep] fraction were evaluated for their effect on THP1 monocyte cells, another important immune cell line. DNA microarray studies showed that [BF/S+L/Ep] and a single purified, phytocompound tested induced similar profiles of complex gene expression patterns (Yang *et al*. 2004; Wang *et al*. 2008; Chiu *et al*. 2008). Both the purified single compound approach and the use of complex crude or fractionated medicinal plant extracts are thus clearly demonstrated to produce meaningful bioactivities toward well-known target molecules involved in important signaling pathways/networks. We therefore advocate that both of the previously mentioned strategies are likely to be useful for future development of phytomedicines. These two very different approaches do not need to be mutually exclusive, but instead should complement each other.

Another Affymetrix gene chip study showed significant upregulation of specific genes encoding for cytokines (IL-8, IL-1β, and IL-18) and chemokines (CACL 2, CCL 5, and CCL 2) within 4 hours after [BF/S+L/Ep] treatment of human immature DCs. Bioinformatic analysis of the upregulated genes revealed a key signaling network involving a number of immunomodulatory molecules leading to the activation of a very important, downstream master gene, *adenylate cyclase 8* (Wang *et al*. 2008). This study is a good example of the use of the relevant bioinformatics packages in conjunction with the results of genomics profiling to quickly provide information about target molecules and molecular signaling of the immune-modulatory and/or anti-inflammatory activities of a traditional medicinal herb, in this case, the Western herb *Echinacea* and its derived phytocompounds. An example of the cellular signaling network information provided by software analysis of DNA microarray data, in this case of the differences in cell response to [BF/S+L/Ep] at 30 minutes and at 12 hours after treatment, is shown in Fig. 20.2.

The identification of a spectrum of gene products, as RNA transcripts, protein/enzymes, and the metabolites produced via related enzymatic activities, in response to treatment with candidate phytomedicines reveals the basic molecular and cellular biology of the test compounds. This information allows us to correlate specific bioactivities with the clinical effects of test phytomedicines at the cellular levels. The effect of alkaloids from Chinese herbal medicine *Tripterygium hypoglaucum* root was examined using DNA microarrays, which shows the involvement of apoptotic genes and, in particular, activation of c-Myc and NF-κB signaling pathways (Zhuang *et al*. 2004). Following the revelation that *Ginko biloba* leaf extract EGb 761 showed antitumor (Li *et al*. 2002; De Feudis *et al*. 2003) and neuroprotective effects (Watanabe *et al*. 2001) *in vivo*, cDNA microarray analyses further revealed that particular groups of apoptosis genes were up- or downregulated in tumor cells when treated with EGb 761 leaf extract (Li *et al*. 2002; Smith *et al*. 2002). The apoptosis genes inhibited in the EGb 761-treated cells, at least in part, overlapped those involved in the neuroprotective effect of EGb 761 (Watanabe *et al*. 2001).

Previously, our laboratory reported the use of herbal extract of *Anoectochilus formosanus* for inhibition on growth of MCF-7 human breast carcinoma cells. cDNA microarray analysis showed the upregulation of genes encoding caspase 8 and cytochrome C (Yang *et al*. 2004). More recently, we characterized the immunomodulatory activities of three anti-inflammatory phytocompounds (emodin, shikonin, and cytopiloyne) and [BF/S+L/Ep], a defined butanol extract from *Echinacea purpurea* by a focused DNA microarray analysis (Chiu *et al*. 2008), using ~230 selected, immune-related genes, on LPS-stimulated THP-1 monocytes. Shikonin and emodin significantly inhibited the early expression (within 0.5 hours) of approximately fifty genes, notably cytokines TNF-α, IL-1β, and IL-4, chemokines CCL4 and CCL8, and inflammatory modulators NFATC3 and PTGS2. In contrast, neither cytopiloyne nor [BF/S+F/Ep] inhibited the early expression of these fifty genes, but rather inhibited most late-stage expression (~12 hours), particularly IL-4, NFATC3, and PTGS2, and cell migration and chemokine molecules CDH1 and ITGA2 (Chiu *et al*. 2008). Interestingly, cytopiloyne as a single phytocompound and [BF/S+L/Ep] as relatively crude plant extract fraction showed very similar patterns on a spectrum of gene expressions in this study.

For research into molecular toxicology, high-quality gene arrays have also been employed as a standard tool, and this allowed several international proof-of-principle tests of the application of global gene expression profiling to study the toxicity of new chemical compounds

(Lettieri 2006). Previous studies had shown that it is possible to identify gene signature patterns after exposure of test tissues to a given toxicant (Tennant 2002; Lettieri 2006). Tumor cell proteins with altered expression profiles in response to phytocompounds have been identified (Zhang *et al*. 2007; MacKeigan *et al*. 2003). The expression of specific proteins involved in cell signaling were changed following genistein treatment (Zhang *et al*. 2007), including down-regulated Rab14, a member of the Rab family and part of a low molecular weight GTPase superfamily, which is known to play an important role in intracellular vesicle trafficking and cytoskeletal rearrangement (Takai *et al*. 2001).

The use of whole plants or their crude extracts as medicines gave way to the reductionist approach of modern medicine with its reliance on pure active compounds, beginning in the early nineteenth century with the isolation of morphine from opium. However, in this reductionistic approach, bioactivity-guided purification steps often resulted in a "loss" of activity, with no single active principle discoverable, perhaps because of their low abundance in plants. Another possible explanation for this is that the spectrum of pharmacological efficacy traditionally observed arose because of the synergistic activity of multiple ingredients present in a single plant or in a combination of multiple plants, as most times formulated by design in Ayurveda or in TCM (Deocaris *et al*. 2008; Williamson 2001). Both of these situations could be addressed by the new technologies being developed for metabolomics. Metabolomics approaches such as GC-MS, LC-MS, 2D-NMR, or other hyphenated or combined techniques are effective tools for quality control and sensitive metabolite profiling of medicinal plant or herb medicinal materials or products (Zeng *et al*. 2007; Ye *et al*. 2007; Yang *et al*. 2006). These metabolomics strategies have proven useful in profiling unique groups of plant secondary metabolites that can be used as index compounds for improved classification of the three *Echinacea* species, and for better quality control of medicinal *Echinacea* extracts (Hou *et al*. 2008). Focused or targeted metabolomics platforms, for example, lipid profiling or lipidomics, are being applied in the drug development process, especially for lipid-related metabolic disorders and inflammatory diseases (Morris & Watkins 2005). Recently, comparative metabolomics approaches using LC-MS-based techniques have been used in drug or xenobiotic metabolism and chemical toxicology research (Chen *et al*. 2007b). In addition, research scientists are also using this to examine the toxicology of specific phytomedicines after metabolism in animals and humans (Ulrich-Merzenich *et al*. 2007).

The omics approach, however, is not without problems. For example, problems in microarray experiments include occasional nonreproducibility in replicate experiments (which often is a "biology" problem, not a gene chip defect), difficulty in statistical relability (due to small number of test samples), the comparison of data from different experiments and research groups, and the inconsistencies in data that may arise from the low specificity of test DNA probes, discrepancy in appropriate use of fold-change in addressing differential gene expression. The technical varia-tions in laboratory skills, probe-labeling, biochemical reactions, scanner and laser use, etc., also contribute to the data variations. Another factor that may limit microarray application is the high cost of this technology. The challenge today is therefore to provide standardized, reproducible microarray platforms, user-friendly databases, and visualization methods for evaluating differ-ential gene expression profiles that are affordable to most biologists. With the development of new uniform and more sophisticated experimental designs, multiple systems for statistical anal-ysis, data mining and management systems, and algorithms for crosstalk and data analysis DNA microarrays approach can be expected to optimize the experimental studies for development of novel phytomedicines.

The application of omics technologies to the systems biology of novel phytomedicines will greatly benefit the development of evidence-based phytotherapeutics, and such research may also lead to a change of paradigm in facilitating the application of complex phytomedicine mixtures and formulations, making a very important contribution to medical biotechnology and modern medicine. Recent developments in computational biology and bioinformatics have already provided scientists with a number of systematic methods for analysis of such molecular networks in biological systems. Generally, in systems biology, metabolomic data are currently being organized with the aim of developing a computer model to simulate the whole system well enough to predict both functional genomic activation and metabolic flows in the complex systems. The metabolic control activity and the functional genomics and proteomics within the test system most often share the same agenda (Kell & Mendes 2000). Therefore, the joint application of various omics technology will be the ultimate genotyping and phenotyping of cells, and this approach has the potential to revolutionize herbal medicinal product or phytodrug research and to advance the adoption of parts of TCM or Ayurveda for use in modern evidence-based medicine (Verpoorte *et al.* 2005; Patwardhan *et al.* 2005). An example of such a future possibility is our study of DCs in response to *Echinacea* phytocompounds (Fig. 20.2). It is our hope that integration of these data into systems biology studies can provide better understanding of biological systems under study and also adequate tools to analyze TCM, Ayurveda, or other traditional medicine systems, aiming at creating evidence-based phytomedicines of the twenty-first century.

CONCLUSION AND FUTURE CHALLENGES

Medicinal plants are an integral part of the Chinese, Ayurvedic, Native American, European, and many other traditional systems of medicine. Clinical experiences, anecdotal observations, and documented data are a starting point for research and development of traditional knowledge-based drug development. Although clinical experience with traditional medicines has spanned 4000 years, it is still very dangerous to extrapolate directly from such traditional use to current practice, because the actual herbs and or formulations previously used may be very different to the commercial herbal products available today. Many other questions and problems remain, including the quality, safety, efficacy, bioavailability, and mechanisms of action of even the most popularly used herbs. These questions and concerns can only be answered by rigorous applications of scientific evidence in high-quality basic, translational, and clinical research. An "-omics" technologies-based approach can facilitate high-throughput, fast track, and whole spectrum/global view-related studies, and can expedite evidence-based botanical drug development.

In conjunction with the functional genomics study (the transcriptomes and proteomes), it is important to overcome the current problems of metabolite complexity, and to allow a deeper mining of the metabolome. The development of combinations of multidimensional chromatographic separations with multidimensional MS or NMR spectrometric analyses is urgently needed to achieve maximal output of the existing technologies. Further optimization of chromatographic separations and better sensitivity of NMR and MS instruments, with higher resolution capability and greater mass accuracy, is generally recognized as the key for future improvements in metabolomics (Griffiths *et al.* 2007). It is therefore essential to establish standard operating protocols (SOPs) to minimize or limit the technical variations in extraction procedures across a broad base of metabolomics datasets. Although the analytic, system biology, and bioinformatics capacities to effectively use the genomics, proteomics, and metabolomics approaches have only been developed for a few years, clinicians, food and nutrition scientists, and systems biology scientists should not sit idle but should act together to set the priorities and agenda for development of novel phytomedicines in future, sooner than later. The integration of omics approaches with systems biology is expected to create a highly complex map, and we need to prepare ourselves to effectively "read" and make use of this map for the development of phytomedicines or other related translational research. A future challenge will also be developing effective crosstalk between clinicians, nutrition and food scientists, and omics biologists, to make practical recommendations to patients or the general public. The methodical and thorough integration of information from genomics, proteomics, and metabolomics will provide a solid foundation and scientific rationale for the evidence-based development of novel phytomedicines, which we believe will prove to be very useful strategy in human health care.

REFERENCES

Anand, P., Sundaram, C., Jhurani, S., Kunnumakkara, A. B. & Aggarwal, B. B. (2008) Cucurmin and cancer: An "old-age" disease with an "age-old" solution. *Cancer Lett 267*:133–164.

Blumenthal, M., Goldberg, A. & Brinckmann, J. (2000) *Herbal medicine: expanded Commission E monographs.* Newton, MA: Intergrative Medicine Communications.

Chang, C. L., Chang, S. L., Lee, Y. M. *et al.* (2007) Cytopiloyne, a polyacetylenic glucoside, prevents type I diabetes in nonobese diabetic mice. *J Immunol 178*, 6984–6993.

Chen, B. J. (2001) Triptolide, a novel immunosuppressive and anti-inflammatory agent purified from a Chinese herb Tripterygium wilfordii Hook F. *Leuk Lymphoma 42*, 253–265.

Chen, C., Gonzalez, F. J. & Idle, J. R. (2007a) LC–MS-based metabolomics in drug metabolism. *Drug Metabol Rev 39*, 581–597.

Chen, C., Ma, X., Malfatti, M. A. *et al.* (2007b) A comprehensive investigation of 2-amino-1-methyl-6-phenylimidazo [4,5-b] pyridine (PhIP) metabolism in the mouse using a multivariate data analysis approach. *Chem Res Toxicol 20*, 531–542.

Chiu, S. C., Tsao, S. W., Hwang, P. I. *et al.* (2008) Comparative functional genomic studies of anti-inflammatory phytocompounds on immune signaling mechanisms in human monocytes (submitted for publication).

Claudino, W. M., Quattrone, A., Biganzoli, L., Pestrin, M., Bertini, I. & Di Leo, A. (2007) Metabolomics: available results, current research projects in breast cancer, and future applications. *J Clin Oncol 25*, 2840–2846.

De Feudis, F. V., Papadopoulos, V. & Drieu, K. (2003) *Ginkgo biloba* extracts and cancer: a research area in its infancy. *Fund Clin Pharmacol 17*, 405–417.

Deocaris, C. C., Widodo, N., Wadhwa, R. & Kaul, S. C. (2008) Merger of Ayurveda and tissue culture-based functional genomics: Inspirations from Systems Biology. *J Trans Med 6*, 14–22.

Dunn, W. B. & Ellis, D. I. (2005) Metabolomics: current analytical platforms and methodologies. *Trend Anal Chem 24*, 285–294.

Ellis, D. I., Dunn, W. B., Griffin, J. L., Allwood, J. W. & Goodacre, R. (2007) Metabolic fingerprinting as a diagnostic tool. *Pharmacogenomics 8*, 1243–1266.

Ellis, D.I. & Goodacre, R. (2006) Metabolic fingerprinting in disease diagnosis: biomedical applications of infrared and Raman spectroscopy. *Analyst 131*, 875–885.

Ferrari, J., Terreaux, C., Sahpaz, S., Msonthi, J. D., Wolfender, J. & Hostettmann, K. (2000) Benzophenone glycosides from Gnidia involucrata. *Phytochemistry 54*, 883–889.

Fritz, P. (2004) Impact of mistletoe lectin binding in breast cancer. *Anticancer Res 24*, 1187–1192.

Gautam, M., Diwanay, S., Gairola, S., Shinde, Y., Patki, P. & Patwardhan, B. (2004) Immunoadjuvant potential of *Asparagus racemosus* in experimental system. *J Ethnopharmacol 91*, 251–255.

Griffin, J. L. & Kauppinen, R. A. (2007) Tumour metabolomics in animal models of human cancer. *J Proteome Res 6*, 498–505.

Griffiths, W. J., Karu, K., Hornshaw, M., Woffendin, G. & Wang, Y. (2007) Metabolomics and metabolite profiling: past heroes and future developments. *Eur J Mass Spect 13*, 45–50.

Han, S. B., Yoon, Y. D., Ahn, H. J. *et al.* (2003) Toll-like receptor-mediated activation of B cells and macrophages by polysaccharide isolated from cell culture of *Acanthopanax senticosus*. *Inter Immunopharmacol 3*, 1301–1312.

Horning, E. C. & Horning, M. G. (1971) Metabolic profiles: gas-phase methods for analysis of metabolites. *Clin Chem 17*, 802–809.

Hou, C. C., Chen, Y. P., Chen, C. H., *et al.* (2008) Species identification and anti-inflammatory bioactivity validation of *Echinacea* plants using metabolomics approach coupled with cell- and gene-based assays (submitted for publication).

Hou, C. C., Chen, Y. P., Wu, J. H. *et al.* (2007) A galactolipid possesses novel cancer chemopreventive effects by suppressing inflammatory mediators and mouse B16 melanoma. *Cancer Res 67*, 6907–6915.

Kell, D. B. (2004) Metabolomics and systems biology: making sense of the soup. *Curr Opin Microbiol 7*, 296–307.

Kell, D. B. (2006a) Systems biology, metabolic modelling and metabolomics in drug discovery and development. *Drug Discov Today 11*, 1085–1092.

Kell, D. B. (2006b) Metabolomics, modelling and machine learning in systems biology – towards an understanding of the languages of cells. *FEBS J 273*, 873–894.

Kell, D. B. & Mendes, P. (2000) Snapshots of systems - metabolic control analysis and biotechnology in the post genomic era. In: Cornish-Bowden, A. & Cardenas, B. L. (eds.), *Technological and Medical Implications of Metabolic Control Analysis*. Dordrecht: Kluwer Academic Publishers, pp. 251–283.

Keusch, G. T. (2006) What do –omics mean for the science and policy of the nutritional sciences? *Am J Clin Nutri 83*, 520–522.

Kim, H. J., Baburin, I., Khom, S., Hering, S. & Hamburger, M. (2008) HPLC-based activity profiling approach for the discovery of GABAA receptor ligands using an automated two microelectrode voltage clamp assay on *Xenopus* oocytes. *Planta Med 74*, 521–526.

Kitano, H. (2002) Systems biology: a brief overview. *Science 295*, 1662–1664.

Lettieri, T. (2006) Recent applications of DNA microarray technology to toxicology and ecotoxicology. *Environ Health Perspect 114*, 4–9.

Li, W., Pretner, E., Shen, L., Drieu, K. & Papadopoulos, V. (2002) Common gene targets of *Ginkgo biloba* extract (EGb 761) in human tumor cells: relation to cell growth. *Cell Mol Biol 48*, 655–662.

Lindon, J. C., Holmes, E. & Nicholson, J. K. (2007) Metabonomics in pharmaceutical R&D. *FEBS J 274*, 1140–1151.

Liu, M., Liu, X. Y. & Cheng, J. F. (2003) Advance in the pharmacological research on matrine. *Zhongguo Zhong Yao Za Zhi 28*, 801–804.

MacKeigan, J. P., Clements, C. M., Lich, J. D., Pope, R. M., Hod, Y. & Ting, J. P. Y. (2003) Proteomic profiling drug-induced apoptosis in non-small cell lung carcinoma: Identification of RS/DJ-1 and RhoGDIα. *Cancer Res 63*, 6928–6934.

McKusick, V. A. & Ruddle, F. (1997) Genomics as it enters its second decade. *Genomics 45*, 243.

Morris, M. & Watkins, S. M. (2005). Focused metabolomic profiling in the drug development process: advance from lipid profiling. *Curr Opin Chem Biol 9*, 407–412.

Oliver, S. G., Winson, M. K., Kell, D. B. & Baganz, F. (1998) Systematic functional analysis of the yeast genome. *Trends Biotechnol 16*, 373–378.

Patwardhan, B., Warude, D., Pushpangadan, P. & Bhatt, N. (2005) Ayurveda and traditional Chinese medicine: a comparative overview. *Evid Based Complement Alternat Med 2*, 465–473.

Pauling, L., Robinson, A. B., Teranishi, R. & Cary, P. (1971) Quantitative analysis of urine vapor and breath by gas–liquid partition chromatography. *PNAS 68*, 2374–2376.

Plaeger, S. F. (2003) Clinical immunology and traditional herbal medicines. *Clin Diagn Lab Immunol 10*, 337–338.

Sauter, H., Lauer, M. & Fritsch, H. (1991) Metabolic profiling of plant: a new diagnostic technique. In: Baker, D. R., Moberg, W. K. & Fenyes, J. G. (eds.), *Synthesis and Chemistry of Agrochemicals II*. Oxford: American Chemical Society Press, pp. 288–299.

Schena, M., Shalon, D., Davis, R. W. & Brown, P. O. (1995) Quantitative monitoring of gene expression patterns with a complementary DNA microarray. *Science 270*, 467–470.

Schena, M., Shalon, D., Heller, R., Chai, A., Brown, P. O. & Davis, R. W. (1996) Parallel human genome analysis: microarray-based expression monitoring of 1000 genes. *PNAS 93*, 10614–10619.

Shyur, L. F. & Yang, N. S. (2008) Metabolomics for phytomedicine research and drug development. *Curr Opin Chem Biol 12*, 66–71.

Smith, J. V., Burdick, A. J., Golik, P., Khan, I., Wallace, D. & Luo, Y. (2002) Anti-apoptotic properties of *Ginkgo biloba* extract EGb 761 in differentiated PC12 cells. *Cell Mol Biol 48*, 699–707.

Song, J. Y., Han, S. K., Son, E. H., Pyo, S. N., Yun, Y. S. & Yi, S. Y. (2002) Induction of secretory and tumoricidal activities in peritoneal macrophages by ginsan. *Inter Immunopharmacol 2*, 857–865.

Sunila, E. S. & Kuttan, G. (2004) Immunomodulatory and antitumor activity of Piper longum Linn and piperine. *J Ethnopharmacol 90*, 339–346.

Takai, Y., Sasaki, T. & Matozaki, T. (2001) Small GTP-binding proteins. *Physiol Rev 81*, 153–208.

Tan, B. K. & Vanitha, J. (2004) Immunomodulatory and antimicrobial effects of some traditional Chinese medicinal herbs: a review. *Curr Topics Med Chem 11*, 1423–1430.

Temple, N. J. (2000) Antioxidants and disease: more questions than answers. *Nutri Res 20*, 449–459.

Tennant, R. W. (2002) The National Centre for Toxicogenomics: using new technologies to inform mechanistic toxicology. *Environ Health Perspect 110*, A8–A110.

Tweeddale, H., Notley-McRobb, L. & Ferenci, T. (1998). Effect of slow growth on metabolism of Escherichia coli, as revealed by global metabolite pool (''metabolome'') analysis. *J Bacteriol 180*, 5109–5116.

Tyler, V. E. & Foster, S. (eds.) (1999) *Tyler's honest herbal: a sensible guide to the use of herbs and related remedies.* (4th ed.). New York: Haworth Herbal Press.

Tyler, V. E. & Robbers, J. E. (eds.) (1999) *Tyler's herbs of choice: the therapeutic use of phytomedicinals.* New York: Haworth Herbal Press.

Ulrich-Merzenich, G., Zeitler, H., Jobst, D., Panek, D., Vetter, H. & Wagner, H. (2007) Application of the ''-omic-'' technologies in phytomedicine. *Phytomed 14*, 70–82.

US Food and Drug Administration (1997). Dietary supplements containing ephedrine alkaloids: proposed rule. *Fed Regist 62*, 30678–30717.

US Food and Drug Administration (2002). Kava and severe liver injury. *FDA Consum 36*, 4.

Vanden, W. E., Beukelman, C. J., Van deen, A. J. J., Kroes, B. H., Labadie, R. P. & Djik, H. V. (2001) Effects of methoxylation of apocynin and analogs on the inhibition of reactive oxygen species production by stimulated human neutrophils. *Euro J Pharmacol 433*, 225–230.

VanEngelenburg, S. B. & Palmer, A. E. (2008) Fluorescent biosensors of protein function. *Curr Opin Chem Biol 12*, 60–65.

Verpoorte, R., Choi, Y. H. & Kim, H. K. (2005) Ethnopharmacology and systems biology: a perfect holistic match. *J Ethnopharmacol 100*, 53–56.

Wang, C. Y., Chiao, M. T., Yen, P. J. et al. (2006) Modulatory effects of *Echinacea purpurea* extracts on human dendritic cells: a cell- and gene-based study. *Genomics 88*, 801–808.

Wang, C. Y., Staniforth, V., Chiao, M. T. et al. (2008) Genomics and proteomics of immune modulatory effects of *Echinacea purpurea* phytocompounds in human dendritic cells. *BMC Genomics* (in press).

Watanabe, C. M., Wolffram, S., Ader, P. et al. (2001) The *in vivo* neuromodulatory effects of the herbal medicine *Ginkgo biloba*. *PNAS 98*, 6577–6580.

Wen, C. C., Kuo, Y. H., Jan, J. T. et al. (2007) Specific plant terpenoids and lignoids possess potent antiviral activities against severe acute respiratory syndrome coronavirus. *J Med Chem 50*, 4087–4095.

Wilkins, M. R., Sanchez, J. C., Gooley, A. A. et al. (1995) Progress with proteome projects: why all proteins expressed by a genome should be identified and how to do it. *Biotechnol Genetic Eng Rev 13*, 19–50.

Willett, W. C. (1994) Diet and health: What should we eat? *Science 254*, 532–537.

Williams, E. A., Coxhead, J. M. & Mathers, J. C. (2003) Anticancer effects of butyrate: use of micro-array technology to investigate mechanisms. *Proc Nutri Soc 62*, 107–115.

Williamson, E. M. (2001) Synergy and other interactions in phytomedicines. *Phytomed 8*, 401–409.

Wishart, D. S., Tzur, D., Knox, C. et al. (2007) HMDB: The Human Metabolome Database. *Nucl Acids Res 35*(Database issue), D521–526.

Yang, N. S., Shyur, L. F., Chen, C. H., Wang, S. Y. & Tzeng, C. M. (2004) Medicinal herb extract and a single-compound drug confer similar complex pharmacogenomic activities of MCF-7 cells. *J Biomed Sci 11*, 418–422.

Yang, S. Y., Kim, H. K., Lefeber, A. W. M. *et al.* (2006) Application of two-dimensional nuclear magnetic resonance spectroscopy to quality control of Ginseng commercial products. *Plant Med 72*, 364–369.

Ye, M., Liu, S. H., Jiang, Z., Lee, Y., Tilton, R. & Cheng, Y. C. (2007) Liquid chromatography/mass spectrometry analysis of PHY906, a Chinese medicine formulation of cancer therapy. *Rapid Comm Mass Spect 21*, 3593–3607.

Zeng, Z. D., Liang, Y. Z., Chau, F. T., Chen, S., Daniel, M. K. & Chan, C. O. (2007). Mass spectral profiling: an effective tool for quality control of herbal medicines. *Anal Chim Acta 604*, 89–98.

Zhang, D., Tai, Y. C., Wong, C. S., Tai, L. K., Koay, E. S. & Chen, C. S. (2007) Molecular response of leukemia HL-60 cells to genistein treatment, a proteomics study. *Leukemia Res 31*, 75–82.

Zhang, L. H., Huang, Y., Wang, L. W. & Xiao, P. G. (1995) Several compounds from Chinese traditional and herbal medicine as immunomodulators. *Phytotherapy Res 9*, 315–322.

Zhuang, W. J., Fong, C. C., Cao, J. *et al.* (2004) Involvement of NF-kappa B and c-myc signaling pathways in the apoptosis of HL-60 cells induced by alkaloids of *Tripterygium hypoglaucum* (levl.). *Hutch Phytomed 11*, 295–302.

21

Contribution of Omics Revolution to Cancer Prevention Research

Nancy J. Emenaker and John A. Milner

Mounting evidence points to diet as an important and modifiable variable that likely influences cancer risk and tumor behavior [1–3]. Historically, about 30% to 40% of all cancers are thought to relate to dietary habits [4]. More recently, the World Cancer Research Fund/American Institute of Cancer Research reported a similar percentage of cancers that were preventable through the consumption of a balanced diet, regular physical activity, and avoidance of obesity [5]. Although a wealth of evidence implicates foods as both preventative and causative agents, there is considerable variability in response [1–5]. This variability may arise from many factors including the genetics of the consumer as briefly discussed in this chapter.

Although developing an ideal diet with specific foods is a laudable, noninvasive, and cost-effective approach for reducing the cancer burden, this is far from being a simple undertaking [5]. The massiveness of the problem in underscored by thousands of compounds within the foods that are consumed each day and therefore the potential for multiple interactions [2, 6]. Furthermore, many of the bioactive food constituents remain largely uncharacterized in terms of the biological changes they produce and thus their possible role in cancer remains largely speculative. Interactions, both synergistic and antagonistic, between the different components within a food may contribute variability in the biological outcomes to the consumption of intact foods versus isolated constituents. Positive or negative interactions arising from constituent interactions may also contribute to the variability in response across studies in overall relationship between eating behaviors and cancer [7]. At least part of this variability likely relates to the amounts of specific bioactive food components consumed in epidemiological studies or the amount provided in controlled intervention studies [1, 6] Likewise, the timing of administration and duration of exposure contribute to some of the variability in the literature about the impact of specific foods or their components on cancer risk and/or tumor behavior [7–9]. Finally, it is becoming increasingly apparent that many anticarcinogenic dietary components when provided in exaggerated amounts stimulate cancer risk [10–12].

GENETICS IS ALSO AN IMPORTANT DETERMINANT OF THE RESPONSE TO DIET

One of the likely major causes of variation in response to foods and their components likely comes from the individuality in response to the food components that are consumed. Evidence continues to surface that genes can influence the response to food components in a similar fashion to that occurring with drugs [13–15].

Advances in these "omics" technologies (genomics, epigenomics, transcriptomics, proteomics, and metabolomics) are increasingly becoming interwoven into all aspects of the

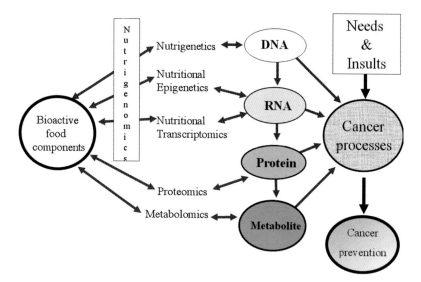

Figure 21.1 The "omic" can influence the response to bioactive food components. The needs for food components for normal growth and development along with specific insults including excess calories, nutrient–nutrient interactions, bacteria, viruses, and environmental contaminants can influence the ability of food components to influence cancer risk and tumor behavior. Because these interrelationships are not linear, effective bioinformatic tools are needed to evaluate the significance of foods and their components and to determine who will benefit and/or be placed at risk due to dietary intervention. For color detail, please refer to the color plate section.

biological sciences and are serving as the basis for significant scientific discoveries. Undeniably, these technologies are fundamental in explaining the phenotypic response that occurs after ingestion of food or their bioactive components [1,2] (Fig. 21.1). This brief commentary highlights some of the proof-of-principles of the importance of the "omic" technologies in influencing the response to foods and their components. This discussion focuses predominately on "omic" approaches in human nutrition research, specifically those that help to identify profiles or characteristics that can be used to identify those who will benefit or be placed at risk due to dietary change.

GENOMICS

Genomics is the study of all DNA sequences encoded by an organism. In humans, the haploid genome contains approximately 20,000–25,000 genes encoding nearly 100,000 proteins in total [16]. Although there is consistency in the human genome, there is also significant variation as evident by the occurrence of millions of polymorphisms, some of which can markedly influence biological processes [17] and variation in copy number [18]. Several studies suggest some of the single nuclear polymorphisms (SNPs) and haplotype blocks may contribute substantially to phenotypic changes adversely impacting individual disease risks including that related to cancer [19–21]. The impact of polymorphisms are not limited to changes in cancer risk as evident by recent findings from the Framingham Heart Study where allelic variants were found to predict renal function traits and chronic renal disease susceptibilities, concomitantly influencing stroke and cardiovascular disease risks [22]. Similar genomic relationships have been reported for the risk of diabetes [23]. It is important to note that if one excludes mutations in cancer susceptibility genes or occupational exposure to chemical carcinogens, dietary habits become one of the most significant risk factors in the etiology of sporadic tumors, yet the impact of diet is dependent on genes that influence digestion, absorption, metabolism, and excretion [14,15].

The field of nutrigenetics encompasses the study of molecular relationships between polymorphisms/copy number and the biological response to foods or their components. Several epidemiological and preclinical studies along with a limited number of clinical interventions suggest individual genetic variations, including SNPs, copy number, and haplotype blocks, contribute to interindividual differences in biologic responses to foods [24–28]. Discordance across epidemiological findings in particular may reflect unreported genotypic variation across participants, altering susceptible individual responses to dietary modification of complex genomic diseases [3,15].

Cancer risks appear to increase as diets shift away from plant-based diets to greater consumption of meats and refined, processed foods [29,30]. The relationship of colorectal cancer was examined as a function of six single nucleotide polymorphisms located in cytochrome P450 genes (c.-163A>C and c.1548T>C in CYP1A2, g.-1293G>C and g.-1053C>T in CYP2E1, c.1294C>G in CYP1B1, and c.430C>T in CYP2C9) [31]. Although separate analyses of these SNPs showed no effect on risk, three allelic variant combinations were found to be associated with a significant increase in risk and interaction with excessive red meat consumption. Collectively these data suggest that by combining genomic information risk for the general population of about 20% [1] was elevated to almost 50-fold [31]. The frequency of these combined polymorphisms was only about 5% in the population examined. These findings raise interesting issues about population findings, which tend to be variable, and may actually reflect the frequency of these particularly vulnerable individuals in study populations. Thus, it is unclear if recommendations to restrain meat consumption should be limited to particularly vulnerable subsets or to the entire population. Furthermore, these significant associations emphasize that the effect of multiple SNPs may be important despite single null single SNP associations.

There is limited evidence that the response to fruits and vegetables depends on genetic polymorphisms. For example, dietary antioxidants are thought to interact with endogenous sources of pro- and antioxidants and thereby reduce breast cancer risk. Recently, a nested case-control study of postmenopausal women in the Cancer Prevention Study-II Nutrition Cohort examined the interaction between oxidative stress-related genes, vegetable/fruit intake, and breast cancer risk [32]. Genetic variations in catalase (CAT) (C-262T), myeloperoxidase (MPO) (G-463A), endothelial nitric oxide synthase (NOS3) (G894T), and heme oxygenase-1 (HO-1) [(GT)(n) dinucleotide length polymorphism] were not associated with breast cancer risk. Interestingly, in women carrying the low-risk CAT CC [odds ratio (OR) = 0.75, 95% confidence interval (CI) 0.50-1.11], NOS3 TT (OR = 0.54, 95% CI = 0.26–1.12, P-trend = 0.10) or HO-1 S allele and MM genotype (OR = 0.56, 95% CI = 0.37–0.55), and consuming high amounts of fruits and vegetables there was a nonsignificantly reduced breast cancer risk. Furthermore, those with > or = 4 putative low-risk alleles in total had significantly reduced risk (OR = 0.53, 95% CI = 0.32–0.88, P-interaction = 0.006) compared with those with < or = 2 low-risk alleles. In contrast, among women with low vegetable and fruit intake (< median), the low-risk CAT CC (OR = 1.33, 95% CI = 0.89–1.99), NOS3 TT (OR = 2.93, 95% CI = 1.38–6.22) and MPO AA (OR = 2.09, 95% CI = 0.73–5.95) genotypes appeared to be associated with elevated risks. These findings and earlier observations [33] suggest that not all individuals will likely respond equally to increased fruits and vegetables. Again, such information raises issues about how best to make dietary recommendations when only target groups appear to be receiving the majority of the anticancer benefits.

In the Tennessee Colorectal Polyp Study, calcium-magnesium balance was assessed in 4623 participants carrying Thr1482Ile mutations (G→A transition) to the magnesium transporter transient receptor potential melastatin 7(TRPM7) [24]. TRPM7 is a unique, ubiquitously expressed ion channel-regulating calcium and magnesium uptake with kinase capacity. Specifically, TRPM7 simultaneously regulates cell depolarization and intracellular calcium signaling as an ion channel while actively phosphorylating cellular substrates via its intrinsic kinase capabilities.

Dai et al. [24] reported non-African American participants carrying one or both mutated alleles were at increasing colorectal cancer risks (OR = 1.35 and OR = 2.32, $P = 0.03$), respectively, regardless of dietary intake considerations. Expectedly, when dietary magnesium levels were elevated (i.e., suggesting diets rich in whole grains, nuts, and beans) decreasing Ca:Mg ratio <2.78, colorectal cancer risk declined to OR = 0.44 (cases $n = 1$; controls $n = 12$). However, for individuals carrying both mutated alleles, dietary intake considerations dramatically increased colorectal cancer risks several/fold (OR = 16.14, $P < 0.01$). These observations point to the need to understand nutrient–nutrient interactions when evaluating diet–SNP interrelationships.

The response to calcium also appears to be influenced by genetic polymorphisms [28]. In the Singapore Chinese Health Study, an association between reduced calcium absorption and increased colorectal cancer risk was reported in individuals carrying either the VDR FokI "Ff" or "ff" polymorphic variants. Individuals carrying either VDR FokI "Ff" or "ff" polymorphic variants and consuming low levels of dietary calcium (<387.74 mg/day) were at 2.1-fold and 2.6-fold (P trend = 0.004), respectively, increased risk for colorectal cancer compared to those carrying the VDR FokI FF genotype [28]. They hypothesized that dietary calcium intake modifies the influence of the VDR on colorectal cancer risk because calcium regulates formation of the active metabolite of vitamin D. Interestingly, colorectal cancer risk was reduced (OR = 1.01; $P_{trend} = 0.97$) in individuals carrying the high-risk VDR FokI Ff or ff polymorphisms and who consumed the greatest amounts of dietary calcium; suggesting personalized approaches to cancer prevention interventions may be beneficial in reducing cancer risks [28].

It is logical to assume that the interrelationships between genes and nutrients will not always be positive. Both protective and harmful effects were observed in a study that examined arachidonic acid metabolism polymorphisms and their relationship to colorectal adenomas [34]. In this Dutch case-control study, fish consumption was found to modify the associations with *COX-2* and *PPARδ* genotypes. Those consuming the highest amount of fish and with SNP *c.-789C→T* in *PPARδ* exhibited the lowest adenoma risk compared with the lowest tertile (OR 0.65, 95% CI 0.41–1.02). However, for those with the CT or TT genotype the increased consumption of fish increased risk (OR 2.22, 95% CI 0.78–6.36). Thus, not only the magnitude of a response to foods can be influenced by genetic polymorphisms; however, in some cases the direction of the response can reverse. Additional research is needed to help identify vulnerable populations to increased intake of selected foods and/or their isolated components.

Intriguing information continues to surface about the importance of nutrigenomics in determining the response to dietary components. Some genetic variations appear to place carriers at enhanced response to the protection of some dietary components whereas in other individuals these same nutrients appear to increase risks. Although a dose–response threshold seems logical, whether the food component is protective or stimulatory, it unfortunately has only rarely been evaluated. It should also be noted that although the evidence continues to suggest genetic variations can markedly alter the physiologic responses to dietary components, the verification of a specific response is exceedingly limited. Undeniably, greater attention needs to be given to the verification of nutrigenomic interactions and how these changes influence cellular process that relate to cancer.

EPIGENETICS

Epigenetics is the study of heritable changes in gene expression that occur without a change in the sequence of DNA. It provides an extra layer of control in gene expression regulation beyond that attributable to SPNs and other genetic shifts. These regulatory processes are critical components for normal development and growth of neoplastic cells. Shifts in epigenetics are increasingly recognized as a contributor to cancer [35]. Several drugs are targeting epigenetic mechanisms as novel cancer therapies. These therapies, including histone deacetylase inhibitors and demethylating agents, show promise with hematological cancer, but their significance with solid tumors remains unresolved. Likewise, several bioactive food components have been reported to influence one or more sites with what constitutes the epigenetic process [36].

Distinct mechanisms within the epigentics domain that are interconnected include DNA methylation, histone modification, noncoding RNA regulation, and polycomb homeostasis. DNA methylation occurring on the 5 position of the pyrimidine ring of cytosines in the dinucleotide sequence CpG forms one of the epigenetic events that controls and modulates genetic expression. DNA methylation is essential for normal mammalian development and imprinting. Aberrant DNA methylation patterns are evident in several diseases, particularly cancer where genome-wide hypomethylation coincides with gene-specific hypermethylation. The relationship between aberrant hypermethylation and hypomethylation on the expression of genes and their relationship to disease risk remains an area of active investigation. Today, DNA methylation patterns may assist with the detection of cancers at very early stages, classify tumors, and predict/monitor the response to antineoplastic treatment [37].

Because epigenetic events can be changed by dietary habits, they offer another explanation about how eating behaviors can influence biological processes and phenotypes (Fig. 21.1). Food components influence this methylation in at least four different ways [38]. First, dietary factors are important in providing and regulating the supply of methyl groups available for the formation of S-adenosylmethionine (SAM), the universal methyl donor. Second, selected dietary factors can modify the utilization of methyl groups by processes including shifts in DNA methyltransferase activity including such food components as selenium. A third plausible mechanism relates to DNA demethylation activity, admittedly this process remains of considerable controversy if it actually occurs. Finally, the DNA methylation patterns may influence the response by regulating genes that influence absorption, metabolism, or the site of action for the bioactive food component.

One of the best models for examining the effects of diet on methylation patterns comes from studies involving the agouti mouse. Providing these yellow mice with supplemental choline, betaine, folic acid, vitamin B-12, methionine, and zinc during pregnancy increases DNA methylation in the promoter region of the agouti gene and causes a change in the color pattern of the hair coat in the offspring to a higher percentage of brownish mice [39]. This phenotypic change coincides with a lower susceptibility to obesity, diabetes, and cancer. It should be noted that this phenotypic change does not occur in all offspring and thus other factors influence the biological change induced by these maternal dietary interventions. The response in phenotype (increased methylation) is not limited to providing enriched methyl donor diets but can also occur by providing supplemental genistein, or can be suppressed (increased hypomethylation) by environmental compounds such as bisphenol A [40]. Although humans do not have the long-term repeating unit found in these mice, these studies serve as a proof-of-principle that diet influences epigenetics and can lead to phenotypic change. Overall, these studies suggest that *in utero* exposure to dietary components may not only influence embryonic development but may also have profound and long-term health consequences including changing the risk of cancer.

In addition to enzymes associated with methylation, the cells have histone-modifying enzymes and accessory proteins, which can facilitate and/or target epigenetic marks [41]. Several lysine and arginine demethylases have been discovered, and the presence of an active DNA demethylase is speculated in mammalian cells [41]. Histone deacetylase (HDAC) inhibitors are thought to reactivate epigenetically silenced genes in cancer cells, triggering cell cycle arrest and apoptosis. A few dietary constituents including butyrate, sulforaphane, and diallyl disulfide have been reported to serve as HDAC inhibitors including isothiocyanates found in cruciferous vegetables and the allyl compounds present in garlic [42]. In a preclinical mouse model, sulforaphane inhibited HDAC activity and induced histone hyperacetylation, which coincided with a depression in the growth of a PC-3 xenograph [43]. It is also important to note that an inhibition of HDAC activity has been observed in circulating peripheral blood mononuclear cells obtained from people who consumed broccoli sprouts [42]. Overall, intriguing studies are emerging about the ability of foods to modify histones, which appears to confer a cancer-protective response.

Within mammalian cells, there exists an abundance of microRNAs (miRNAs). MicroRNAs (miRNAs) are a family of naturally occurring, evolutionary conserved, small (approximately 19–23 nucleotides), non protein-coding RNA molecules that generally negatively regulate post-transcriptional gene expression. miRNAs are estimated to account for about 3% of all human genes and to control expression of thousands of target mRNAs. Multiple miRNAs are thought to target each mRNA. These noncoding RNAs are important in posttranslational gene regulation, including regulation of cell proliferation, apoptosis, and differentiation processes. Evidence is mounting that miRNAs are involved in cancer initiation and progression and their expression patterns may serve as phenotypic signatures of different cancers. Several dietary components including folate, retinoids, and curcumin appear to exert cancer-protective effects at least in part by altered miRNA expression [44]. Recently, dietary fat has been reported to influence miRNA expression [45]. The ability to upregulate the expression of miR-143 in the mesenteric fat of high-fat diet-induced obese mice may contribute to the regulated expression of adipocyte genes involved in the pathophysiology of obesity and the observed linkage between obesity and cancer [46].

Various regulatory factors are involved with epigenetic regulation. For example, Polycomb group (PcG) and Trithorax group (TrxG) proteins have emerged as players in gene regulation and are thought to function coordinately to orchestrate DNA accessibility throughout development. These epigenetic regulators act antagonistically to either promote (TrxG) or repress (PcG) transcription through regulation of specific amino acid modifications of critical histones. It is not known how the PcG and TrxG proteins switch and balance between transcriptionally silenced heterochromatin. However, there is evidence that dietary components such as the polyphenols can influence the polycomb homeostasis in skin and possibly at other sites [47].

TRANSCRIPTOMICS

Genomic and epigenomic fluctuations do not entirely account for the influence that dietary factors can have on phenotypes because changes in the rate of transcription of genes (transcriptomics) can also be exceedingly important [48]. Several bioactive food components have been reported to be important regulators of gene expression patterns both *in vitro* and *in vivo*. Vitamins, minerals, and a host of phytochemicals are reported to influence significantly gene transcription and translation in a dose- and time-dependent manner [1]. These changes are fundamental to a change in cellular events and thereby likely influence one or more biological processes associated with cancer including cellular energetics, cell growth, apoptosis, and differentiation.

Transcriptomics analysis allows for a genome-wide monitoring of expression for the simultaneous assessment of tens of thousands of genes and of their relative expression. Although microarray technologies provide an important tool to discover expression changes, it must be remembered that any response is likely cellular dependent and varies between healthy and diseased conditions. Studies using animals are beginning to identify specific sites of action of bioactive food components. For example, the nuclear factor E2 p45-related factor 2 (Nrf2) and the Kelch domain-containing partner Keap1, which are modified by sulforaphane and allyl sulfur [49–51]. Gene expression profiles from wild-type and Nrf2-deficient mice fed sulforaphane have shown several novel downstream events and thus more clues about the true biological response to this food component. The upregulation of glutathione s-transferase, nicotinamide adenine dinucleotide phosphate:quinone reductase, gamma-glutamylcysteine synthetase, and epoxide hydrolase, occurring because of release of Nrf2 from its cytosolic complex, likely explains the ability of sulforaphane to influence multiple processes including those involving xenobiotic metabolizing enzymes, antioxidants, and biosynthetic enzymes of the glutathione and glucuronidation conjugation pathways.

Mammals adapt to limitation and excess exposure to foods and their components through shifts in absorption, metabolism, or excretion. Thus, the quantity and duration of exposure must be considered when translating the response in gene expression patterns following exposure to a given food or components. Because microarray technologies provide only a single snapshot, overinterpretation of their physiological significance is possible. Although mRNA microarray technology continues to provide a powerful tool for examining potential sites of action of food components, their usefulness for population studies remains limited by cost. Transcriptomic technologies have been used to examine the relationship between diet and prostate cancer among native Japanese and second-generation Japanese American men as a function of consumption of animal fat and soy [52]. This technology was able to discriminate differences between those men with cancer and those who were cancer-free. Likewise, these studies detected changes associated with body mass and metabolism. Weight loss caused by caloric restriction has been reported to be associated with changes in the expression of several inflammatory related genes as discovered by transcriptomics [53]. To date, relatively few human studies have used transcriptomics to characterize the response to specific dietary components and thus it is hard to mark firm conclusions about the utility of this technology. Nevertheless, much of the current evidence suggests that mRNA abundance often provides little clue about protein activity and thus cannot substitute for functional and ecological analyses of candidate genes [48]. It is possible that more select arrays maybe be useful if targeted to some cellular process.

Transcriptomics examines the expression level of mRNAs in specific cell populations and is fundamental to the identification of cellular mechanisms involved with cancer. The transcriptome is regulated by the activity of transcription factors. A transcription factor comprises one or more proteins that bind to a specific gene DNA sequence and either initiate or repress transcription thereby leading to increased or decreased mRNA production. DNA binding is induced by protein kinases and transcription factors that act downstream of these signaling cascades that ultimately influence cellular processes that determine survival, proliferation, apoptosis, or differentiation. Thus, one of the mechanisms by which food components may influence the transcriptome and thereby cellular processes is by interacting with members of the nuclear receptor superfamily of transcription factors. This family is recognized to comprise many members that bind with varying affinities to hormones, nutrients, and metabolites and thereby change phenotypic expressions. These receptors function as sensors and transmit signals to specific molecular targets to induce an adaptive response and changes in transcript levels. Receptors are binding proteins that translocate from the cytoplasm to the nucleus and include such receptors as the glucocorticoid (GR), estrogen (ERα and β) receptors, and peroxisome proliferator-activated receptor gamma (PPAR gamma). PPAR gamma is a nuclear receptor that regulates intestinal inflammation and is known to be influenced by a number of dietary components [54]. Diurnal variation in various nuclear transcription factors may be a result of shifts in the concentrations of nutrients or their metabolites [55].

PROTEOMICS

It can be argued that gene expression parts are only part of the issue when it comes to the regulation of cellular processes. The transcriptome must be converted to proteins to fulfill their biological roles. New and exciting technologies in protein chemistry are being developed that assist with biomarker development and assessment of the impact of diet [56, 57]. Classical two-dimensional (2D) gel electrophoresis, coupled with spot analysis by mass spectrometry, is still the most widely used technical approach. Although the results from these approaches are impressive, they can also be subject to overinterpretation due to inadequate identification of the protein in question. Unfortunately, the literature tends to be enriched with reviews than actual experimentally derived data.

Since significant post-translational modification can occur on this protein its formation is only one site of regulation. Many proteins have N-terminal signaling peptides that are removed once the protein is formed. Most proteins undergo extensive posttranslational modification by addition of phosphates, carbohydrates, methylation, etc. On average, these modifications occur in proteins in about every 8–12 amino acid residues [58]. Likewise, oxidative stress can influence these proteins once they are formed. Bioactive food components can influence not only the rate of formation of proteins but also their posttranslational modification.

Several food components are reported to influence the formation or characteristics of cellular proteins. For example, genistein has been demonstrated to influence a number of proteins in mammary gland tissue [59]. GTP cyclohydrolase 1 was one of six proteins whose abundance was changed in the mammary gland following genistein administration early in life, whereas tyrosine hydroxylase was upregulated later in the development of the rodent [59].

Human and animal studies suggest quercetin may be involved in preventing or inhibiting oncogenesis, but the underlying mechanism remains unclear. Recent studies evaluated the effect(s) of quercetin on normal and malignant prostate cells in an attempt to identify proteins that were being influenced [60]. Quercetin was found to promote cancer cell apoptosis by down-regulating the levels of heat shock protein (Hsp) 90. Knockdown of Hsp90 by short interfering RNA also resulted in induction apoptosis similar to quercetin. It should be noted that Hsp 90 is considerably lower in normal nonneoplastic cells and thus a response in control cultures was not detected.

The area of nutritional proteomics is an exciting but clearly emerging area. Most studies to date have been conducted with cells in culture with animal models. Proteomics has not been as universally accepted in human studies as has transcriptomics, which also continues to be sparsely utilized. Regardless, it is unlikely that a single protein by itself clearly defines a specific response to a food component or to a change in cancer risk or tumor behavior. Thus, multiple protein biomarkers may allow a more definite evaluation of the benefits or risk associated with a change in dietary habits. A combination of proteomics data with other physiologically relevant indicators and employing appropriate statistical tools for developing predictive models is likely in the future. ·

METABOLOMICS

Metabolomics holds promise to identify pathways involved with the maintenance of normal and/or pathophysiologic conditions including cancer [57, 61, 62]. The metabolome can best be described as the full set of endogenous or exogenous low molecular weight metabolic entities of approximately <1000 Da (metabolites) and the small pathway motifs that are present in a biological system (cell, tissue, organ, organism, or species). The most common metabolites are amino acids, lipids, vitamins, small peptides, or carbohydrates. These metabolites are the ultimate endpoint of gene expression and of any physiological regulatory processes. Whereas these metabolites may be more sensitive indicators of disease status, they are also subject to enormous variation due to dosage and temporal effects of foods and food components [3, 61, 62].

Typically, metabolomic approaches use ^{1}H and ^{13}C nuclear magnetic resonance (^{1}H NMR, ^{13}C NMR) or gas chromatography combined with mass spectrometry (GC-MS). Whereas ^{1}H NMR detection limits range from 20–40 metabolites in extracted tissues and 100–200 metabolites in urine samples, GC-MS techniques may have higher detection rates across these samples. Despite this possible difference, NMR techniques are sensitive detection methods capable of discerning metabolites signaling alterations in tumor glucose metabolism [63] and hopefully will allow for the detection of subtle changes induced by dietary change in both normal and neoplastic tissue(s).

Undeniably, measuring metabolic changes in targeted single cells or across single tissues and entire organ systems is a daunting task because it is unclear what surrogate cells/fluids are best utilized. Although a host of repository tissues exists, it is unclear what special handling procedures are required to ensure data obtained from these will be meaningful. Without question genomics, transcriptomics, and proteomics measure potential human response to exposure of particular stimuli (e.g., dietary constituents), whereas metabolomics attempts to measure all metabolic end- and by-products resulting in response to dietary exposures, thus yielding a comprehensive yet extremely complex systems-biology snapshot of events at a given moment. How these events relate to health and wellness is a matter of continued investigation.

Solanky et al. [64] used ^{1}H NMR spectroscopy of blood plasma samples to analyze metabolic changes associated with a change in soy intake in premenopausal women. Consumption of soy was linked to increased concentrations of 3-hydroxybutyrate, N-acetyl glycoproteins and lactate, and a concomitant decrease in plasma sugars. These changes were suggestive to reflect an alteration in carbohydrate and energy metabolism. These changes were also accompanied by changes in plasma lipoproteins as well as glycoproteins. Although these observations point to the utility of using metabolomics profiles to detect shifts in energy metabolism and consequential changes in the lipoproteins, more research is needed to determine the biological significance of these changes. Because soy has frequently been associated with a reduction in cancer risk [65], it will be important to determine if subtle energetic changes may be at the root of the protection.

The Human Metabolome Database is an impressive collection of human metabolite and human metabolism data [66]. It contains records for more than 2180 endogenous metabolites with information gathered from thousands of books, journal articles, and electronic databases. In addition to its comprehensive literature-derived data, it also contains an extensive collection of experimental metabolite concentration data compiled from hundreds of mass spectra (MS) and nuclear magnetic resonance (NMR) metabolomic analyses performed on urine, blood, and cerebrospinal fluid samples.

CONCLUSIONS

Each of the "omic" technologies offers unique insights into unraveling the role of diet in modifying the cancer process. Research in each of the "omics" is beginning to yield a better understanding of the carcinogenic process and how environmental agents including dietary components might best be used. This research is identifying targets likely influenced by bioactive food components. It should be noted that most food components influence multiple sites simultaneously and thus may be one of the best attributes to the use of diet as an effective intervention strategy, especially because they are often less toxic than fabricated drugs. Undeniably, understanding how the "omics" influence these sites of action is of paramount importance to human health.

Many dietary factors may be effective preventive agents by inhibiting or reversing premalignant lesions and/or reducing tumor incidence. Some compounds that show promise include carotenoids, isothiocyanates, allyl sulfurs, butyrate, and flavonoids, along with a host of essential nutrients including vitamin D, calcium, selenium, and folate. Rigorous investigations are needed to determine the molecular actions of these and other bioactive food components, their long-term effectiveness, and safety. Preclinical studies are needed to address the bioavailability, toxicity, molecular target, signal transduction pathways, and side effects of bioactive food components prior to entering into long-term high-risk interventions in humans. Clinical trials based on established mechanistic approaches that use target populations rather than the masses are also essential to unraveling the mysteries associated with diet and cancer prevention.

REFERENCES

1. Milner, JA (2004) Molecular targets for bioactive food components. *J Nutr. 134*, 2492s–2498s.
2. Davis CD, Milner JA (2007) Molecular targets for nutritional preemption of cancer. *Curr Cancer Drug Targets 7*(5):410–415.
3. Jenab M, Slimani N, Bictash M, Ferrari P, Bingham SA (2009) Biomarkers in nutritional epidemiology: applications, needs and new horizons. *Hum Genet. 125*(5–6):507–525.
4. Doll R, Peto R (1981) The causes of cancer: quantitative estimates of avoidable risks of cancer in the United States today. *J Natl Cancer Inst. 66*(6):1191–1308.

5. World Cancer Research Fund/American Institute for Cancer Research (2007) *Food Nutrition, Physical Activity and the Prevention of Cancer: A Global Perspective.* AICR, Washington, DC.
6. Thiébaut AC, Chajès V, Gerber M, Boutron-Ruault MC, Joulin V, Lenoir G, Berrino F, Riboli E, Bénichou J, Clavel-Chapelon F (2009) Dietary intakes of omega-6 and omega-3 polyunsaturated fatty acids and the risk of breast cancer. *Int J Cancer 124*(4):924–931.
7. Davis, CD (2007) Nutritional interactions: credentialing of molecular targets for cancer prevention. *Exp Biol Med. 232*, 176–183.
8. Zhang M, Holman CD, Huang JP, Xie X (2007) Green tea and the prevention of breast cancer: a case-control study in Southeast China. *Carcinogenesis 28*(5):1074–1078.
9. Tomar RS, Shiao R (2008) Early life and adult exposure to isoflavones and breast cancer risk. *J Environ Sci Health C Environ Carcinog Ecotoxicol Rev. 26*(2):113–173.
10. Virtamo J, Pietinen P, Huttunen JK, Korhonen P, Malila N, Virtanen MJ, Albanes D, Taylor PR, Albert P; ATBC Study Group (2003) Incidence of cancer and mortality following alpha-tocopherol and beta-carotene supplementation: a postintervention follow-up. *JAMA 290*(4):476–485.
11. Waters DJ, Shen S, Cooley DM, Bostwick DG, Qian J, Combs GF Jr, Glickman LT, Oteham C, Schlittler D, Morris JS (2003) Effects of dietary selenium supplementation on DNA damage and apoptosis in canine prostate. *J Natl Cancer Inst. 95*(3):237–241.
12. Toyokuni S (2009) Role of iron in carcinogenesis: cancer as a ferrotoxic disease. *Cancer Sci. 100*(1):9–16.
13. Goldstein DB, Need AC, Singh R, Sisodiya SM (2007) Potential genetic causes of heterogeneity of treatment effects. *Am J Med. 120*(4 Suppl 1):S21–25.
14. El-Sohemy A (2007) Nutrigenetics. *Forum Nutr. 60*:25–30.
15. Lampe JW (2009) Interindividual differences in response to plant-based diets: implications for cancer risk. *Am J Clin Nutr. 89*(5):1553S–1557S.
16. International Human Genome Sequencing Consortium. 2004. Finishing the euchromatic sequence of the human genome. *Nature 431*(7011):931–945.
17. Voisey J, Morris CP (2008) SNP technologies for drug discovery: a current review. *Curr Drug Discov Technol. 5*(3):230–235.
18. Clay Montier LL, Deng JJ, Bai Y (2009) Number matters: control of mammalian mitochondrial DNA copy number. *J Genet Genomics 36*(3):125–131.
19. Thomas G, Jacobs KB, Yeager M, Kraft P, Wacholder S, Orr N, Yu K, Chatterjee N, Welch R, Hutchinson A, Crenshaw A, Cancel-Tassin G, Staats BJ, Wang Z, Gonzalez-Bosquet J, Fang J, Deng X, Berndt SI, Calle EE, Feigelson HS, Thun MJ, Rodriguez C, Albanes D, Virtamo J, Weinstein S, Schumacher FR, Giovannucci E, Willett WC, Cussenot O, Valeri A, Andriole GL, Crawford ED, Tucker M, Gerhard DS, Fraumeni JF Jr, Hoover R, Hayes RB, Hunter DJ, Chanock SJ (2008) Multiple loci identified in a genome-wide association study of prostate cancer. *Nat Genet. 40*(3):310–315.
20. Zheng SL, Sun J, Cheng Y, Li G, Hsu FC, Zhu Y, Chang BL, Liu W, Kim JW, Turner AR, Gielzak M, Yan G, Isaacs SD, Wiley KE, Sauvageot J, Chen HS, Gurganus R, Mangold LA, Trock BJ, Gronberg H, Duggan D, Carpten JD, Partin AW, Walsh PC, Xu J, Isaacs WB (2007) Association between two unlinked loci at 8q24 and prostate cancer risk among European Americans. *J Natl Cancer Inst. 99*(20):1525–1533.
21. Gold B, Kirchhoff T, Stefanov S, Lautenberger J, Viale A, Garber J, Friedman E, Narod S, Olshen AB, Gregersen P, Kosarin K, Olsh A, Bergeron J, Ellis NA, Klein RJ, Clark AG, Norton L, Dean M, Boyd J, Offit K (2008) Genome-wide association study provides evidence for a breast cancer risk locus at 6q22.33. *Proc Natl Acad Sci U S A 105*(11): 4340–4345.
22. Hwang SJ, Yang Q, Meigs JB, Pearce EN, Fox CS (2007) A genome-wide association for kidney function and endocrine-related traits in the NHLBI's Framingham Heart Study. *BMC Med Genet. 8* Suppl 1–10.
23. Florez JC, Burtt N, de Bakker PI, Almgren P, Tuomi T, Holmkvist J, Gaudet D, Hudson TJ, Schaffner SF, Daly MJ, Hirschhorn JN, Groop L, Altshuler D (2004) Haplotype structure and

genotype-phenotype correlations of the sulfonylurea receptor and the islet ATP-sensitive potassium channel gene region. *Diabetes 53*(5):1360–1368.

24. Dai Q, Shrubsole MJ, Ness RM, Schlundt D, Cai Q, Smalley WE, Li M, Shyr Y, Zheng W (2007) The relation of magnesium and calcium intakes and a genetic polymorphism in the magnesium transporter to colorectal neoplasia risk. *Am J Clin Nutr.* 2007 Sep;*86*(3):743–751.

25. Siezen CL, van Leeuwen AI, Kram NR, Luken ME, van Kranen HJ, Kampman E (2005) Colorectal adenoma risk is modified by the interplay between polymorphisms in arachidonic acid pathway genes and fish consumption. *Carcinogenesis 26*(2):449–457.

26. Guerreiro CS, Cravo ML, Brito M, Vidal PM, Fidalgo PO, Leitão CN (2007) The D1822V APC polymorphism interacts with fat, calcium, and fiber intakes in modulating the risk of colorectal cancer in Portuguese persons. *Am J Clin Nutr 85*(6):1592–1597.

27. Herron KL, McGrane MM, Waters D, Lofgren IE, Clark RM, Ordovas JM, Fernandez ML (2006) The ABCG5 polymorphism contributes to individual responses to dietary cholesterol and carotenoids in eggs. *J Nutr. 136*(5):1161–1165.

28. Wong HL, Seow A, Arakawa K, Lee HP, Yu MC, Ingles SA (2003) Vitamin D receptor start codon polymorphism and colorectal cancer risk: effect modification by dietary calcium and fat in Singapore Chinese. *Carcinogenesis 24*(6):1091–1095.

29. Norat T, Bingham S, Ferrari P, Slimani N, Jenab M, Mazuir M, Overvad K, Olsen A, Tjønneland A, Clavel F, Boutron-Ruault MC, Kesse E, Boeing H, Bergmann MM, Nieters A, Linseisen J, Trichopoulou A, Trichopoulos D, Tountas Y, Berrino F, Palli D, Panico S, Tumino R, Vineis P, Bueno-de-Mesquita HB, Peeters PH, Engeset D, Lund E, Skeie G, Ardanaz E, González C, Navarro C, Quirós JR, Sanchez MJ, Berglund G, Mattisson I, Hallmans G, Palmqvist R, Day NE, Khaw KT, Key TJ, San Joaquin M, Hémon B, Saracci R, Kaaks R, Riboli E (2005) Meat, fish, and colorectal cancer risk: the European Prospective Investigation into cancer and nutrition. *J Natl Cancer Inst. 97*(12):906–916.

30. González CA, Jakszyn P, Pera G, Agudo A, Bingham S, Palli D, Ferrari P, Boeing H, del Giudice G, Plebani M, Carneiro F, Nesi G, Berrino F, Sacerdote C, Tumino R, Panico S, Berglund G, Simán H, Nyrén O, Hallmans G, Martinez C, Dorronsoro M, Barricarte A, Navarro C, Quirós JR, Allen N, Key TJ, Day NE, Linseisen J, Nagel G, Bergmann MM, Overvad K, Jensen MK, Tjonneland A, Olsen A, Bueno-de-Mesquita HB, Ocke M, Peeters PH, Numans ME, Clavel-Chapelon F, Boutron-Ruault MC, Trichopoulou A, Psaltopoulou T, Roukos D, Lund E, Hemon B, Kaaks R, Norat T, Riboli E (2006) Meat intake and risk of stomach and esophageal adenocarcinoma within the European Prospective Investigation Into Cancer and Nutrition (EPIC). *J Natl Cancer Inst. 98*(5):345–354.

31. Küry S, Buecher B, Robiou-du-Pont S, Scoul C, Sébille V, Colman H, Le Houérou C, Le Neel T, Bourdon J, Faroux R, Ollivry J, Lafraise B, Chupin LD, Bézieau S (2007) Combinations of cytochrome P450 gene polymorphisms enhancing the risk for sporadic colorectal cancer related to red meat consumption. *Cancer Epidemiol Biomarkers Prev. 16*(7):1460–1467.

32. Li Y, Ambrosone CB, McCullough MJ, Ahn J, Stevens VL, Thun MJ, Hong CC (2009) Oxidative stress-related genotypes, fruit and vegetable consumption and breast cancer risk. *Carcinogenesis 30*(5):777–784.

33. Ahn J, Gammon MD, Santella RM, Gaudet MM, Britton JA, Teitelbaum SL, Terry MB, Neugut AI, Josephy PD, Ambrosone CB (2004) Myeloperoxidase genotype, fruit and vegetable consumption, and breast cancer risk. *Cancer Res. 64*(20):7634–7639.

34. Siezen CL, van Leeuwen AI, Kram NR, Luken ME, van Kranen HJ, Kampman E (2005) Colorectal adenoma risk is modified by the interplay between polymorphisms in arachidonic acid pathway genes and fish consumption. *Carcinogenesis 26*(2):449–457.

35. Lopez J, Percharde M, Coley HM, Webb A, Crook T (2009) The context and potential of epigenetics in oncology. *Br J Cancer 100*(4):571–577.

36. Ross SA (2007) Nutritional genomic approaches to cancer prevention research. *Exp Oncol. 29*(4):250–256.

37. Tost J (2008) DNA methylation: an introduction to the biology and the disease-associated changes of a promising biomarker. *Methods Mol Biol. 2009*(507):3–20.

38. Ross SA (2003) Diet and DNA methylation interactions in cancer prevention. *Ann N Y Acad Sci.* *983*:197–207.
39. Dolinoy DC (2008) The agouti mouse model: an epigenetic biosensor for nutritional and environmental alterations on the fetal epigenome. *Nutr Rev. 66* Suppl 1:S7–11.
40. Dolinoy DC, Huang D, Jirtle RL (2007) Maternal nutrient supplementation counteracts bisphenol A-induced DNA hypomethylation in early development. *Proc Natl Acad Sci U S A.* *104*(32):13056–13061.
41. Kim JK, Samaranayake M, Pradhan S (2009) Epigenetic mechanisms in mammals. *Cell Mol Life Sci. 66*(4):596–612.
42. Dashwood RH, Ho E (2007) Dietary histone deacetylase inhibitors: from cells to mice to man. *Semin Cancer Biol. 17*(5):363–369.
43. Myzak MC, Tong P, Dashwood WM, Dashwood RH, Ho E (2007) Sulforaphane retards the growth of human PC-3 xenografts and inhibits HDAC activity in human subjects. *Exp Biol Med (Maywood). 232*(2):227–234.
44. Davis CD, Ross SA. (2008) Evidence for dietary regulation of microRNA expression in cancer cells. *Nutr Rev. 66*(8):477–482.
45. Takanabe R, Ono K, Abe Y, Takaya T, Horie T, Wada H, Kita T, Satoh N, Shimatsu A, Hasegawa K (2008) Up-regulated expression of microRNA-143 in association with obesity in adipose tissue of mice fed high-fat diet. *Biochem Biophys Res Commun. 376*(4):728–732.
46. Calle EE, Thun MJ (2004) Obesity and cancer. *Oncogene 23*(38):6365–6378.
47. Balasubramanian S, Lee K, Adhikary G, Gopalakrishnan R, Rorke EA, Eckert RL (2008) The Bmi-1 polycomb group gene in skin cancer: regulation of function by (-)-epigallocatechin-3-gallate. *Nutr Rev. 66* Suppl 1:S65–68.
48. Feder ME, Walser JC (2005) The biological limitations of transcriptomics in elucidating stress and stress responses. *J Evol Biol. 18*(4):901–910.
49. Juge N, Mithen RF, Traka M (2007) Molecular basis for chemoprevention by sulforaphane: a comprehensive review. *Cell Mol Life Sci. 64*(9):1105–1127.
50. Hu R, Xu C, Shen G, Jain MR, Khor TO, Gopalkrishnan A, Lin W, Reddy B, Chan JY, Kong AN (2006) Gene expression profiles induced by cancer chemopreventive isothiocyanate sulforaphane in the liver of C57BL/6J mice and C57BL/6J/Nrf2 (-/-) mice. *Cancer Lett. 243*(2):170–192.
51. Kalayarasan S, Sriram N, Sureshkumar A, Sudhandiran G (2008) Chromium (VI)-induced oxidative stress and apoptosis is reduced by garlic and its derivative S-allylcysteine through the activation of Nrf2 in the hepatocytes of Wistar rats. *J Appl Toxicol. 28*(7):908–919.
52. Marks LS, Kojima M, Demarzo A, Heber D, Bostwick DG, Qian J, Dorey FJ, Veltri RW, Mohler JL, Partin AW (2004) Prostate cancer in native Japanese and Japanese-American men: effects of dietary differences on prostatic tissue. *Urology 64*(4):765–771.
53. Viguerie N, Poitou C, Cancello R, Stich V, Clément K, Langin D (2005) Transcriptomics applied to obesity and caloric restriction. *Biochimie 87*(1):117–123.
54. Marion-Letellier R, Déchelotte P, Iacucci M, Ghosh S (2009) Dietary modulation of peroxisome proliferator-activated receptor gamma. *Gut 58*(4):586–593.
55. Yang X, Lamia KA, Evans RM (2007) Nuclear receptors, metabolism, and the circadian clock. *Cold Spring Harb Symp Quant Biol. 72*:387–394.
56. de Roos B (2008) Proteomic analysis of human plasma and blood cells in nutritional studies: development of biomarkers to aid disease prevention. *Expert Rev Proteomics 5*(6):819–826.
57. Trujillo E, Davis C, Milner J (2006) Nutrigenomics, proteomics, metabolomics, and the practice of dietetics. *J Am Diet Assoc. 106*(3):403–413.
58. Nielsen ML, Savitski MM, Zubarev RA (2006) Extent of modifications in human proteome samples and their effect on dynamic range of analysis in shotgun proteomics. *Mol Cell Proteomics 5*:2384–2391.
59. Rowell C, Carpenter DM, Lamartiniere CA (2005) Chemoprevention of breast cancer, proteomic discovery of genistein action in the rat mammary gland. *J Nutr. 135*:2953S–2959S.

60. Aalinkeel R, Bindukumar B, Reynolds JL, Sykes DE, Mahajan SD, Chadha KC, Schwartz SA (2008) The dietary bioflavonoid, quercetin, selectively induces apoptosis of prostate cancer cells by down-regulating the expression of heat shock protein 90. *Prostate 68*(16):1773–1789.
61. Barnes S (2008) Nutritional genomics, polyphenols, diets, and their impact on dietetics. *J Am Diet Assoc. 108*(11):1888–1895.
62. Kim YS, Maruvada P, Milner JA (2008) Metabolomics in biomarker discovery: future uses for cancer prevention. *Future Oncol. 4*(1):93–102.
63. Griffin, Julian L and Shockcor, John P (2004) Metabolic profiles of cancer cells. *Nature Reviews Cancer 4*:551–561.
64. Solanky KS, Bailey NJ, Beckwith-Hall BM, Davis A, Bingham S, Holmes E, Nicholson JK, Cassidy A (2003) Application of biofluid ^{1}H nuclear magnetic resonance-based metabonomic techniques for the analysis of the biochemical effects of dietary isoflavones on human plasma profile. *Anal Biochem. 323*:197–204.
65. Kim J (2008) Protective effects of Asian dietary items on cancers – soy and ginseng. *Asian Pac J Cancer Prev. 9*(4):543–548.
66. Wishart DS, Knox C, Guo AC, Eisner R, Young N, Gautam B, Hau DD, Psychogios N, Dong E, Bouatra S, Mandal R, Sinelnikov I, Xia J, Jia L, Cruz JA, Lim E, Sobsey CA, Shrivastava S, Huang P, Liu P, Fang L, Peng J, Fradette R, Cheng D, Tzur D, Clements M, Lewis A, De Souza A, Zuniga A, Dawe M, Xiong Y, Clive D, Greiner R, Nazyrova A, Shaykhutdinov R, Li L, Vogel HJ, Forsythe I (2009) HMDB: a knowledge base for the human metabolome. *Nucleic Acids Res. 37*(Database issue):D603–D610.

Index

Adipocytes, 249–250
Advanced glycation end products (AGE), 226, 228, 236
Advisory Committee on Novel Foods and Processes (ACNFP), 26
Affymetrix GeneChips, 13
Affymetrix Microarray Suite 5.0, 78
Allyl sulfides, 185
Amino acids, unbalance and excess of, 133, 134 *f*
Anthocyanins, 12
ApoAI gene, 3
Apolipoprotein (apo) E gene, 3
Apple flavonoids, 108*t*
Apple polyphenols, 145–146, 146 *f*
Arginine, 12
Array platforms, 97
Association network analysis tools, 64
Autophagy, 88–89

Bacteroidetes, 86
Beggar's ticks, 302*t*
BE-30, 158, 168–169
Bioanalyzer 2100, 77
Biobanks, 25–26
Bioconductor, 62–63
Bioinformatics. *See* Functional foods, application of genomics and bioinformatics in study of
Biomarker proteins from blood cells, 216–217
Black cohosh, 301*t*
Boswellia serrata (frankincense), genetic basis of anti-inflammatory properties of, 155–172
  human clinical studies, 169–171, 170 *f*
  mechanism of action, 157–160, 163–166
    GeneChip analysis, 158–160, 159 *f*–160 *f*, 161*t*–163*t*
    genetic basis for efficacy of BE-30, 158
    5-LOXIN (BE-30), 158
    proteomics, 160, 163–166, 164 *f*–166 *f*
  overview, 76, 77 *f*–79 *f*, 155–157, 156 *f*, 171–172, 171 *f*
  safety of BE-30, 168–169
  in vivo studies, 167–168, 167 *f*–168 *f*
Butyrate, 6

Cafestol, chemical structure of, 177 *f*
Caffeic acid phenethyl ester, 111*t*
  chemical structure, 177 *f*
Cancer prevention research, 74–75
  contribution of omics revolution to, 315–323
    epigenetics, 318–320
    genetics and response to diet, 315–316, 316 *f*
    genomics, 316–318
    metabolomics, 322–323
    overview, 315, 323
    proteomics, 321–322
    transcriptomics, 320–321
  phytochemicals. *See* Chemoprevention of cancer with anti-inflammatory and antioxidant phytochemicals
Capillary electrophoresis (CE), 241, 275
Capsaicin, 180
  chemical structure, 176 *f*
Cardiomyopathy, diabetic, 257
Cardiovascular system and grape seed proanthocyanidins (GSP), 227–228
Carnosol, 186
  chemical structure, 177 *f*
Carotenes, 288 *f*
Carotenoids, 288 *f*
cDNA (complementary DNA), 12
CE (capillary electrophoresis), 241, 275
Cell mapping proteomics, 231
Chalcones, 187
Chamomile tea, 6, 16
Chemokine receptor 6 (CCR6), 89
Chemoprevention of cancer with anti-inflammatory and antioxidant phytochemicals, 175–189
  interaction/crosstalk between NRF2 and NF-κB signaling pathways, 187–189, 188 *f*
  molecular-based cancer chemoprevention, 177
  nuclear factor erythroid 2 P45 (NF-E2)-related factor (NRF2), 181–187, 183 *f*
    allyl sulfides, 185
    carnosol, 186
    chalcones, 187

Chemoprevention of cancer with
     anti-inflammatory and antioxidant
     phytochemicals (*Continued*)
  coffee-derived diterpenes, 186
  curcuminoids, 184
  EGCG, 184–185
  lycopene, 186
  pungent vanilloids, 186
  resveratrol, 185–186
  sulforaphane, 182–184
  xanthohumol, 187
  zerumbone, 187
  nuclear factor-kappa B (NF-κB), 89,
     177–181, 179 *f*
  capsaicin, 180
  curcumin, 179–180
  genistein, 181
  quercetin, 181
  resveratrol, 180–181
  [6]-gingerol, 180
  sulforaphane, 181
  overview, 175, 176 *f*–177 *f*, 189
Chocolate, 17
"Choosing Health," 30–31
Chronic inflammation in disease, 73–75
  cancer, 74–75
  gastrointestinal tract, 74
  neurodegenerative diseases, 74
  neutraceuticals in management of, 75–76
  respiratory disorders, 74
  rheumatic diseases, 75
  vascular disorders, 73
Cinnamaldehyde, 107*t*
Circumin, chemical structure of, 176 *f*
Cocoa, 142–143, 143 *f*
Coding genes, 35, 39
Coffee-derived diterpenes, 186
Comfrey, 301*t*
Complementary DNA (cDNA), 12
Cone flower, 301*t*
Conjugated linoleic acid (CLA), 217–218
Crohn's disease, 87–90
  dietary modulation in response to specific
     genetic variants, 89–90
  influence of transporter and barrier functions,
     87
  innate and adaptive immune responses to
     microbial pathogens, 87–89
Curcumin and curcuminoids, 179–180, 184

DASH (Dietary Approaches to Stop
     Hypertension) Eating Plan, 4
Data Mining Tool (DMT) 2.0, 78
dChip software, 78
Devil's bush, 301*t*

DGAT2 (diacylglycerol transferase 2), 13
Diabetes, grape seed proanthocyanidins (GSP)
     and, 228–229
Diabetes, proteomics for elucidating insulin
     dysregulation, 241–258
  diabetes mellitus, 243
  overview, 241, 242 *f*, 257–258
  proteomics in diabetes, 243–251
     adipocytes, 249–250
     liver, 248
     pancreas, pancreatic islets, and β-cells,
        243–248, 244 *f*–246 *f*
     serum and plasma from diabetes, 250–251
     skeletal muscle, 248–249
  proteomics of diabetic complications,
     251–257
     diabetic cardiomyopathy, 257
     diabetic nephropathy, 234–236, 252–254,
        254 *f*, 255*t*
     diabetic retinopathy, 254–257
Diacylglycerol transferase 2 (DGAT2), 13
Diallyl sulfide, chemical structure of, 176 *f*
Diallyl trisulfide, 107*t*
  chemical structure, 176 *f*
Diet
  disease and, 277–278
  genetics and response to, 315–316, 316 *f*
Dietary antioxidants, demonstrating safety and
     efficacy of with genomics, 95–116
  antioxidants and gene expression, 99–100,
     103–105, 112
     conventional methods available to measure
        oxidative stress, 104
     genomics-based approaches to identify the
        antioxidant effect, 104–105
     oxidative stress, antioxidants, and cell
        signaling, 100, 101*t*–103*t*, 103–104
     in vitro studies, 105, 106*t*–111*t*
     in vivo studies, 105, 112
  array platforms, 97
  functional genomics, 96
  nutrigenetics and nutrigenomics, 112–116,
     113 *f*
     safety and efficacy assessment of dietary
        nutrients/antioxidants, 113–114,
        115 *f*
     toxicogenomics as a predictive tool,
        115–116
     toxicological genomic approach in the
        assessment of dietary
        nutrients/antioxidants, 114–115
  overview, 95–96, 116
  performing a microarray experiment, 97–99
     differential expression, 98–99
     experimental design, 97

hybridization, 98
    image analysis and data extraction, 98
    validation by real-time PCR, 99
Dietary Approaches to Stop Hypertension
        (DASH) Eating Plan, 4
Dietary Guidelines for Americans (2005), 4
Dietary protein, 131–132, 132*t*
Dietary supplements, safety and efficacy of,
        47–53
    HCA-SX
        efficacy in weight loss, 49–50
        regulation of transcriptomic expression,
            51–53
        safety studies, 48–49
    NBC
        safety evaluations, 47
        transcriptomic changes, 50–51
        in weight loss, 47–48
    nutrigenomics in weight management
        supplements, 50
Difference gel electrophoresis (DIGE), 14–15,
        215
DMT (Data Mining Tool) 2.0, 78

EGCG, 184–185
    chemical structure, 176*f*
Eicosanoid biosynthetic pathways, 155, 156*f*
Electrospray ionization (ESI), 14, 214, 241
Endothelial lipase (LIPG) gene, 53
Enolase-3 (ENO3) gene, 51
Entrez Gene search, 64, 66, 67*t*
Epigallocatechin gallate, 106*t*–107*t*, 110*t*
Epigenetics, 318–320
Ethanol metabolism genes, 65–66, 67*t*, 68–69,
        68*f*–69*f*
Ethical questions concerning nutrigenomics and
        statistical power, 25–32
    ethics and advice, 29–32
    nutrigenomics versus pharmacogenomics,
        27
    overview, 25–26, 32
    personalization, 26–27
    prerequisites, 28–29
    skepticism, 27–28
European mistletoe, 301*t*
EXPANDER, 62–63
Expression Profiler, 62–63
Expression proteomics, 231

Fasting, 135–136, 136*f*
Ferulic acid, 105
Firmicutes, 86
Fish oil supplements, 215, 218
5-LOXIN (BE-30), 158, 168–169
Flavonoids, 6, 288*f*

Flaxseed, 215
Fluorescence 2-D difference gel electrophoresis
        (DIGE), 231
Fluxomics, 7
Food allergy, 136–138, 137*f*
Food Guide Pyramid, 3, 50
Food-related systems biology, 149, 150*f*
Foods for specified health use (FOSHU), 128,
        129*t*–130*t*, 130
Fourier transform infrared (FT-IR)
        spectroscopy, 275
Fourier transform ion cyclotron resonance MS
        (FTICR-MS), 14
Fox tail, 301*t*
Frankincense. *See Boswellia serrata*
        (frankincense), genetic basis of
        anti-inflammatory properties of
Fructooligosaccharide, 142, 143*t*
Functional foods
    application of genomics and bioinformatics
        in study of, 61–70
        analysis tools, 62–63
            bioconductor, 62
            EXPANDER, 62–63
            Expression Profiler, 62–63
            GeneSpring GX, 62
            GEPAS, 62–63
            TM4, 62–63
        animal study and DNA microarray
            analysis, 65–66
        application example (kale), 64–65, 65*f*
        association network analysis tools, 64
        data analysis, 66, 68–69
            Entrez Gene search, 66, 67*t*
            expression data, 66
            filtering, 66, 69*f*–70*f*
            result, 66, 68–69, 70*f*
            STRING search, 66, 68*f*
        Gene Oncology (GO) project tools, 63
        interpretation tools, 63
        overview, 61–62, 69–70
        pathway analysis tools, 63–64
    applications of genomics in Japan, 138–140,
        142–149
        apple polyphenols, 145–146, 146*f*
        cocoa, 142–143, 143*f*
        fructooligosaccharide, 142, 143*t*
        lycopene, 146–147, 147*f*
        neoculin, 147–149, 148*f*
        proanthocyanidin, 144–145, 145*f*
        royal jelly, 143–144, 144*t*
        sesamin, 140, 141*f*, 142
        soy protein isolate, 138–140
    definition of, 3
    metabolomics and, 278–279

Functional foods (*Continued*)
  novel omics technologies in. *See* Omics
    technologies, nutraceutical and
    functional food research
  recent advances in
    nutritional genomics, 4–7
      nutrigenetics, 4
      nutrigenomics, 5–7
    omics in future nutrition research, 7
    overview, 3–4, 7
Functional genomics, 96

Gallic acid, 107*t*
Garlic, 301*t*
Gas chromatography (GC), 14
Gastrointestinal tract, 74, 84–86
Gel-based techniques, 5
"Gel-free" proteomics, 232–233
GeneChip, 12, 76–80, 77*f*–79*f*
  analysis of *Boswellia* extract, 158–160,
    159*f*–160*f*, 161*t*–163*t*
Gene–diet interactions, 85*f*, 315–316,
  316*f*
  in Crohn's disease, 87–90
  in gastrointestinal health, 87
Gene Ontology (GO) project tools, 63
GeneSpring GX, 62–64, 66
Genistein, 6, 12, 106*t*, 109*t*, 111*t*, 181
  chemical structure, 176*f*
Genome-wide association studies (GWAs),
  87–88
Genomics, 302–304, 303*f*, 305*f*
  antioxidants, demonstrating safety and
    efficacy of. *See* Dietary antioxidants,
    demonstrating safety and efficacy of
    with genomics
  application in study of functional food. *See*
    Functional foods, application of
    genomics and bioinformatics in
    study of
  applied to nutrients and functional foods in
    Japan. *See* Japan, application of
    genomics to nutrients and functional
    foods in
  Boswellia extract, genetic basis of
    anti-inflammatory properties of. *See*
    Boswellia serrata (frankincense),
    genetic basis of anti-inflammatory
    properties of
  cancer prevention research. *See* Cancer
    prevention research;
    Chemoprevention of cancer with
    anti-inflammatory and antioxidant
    phytochemicals
  definition of, 11, 302, 316

  nutrigenomics, application in gastrointestinal
    health. *See* Nutrigenomics,
    application in gastrointestinal health
  nutrigenomics and statistical power, ethical
    questions concerning. *See*
    Nutrigenomics and statistical power,
    ethical questions concerning
  nutrimiromics. *See* Nutrimiromics
  as tool to characterize anti-inflammatory
    nutraceuticals. *See* Nutraceuticals,
    anti-inflammatory, genomics as tool
    to characterize
  weight loss nutraceuticals. *See* Weight loss
    nutraceuticals, genomics in
GEPAS, 62–63
GEPAT (Genome Expression Pathway Analysis
  Tool), 64
Ginkgo biloba, 114–115, 301*t*
Ginseng, 301*t*
Glucose phosphate isomerase (GPI) gene, 51
Glucose tolerance factor (GTF), 50
Glucosinolates, 289*f*
GO (Gene Ontology) project tools, 63
Grape seed extract (GSE), 6, 144–145, 145*f*
Grape seed proanthocyanidins (GSP),
    proteomics approach to assess
    potency of, 225–237
  chemoprotective properties of GSP, 225–230
    antiaging effects, 229
    antinonenzymatic glycation and
      anti-inflammation effects, 226–227
    antioncogenesis effects, 229–230
    antiosteoporosis, 230
    antioxidant effects, 226
    components and molecules, 225–226
    protective effects of diabetes and its
      complications, 228–229
    protective effects on cardiovascular
      system, 227–228
    wound healing, 230
  overview, 225, 236–237
  proteomic platform, 231–233
    "gel-free" proteomics, 232–233
    protein chips, 233
    two-dimensional gel electrophoresis
      (2-DE) proteomics, 231–232
  proteomics analysis of GSP actions, 233–237
    in brain of normal rats, 233–234
    functional confirmation of proteins, 236
    future perspective, 236–237
    in rats diabetic nephropathy, 234–236,
      234*f*–235*f*
Green tea catechins, 12
GSE (grape seed extract), 6, 144–145, 145*f*
GSE (*Gymnema sylvestre* extract), 49

GSP. *See* Grape seed proanthocyanidins, proteomics approach to assess potency of
GSTM, 235
GTF (glucose tolerance factor), 50
Gut microflora, metabolomics and, 17, 278
GWAs (genome-wide association studies), 87–88
Gymnema sylvestre extract (GSE), 49

HCA-SX
   efficacy in weight loss, 49–50
   regulation of transcriptomic expression, 51–53
   safety studies, 48–49
Heat shock proteins (HSPs), 205
Herbal medicines, 76
High-density oligonucleotide arrays (gene chips), 97
High-resolution magic angle spinning (HRMAS), 274, 278
HPLC systems, 16
HPLC-TOF-MS, 16
Human Genome Project (HGP), 4, 39
Human umbilical vein endothelial cells (HUVEC), 226–227

ICAT (isotope-coded affinity tag), 15, 232
IL-12–IL-23 pathway, 89
Illumina BeadArrays, 13
Indole-3-carbinol, chemical structure of, 177 *f*
Ingenuity Pathways Analysis tool, 64
Insulation dysregulation. *See* Diabetes, proteomics for elucidating insulin dysregulation
Intelectin-1 (ITLN-1), 88
Interference RNA (RNAi), 36
Isolation stress, 138–139 *f*
Isoliquiritigenin, chemical structure of, 176 *f*
Isomalt, 13
Isotope-coded affinity tag (ICAT), 15, 232
iTRAQ (isobaric tagging for relative and absolute quantification), 232, 247

Janus kinase 22 (JAK2), 89
Japan, application of genomics to nutrients and functional foods in, 127–150
   functional food science application, 138–140, 142–149
      apple polyphenols, 145–146, 146 *f*
      cocoa, 142–143, 143 *f*
      fructooligosaccharide, 142, 143 *t*
      lycopene, 146–147, 147 *f*
      neoculin, 147–149, 148 *f*
      proanthocyanidin, 144–145, 145 *f*
      royal jelly, 143–144, 144 *t*
      sesamin, 140, 141 *f*, 142
      soy protein isolate, 138–140
   nutritional science application, 131–138
      dietary protein, 131–132, 132 *t*
      fasting, 135–136, 136 *f*
      food allergy, 136–138, 137 *f*
      isolation stress, 138–139F
      mineral deficiency, 133–135, 135 *f*
      unbalance and excess of amino acids, 133, 134 *f*
   overview, 127–131, 149–150
      food as a heterogeneous system, 127–128, 128 *f*
      foods that have a variety of components, 128, 129 *f*–130 *f*, 130
      genomics as means of quality assessment, 130–131, 131 *f*
   perspectives, 149–150
      food-related systems biology, 149, 150 *f*
      personalized nutrition for antiobesity and potential utility of made-to-order foods, 149–150, 150 *t*
Japan Science and Technology Agency (JST), 128

Kahweol, chemical structure of, 177 *f*
Kale, application of genomics and bioinformatics in study of, 64–65, 65 *f*
Kaposi's sarcoma (KS), 75
Kava-kava, 301 *t*
Kutki, 302 *t*
Kyoto Encyclopedia of Genes and Genomes (KEGG), 64, 78

Lactobacillus GG, 5
Lei gong teng, 301 *t*
Leptin gene, 52
Lifestyle modification and weight management, 46–47
Lignans and lignins, 288 *f*
LIPG (endothelial lipase) gene, 53
Liponomics, 7
Liquid chromatography (LC), 14, 241
Liver, 248
Long pepper, 301 *t*
Lycopene, 108 *t*, 146–147, 147 *f*, 186
   chemical structure, 176 *f*

Macroarrays, 97
Madison Metabolomics Consortium Database, 276
Magnesium, 133–135, 135 *f*
Ma huang, 301 *t*

MAP kinase signaling, 100
Marine sponge, 302*t*
Mass spectrometry (MS), 5–6, 14, 16, 214,
    214*f*, 241, 242*f*, 274–275
Matrix-assisted laser desorption ionization
    (MALDI), 214, 214*f*
Matrix-assisted laser desorption/ionization-time
    of flight mass spectrometry
    (MALDI-TOF MS), 6, 14, 231, 241,
    247
Mechanistic biomarkers from animal studies,
    217–220, 217*f*
Messenger RNA (mRNA), 5, 11
Metabolic variability in populations, 16
Metabolites, 15, 276
    profiling
        global, 292
        without identification of detected
            compounds, 292
Metabolome, definition of, 6
Metabolomics, 6–7, 15–17, 149*f*, 271–279,
    303*f*, 306, 316*f*
    cancer prevention research, 322–323
    definition of, 6, 11–12
    experimental design and metabolomic
        technologies, 272–276, 273*f*
        analysis of biofluids and tissues, 272–274
        capillary electrophoresis (CE), 275
        data processing, 275
        Fourier transform infrared (FT-IR)
            spectroscopy, 275
        identification of low molecular weight
            metabolites, 276
        mass spectrometry, 274–275
        NMR spectroscopy, 274
        statistical analysis, 275–276
    future directions, 279
    NMR-based strategy. *See* NMR-based
        metabolomics strategy for
        classification and quality control of
        nutraceuticals and functional foods
    nutritional sciences and, 276–279
        diet and disease, 277–278
        dietary influences on metabolite profiles,
            276–277
        functional foods, 278–279
        gut microflora, 278
    overview, 271–272, 273*f*, 279
    phytonutrients, evaluation of beneficial
        effects of. *See* Phytonutrients,
        evaluation of beneficial effects by
        metabolomics
    terminology, 272, 272*t*
Metabonomics, definition of, 12
MetaCore, 64

Metalloproteinases (MMP), 52–53, 79–80
Methyltetrahydrofolate reductase (MTHFR)
    gene, 4
METLIN Metabolite Database, 276
MIAME (minimum information about a
    microarray experiment), 13
Microarray analysis, 5, 12–13, 62–63, 97, 130,
    131*f*
    dietary antioxidants, 97–99
        differential expression, 98–99
        experimental design, 97
        hybridization, 98
        image analysis and data extraction, 98
        validation by real-time PCR, 99
    standards for reporting, 13
Micro RNA (miRNA, miR), 36–40, 36*f*
Mineral deficiency, 133–135, 135*f*
Minimum information about a microarray
    experiment (MIAME), 13
miRNomics, 7
Misteltoe, European, 301*t*
Mitochondrial uncoupling protein 1 (UCP1)
    gene, 51
MMP (metalloproteinases), 52–53, 79–80
mRNA (messenger RNA), 5, 11
MS (mass spectrometry), 5–6, 14, 16, 214,
    214*f*, 241, 241*f*, 274–275
MTHFR (methyltetrahydrofolate reductase)
    gene, 4
Multiple reaction monitoring, 5
Multiplexed immunoassays, 5
MyPyramid, 3–4, 50

NBC (niacin-bound chromium(III)), 45–51
    safety evaluations, 47
    transcriptomic changes, 50–51
    in weight loss, 47–48
ncRNA (noncoding RNA), 35–36
Neoculin, 147–149, 148*f*
Nephropathy, diabetic, 234–236, 252–254,
    254*f*, 255*t*
Neurodegenerative diseases, 74
NF-κB pathway. *See* Chemoprevention of
    cancer with anti-inflammatory and
    antioxidant phytochemicals
Nitrigenetics, 25
Nitrigenomics, 25
NMR (nuclear magnetic resonance)-based
    metabolomics strategy for
    classification and quality control of
    nutraceuticals and functional foods,
    265–269
    data analysis, 266–267
        supervised method, 267
        unsupervised analysis, 267

NMR-based metabolic profiling techniques, 266
overview, 265–266
quality control, 267–269
NMR (nuclear magnetic resonance) spectroscopy, 274
Nodding burnweed, 302*t*
Noncoding genes, 35, 39
Noncoding RNA (ncRNA), 35–36
Northern blotting, 12
Novel proteins, 5
NRF2-signaling pathway. *See* Chemoprevention of cancer with anti-inflammatory and antioxidant phytochemicals
Nutraceuticals
definition of, 3
novel omics technologies in. *See* Omics technologies, nutraceutical and functional food research
recent advances in, 3–7
nutritional genomics, 4–7
nutrigenetics, 4
nutrigenomics, 5–7
omics in future nutrition research, 7
overview, 3–4, 7
Nutraceuticals, anti-inflammatory, genomics as tool to characterize, 73–80
chronic inflammation in disease, 73–75
cancer, 74–75
gastrointestinal tract, 74
neurodegenerative diseases, 74
respiratory disorders, 74
rheumatic diseases, 75
vascular disorders, 73
GeneChip, 76–80, 77*f*–79*f*
neutraceuticals in management of chronic inflammation, 75–76
overview, 73
Nutrigenetics, 4, 316*f*
definition of, 4, 11
Nutrigenetics and nutrigenomics, 112–116, 113*f*
safety and efficacy assessment of dietary nutrients/antioxidants, 113–114, 115*f*
toxicogenomics as a predictive tool, 115–116
toxicological genomic approach in the assessment of dietary nutrients/antioxidants, 114–115
Nutrigenomics, 5–7, 130
cancer prevention. *See* Cancer prevention research, contribution of omics revolution to
definition of, 11
metabolomics, 6–7

miR and, 39–40
omics for the development of novel phytomedicines. *See* Omics technologies, novel phytomedicines
proteomics, 5–6
transcriptomics, 5
weight management supplements, 50
Nutrigenomics, application in gastrointestinal health, 83–91
gene–diet interactions
in Crohn's disease, 87–90. *See also* Crohn's disease
in gastrointestinal health, 87
human gastrointestinal ecosystem, 86
human gastrointestinal tract, 84–86
-omics approaches to optimizing gastrointestinal health, 90
overview, 83–84, 85*f*, 90–91
Nutrigenomics and statistical power, ethical questions concerning, 25–32
ethics and advice, 29–32
nutrigenomics versus pharmacogenomics, 27
overview, 25–26, 32
personalization, 26–27
prerequisites, 28–29
skepticism, 27–28
Nutrimiromics
miromics, 38–39
nutrigenomics and miR, 39–40
overview, 35–38, 36*f*
Nutritional epigenetics, 316*f*
Nutritional genomics, 4–7
definition of, 4
nutrigenetics, 4
nutrigenomics, 5–7
Nutritional science
applications of genomics in Japan, 131–138
metabolomics and, 276–279
Nutritional transcriptomics, 316*f*
Nutrition research application of proteomics, 213–221
biomarker proteins from blood cells, 216–217
mechanistic biomarkers from animal studies, 217–220, 217*f*
overview, 213, 220–221, 220*f*
proteomics technology in nutrition sciences, 213–215, 214*f*
search for plasma biomarkers, 215–216

Obesity, 39–40, 86
antiobesity mechanisms, 150*t*
epidemic and its consequences, 46–47
lifestyle modification and weight management, 46–47

Oligomeric proanthocyanidin (OPC), 225
Olive leaf extract, 114
Omics technologies
  future nutrition research, 7
  gastrointestinal health, 90
  novel phytomedicines, 299–310
    future challenges, 310
    genomics, 302–304, 305 f
    metabolomics, 306
    omics in phytomedicine, 307–310
    overview, 299–300, 301t–302t, 302, 303 f,
      310
    proteomics, 304–306
    systems biology, 307
  nutraceutical and functional food research,
    11–18
    metabolomics, 15–17
    overview, 11–12, 18, 18 f
    proteomics, 14–15
    systems biology, 17–18
    transcriptomics, 12–13
OPC (oligomeric proanthocyanidin), 225
Orotic acid, 18
Oxidative stress, 99–100, 101t–103t,
  103–104

Pancreas, pancreatic islets, and β-cells,
  243–248, 244 f–246 f
Panorama protein microarray, 15
PathArt, 64
Pathway analysis tools, 63–64
Pattern recognition receptors (PRRs), 88
Perilipin gene, 53
Peroxisomal trans-2-enoyl-CoA reductase
  (PECR) gene, 53
Peroxisome proliferator-activated
  receptor-gamma (PPARγ) gene, 3
Peroxisome proliferator-activated receptor
  (PPAR) gamma coactivator 1-alpha
  (PGC1-alpha) gene, 53
Personalized nutrition, 149–150, 150t
Phenethylisothiocyanate (PEITC), 109t
Phenolic acids, 288 f
Phytochemicals, 175
Phytomedicine, omics in. See Omics
  technologies, novel phytomedicines
Phytonutrients, evaluation of beneficial effects
  by metabolomics, 287–294
  effect of gut microflora, 290–291, 291 f
  health benefits of phytonutrients or
    phytochemicals, 289–290
  limitation of assessing health benefits, 290
  metabolomics tools for phytonutrients
    research, 291–292
  overview, 287

phytochemical research requires
  "metametabolomics" approach,
  293–294, 293 f
phytonutrients or phytochemicals, 287,
  288t–289t
Plasma biomarkers, 215–216
Polyphenols, 288 f
Polyunsaturated fatty acids, 12
Posttranscriptional gene silencing (PTGS), 35
PPAR (proliferator-activated receptors),
  226–227
PPARγ (peroxisome proliferator-activated
  receptor-gamma) gene, 3
Precursor miRNA (pre-miRNA), 37
Proanthocyanidin, 144–145, 145 f
Protein, dietary, 131–132, 132t
Protein chips, 15, 233
Proteobacteria, 86
Proteome, definition of, 5, 201
Proteomics, 5–6, 14–15, 149 f, 303 f, 304–306,
  316 f
  analysis of functionality of Toona sinensis by
    2D-gel electrophoresis. See Toona
    sinensis proteomics analysis by
    2D-gel electrophoresis
  cancer prevention research, 321–322
  cell mapping, 231
  definition of, 5, 11
  in diabetes, 243–251
    adipocytes, 249–250
    liver, 248
    pancreas, pancreatic islets, and β-cells,
      243–248, 244 f–246 f
    serum and plasma from diabetes,
      250–251
    skeletal muscle, 248–249
  dietary grape seed proanthocyanidin potency
    assessment. See Grape seed
    proanthocyanidins, proteomics
    approach to assess potency of
  expression proteomics, 231
  insulin dysregulation in diabetes. See
    Diabetes, proteomics for elucidating
    insulin dysregulation
  nutrition research. See Nutrition research
    application of proteomics
Proton nuclear magnetic resonance (NMR)
  technology, 16
PRRs (pattern recognition receptors), 88
PTGS (posttranscriptional gene silencing),
  35
Pungent vanilloids, 186

Quercetin, 12, 105, 106t, 109t, 181
  chemical structure, 176 f

Reactive nitrogen species (RNS), 177, 181
Reactive oxygen species (ROS), 177, 181
Receptor of advanced glycation end products
     (RAGE) protein, 226, 229
Red wine polyphenolics (RWP), 110*t*
Respiratory disorders, 74
Resveratrol, 105, 108*t*, 110*t*, 180–181,
     185–186
  chemical structure, 176*f*
Retinopathy, diabetic, 254–257
Reverse transcription coupled polymerase chain
     reaction (RT-PCR), 12, 96
Rheumatic diseases, 75
Ribosomal RNA (rRNA), 36
RNAi, 35
RNA-induced silencing complex (RISC),
     36–37, 36*f*
RNeasy kit, 77
RNomics, 7
RNS (reactive nitrogen species), 177, 181
ROS (reactive oxygen species), 177, 181
Royal jelly, 143–144, 144*t*
RT-PCR (reverse transcription coupled
     polymerase chain reaction), 12, 96
RWP (red wine polyphenolics), 110*t*

S-allyl cysteine, chemical structure of, 176*f*
Satavar, 302*t*
Saw palmetto, 301*t*
SCD (stearoyl-coenzyme A desaturase), 13
Search Tool for Retrieval of Interacting
     Genes/Protein (STRING), 64, 66,
     68*f*
SELDI-TOF-MS (surface-enhanced laser
     desorption ionization with a
     time-of-flight mass spectrometry), 14,
     233, 241
Sesamin, 140, 141*f*, 142
Short interfering RNA (siRNA), 36
"Shotgun" proteomics, 232
Signal transducer and activator of transcription
     3 (STAT3), 89
SILAC (stable isotope labeling by amino acids
     in cell culture), 232
Single nucleotide polymorphisms (SNPs), 4,
     316
[6]-gingerol, 180
  chemical structure, 176*f*
16S rRNA gene, 86
Skeletal muscle, 248–249
SLC22A4 and SLC22A5 genes, 87
Small nuclear RNA (snRNA), 36, 38
Small nucleoloar RNA (snoRNA), 36
Smoking, 30, 147, 147*f*
Snail seed, 301*t*

Soy isoflavones, 6, 12–13, 16, 129*t*, 150*t*
  biotransformation by gut microorganisms,
     290, 291*f*
Soy protein isolate, 138–140
Sperm functions and TSL extracts, 206–208
  adenosine triphosphate (ATP) level, 206–207
  cytotoxicity, 207
  human sperm functions, 207–208
  intracellular ROS levels, 206
  mitochondrial membrane potential (MMP),
     206–207
  sperm chromatin structure assay (SCSA), 207
  sperm motility, 206–207
St. John's wort, 301*t*
Stable isotope labeling by amino acids in cell
     culture (SILAC), 232
Statistical analysis, 275–276
  ethical questions concerning nutrigenomics
     and statistical power. *See* Ethical
     questions concerning nutrigenomics
     and statistical power
STATs (signal transducer and activator of
     transcription 3), 89
Stearoyl-coenzyme A desaturase (SCD), 13
Stilbenes, 288*f*
STRING (Search Tool for Retrieval of
     Interacting Genes/Protein), 64, 66,
     68*f*
Sucrose, 13
Sulforaphane, 181–184
  chemical structure, 177*f*
Super CitriMax (HCA-SX), 48–53
Supplements, dietary, safety and efficacy of,
     47–53
  HCA-SX
     efficacy in weight loss, 49–50
     regulation of transcriptomic expression,
        51–53
     safety studies, 48–49
  NBC
     safety evaluations, 47
     transcriptomic changes, 50–51
     in weight loss, 47–48
  nutrigenomics in weight management
     supplements, 50
Surface-enhanced laser desorption ionization
     with a time-of-flight mass
     spectrometry (SELDI-TOF-MS), 14,
     233, 241
Sweet wormwood, 301*t*
Systems biology, 17–18, 307

(10)-Shogaol, chemical structure of, 176*f*
Tissue-type plasminogen activator (PLAT)
     gene, 52

TM4, 62–64

TNFα, 75–79, 77 f–79 f, 158–160, 159 f, 161t–163t, 164–165, 164 f–166 f

Toona sinensis proteomics analysis by 2D-gel electrophoresis, 201–208

  antioxidant enzymes, 208

    under normal physiological condition, 208

    under oxidative stress, 208

  overview, 201, 208

  Toona sinensis as a novel antioxidant, 201–206

    antioxidant activity in leaf extracts, 201–202

    functionality by 2D-gel electrophoresis, 202

    functions of TSL-2, 202–204

    functions of TSL-6, 204–206

    functions of TSL-2P, 204

    possible active compounds in TSL extracts, 202

  Toona sinensis leaf (TSL) extracts and human sperm functions, 206–208

    adenosine triphosphate (ATP) level, 206–207

    cytotoxicity, 207

    human sperm functions, 207–208

    intracellular ROS levels, 206

    mitochondrial membrane potential (MMP), 206–207

    sperm chromatin structure assay (SCSA), 207

    sperm motility, 206–207

Toxicogenomics, 115–116

Toxicological genomic approach in the assessment of dietary nutrients/antioxidants, 114–115

Toxigenomics, 7

Transcriptome, definition of, 5

Transcriptomics, 5, 12–13, 149 f, 303

  cancer prevention research, 320–321

  definition of, 11

  problems or limitations, 13

Transfer RNA (tRNA), 36

TSL-2, 202–204

TSL-2P, 204

TSL-6, 204–206

Turmeric supplements, 110t

Two-dimensional (2D) gel electrophoresis, 14, 214, 214 f, 231–232, 241, 242 f, 244 f, 246 f, 254 f. *See also Toona sinensis* proteomics analysis by 2D-gel electrophoresis

UCP1 (mitochondrial uncoupling protein 1) gene, 51

UPLC technology, 16

Vanillin, 107t

Vanilloids, pungent, 186

Vascular disorders, 73

VEGF, 100

Vitamins

  C, 99, 111t, 226

  D, 12

  E, 12, 16, 99–100, 108t–109t, 226

Weight loss nutraceuticals, genomics in, 45–54

  obesity epidemic and its consequences, 46–47

    lifestyle modification and weight management, 46–47

  overview, 45–46, 53–54

  safety and efficacy of dietary supplements, 47–53

    HCA-SX efficacy in weight loss, 49–50

    HCA-SX regulation of transcriptomic expression, 51–53

    HCA-SX safety studies, 48–49

    NBC-induced transcriptomic changes, 50–51

    NBC in weight loss, 47–48

    NBC safety evaluations, 47

    nutrigenomics in weight management supplements, 50

Western blotting, 96

Wheat flour, hypoallergenic, 12

World Health Organization, 26

Wound healing and grape seed proanthocyanidins (GSP), 230

Xanthohumol, 187

  chemical structure, 176 f

Xanthophylls, 288 f

Yellow cypress, 302t

Yohimbe, 302t

Zerumbone, 187

  chemical structure, 176 f

Zinc, 218–219